FLUID-STRUCTURE INTERACTIONS

Structures in contact with fluid flow, whether natural or man-made, are inevitably subject to flow-induced forces and flow-induced vibration: from plant leaves to traffic signs and to more substantial structures, such as bridge decks and heat exchanger tubes. Under certain conditions the vibration may be self-excited, and it is usually referred to as an instability. These instabilities and, more specifically, the conditions under which they arise are of great importance to designers and operators of the systems concerned because of the significant potential to cause damage in the short term. Such flow-induced instabilities are the subject of this book. In particular, the flow-induced instabilities treated in this book are associated with cross-flow, that is, flow normal to the long axis of the structure, although the book does not aim to cover every possible type. Instead it treats a specific set of problems that are fundamentally and technologically important: galloping, vortex-shedding oscillations under lock-in conditions, and rain-and-wind-induced vibrations, among others. The emphasis throughout is on providing a physical description of the phenomena that is as clear and up-to-date as possible.

Michael P. Païdoussis is a Mechanical Engineering professor at McGill University. He is a Fellow of the Royal Society of Canada, the Canadian Academy of Engineering, the Canadian Society for Mechanical Engineering (CSME), the Institution of Mechanical Engineers (IMechE), the American Society of Mechanical Engineers (ASME), and the American Academy of Mechanics (AAM). Professor Païdoussis is the founding editor of the *Journal of Fluids and Structures*. His principal research interests are in fluid-structure interactions, flow-induced vibrations, aero- and hydroelasticity, dynamics, nonlinear dynamics, and chaos.

Stuart J. Price is a Mechanical Engineering professor at McGill University. His research is focused on the dynamics and stability of structures exposed to a fluid flow. The topics studied are inspired by, and often directly related to, real engineering problems – for example, heat exchangers, offshore structures, overhead transmission lines, and aircraft structures.

Emmanuel de Langre is a professor of Mechanics in École Polytechnique. He is an associate editor of the *Journal of Fluids and Structures*. He has worked as an engineer in the French nuclear industry. His principal research interests are in fluid-structure interactions and vibrations of engineering systems, such as heat exchangers and underwater offshore risers, but also of natural systems such as crops and trees moving under wind load.

Fluid-Structure Interactions

CROSS-FLOW-INDUCED INSTABILITIES

Michael P. Païdoussis
McGill University

Stuart J. Price
McGill University

Emmanuel de Langre
École Polytechnique

CAMBRIDGE UNIVERSITY PRESS
Cambridge, New York, Melbourne, Madrid, Cape Town,
Singapore, São Paulo, Delhi, Mexico City

Cambridge University Press
32 Avenue of the Americas, New York NY 10013-2473, USA

Published in the United States of America by Cambridge University Press, New York

www.cambridge.org
Information on this title: www.cambridge.org/9781107652958

First published 2011
First paperback edition 2013

A catalogue record for this publication is available from the British Library

Library of Congress Cataloguing in Publication Data
Païdoussis, M. P.
Fluid-structure interactions : cross-flow-induced instabilities / Michael Païdoussis, Stuart Price, Emmanuel de Langre.
   p.   cm.
Includes bibliographical references and index.
ISBN 978-0-521-11942-9
1. Fluid-structure interaction.   2. Unsteady flow (Fluid dynamics)   I. Price, Stuart, 1951–   II. Langre, Emmanuel de, 1958–.   III. Title.
TA356.5.F58P346   2010
624.1′7–dc22

2010031766

ISBN 978-0-521-11942-9 Hardback
ISBN 978-1-107-65295-8 Paperback

Contents

Preface

Structures in contact with fluid flow, whether natural (e.g. wind and ocean currents) or man-made, are inevitably subject to flow-induced forces and flow-induced vibration: from plant leaves to traffic signs, to more substantial structures, such as bridge decks and heat-exchanger tubes. These vibrations may be of small or large amplitude, and they may be inconsequential, or of mild or even grave concern.

Consider overhead transmission lines, bridges, tall buildings, and chimneys subjected to wind, offshore risers and umbilicals in ocean currents, cylinders and cylindrical tube arrays in power-generating and chemical plants, for example. Such structures vibrate to some extent at any flow velocity, e.g. due to turbulence or vortex shedding. If the vibrations are of small amplitude, they may lead to fatigue or fretting wear *in the long term*. However, under certain circumstances, the vibration amplitude is large, and damage may occur *in the short term*, in hours or weeks. Moreover, the vibration may be *self-excited*. Typically, but not universally, such vibration is associated with a threshold of flow velocity: on one side of the threshold, oscillations due to some perturbation imparted to the system die out; on the other side, oscillations grow. More generally, we may define self-excited vibration simply as one that grows exponentially with time until it settles down to a limit-cycle motion. Clearly, such phenomena, more specifically the conditions under which they arise, are of great importance to designers and operators of the systems concerned, because of the great potential to cause damage in the short term. Such *flow-induced instabilities* are the subject of this book.

In particular, the flow-induced instabilities treated in this book are associated with cross-flow, i.e. flow normal to the long axis of the structure(s), presuming the geometry is such; i.e. we are mostly concerned with more or less slender structures in cross-flow. These *cross-flow-induced instabilities* are known to be very severe and to occur in the range of natural or engineering flow velocities. Axial-flow-induced instabilities are treated in other books.

Cross-flow about slender structures generally involves flow separation, making its modelling quite difficult compared with axial-flow-related analyses. For that reason, cross-flow-induced instability models inevitably involve some degree of empiricism, and progress in their understanding has only been possible via the intimate interweaving of theory and experiment.

It is not the aim of this monograph to cover every possible type of cross-flow-induced instability. Instead, a specific set of problems is treated which are fundamentally and technologically important: galloping, vortex-shedding oscillations (VIV) under lock-in conditions, rain-and-wind-induced vibrations (RWIV), wake-induced vibrations of small groups of cylinders and flow-induced instabilities of arrays of cylinders, galloping and flutter of bridge decks, and ovalling oscillations of cylindrical shells (e.g. chimney stacks). Flutter in aeronautical structures is not treated in this book.

The emphasis throughout is to provide as clear a physical description of the phenomena as possible, as well as the state of the art in each topic, both in terms of physical understanding and means of prediction. Each topic is first treated in a simplified way, affording an easy grasp of the fundamentals; this is followed by more sophisticated treatment, ending with the most up-to-date state of the art. Care is taken to provide a coherent and fair account of the historical aspects of each topic. An extensive list of references is provided to direct the interested reader to the primary sources, if desired.

This is a monograph for engineers and applied scientists interested in these topics – researchers and practicing professionals alike, working in wind and ocean engineering and in the power-generating industry. It can also serve as a textbook in a graduate-level course.

We are grateful to the Natural Sciences and Engineering Research Council (NSERC) of Canada and Fonds Québécois de la Recherche sur la Nature et les Technologies (FQRNT) of Québec, as well as l'Institut Français du Pétrole, le Centre National de la Recherche Scientifique (CNRS), École Polytechnique, and McGill University.

Writing a book involves hard work and necessitates relentless concentration of effort from time to time, which has meant "disappearances" into our respective offices. We are very grateful to our respective spouses Vrisseis (*Βρισηΐς*), Carol, and Sabine for their forbearance and relentless support.

Finally, we express our unbounded gratitude to Mary Fiorilli for helping in the overall organization and for typing large tracts of the text with such loving care and exemplary commitment.

Michael P. Païdoussis
Stuart J. Price
Emmanuel de Langre

1 Introduction

1.1 General Overview

Cross-flow-induced vibration of bluff bodies, i.e. bodies whose aspect is not small compared with the streamwise dimension, are ubiquitous, in nature as well as in man-made constructions. The wind-induced fluttering of leaves and tree branches and the waving motions in wheat fields are examples of the former. The Aeolian harp, going back perhaps 3000 years, is an example of the earliest realization and/or exploitation of the existence of these vibrations made by man.

Perhaps the first documented and surviving realization of the existence of vortex shedding as such goes back to two Renaissance paintings in Bologna and a sketch by Leonardo da Vinci, thus, to the 14th and 15th centuries.* The modern study of vortex shedding began in the late 19th century, with Strouhal (1878), Bénard (1908) and von Kármán (1912). Studies on vortex-induced vibrations followed soon after; lock-in, or shedding frequency synchronization, was first documented by Bishop & Hassan (1964).

With such a venerable and long pedigree, it is not surprising that the topic of cross-flow-induced vibrations and instabilities of bluff bodies, notably cylinders or groups of cylinders, is truly vast. To make any headway in this topic, one must first understand the fluid mechanics of the flow around bluff bodies, while stationary or in motion, and the forces generated thereon. Because these depend on the Reynolds number, roughness, flow confinement, aspect ratio, amplitude of motion and many other factors, the task of documenting, categorizing and making sense of the voluminous amount of research done over the past 100 years or so is truly Herculean. In this regard, one must pay tribute to the excellent work done by Zdravkovich (1997, 2003); the task is so huge that the work covered in the first two volumes already published, involving 1264 pages, has not yet reached the point of considering

---

* One painting from the end of the 14th century, attributed to Giovanni da Modena and found in Bologna, depicts the Christ-bearing ($X\rho\iota\sigma\tau\acute{o}\phi o\rho o\varsigma$) San Cristoforo crossing a stream and shows an alternating pattern of vortices downstream of his legs (Tokaty 1971; Sumner 1999). Another, entitled *Madona col Bambino tra i Santo Domenico, Pietro Martire e Cristoforo* is a 15th-century mural in the Basilica de San Domenico in Bologna, again showing vortices from the foot of St. Christopher crossing a stream (von Kármán 1954; Zdravkovich 2003). The drawing by Leonardo da Vinci from roughly the same period shows vortices in the wake of a pile in a stream (Lugt 1983; Blevins 1990; Zdravkovich 1997; Mizota *et al.* 2000).

fluid-coupled, self-excited motions. The difficulty of this task is exacerbated by the fact that, routinely, for decades now, there is hardly an issue of the *Journal of Fluid Mechanics* or the *Journal of Fluids and Structures*, or indeed the *Journal of Sound and Vibration*, the *Journal of Wind Engineering and Industrial Aerodynamics* or the *Journal of Fluids Engineering*, without one or several papers related to cross-flow about bluff bodies, the forces and motions induced thereby and so on. Thus, it is not only that the accumulated knowledge is vast, but also that the accretion of knowledge and experience on the topic continues to grow unabated, perhaps exponentially.[*]

Of course, other books exist in which chapters may be found on cross-flow-induced vibrations and instabilities, published over the past 25 years: by Blevins (1977, 1990), Sarpkaya & Isaacson (1981), Chen (1987), Naudascher & Rockwell (1994, 2005), Gibert (1998), Sumer & Fredsøe (1997), Au-Yang (2001), Axisa (2001), de Langre (2001), Kaneko *et al.* (2008) and others. However, these books cover several topics other than cross-flow-induced vibrations and instabilities, and in Zdravkovich (1997, 2003) and the forthcoming Volume 3 of that work it is attempted to cover the whole field of cross-flow about bluff bodies. In contrast, the present book is more modest in scope and its aim more focussed.

In this book, the focus is on the interaction of the cross-flow with motions of the bluff structure, presuming that the flow field and the forces associated with prescribed motions of the structure are known *a priori*. Furthermore, the vista is further limited by excluding extraneously induced excitation (EIE) and instability-induced excitation (IIE) in Naudascher & Rockwell's (1980, 1994) classification of flow-induced oscillation phenomena. The subject matter in this book is therefore broadly associated with *movement-induced excitation* (MIE) phenomena, in which the excitation is intimately coupled with, indeed caused by, movements of the body. Hence, the phenomena are *self-excited*. In the linear sense, these phenomena are *instabilities*; i.e., as a parameter is incremented, a system hitherto in a quiescent state becomes subject to self-excited oscillation – as discussed further in Section 1.2. Hence, the topic is: *self-excited oscillations involving bluff bodies in cross-flow.*[†]

Why so much interest in bluff-body/flow interactions, indeed, in the subject of this book? The immediate answer is that (i) bluff bodies, in particular, cylinders and prisms, are ubiquitous in engineering structures, typically as components of larger systems; (ii) in many cases these bluff bodies are subjected to flow (cross-flow); and (iii) frequently problems arise, often in the form of self-excited oscillations, the solution if not prevention of which necessitates understanding the fluid-structure interaction mechanisms involved. Every engineering student learns about the Tacoma Narrows Bridge disaster and may have seen the spectacular ciné-film of its collapse. However, apart from bridges, cross-flow-induced vibrations occur in (i) heat exchangers and other power-generation components; (ii) offshore structures, including risers and submerged pipelines; (iii) high-rise buildings, silos and chimneys;

---

[*] In this respect, one has to marvel at Mickey Zdravkovich's tenacity. As he has told the first author, the main difficulty in writing his books was that, no sooner was a particular chapter closed and the writing progressed to the next and subsequent ones, that it had to be reopened because interesting and pertinent new information had been published in the meantime. And, of course, one cannot cry "stop!" anymore than one can ignore the new knowledge.

[†] Here, of course, the definition of "bluff-bodies in cross-flow" is pleonastic, just as "slender body in axial flow" is; however, the redundancy enhances the clarity of the definition.

(iv) overhead transmission lines and cables; and within (v) fluid-manipulating machinery in mechanical and chemical plants, to give but a partial list. Thus, the flows involved are either contained gas or liquid flows, or generally unconfined flows due to wind and water currents.

In the long list of engineering applications just mentioned, "problems" arise associated with self-excited oscillations or cross-flow-induced instabilities. These problems range from short-term destruction of the structure to unacceptable long-term wear (fretting) problems and fatigue. Some examples may be found in Païdoussis (1980, 2006), Axisa (1993), Au-Yang (2001) and Kaneko *et al.* (2008). Many of these are related to the power-generating industry, in particular, to nuclear plants, where disclosure of all types of problems, including flow-induced problems, is mandatory in many states. Other incidents, however, remain hidden from public view, their existence being surmised only by sudden upsurges in research funding; or, at the very least, they are incompletely reported,* e.g. in the offshore industry.

It is opportune to contrast the research on cross-flow-induced instabilities to that on axial-flow-induced ones. In the latter, much, though by no means all, of the research work was curiosity-driven (Païdoussis 1998, 2004), with many of the applications emerging 10 or 20 years later (Païdoussis 1993). For cross-flow, on the other hand, much work was inspired by, or necessitated for, concrete applications. This reflects the fact that, with the exception of some classes of axial-flow-induced vibration, notably involving annular and leakage flows, catastrophic failure is rather rare. For cross-flow situations, however, problems have abounded and are not all that rare even today. In one subtopic alone, that of fluidelastic instabilities of cylinder arrays in cross-flow, the cumulative damages incurred over a decade were estimated at 1000 M$ (Païdoussis 2006).

Something that ought to be stressed is that flow-induced vibrations of structures subject to cross-flow are inevitable and often innocuous. It is only when the amplitudes become large enough, as is often the case with flow-induced instabilities, that they become worrisome. The main task of this book is to elucidate the mechanisms underlying these instabilities and to provide means for predicting their occurrence.

It should also be pointed out that flow-induced vibrations and instabilities are not always undesirable. For instance, naturally occurring flow-induced vibrations help in promoting the dispersion of plant seeds (de Langre 2008). In addition, they can be exploited for engineering purposes, e.g. in ocean-current-driven energy-harvesting devices.

As stated in the Preface, it is here emphasized that the treatment in this book is not exhaustive. Rather, the emphasis is very much on the fundamentals and on a physical understanding of the mechanisms involved to the extent possible. Beyond that, a full list of references guides the reader to the available literature in each subtopic.

1.2 Concepts and Mechanisms

The purpose of this section is to clarify some of the terms and concepts referred to in the foregoing and used extensively in this book, e.g. the concepts of instability

---

* Mainly to protect the corporate image on a trade mark, or for fear of litigation.

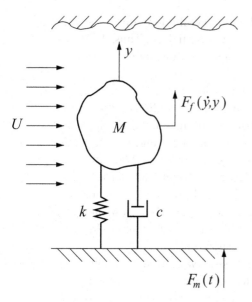

Figure 1.1. A flexibly supported bluff body of mass $M \equiv ml$ in cross-flow.

and self-excited vibrations, in the process clarifying also some of the underlying mechanisms.

1.2.1 Self-excited oscillations and instabilities

The truth about self-excited oscillations is that they are not truly self-excited. That is, a mechanical system does not by itself spontaneously break into oscillation, unless the definition of the system includes the source of energy, e.g. a fluid flow, which is responsible for the oscillation (Den Hartog 1956, chapter 7; Magnus 1965, chapter 3). As we shall see, however, the governing equation of motion may be written in a way that the resulting oscillation *appears* to be self-excited.

Consider, for example, a flexibly supported bluff body which could be modelled as a mass-dashpot-spring system, as shown in Figure 1.1, free to move in the direction transverse to the flow; the cross-section of the body is uniform along its length l (normal to the plane of the paper), so that its total mass $M = ml$, where m is the mass per unit length. The bluff body is subjected to a fluid-dynamical force $F_f(\dot{y}, y)$, as well as a mechanical force $F_m(t)$, e.g. a base excitation; y is the transverse displacement and $\dot{y}$ the corresponding velocity. Thus, we have

$$ml\ddot{y} + c\dot{y} + ky = F_f(\dot{y}, y) + F_m(t), \qquad (1.1)$$

where the overdot denotes differentiation with respect to t. Here, l is assumed to be sufficiently large for the flow around the body to be sensibly two-dimensional. Suppose further that F_f may be expressed as $\frac{1}{2}\rho U^2 hl C_{f_1}(\dot{y}, U) + \frac{1}{2}\rho U^2 hl C_{f_2}(y, U)$, where h is a characteristic length (typically the diameter for a cylindrical body, or the frontal height of the cross-section *vis-à-vis* the flow), and C_{f_1} and C_{f_2} are fluid-dynamic force coefficients, respectively functions of $\dot{y}$ and y, and weakly of the mean flow velocity U. Velocity dependence may arise because the instantaneous angle of attack of the flow on the body as the body oscillates is $\theta = \tan^{-1}(\dot{y}/U)$. Position dependence may arise through proximity of the bluff body to, say, a wall, so that the fluid forces depend on the distance from the wall. In equation (1.1), it is presumed that $m = m_s + m_a$, m_s being the structural mass and m_a the added or virtual fluid-dynamic mass per unit length. For an oscillating body, the acceleration-related pressure field gives rise to a force which may be written as $m_a l \ddot{y}$, thus defining m_a; for a dense fluid, m_a/m_s is not negligible (e.g., for a circular cross-section, $m_a = \rho(\pi/4)D^2$ according to potential flow theory, D being the diameter) – see Section 3.4.1. Thus, this could have been incorporated in (1.2) as a third fluid-dynamical force, linearly dependent on $\ddot{y}$ and independent of U, while taking $m = m_s$.

Succinctly, the difference between F_f and F_m is that for $y = 0$ and $\dot{y} = 0$, F_f will be zero (or a constant force that can be eliminated by a change of variable from y to $y^* = y - y_0$). Thus, writing the equation in terms of y^* and suppressing the asterisk, we have

$$m l \ddot{y} + c \dot{y} + k y = \tfrac{1}{2}\rho U^2 h l C_{f_1}(\dot{y}, U) + \tfrac{1}{2}\rho U^2 h l C_{f_2}(y, U) + F_m(t), \qquad (1.2)$$

where it is understood that if $y = 0$, $\dot{y} = 0$, the first two forcing functions (representing $F_f(\dot{y}, y)$) vanish; thus, they only arise because of motion, whereas $F_m(t)$ is not affected by the motion. Having served to clarify the distinction between itself and F_f, we shall from now on ignore $F_m(t)$.

We can next write equation (1.2) in dimensionless form by defining $c/ml = 2\zeta\omega_n$ and $k/ml = \omega_n^2$, as well as

$$\eta = y/l, \quad \tau = \omega_n t, \quad U_r = U/(\omega_n h), \quad \mu_r = \rho h^2/m, \qquad (1.3)$$

where U_r is the so-called *reduced flow velocity* and μ_r is a *mass ratio*, obtaining*

$$\ddot{\eta} + 2\zeta\dot{\eta} + \eta = \tfrac{1}{2}\mu_r U_r^2 C_{f_1}(\dot{\eta}, U_r) + \tfrac{1}{2}\mu_r U_r^2 C_{f_2}(\eta, U_r), \qquad (1.4)$$

where the overdot is now $d(\ )/d\tau$. It is assumed next, for simplicity, that C_{f_1} may be expressed as a function of $\dot{\eta}/U_r$ and C_{f_2} as a function of η alone. For the purposes of this illustrative example, let

$$\tfrac{1}{2}\mu_r U_r^2 C_{f_1}(\dot{\eta}, U_r) = \beta_1(U_r)\dot{\eta} - \beta_3(U_r)\dot{\eta}^3, \quad \tfrac{1}{2}\mu_r U_r^2 C_{f_2}(\eta, U_r) = \gamma_1(U_r)\eta - \gamma_3(U_r)\eta^3,$$

and hence equation (1.4) is written as

$$\ddot{\eta} + \left[2\zeta - \beta_1(U_r) + \beta_3(U_r)\dot{\eta}^2\right]\dot{\eta} + \left[1 - \gamma_1(U_r) + \gamma_3(U_r)\eta^2\right]\eta = 0. \qquad (1.5)$$

Thus, at first glance, considering the quantities in square brackets as an effective damping and an effective stiffness, the source of energy input in this autonomous system is "hidden".

Let us further assume that β_1, β_3, γ_1 and γ_3 are positive, monotonically increasing functions of U_r in view of the weak dependence of C_{f_1} and C_{f_2} on U_r, and let us consider the dynamics displayed by equation (1.5).

First, taking $\gamma_1 = \gamma_3 = 0$ for the moment, it is clear that for arbitrarily small $|\dot{\eta}|$ the dynamics is controlled by the linear terms and hence by the sign of $2\zeta - \beta_1$: if it is positive, as it must be for sufficiently small U_r, the damping is positive and the oscillations will be damped; for higher U_r, however, it becomes negative, which

---

* Equation (1.4) holds true also if equation (1.1) is written in two-dimensional or "sectional" form. In that case, m would replace ml in equation (1.2); similarly, k and c could represent distributed quantities per unit length, or the total acting on the bluff body divided by l; also, l would be absent from the right-hand side of (1.2). Equations (1.3) would be the same.

means *negative damping* and *self-excited oscillations*. Thus, the threshold of linear instability of the system, U_{rc}, occurs at $\beta_1(U_{rc}) = 2\zeta$; according to linear theory, the amplitude of the self-excited oscillation will grow indefinitely. However, taking the $\beta_3(U_r)\dot{\eta}^3$ term into account, it is clear that for sufficiently large $|\dot{\eta}^3|$, the damping ceases to be negative; indeed, the quantity in square brackets becomes zero on the average, and one obtains *limit-cycle oscillation*. Thus, the growth of amplitude is self-limiting. In this case, the limit cycle is stable, as both positive and negative increments (perturbations) to $\dot{\eta}$ are damped, returning the system to the limit cycle; the case of an unstable limit cycle is discussed later.

From the nonlinear perspective, the linear threshold of instability is referred to as a *bifurcation*, leading in this case from one stable state, the trivial equilibrium, to another stable state, the limit-cycle oscillation. However, in engineering terms, the linear threshold of instability at U_{rc} is commonly called the threshold of instability, regardless.

Bifurcation has a broader meaning than for the situation just discussed: it is associated with any qualitative change in the state or dynamical behaviour of a system, e.g. from periodic to quasiperiodic oscillation, or from quasiperiodic to chaotic.

Let us next consider the statics of the system, clearly governed by the last term of equation (1.5). For small $|\eta|$, the effective linear stiffness, $1 - \gamma_1(U_r)$, is positive, provided U_r is sufficiently small; for any small departure from $\eta = 0$, the stiffness force restores the original static equilibrium. For higher U_r, however, we may have a *negative stiffness*, $1 - \gamma_1(U_r) < 0$, leading to static instability, a *static divergence*, implying a nonoscillatory amplification, without limit, of any small departure from the now unstable trivial equilibrium. Taking the nonlinear term into account, however, it is clear that two new equilibria are born for sufficiently large $|\eta|$: at $\eta_{st} = \pm\{[1 - \gamma_1(U_r)]/\gamma_3(U_r)\}^{1/2}$, which may, in general, be stable or unstable (in the sense of the equilibria of a pendulum at $\theta = 0$ and π, respectively) – but for the form of the stiffness term here always unstable.

The dynamics of the system of equation (1.5) could be displayed as a three-dimensional plot of $(\eta, \dot{\eta}, U_r)$. Any "cut" thereof along the U_r-axis would yield a phase-plane plot $(\eta, \dot{\eta})$. To make things more interesting and instructive, we henceforth relax the requirement that $\beta_3(U_r)$ and $\gamma_3(U_r)$ be positive. Thus, consider the system at $U_r = U_{r1}$, such that equation (1.5) becomes

$$\ddot{\eta} - 0.02(1 - \dot{\eta}^2)\dot{\eta} + (0.95 - 0.018\eta^2)\eta = 0. \qquad (1.6)$$

In view of the foregoing, this represents a system just beyond the onset of linear self-excited oscillation; i.e. here $2\zeta - \beta_1(U_{r1}) = -0.02$, while $\beta_3(U_{r1}) = 0.02$. Also, $1 - \gamma_1(U_{r1}) = 0.95$ and $\gamma_3(U_{r1}) = -0.018$.

The dynamics is displayed in Figure 1.2(a). It is seen that the origin (trivial equilibrium) is unstable, and that a stable limit cycle exists at $|\eta| \simeq 1.1$ (the oval region purposely left blank for clarity). Trajectories for $|\eta| < 1.1$ and $|\eta| > 1.1$ but not too far away, spiral outwards and inwards, respectively, towards the limit cycle. There are also two new *fixed points*, i.e. points of static equilibrium, at $|\eta| = (0.95/0.018)^{1/2} \simeq 7.26$. They are unstable; specifically, they are *saddle points*. In this case, the *basin of attraction* of the limit cycle is the diagonal swath from the upper left of the figure to the lower right, within the area delimited by the trajectories going through the saddle points.

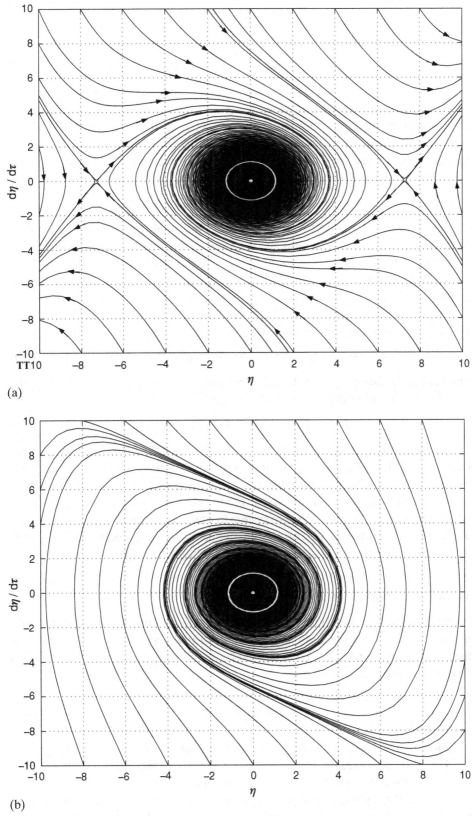

(a)

(b)

Figure 1.2. Phase-plane diagrams for the system of equation (1.4) for different system parameters: (a) for the system $\ddot{\eta} - 0.02(1 - \dot{\eta}^2)\dot{\eta} + (0.95 - 0.018\eta^2)\eta = 0$; (b) for the system $\ddot{\eta} - 0.02(1 - \dot{\eta}^2)\dot{\eta} + (0.95 + 0.018\eta^2)\eta = 0$.

For $\gamma_3(U_{r1}) = +0.018$ in equation (1.5), i.e. when the equation of motion is

$$\ddot{\eta} - 0.02(1 - \dot{\eta}^2)\dot{\eta} + (0.95 + 0.018\,\eta^2)\eta = 0, \qquad (1.7)$$

the saddle points disappear, as shown in Figure 1.2(b), and the basin of attraction of the limit cycle covers the whole figure. Taking a global view of the dynamics, we can say that the system is *unstable in the small* (i.e. in the region within the limit cycle close to the origin), and *stable in the large*. In general, "small" and "large" are suggested by the physics of the system, but may be subjective.

For $\gamma_3(U_{r1}) = 0.018$ and $\beta_3(U_{r1}) = -0.02$, i.e. for the equation

$$\ddot{\eta} - 0.02(1 + \dot{\eta}^2)\dot{\eta} + (0.95 + 0.018\,\eta^2)\eta = 0, \qquad (1.8)$$

the stable limit cycle disappears also. This does not imply that the physical system is unstable at all nonzero amplitudes. It simply means that the nonlinear model of equation (1.8) is not accurate enough. A more accurate representation, e.g. involving a positive $\beta_5(U_{r1})\dot{\eta}^5$, could again give rise to a stable limit cycle (see, e.g., Païdoussis (1998, section 2.3)).

Next, consider the system of equation (1.5) at $U_{r2} < U_{r1}$, such that $2\zeta - \beta_1(U_{r2}) > 0$. The equation of motion is now

$$\ddot{\eta} + 0.02(1 + \dot{\eta}^2)\dot{\eta} + (0.95 + 0.018\eta^2)\eta = 0. \qquad (1.9)$$

The limit cycle disappears and the trivial equilibrium becomes a stable fixed point.

Consider next another system, governed by

$$\ddot{\eta} + 0.02(1 - \dot{\eta}^2 + 0.05\dot{\eta}^4)\,\dot{\eta} + (0.95 - 0.018\eta^2)\eta = 0. \qquad (1.10)$$

The phase-plane plot is shown in Figure 1.3. In this case the blank oval at $|\,\eta\,| \simeq 1.1$ is an *unstable* limit cycle, nesting within a stable limit cycle going through $|\,\eta\,| \simeq 6.2$. Thus, trajectories within the unstable limit cycle spiral towards the origin, and those on the outside spiral towards the stable limit cycle – as do trajectories outside the latter. This represents a not-too-rare system in practice: a system stable at the origin which if lightly perturbed will return to the origin; but, if strongly perturbed to beyond the unstable limit cycle, it will develop large-amplitude limit-cycle oscillations.

1.2.2 Argand diagrams and bifurcations

For an N-degree-of-freedom system or an N-mode discretization of a distributed parameter system, let the N dimensionless eigenfrequencies be denoted by ω_r and the eigenvectors by $\{A\}_r$, $r = 1, \ldots, N$, and let the linear solution of the autonomous system be expressed as

$$\{q\} = \sum_{r=1}^{N} \{A\}_r\, e^{i\omega_r \tau}. \qquad (1.11)$$

In general, $\omega_r = \mathcal{R}e(\omega_r) + i\,\mathcal{I}m(\omega_r)$. It is clear that if for one of the ω_r, say for ω_s, $\mathcal{I}m(\omega_s)$ is negative, the system is linearly unstable, since the solution will then involve a term $\exp(\alpha_s \tau)$, where $\mathcal{I}m(\omega_s) = -\alpha_s$ and $\alpha_s > 0$.

As one of the system parameters is varied, say the dimensionless flow velocity u, the evolution of the ω_r is often displayed as an Argand diagram, in which $\mathcal{I}m(\omega_r)$ is plotted versus $\mathcal{R}e(\omega_r)$ with u as parameter.

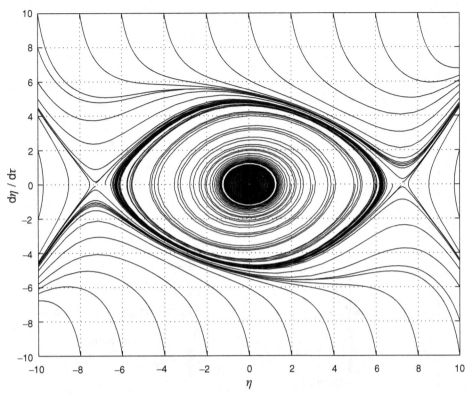

Figure 1.3. Phase-plane diagram for the version of the system of equation (1.4) described by $\ddot{\eta} + 0.02(1 - \dot{\eta}^2 + 0.05\dot{\eta}^4)\dot{\eta} + (0.95 - 0.018\eta^2)\eta = 0$.

Figure 1.4 shows such diagrams, illustrating several ways in which the frequency loci may cross from the stable $+\mathcal{I}m(\omega_r)$ half of the frequency plane to the unstable $-\mathcal{I}m(\omega_r)$ half.

Figure 1.4(a) shows the onset of *divergence* via a *pitchfork bifurcation** in the first mode of a conservative system. As ω_1 is purely imaginary for $u > u_c$, this is clearly a static instability.

Figure 1.4(b) illustrates loss of stability via *Hopf bifurcation* for a nonconservative system with zero structural damping ($\mathcal{I}m(\omega_2) = 0$ at $u = 0$). Clearly, as $\mathcal{R}e(\omega_2) \neq 0$ at $u = u_c$, this is an oscillatory instability, signifying *single-mode amplified oscillations* or *flutter*.

Flutter can also arise through coalescence of two modes in the form of *coupled-mode flutter*, as shown in Figure 1.4(c, d), again for systems with zero structural damping. The fact that the eigenfrequencies are purely real prior to instability is indicative of the system being conservative. The coupled-mode flutter displayed in Figure 1.4(c) is via a so-called *Hamiltonian Hopf bifurcation*.

In Figure 1.4(d) the loci of the modes lie either on the $\mathcal{R}e(\omega)$ or the $\mathcal{I}m(\omega)$ axis, but they are drawn just off the axes for clarity. The coupled-mode flutter in this case is via a *secondary bifurcation*, i.e. after the system has lost stability by divergence – in

---

* Strictly speaking, the type of bifurcation involved is defined by the nonlinear terms in the equation of motion. In this case, the flow-related nonlinearities in the stiffness term are cubic and similar to those in a softening cubic spring. This is what gives rise to two stable static equilibria for $u > u_c$.

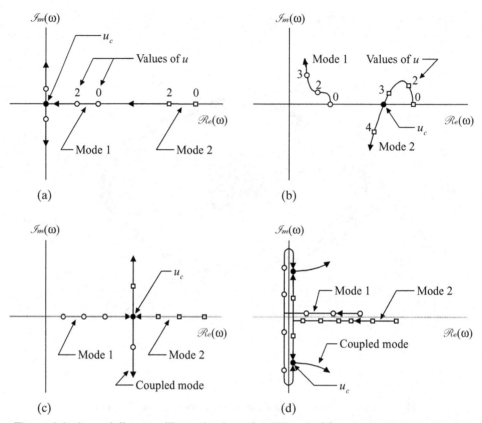

Figure 1.4. Argand diagrams illustrating loss of stability via (a) a pitchfork bifurcation of a conservative system, leading to static divergence, (b) a Hopf bifurcation of a nonconservative system, (c) a Hamiltonian Hopf bifurcation, leading to coupled-mode flutter, (d) a so-called Païdoussis coupled-mode flutter, all for nondissipative systems; u_c denotes the critical dimensionless flow velocity.

this case, in both the first and second modes.[*] To distinguish it from the Hamiltonian coupled-mode flutter, Done & Simpson (1977) christened it as *Païdoussis coupled-mode flutter*, because it was first documented in a paper by Païdoussis & Issid (1974); its principal characteristic is that, at onset, the frequency of oscillation is zero, but it becomes finite as u is increased.

It is instructive to consider how these bifurcations are affected by the presence of dissipative effects and nonlinearities. Figure 1.5 shows the effect of dissipation on the bifurcations. It is clear that the bifurcations in Figure 1.5(a, b) are not qualitatively different from those in Figure 1.4(a, b).

Figure 1.5(c) is distinctly different, however. The two modes nearly collide and then veer away from each other (*mode-veering* phenomenon), and one of them crosses the $\mathcal{Re}(\omega)$-axis to the unstable domain; thus, the coupled-mode flutter devolves to a form of single-mode flutter. In a sense, something similar is shown in Figure 1.5(d), although the coupled-mode flutter in this case survives, but involves the coalescence of two branches of the first mode.

---

[*] Another form involves the coalescence of the two branches of the same mode [see, e.g., Païdoussis (1998, fig. 3.14); also Figure 1.5(d)].

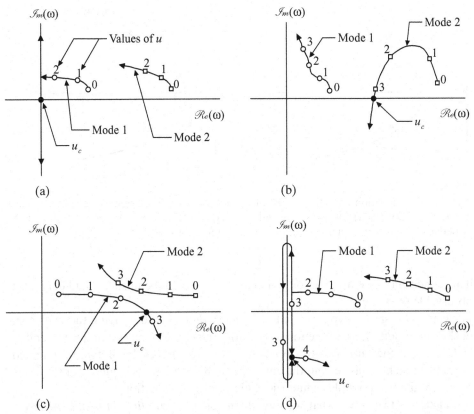

Figure 1.5. The corresponding Argand diagrams to those of Figure 1.4 for dissipative systems: (a) pitchfork bifurcation, (b) Hopf bifurcation, (c) degenerate form of the Hamiltonian Hopf bifurcation, and (d) Païdoussis-type coupled-mode flutter bifurcation.

In terms of nonlinear theory, the evolution of the modes with increasing u is a little or a great deal different. First, it should be emphasized that linear theory is reliable only in the prediction of the threshold of the first loss of stability of a system; beyond that, the dynamics can only be reliably predicted via nonlinear theory. Thus, subsequent (secondary *et seq.*) instabilities predicted by linear theory may or may not materialize;* and, if they do, they will not arise in the same way as linear theory predicts. This is mainly because, once the first instability has occurred, the system is no longer in its trivial equilibrium state; hence, in general, secondary bifurcations emanate from the stable state beyond the first instability, whether this is a buckled state or a limit cycle, and not from the trivial equilibrium (as implicitly presumed by linear theory). Clearly, this has the greatest effect on the coupled-mode flutter of Figure 1.5(d).

Of course, linear theory cannot predict the amplitude of divergence or flutter that emerges from the corresponding bifurcation. Nonlinear theory does. Furthermore,

---

* For the case where linear predictions are contradicted by nonlinear theory, see Païdoussis (1998, sections 3.4.1 and 3.5.2). For the opposite, see Païdoussis (2004, section 8.11.2) and Modarres-Sadeghi (2005); in this case, although the origination of the post-divergence flutter is different, linear and nonlinear predictions of the critical flow velocity are not too different; the phenomenon also materializes in experiments.

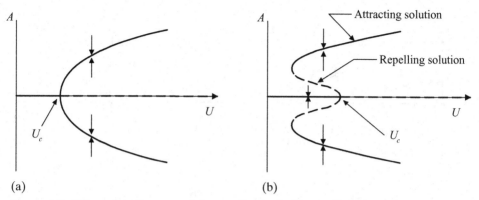

Figure 1.6. Bifurcation diagrams of the amplitude A of static divergence or flutter versus the flow velocity U: (a) for a supercritical bifurcation, and (b) for a subcritical bifurcation displaying hysteresis; —, stable solutions; – – –, unstable solutions.

it also finds that the emerging solution may in some cases be unstable, and hence not observable physically.

Figure 1.6 shows two so-called *bifurcation diagrams*, displaying the evolution of the amplitude, A, as a function of the flow velocity U. The system loses stability at $U = U_c$, at the same point as predicted by linear theory. Here, the positive and negative branches may delimit the envelope of the oscillations in flutter, or they may represent the two possible symmetric static divergence solutions.

Figure 1.6(a) shows what we now define as a *supercritical bifurcation*. This is characterized by the fact that, for $U = U_c + \varepsilon$, the amplitude of divergence or flutter, as the case may be, is vanishingly small; it then grows steadily with increasing U. Beyond U_c, the trivial equilibrium is unstable; the arrows therefore show that if the system is perturbed to either higher or lower amplitudes, it will return to the stable solution (full line). For flutter, the situation corresponds to that of the limit cycle in Figure 1.2(a, b) which would correspond to a cut of Figure 1.6 at one particular U (or U_r), specifically at U_{r1}.

Figure 1.6(b) shows a *subcritical bifurcation*. Here, the emerging solution is unstable. For small perturbations about the stable trivial equilibrium at $U < U_c$, the system returns to its original state. However, for large enough perturbations, larger than the unstable ("repelling") solution, the system may end up on the larger "attracting solution". For flutter, this is the situation in Figure 1.3, involving an unstable (repelling) limit cycle nesting within a stable (attracting) one.

1.2.3 Energy considerations

The conditions of stability/instability may be determined from solutions of the partial or ordinary differential equations governing free motions of the system. However, they may sometimes be obtained by calculation of the work done by all fluid-dynamic forces, $F_f(y, \dot{y}, \ddot{y})$ on the system, together with the work done (energy lost) by all dissipative forces, say $F_d(\dot{y})$. Even if the conditions of instability are not determined thereby, some valuable insights are often gained.

Consider self-excited oscillations of a one-dimensional system in a spatial domain $[0, L]$ subjected to a flow velocity U. The work done in a cycle of putative oscillation is

$$\Delta W = \int_0^T \int_0^L [F_f(U, y, \dot{y}, \ddot{y}) + F_d(\dot{y})] \dot{y} \, dxdt, \qquad (1.12)$$

where $y = y(x, t)$ and T is the period of oscillation. Clearly, if the system is stable, then $\Delta W < 0$; the threshold of instability occurs as ΔW passes through zero to positive values. If this cannot be found by inspection or simple manipulation of (1.12), this expression may nevertheless reveal some essential conditions for $\Delta W > 0$.[*]

For a one-degree-of-freedom system, the integral over x vanishes. For example, for the system equation (1.5), we have

$$\Delta W = \int_0^T -\left[(2\zeta - \beta_1(U_r)) \dot{\eta} + \beta_3(U_r)\dot{\eta}^3\right] \dot{\eta} \, dt. \qquad (1.13)$$

The terms involving γ_1 and γ_3 do not contribute, appropriately, as they are conservative.[†] Rewriting the equation as

$$\Delta W = -\int_0^T \left[(2\zeta - \beta_1)\dot{\eta}^2 + \beta_3\dot{\eta}^4\right] dt.$$

for convenience, the following insights may be obtained:

(i) For $2\zeta - \beta_1 > 0$ and $\beta_3 > 0$, the system is unconditionally stable. If $\beta_3 < 0$, however, for sufficiently large $\dot{\eta}^2$, an unstable limit cycle may arise.
(ii) For $2\zeta - \beta_1 < 0$ and $\beta_3 > 0$, one may have a stable limit cycle for sufficiently large $\dot{\eta}^2$; e.g. in the case of equation (1.6), one has $2\zeta - \beta_1 = -0.02$ and $\beta_3 = 0.02$; hence, a limit cycle may exist at $|\dot{\eta}| \sim \mathcal{O}[-(2\zeta - \beta_1)/\beta_3]^{1/2}$, in this case $\mathcal{O}(1)$, which agrees with Figure 1.2(a).

One cannot go much farther in this discussion without considering specific physical systems, and this is not the appropriate place for that.

1.3 Notation

Because of the wide-ranging subject matter in this book, the notation used differs from one chapter to the next. In this respect, it was considered preferable to keep close to the coventional notation firmly embedded in each subject area, thus familiar to those acquainted with the subject and close to that used in primary references in each area, rather than enforcing uniformity across the board.

Nevertheless, to help the reader, a glossarial listing is included at the beginning of each chapter, defining some of the nomenclature and focussing on the symbols that are different in other chapters.

---

[*] For a simple example, that of a pipe conveying fluid (axial flow), refer to Païdoussis (1998, section 3.2).
[†] E.g. $\int_0^T -\gamma_3\eta^3\dot{\eta} \, dt = -\frac{1}{4}\gamma_3\int_0^T d(\eta^4)/dt = -\frac{1}{4}\gamma_3\eta^4\big|_0^T = 0$.

1.4 Contents of the Book

Apart from the Introduction and Epilogue, there are six chapters in this book. Each chapter begins with a simplified presentation of the topic, not necessarily the most modern or sophisticated, but one that helps elucidate the mechanism involved, and, moreover, one that is easy to understand at the physical level.

Chapter 2 deals with galloping of prismatic bodies, as well as galloping and flutter of structures with elongated cross-sections. Chapter 3 is on vortex-induced vibrations, in particular, aspects of lock-in vibration; under lock-in resonance conditions, these vibrations are fundamentally similar to the rest described in this book.

Chapters 2 and 3 deal with just a single bluff body, while Chapters 4 and 5 deal with small or large groups of cylinders. Specifically, Chapter 4 deals with so-called wake flutter of pairs or small groups of cylinders, with special application to electrical transmission lines and multitube offshore risers. Chapter 5 deals with the so-called fluidelastic instabilities of arrays of cylinders, with particular application to cylinder arrays in heat exchangers. Chapter 7 deals with a special, intriguing topic: rain-and-wind-induced vibration of cables (RWIV) and cable-stayed bridges.

In all the foregoing, the bluff body cross-section is nondeformable, but in Chapter 6 we discuss the case of a deformable cross-section, specifically the ovalling oscillations of cylindrical shells in cross-flow, with application to the ovalling oscillation of tall, thin-walled chimney stacks.

Finally, some concluding remarks are presented in the Epilogue.

2 Prisms in Cross-Flow – Galloping

2.1 Introductory Comments

In this chapter, we use the word *prism* in a general sense to denote a structure of noncircular, not necessarily polygonal, cross-section. We purposely avoid such expressions as "rectangular cylinders", preferring instead "prisms of rectangular cross-section" or simply "rectangular prisms".*

Consider a slender flexible beam or cable, cross-sectionally bluff, submitted to flow normal to its long axis; or alternatively, a flexibly supported bluff body, as in Figure 1.1.[†] We define *galloping* as a velocity-dependent, damping-controlled instability,[‡] giving rise to transverse or torsional motions – for the present, considering it as a one-degree-of-freedom (1-dof) instability. Parkinson (1971) finds the name "rather appropriate", "because of the visual impression given" when it occurs in transmission lines: typically a low-frequency ($\sim$1 Hz), high-amplitude (as much as 3 m) oscillation, reminiscent of a galloping horse – in contrast, on both counts, to the vortex-shedding related Aeolian vibration.[§] For the same reasons presumably, in the early days, galloping was also referred to as "dancing vibrations", e.g. by Davison (1930) who was among the first to describe the phenomenon in detail.

A circular cylinder in cross-flow is immune to galloping. As illustrated in Figure 2.1(a), the flow-related force does not change magnitude and is always in the direction of the flow. Hence, when the body is in motion, the cylinder velocity

---

* A cylinder ($\kappa\acute{\upsilon}\lambda\iota\nu\delta\rho o s$) is something that rolls, from $\kappa\upsilon\lambda\acute{\iota}\omega$ = to roll, and the Greeks have known for quite a while that square wheels do not roll. From that perspective a "square cylinder" is rather oxymoronic. Nevertheless, thanks to American unbridled optimism, such expressions as "square" and "rectangular cylinders" are commonplace in the technical literature nowadays, supplanting the word "prism".

[†] In experiments, a flexibly supported rigid prism system is referred to as a *sectional model*, and the flexible beam or cable as a *full model*.

[‡] We use the expression *velocity-dependent, damping-controlled*, in preference to the widely used *negative-damping instability*, since (i) ultimately all oscillatory instabilities involve negative damping and (ii) the expression we use describes more closely the evolution in the dynamics and, hence, origination of the instability.

[§] For transmission-line conductors, the amplitude of galloping oscillations typically ranges from a few centimeters to 6 m and the frequency from $\frac{1}{4}$ to $1\frac{1}{4}$ Hz. Aeolian vibrations are typically of no more than 2 to 3-cm amplitude and of 10 to 100-Hz frequency (Edwards & Madeyski 1956).

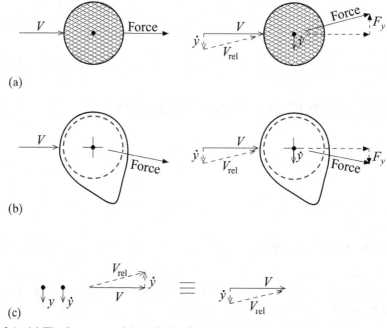

Figure 2.1. (a) The force on a plain cylinder (electrical wire conductor) due to cross-flow; (b) the force on an iced conductor. Diagrams on the left are for a stationary conductor, and on the right for a downwards moving one [after Den Hartog (1956)]. (c) Two equivalent ways of determining V_{rel}: the conventional one (left) and a nonstandard one (right) often used in the literature.

and the component of force in that direction oppose each other, thus negating the possibility of sustenance of the vibration by the flow.

On the other hand, for certain cross-sectional forms, e.g. as in Figure 2.1(b), it is possible for the instantaneous $\dot{y}$ and F_y to be in the same direction and thus for energy to flow into the structure, as discussed by Den Hartog (1932, 1956).

In this chapter, following the classical work on this topic, y and $\dot{y}$ are taken positive downwards, as in Figure 2.1. For the same reason, the nonstandard diagram for obtaining the relative velocity, V_{rel}, is used; as illustrated in Figure 2.1(c), it is wholly equivalent to the conventional one.

Perhaps the description of galloping of a D-section prism with the flat surface facing the flow by Lanchester (1907) was the first ever. This very same system was proposed by Den Hartog (1956) as a benchtop experiment to demonstrate galloping (Figure 2.2(a)).* If, instead of mounting the D-section on springs, it is pivoted in the middle and placed before a fan, then, instead of vibration, we have *autorotation* (Riabouchinsky 1935; Lugt 1978); i.e., if the system is given a starting rotation, it keeps rotating. This is Lanchester's *aerial tourbillion* (Figure 2.2(b)), another effective benchtop teaching aid.

---

* And what an effective teaching aid it is, to demonstrate not only galloping but also the existence of limit-cycle motion! As we shall see later, however, the D-section is prone to galloping only in highly turbulent flow or if subjected to substantial perturbation; but this requirement is perfectly satisfied by the unsteadiness and nonuniformity of the flow in the apparatus shown in Figure 2.2(a).

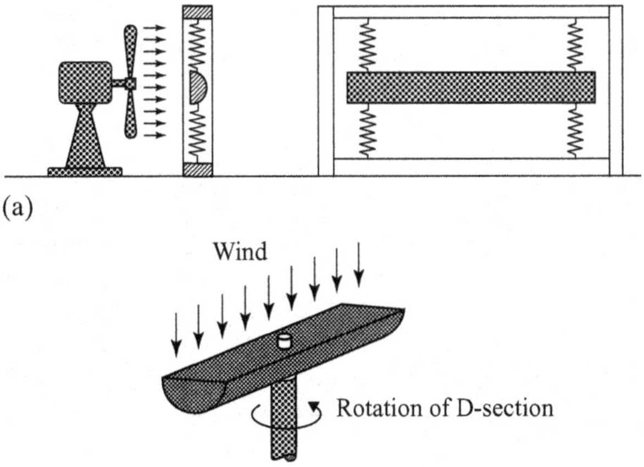

Figure 2.2. (a) Benchtop experiment for demonstrating galloping of a D-section; left: side view; right: front view [from Den Hartog (1956)]. (b) The Lanchester aerial tourbillion [Lanchester (1907)].

A cross-sectional shape as in Figure 2.1(b) may arise because of the accretion of ice on electric transmission lines due to freezing rain or sleet. Such *icing* depends on atmospheric conditions and hence on geography; the ice can be thickest windward or leeward, towards the top or the bottom of the conductor (Rawlins 1979); some iced shapes can gallop if the wind is sufficiently strong, while others cannot. Galloping of transmission lines is a serious problem. Large amplitudes may cause conductor

Notation used in Chapter 2	
Frontal height of prism (normal to flow):	h
"Depth" of prism (in flow direction):	d
Free-stream flow velocity:	V
Reduced flow velocity:	$U = V/\omega h$, ω being the radian frequency $U^* = V/fh$, f being the frequency in Hz $\overline{U} = (\overline{n}/\zeta)U$; $\overline{U}^* = (\overline{n}/\zeta)U^*$
Mass per unit length (span):	$m \equiv M/l$, M being the total mass
Mass parameter:	$n = \rho h^2/2m$; $\overline{n} = \rho h^2/4m$ defined in equations (2.18) and (2.30)
Lateral displacement of prism:	y (positive downwards)
Dimensionless lateral displacement:	$Y = y/h$; $\overline{Y} = (\overline{n}/\zeta)Y$
Angular displacement of prism:	θ (positive clockwise)
Other symbols are defined in the text.	Distinct notation is used in Section 2.11, specific to bridge-deck work.

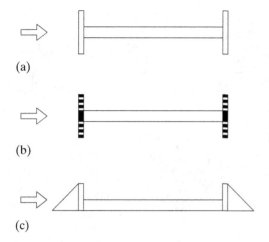

(a)

(b)

Figure 2.3. (a) Typical bridge deck section in cross-wind. (b) Perforated bridge deck sides and (c) bridge deck fitted with fairings.

(c)

clashing, resulting in interruption of service for hours or days, with the costs ranging from thousands to millions of dollars (Rawlins 1979).

Bridge decks of certain cross-sectional shapes are also prone to galloping (Figure 2.3), but the mechanism in this case may be more complex (see, e.g., Naudascher (1974, Session D), Scanlan (1990)) and, almost needless to say, the damage prohibitively expensive. Hence, a great deal of effort has gone into developing modified bridge-deck cross-sectional shapes to elude galloping, e.g. by introducing perforations for the wind to go straight through or by adding fairings.

Parkinson (1971) recognizes four classes of galloping. First, the *transverse* or *translational galloping* discussed in the foregoing in conjunction with iced electric conductors. Second, *torsional galloping* that often arises in bridge decks. A third type involves essentially translational galloping of stranded conductors when the wind is directed a few degrees off the normal to the span (sometimes referred to as a *quartering wind* (Scruton 1971)) and is associated with different separation points on the upper and lower surfaces in the helically wound strands. This phenomenon "occurred spectacularly" circa 1960 on transmission lines crossing the rivers Severn and Wye in the United Kingdom, and it has been studied thoroughly by Davis, Richards & Scriven (1963). A simple remedy was possible here: to wrap tape around the conductors to give them a fairly smooth cross-section (Davis *et al.* 1963; Scruton 1971). A fourth form of galloping may be referred to as *stall flutter*, similar to stall flutter of airfoils oscillating about their stalling angle. Its mechanism involves hysteresis in the separation and reattachment of the flow during oscillation, e.g. in radio telescope and radar reflector bowls (Scruton 1960, 1971).

The most famous occurrence of galloping was that leading to the catastrophic oscillation and collapse of the original, elegant Tacoma Narrows Bridge on 7 November 1940 in 68 km/h wind (Figure 2.4), luckily captured for posterity on ciné-film, thanks to the rather long time it took before catastrophic failure. As glimpsed from the figure, the torsional motion was quite enormous, the cross-section at the quarter points twisting with amplitudes of ± 35° approximately! More will be said about this, later in this chapter; see Sections 2.11 and 2.12.

Figure 2.4. Instantaneous photograph of the galloping Tacoma Narrows bridge in the anti-symmetric torsional mode before collapse [from Scruton (1971)].

Wind-induced galloping of buildings, bridges, cables and traffic signs, for example, has been reported extensively. Less widely known is ocean-current-induced galloping of marine structures, such as box piles and cables (Bokaian & Geoola 1984c); in this case also, cables or bundles thereof are wrapped in waterproof tape or encased in cylindrical containers to prevent galloping (Hallam, Heaf & Wootton 1977).

There is a very rich literature on galloping. A fairly extensive treatment is provided by Blevins (1990), wherein many of the pertinent references on the topic may be found. Most of the important references may also be found in the excellent reviews of the topic by one of the pioneers in research on galloping, G. V. Parkinson (1971, 1972, 1989),* as well as in the compendium of papers by another of the principal researchers in the field, Y. Nakamura (Ohya 1997).

The mechanism of galloping will be elucidated in Section 2.2. This is followed in subsequent sections by discussion of some of the many complexities of the phenomenon, e.g. due to the effects of turbulence, interference by vortex shedding, and reattachment of the separated shear layers, as well as the treatment of torsional and multi-degree-of-freedom galloping.

2.2 The Mechanism of Galloping

In this section, the simplest possible treatment of single-degree-of-freedom galloping is presented, aiming to convey to the reader an appreciation of the underlying

---

* Appropriately, perhaps, Geoff Parkinson was bluff in aspect, and so, when he presented a superb General Lecture on "Mathematical models on flow-induced vibration of bluff bodies", his *forte*, he was introduced by the chairman as "the biggest bluff body of them all" (Bearman *et al.* 2006; p. 993).

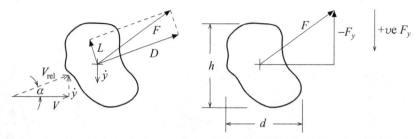

Figure 2.5. A prism, of frontal characteristic dimension ("height") h and depth d, subjected to cross-flow of velocity V and undergoing transverse oscillation in the vertical direction, with downwards velocity v at the instant shown. At that instant, the resultant lift L and drag D are shown in the diagram on the left, and the vertical component of the fluid-dynamic force, F_y, in the diagram on the right.

mechanism. The presentation is based overwhelmingly on the seminal work by Den Hartog (1932, 1956) and Parkinson & Smith (1964).

2.2.1 The linear threshold of galloping

Consider a prismatic bluff body in steady cross-flow of velocity V, as shown in Figure 2.5. The bluff body is assumed to be flexibly mounted as in Figure 1.1, such that only transverse motion, $y(t)$, is possible. It is presumed that the bluff body (prism) oscillates about its static equilibrium position. At a particular instant, its velocity is $v \equiv \dot{y}$ downwards, and hence the effective flow velocity relative to the prism is V_{rel}, at an angle

$$\alpha = \tan^{-1}(\dot{y}/V), \tag{2.1}$$

as shown in Figure 2.5. Hence, the instantaneous lift, L, and drag, D, are in the directions shown in the figure. As motion takes place in the y-direction, we are clearly most interested in the component of the fluid-dynamic force in that direction, F_y, which we define to be *positive downwards*, i.e. in the same direction as $+y$ and $+\dot{y}$; hence, we have $-F_y = L\cos\alpha + D\sin\alpha$ in Figure 2.5, so that

$$F_y = -L\cos\alpha - D\sin\alpha. \tag{2.2}$$

More specifically, in order to assess stability of the system, we are interested in $\mathrm{d}F_y/\mathrm{d}\alpha$. Thus, referring to Figure 2.5, consider an increase in v, $\Delta v \equiv \Delta\dot{y} > 0$; this results in $\Delta\alpha > 0$. If $\mathrm{d}F_y/\mathrm{d}\alpha \simeq \Delta F_y/\Delta\alpha < 0$ and $\Delta\alpha > 0$, then $\Delta F_y < 0$.* From the figure, a positive $\Delta\dot{y}$ is downwards, whereas a negative ΔF_y is upwards, i.e. in the opposite direction, and hence the oscillation will not grow; it will decay. On the other hand, the system is unstable if

$$\frac{\mathrm{d}F_y}{\mathrm{d}\alpha} > 0; \tag{2.3}$$

for in this case $\Delta\dot{y}$ and ΔF_y will be in the same direction.

---

* This cannot be "seen", geometrically, from the figure. ΔF_y being positive (or negative) depends on how L and D vary with α.

Utilizing equation (2.2), we may write

$$\frac{\mathrm{d}F_y}{\mathrm{d}\alpha} = \left(+L - \frac{\mathrm{d}D}{\mathrm{d}\alpha}\right)\sin\alpha + \left(-\frac{\mathrm{d}L}{\mathrm{d}\alpha} - D\right)\cos\alpha. \tag{2.4}$$

For small enough $\dot{y}$, hence small enough α, we have $\sin\alpha \sim \mathcal{O}(\epsilon)$. Therefore, provided that L and $\mathrm{d}D/\mathrm{d}\alpha$ are of a similar order of magnitude as $\mathrm{d}L/\mathrm{d}\alpha$ and D (in fact, they are likely to be smaller), equation (2.4) may be written as

$$\frac{\mathrm{d}F_y}{\mathrm{d}\alpha} \simeq -\left(\frac{\mathrm{d}L}{\mathrm{d}\alpha} + D\right). \tag{2.5}$$

Thus, the criterion for galloping instability, commonly attributed to Den Hartog, is

$$\frac{\mathrm{d}L}{\mathrm{d}\alpha} + D < 0. \tag{2.6}$$

Here, a parenthesis is in order. The same criterion was obtained considerably earlier by Glauert (1919) in connection with the autorotation of a stalled airfoil about a fixed axis.* It is important to stress that here we have taken both y and F_y to be positive downwards, whereas in Glauert's (1919) and Den Hartog's (1932) work F_y is taken to be positive upwards, which results in $\mathrm{d}F_y/\mathrm{d}\alpha < 0$ for galloping; but as $F_y = L\cos\alpha + D\sin\alpha$ in that case, one still obtains $(\mathrm{d}L/\mathrm{d}\alpha) + D < 0$ for galloping. In any case, both sign conventions are widely used in the literature; hence, the reader should beware!

An alternative derivation of inequality (2.6) may be obtained as follows. Rewriting (2.2) for small α as $F_y \simeq -L - D\alpha$ and making Taylor expansions $L \simeq L_0 + (\partial L/\partial\alpha)\alpha$, $D \simeq D_0 + (\partial D/\partial\alpha)\alpha$ leads to

$$F_y = -L_0 + \alpha\left(-\frac{\partial L}{\partial\alpha} - D_0\right) + \mathcal{O}(\alpha^2); \tag{2.7}$$

or, dropping the subscript 0 and utilizing (2.1),

$$F_y \simeq -L - \frac{\dot{y}}{V}\left(\frac{\partial L}{\partial\alpha} + D\right).$$

Now, for the flexibly supported bluff body, recalling that both y and F_y are positive downwards, the equation of motion is $M\ddot{y} + ky = F_y = -L - (\dot{y}/V)[(\partial L/\partial\alpha) + D]$, which can be written as

$$M\ddot{y} + \left[\left(\frac{\partial L}{\partial\alpha} + D\right)\left(\frac{1}{V}\right)\right]\dot{y} + ky = -L, \tag{2.8}$$

where the right-hand side would typically be zero for motions about the static equilibrium. It is seen that the second term is a damping term. It becomes negative, leading to oscillatory instability, if $(\partial L/\partial\alpha) + D < 0$, i.e. if criterion (2.6) is satisfied.

Defining $L = \frac{1}{2}C_L\rho V^2 hl$ and $D = \frac{1}{2}C_D\rho V^2 hl$, and similarly for F_y, ρ being the fluid density, h a characteristic frontal dimension, and l the length of the prism

---

* It is both curious and significant that Scruton (1960, 1971) never cites Den Hartog's work, attributing the galloping criterion to Glauert instead; it is even more curious that Den Hartog does not mention Glauert's work, even though he was familiar with the ARC Reports and Memoranda. Parkinson & Smith (1964) cite both, making it clear that Glauert's work has precedence.

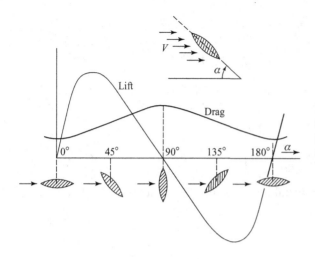

Figure 2.6. Lift and drag as a function of the angle of attack α for a lenticular cross-section, over the whole possible range for α [from Den Hartog (1956)].

perpendicular to the plane of the paper (the span), equation (2.6) may be rewritten in terms of the lift, drag and transverse force coefficients:

$$\frac{\partial C_{F_y}}{\partial \alpha} = -\frac{\partial C_L}{\partial \alpha} - C_D > 0 \quad \text{or} \quad \frac{\partial C_L}{\partial \alpha} + C_D < 0. \tag{2.9}$$

As was made clear in Section 1.2.1 in general terms and as shown in the derivation leading to equation (2.8), galloping is a velocity-dependent damping-controlled instability.

To use equation (2.9), one clearly needs to know C_L and C_D as a function of α. For a given section shape, this data may be obtained from static wind-tunnel measurements, in which the orientation of the section is varied relative to the wind, generating results such as those in Figure 2.6. One then invokes the quasi-steady assumption, according to which C_L and C_D in the course of oscillation are the same at each α as the values measured at the corresponding steady angle of attack α in wind-tunnel experiments.

Clearly, quasi-steady fluid dynamics is applicable if the motion is "quite slow". This is generally true for galloping (see Section 2.1). However, "slow" needs to be quantified properly. Two different criteria have been proposed concerning the applicability of quasi-steady fluid dynamics (Price *et al.* 1988): one by Fung (1955) and the other by Blevins (1977, 1990).

According to Fung (1955), the criterion is this: any disturbance experienced by the oscillating body at a certain point in its oscillatory motion must be swept downstream sufficiently far, by the time the body comes back to that same point (one period later), for the disturbance to no longer affect the flow around the body. Assuming that disturbances are carried downstream with a velocity equal to the free-stream velocity V, then during a period of oscillation they will be swept downstream a distance V/f_n, f_n being the frequency of the oscillation. Fung proposes that this should be at least 10 times the body "diameter" or the characteristic length h (or d), giving $V/f_n h \geq 10$.

According to Blevins (1977), the criterion is that the frequency of shed vortices (the disturbance to the mean flow), that is the Strouhal frequency f_S (see Chapter 3), must be at least twice as large as the oscillation frequency f_n. Assuming a Strouhal number $S = 0.2$, it follows that $V/f_n h \geq 2(V/f_s h) = 2S^{-1} \simeq 10$. Thus, it is interesting to note that, although different physical reasoning was employed, the same end result was obtained by Blevins and Fung for quasi-steady fluid dynamics to be applicable, namely

$$\frac{V}{f_n h} > 10. \tag{2.10a}$$

Later, without further explanation, Blevins (1990) revised the criterion to $V/f_n h$ or $V/f_n D > 20$, D denoting the cylinder diameter – perhaps counting each vortex shed, rather than pairs, or perhaps because for prisms S is generally rather smaller than 0.2 (Section 3.2). In any event, this criterion is usually easily satisfied in most galloping situations. The limits of successful prediction of galloping via equation (2.9) were studied by Nakamura & Mizota (1975a), finding, for example, that one must have $V/f_n h \geq 2(V/f_n h)_{cr}$ for a square prism, where $(V/f_n h)_{cr}$ is the critical reduced flow velocity for vortex-shedding resonance; otherwise the occurrence of galloping is disrupted by vortex shedding (refer to Section 2.10).

Bearman, Gartshore, Maull & Parkinson (1989), on the basis of their work with square prisms and interference by vortex shedding, conclude that it is safe to use the quasi-steady theory of galloping only if $V/\omega_n h \geq 4(2\pi S)^{-1} \simeq 5$, which translates to

$$\frac{V}{f_n h} > 30. \tag{2.10b}$$

Hence, caution should be exercised as to which version of the criterion it is safe to use.

Furthermore, as discussed by van Oudheusden (1995), there is a second requirement for the quasi-steady assumption to be applicable, namely that it is necessary to be able "to define a steady situation (in which the structure is in rest with regard to some suitably chosen reference frame) which is aerodynamically equivalent to the unsteady situation". This innocuous-sounding requirement is sometimes difficult to satisfy; see, e.g., Section 2.7.1.

A very useful nondimensional form of the linear threshold for galloping may be obtained by considering the prism as a damped oscillator subjected to cross-flow, the equation of motion of which is

$$m\ddot{y} + c\dot{y} + ky = F_y \equiv \tfrac{1}{2}\rho V^2 h\, C_{F_y}, \tag{2.11}$$

where m is the section mass (or mass per unit length of the bluff body), and similarly c and k are the damping coefficient and spring constant for the section – thus, the length l, or "span", of the prism (the dimension normal to the plane of the paper) does not appear in (2.11); h is the characteristic length, say the frontal height of the bluff body (Figure 2.5). Expanding C_{F_y} in a Taylor series in α, we have

$$C_{F_y} = C_{F_y}\big|_{\alpha=0} + \big[(\partial C_{F_y}/\partial\alpha)\big|_{\alpha=0}\big]\alpha + \cdots \tag{2.12}$$

Ignoring the equilibrium component, zero for a symmetric section, and recalling that $\alpha \simeq \dot{y}/V$, we can rewrite (2.11) as

$$m\ddot{y} + \big[c - \tfrac{1}{2}\rho V h\, (\partial C_{F_y}/\partial\alpha)\big]\dot{y} + ky = 0, \tag{2.13}$$

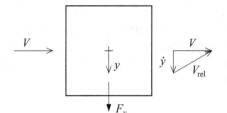

Figure 2.7. A square prism in cross-flow, on an elastic dissipative support (not shown), free to oscillate in the y-direction.

where it is understood that the partial derivative is evaluated at $\alpha = 0$. Further, expressing $c/m = 2\zeta\omega_n$ and $k/m = \omega_n^2$, equation (2.13) leads to

$$\ddot{y} + 2\omega_n \left[\zeta - \frac{\rho V h}{4m\omega_n} \frac{\partial C_{F_y}}{\partial \alpha} \right] \dot{y} + \omega_n^2 y = 0. \tag{2.14}$$

The threshold of galloping occurs when the quantity in square brackets vanishes, and so $V_{\text{crit}} = 4m\omega_n\zeta / \left[\rho h (\partial C_{F_y}/\partial \alpha) \right]$. This may be written in dimensionless form as

$$\frac{V_{\text{crit}}}{f_n h} = \frac{4}{\left[\partial C_{F_y}/\partial \alpha \right]} \frac{m\delta}{\rho h^2} = \frac{4}{\left[-(\partial C_L/\partial \alpha) - C_D \right]} \frac{m\delta}{\rho h^2}, \tag{2.15}$$

where $f_n = \omega_n/2\pi$ and, assuming a lightly damped system, $\delta \simeq 2\pi\zeta$; the product of the mass ratio $m/\rho h^2$ and the logarithmic decrement δ, i.e. $m\delta/\rho h^2$, is usually referred to as the mass-damping parameter. The bracketed expression, in both forms above, must be positive for galloping to occur.

Variants of this expression recur throughout this book – cf. equations (4.9) and (5.10) with $h = D$. They differ from (2.15) because a time delay may exist between the motion and the fluid-dynamic forces causing it as well as different sign conventions.

2.2.2 Nonlinear aspects

In the foregoing section, the critical flow velocity (wind speed) for galloping has been predicted according to linear theory. This critical velocity is the same as that predicted by nonlinear theory, only so long as the bifurcation is supercritical (see Figure 1.6); but, as we shall see, in the case of galloping the dynamical behaviour can be more complex. Nonlinear theory, of course, additionally gives an estimate of the amplitude and frequency of galloping. Here, we shall illustrate the analysis and results obtainable by nonlinear considerations, following Parkinson & Smith's (1964) seminal work, although a different method of solution is adopted.

Consider a flexibly mounted square prism in cross-flow, as shown in Figure 2.7, free to oscillate transversally, in the vertical direction. The equation of motion may be written as

$$ml\frac{d^2y}{dt^2} + r\frac{dy}{dt} + ky = \tfrac{1}{2}C_{F_y}\rho V^2 hl, \tag{2.16}$$

where m is the mass of the prism per unit length, h its height (and depth), l its length normal to the plane of the paper, k and r are, respectively, the stiffness and damping of the flexible support (not shown in the figure), and C_{F_y} is the coefficient of the vertical component of the fluid-dynamic force, F_y (see Figure 2.5). It is important to note that, as in Section 2.2.1, F_y here is taken *positive downwards* (the same as

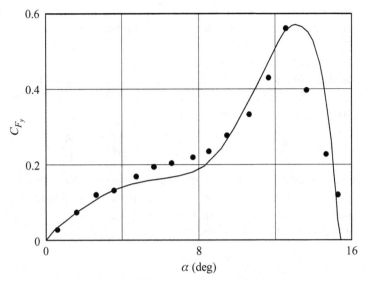

Figure 2.8. Transverse force coefficient C_{F_y} as a function of the angle of attack α for a square prism: •, experimental points (Smith 1962) at Re = 22 300; —, polynomial approximation, equation (2.20) with $A = 2.69$, $B = 168$, $C = 6270$ and $D = 59\,900$ (Parkinson & Smith 1964).

for y). Thus, invoking (2.12) with $C_{F_y}|_{\alpha=0}$ suppressed and (2.9), we must have

$$\frac{\partial C_{F_y}}{\partial \alpha} \equiv -\frac{\partial C_L}{\partial \alpha} - C_D > 0 \qquad (2.17)$$

for galloping.

Next, let

$$y = Yh, \quad \tau = \omega t, \quad V = U\omega h, \quad n = \frac{\rho h^2}{2m}, \quad \omega^2 = \frac{k}{ml}, \quad 2\zeta\omega = \frac{r}{ml}, \qquad (2.18)$$

where n is dimensionless mass parameter and ζ is the damping ratio. Substituting these relationships in equation (2.16), we obtain the dimensionless equation

$$\ddot{Y} + 2\zeta\dot{Y} + Y = n\,U^2\,C_{F_y}, \qquad (2.19)$$

where the overdot denotes a derivative with respect to τ.

For the square prism, Smith (1962) measured C_{F_y} as a function of α in a wind tunnel, and the results are shown in Figure 2.8. Parkinson & Smith (1964) obtained a polynomial approximation to these results (again suppressing the $C_{F_y}|_{\alpha=0}$ term, which changes the static equilibrium but does not affect the dynamics),

$$C_{F_y} = A\left(\frac{dy/dt}{V}\right) - B\left(\frac{dy/dt}{V}\right)^3 + C\left(\frac{dy/dt}{V}\right)^5 - D\left(\frac{dy/dt}{V}\right)^7$$

$$= A\left(\frac{\dot{Y}}{U}\right) - B\left(\frac{\dot{Y}}{U}\right)^3 + C\left(\frac{\dot{Y}}{U}\right)^5 - D\left(\frac{\dot{Y}}{U}\right)^7, \qquad (2.20)$$

also shown in the figure as a continuous line. Equation (2.19) may now be written as

$$\ddot{Y} + Y = nA\left\{\left(U - \frac{2\zeta}{nA}\right)\dot{Y} - \left(\frac{B}{AU}\right)\dot{Y}^3 + \left(\frac{C}{AU^3}\right)\dot{Y}^5 - \left(\frac{D}{AU^5}\right)\dot{Y}^7\right\}, \qquad (2.21)$$

where the first term on the right-hand side represents the linear damping, both structural and fluid-dynamic, and the rest are higher-order velocity-dependent terms.

Parkinson & Smith obtained solutions to this equation by means of the Krylov & Bogoliubov method (Minorsky 1962). Here, however, we use the method of multiple scales (see, e.g., Nayfeh (1981)), outlined for a simpler problem in Appendix A. We rewrite equation (2.21) in the following form:

$$\ddot{Y} + Y = \varepsilon \left\{ \left[UA_1 - \frac{2\zeta}{n} \right] \dot{Y} - \frac{A_3}{U} \dot{Y}^3 + \frac{A_5}{U^3} \dot{Y}^5 - \frac{A_7}{U^5} \dot{Y}^7 \right\}, \qquad (2.22)$$

where we have replaced the n multiplying the expression in curly brackets by ε; in view of the definition of n in (2.18), for airflow at least, $\varepsilon \equiv n$ is indeed small,* and hence this equation is now in a form that may be analysed by the method adopted.

Defining ordinary ("fast") time and slow time by

$$\xi = \tau \quad \text{and} \quad \eta = \varepsilon \tau, \qquad (2.23)$$

respectively, we next expand Y in a Lindstedt-type perturbation series

$$Y = Y_0 + \varepsilon Y_1 + \varepsilon^2 Y_2 + \cdots \qquad (2.24)$$

Substituting into (2.22) and collecting terms in powers of ε, we find the generating solution to be

$$Y_0 = \overline{A}(\eta) \cos \xi + \overline{B}(\eta) \sin \xi = R(\eta) \cos(\xi + \phi(\eta)), \qquad (2.25)$$

where the overbar is used to differentiate these quantities from those in (2.21).

Proceeding with the first-order solution as in Appendix A leads to

$$\frac{dR}{d\eta} = \frac{1}{2} \varepsilon \left\{ \left[UA_1 - \frac{2\zeta}{n} \right] R - \frac{3}{4} \frac{A_3}{U} R^3 + \frac{5}{8} \frac{A_5}{U^3} R^5 - \frac{35}{64} \frac{A_7}{U^5} R^7 \right\}, \qquad (2.26)$$

after elimination of the secular terms. The amplitude and stability of any possible limit-cycle solutions may be determined from this equation. Specifically, the limit-cycle amplitude is determined by letting $dR/d\eta = 0$, yielding

$$f(R) = \left[UA_1 - \frac{2\zeta}{n} \right] R - \frac{3}{4} \frac{A_3}{U} R^3 + \frac{5}{8} \frac{A_5}{U^3} R^5 - \frac{35}{64} \frac{A_7}{U^5} R^7 = 0, \qquad (2.27)$$

which has seven roots. However, only finite real roots correspond to limit cycles. There is also the trivial root $R = 0$, which governs the linear stability of the system. Making reference to Section 1.2.1, or just looking at equation (2.26), or indeed (2.22), it is clear that the threshold of galloping corresponds to the point where the linear damping becomes negative, i.e.

$$U_{cr} = \frac{2\zeta}{nA_1}, \qquad (2.28)$$

which corresponds to the result obtainable by linear theory.

However, to relate to the work presented in Section 2.2.1, it is best to look at equation (2.22) and to consider the *mechanically undamped system* ($\zeta = 0$), whereupon it becomes clear that the system is subject to negative damping

---

* In Smith's (1962) experiments, $n = 4.3 \times 10^{-4}$ (Parkinson & Smith 1964).

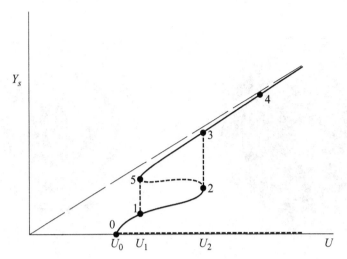

Figure 2.9. Galloping amplitude versus flow velocity, showing the various bifurcations and the "oscillation hysteresis" [from Parkinson & Smith (1964)]; $U_o = U_{cr}$.

provided that $A_1 > 0$; moreover, from (2.20), it is equally clear that $A_1 = [dC_{F_y}/d(\dot{y}/V)]_{\dot{y}/V=0} \equiv (dC_{F_y}/d\alpha)|_{\alpha=0}$. Hence, we retrieve the Den Hartog criterion of equations (2.3) and (2.9) for galloping, as well as the earlier version by Glauert (1919) for autorotation – the only difference being that, here, F_y is defined positive downwards, whereas in Den Hartog and Glauert it is defined positive upwards; see also Parkinson & Smith (1964, p. 231).* With mechanical damping included, equation (2.28) can readily be transformed into the first form of equation (2.15).

Returning now to equation (2.27) and possible real roots, the following was found by Parkinson & Smith:

(i) for $U < U_{cr} \equiv U_0$, only the trivial $R = 0$ solution exists, and the system is stable ($dR/d\eta < 0$);
(ii) for $U_0 < U < U_1$ and for $U > U_2$ two roots are found, one of which corresponds to a stable limit cycle, whereas the trivial solution is now unstable;
(iii) for $U_1 < U < U_2$ four roots are found, involving (a) the unstable source at the origin of the phase plane, (b) a stable limit cycle nesting within (c) an unstable one, which in turn nests within (d) a larger stable limit cycle.

The overall picture, including the obvious meaning of U_1 and U_2, is seen in Figure 2.9. Thus, the bifurcation is supercritical, but one can have (a) a single, "small", stable limit cycle; (b) two, one small and the other "large", with an unstable one in-between; or (c) only a "large" stable limit cycle.

Phase-plane diagrams for the same parameters as in Parkinson & Smith (1964) have been constructed, as shown in Figure 2.10(a–c); i.e. for the same values of A, B, C, D as in the caption of Figure 2.8 and $n = 4.3 \times 10^{-4}$; ζ was arbitrarily taken as $\zeta = 4.28 \times 10^{-2}$, so that $nA_1/2\zeta = 1.35 \times 10^{-2}$. The values of $(nA_1/2\zeta)U$ – cf.

---

* It is reiterated that this is an important source of confusion in the literature: whether y is defined positive upwards or downwards, and similarly for F_y. Both versions are used extensively; hence, care should be exercised by the reader to ascertain definitions in each paper. The extent of the problem is illustrated by the fact that different conventions have been used in the first and second editions of Blevins' (1977, 1990) book.

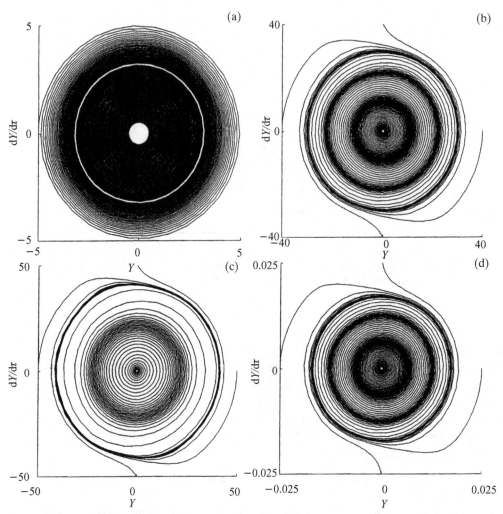

Figure 2.10. Phase-plane diagrams for the dynamical behaviour of the system (a) for $U_0 < U < U_1$, (b) and (d) $U_1 < U < U_2$ and (c) $U > U_2$, where U_0, U_1 and U_2 are defined in Figure 2.9. Specifically the results shown are for a system with A to D as in Figure 2.8, and $\zeta = 4.28 \times 10^{-2}$. In (a) $n = 4.3 \times 10^{-4}$, $(nA_1/2\zeta)U = 1.08$; (b) $n = 4.3 \times 10^{-4}$, $(nA_1/2\zeta)U = 1.50$; (c) $n = 4.3 \times 10^{-4}$, $(nA_1/2\zeta)U = 2.00$. In (d) $n = 7.5 \times 10^{-1}$ and $(nA_1/2\zeta)U = 1.50$.

Figure 2.11 – are as follows: in (a) 1.08 (corresponding to $U \simeq 80$); in (b) 1.50 ($U \simeq 111$), in (c) 2.00 ($U \simeq 148^*$). We see in (a) one limit cycle with an amplitude $Y = Y_s \simeq 3.19$ – the white circle (purposely left blank in the spiralling trajectories for easy visual identification). In (b) there is a stable limit cycle at $Y_s \simeq 10.6$ (the inner dark circle), nesting within an unstable one at $Y_s \simeq 22.3$ (the intermediate dark circle), which in turn is within a larger stable one at $Y_s \simeq 30$ (the outer dark circle); i.e. we are between U_1 and U_2 in Figure 2.9. In (c) we once more have but one stable limit cycle with $Y_s \simeq 41.2$; i.e. we are beyond U_2 in Figure 2.9.

---

* For a body with $h \simeq 16$ cm and galloping at 1 Hz ($\omega = 2\pi$ rad/s), these values of U would be approximately the same as for V in m/s. If $h \simeq 1.6$ cm, V (m/s) would be ten times smaller; cf. equations (2.15).

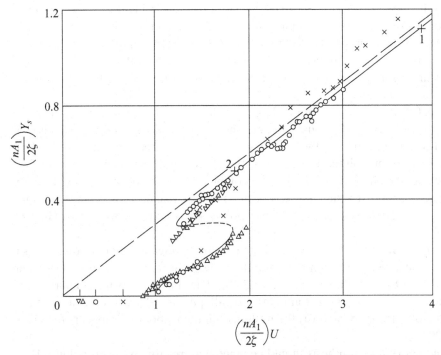

Figure 2.11. Collapsed amplitude-velocity characteristics of galloping of a square prism. — , Theoretical stable limit cycle; – – – , experimental unstable limit cycle. Experimental data points: ×, $\zeta = 0.00107$; ○, $\zeta = 0.00196$; △, $\zeta = 0.00364$; ▽, $\zeta = 0.00372$; +1, $\zeta = 0.0012$; +2, $\zeta = 0.0032$; Reynolds number range $4 \times 10^3 - 20 \times 10^3$; $n = 4.3 \times 10^{-4}$ [from Parkinson & Smith (1964)].

In Figure 2.10(d) we use the same parameters, except for $n = 7.5 \times 10^{-1}$, which might be representative of waterflow rather than airflow (but not so small in terms of setting $n = \varepsilon$), and $nA_1/2\zeta = 23.57$. In this case, the results are similar to those in Figure 2.10(b): stable, unstable and larger stable limit cycles at: $Y_s \simeq 6.2 \times 10^{-3}$, 12.7×10^{-3} and 16.8×10^{-3}. These amplitudes are much smaller, three orders of magnitude smaller, than those in (b). Nevertheless, the values of $(nA_1/2\zeta)Y_s$ in the two cases are sensibly the same – cf. Figure 2.11. It should also be remarked in conjunction with the "waterflow" results in (d) that, not only the amplitudes are smaller than for airflow, but the dimensionless critical flow velocities U are also correspondingly three orders of magnitude smaller; cf. Figure 2.11 again.

Explicit expressions for the amplitude of galloping, both outside and within the hysteresis region, are given by Novak (1969).

Parkinson & Smith (1964) also conducted experiments for the square prism, with $n = 4.3 \times 10^{-4}$ and variable ζ. By plotting the results in a $(nA_1/2\zeta)\, Y_s$ versus $(nA_1/2\zeta)\, U$ plot, where Y_s is the galloping amplitude, the results for various ζ collapse nicely onto effectively a single "universal" curve, as shown in Figure 2.11 – but see also Section 2.3.2. The experimental results agree remarkably well with the theoretical curve. It should also be mentioned that all the values of Y_s for stable and unstable limit cycles in the previous two paragraphs related to Figure 2.10, once converted to $(nA_1/2\zeta)Y_s$, fall almost exactly on the curve generated by Parkinson & Smith, shown in Figure 2.11.

As the flow velocity is increased to beyond $U = U_2$ in Figure 2.9, the galloping amplitude should jump from point 2 to point 3, and then continue on to point 4; whereas starting from point 3 and reducing U to less than U_2, the galloping follows the path from 3 to 5, and then jumps down to point 1. Remarkably, even these jumps, indeed the whole hysteresis phenomenon, are well followed by the experimental data, as seen in Figure 2.11.

Luo, Chew & Ng (2003) studied numerically the hysteresis phenomenon in the galloping oscillation of a square prism. They confirmed that the phenomenon is related to the inflection in the C_{F_y} versus α curve (Figures 2.8 and 2.12(b)). Furthermore, they pinpointed the cause of the inflection as due to intermittent flow reattachment at alternate shedding cycles on one side of the prism. The existence of intermittent flow reattachment was confirmed by flow visualization in a water tunnel. Further work on the role of the inflection points was conducted by Barrero-Gil *et al.* (2009a).

Barrero-Gil *et al.* (2009b) studied the possibility of galloping of a square prism at low enough Reynolds numbers for the flow to be laminar (Re < 200) using Re-dependent coefficients in equation (2.20), determined from the simulations of Sohankar *et al.* (1998). It was found that (i) galloping is indeed possible at such low Re, but not if Re < 159 and (ii) the hysteresis in the response disappears for $159 \leq$ Re ≤ 200.

An important parenthesis should be made here regarding the grouping $nA_1/2\zeta$ utilised in Figure 2.11. In view of (2.18), this may be written as $A_1(\rho h^2/2m)(1/2\zeta) \simeq A_1\pi(2\mu\delta)^{-1}$, where $\mu = (2n)^{-1} = m/\rho h^2$ and $\delta = 2\pi\zeta/(1-\zeta^2)^{1/2} \simeq 2\pi\zeta$ for small damping. At this point we recognize $2\mu\delta$ as the *Scruton number*, Sc (Zdravkovich 1982).

Finally, a recent study by Vio, Dimitriadis & Cooper (2007) should be cited, in which a variety of methods were used to determine the limit-cycle oscillation amplitude: (i) cell mapping, (ii) harmonic balance, (iii) higher-order harmonic balance, (iv) centre manifold linearization, (v) normal form reduction and (vi) numerical continuation methods; their advantages and disadvantages were discussed.

2.3 Further Work on Translational Galloping

A great deal of work has been done beyond Parkinson & Smith's (1964) seminal paper on translational galloping of square prisms and the other work presented in Section 2.2. Some of it is discussed here, and some in the sections that follow.

2.3.1 The effect of sectional shape

As discussed in the excellent reviews by Parkinson (1974, 1989) and Bearman *et al.* (1987), the importance of *afterbody* length and shape on galloping cannot be overemphasised; indeed, as succinctly put by Parkinson, in the absence of an afterbody there is no galloping. Here by "afterbody" we understand the part of the body downstream of the points of separation; thus, for square-section and D-section prisms, it is the whole of the body behind the front face.

A very useful compilation of data as to which shapes are prone to galloping is provided by Blevins (1990), based on data from Parkinson & Brooks (1961), Slater (1969), Nakamura & Mizota (1975a) and Nakamura & Tomonari (1977), reproduced in Table 2.1. It gives the values of $\partial C_{F_y}/\partial\alpha$, defined in equation (2.9), for various section shapes and for both smooth and turbulent flow. It is recalled that, according to (2.17), galloping will occur if $\partial C_{F_y}/\partial\alpha > 0$. The section shapes are characterised by their depth, d, and height, h (cf. Figure 2.5), both expressed in terms of a single parameter D in the diagrams in the table. Thus, for smooth flow, it is seen that several sections will not gallop according to this criterion, including the oblong rectangular sections with small height-to-depth ratio ($h/d = \frac{1}{4}$ or less), as well as the D-section (refer to the fifth footnote of Section 2.1). It is also noted that some sections, which according to Table 2.1 will not gallop in smooth flow, may do so in turbulent flow, and *vice versa* (see Section 2.9).

In the simplified treatment of galloping of Section 2.2.1 it was presumed that the separated shear layer, typically from the front of the body, does not reattach to it farther downstream, which could complicate matters. This (no reattachment) is generally true if the depth d of the section, is not much larger than h. As discussed in Section 2.3.5, reattachment plays a central role in the characteristics of galloping.

The most important physical parameters for galloping (and, indeed, for vortex shedding) are the size, relative to the height h, and shape of the afterbody (Parkinson 1974). The pressure loading associated with galloping acts principally on the afterbody surface. Hence, a very short afterbody will not oscillate from rest* (e.g. for $h/d = 4$ or 2 in smooth flow), but a square prism will. Under appropriate conditions, a D-section with the flat surface facing the wind will also gallop (Figure 2.2(a)), as $\partial C_{F_y}/\partial\alpha \simeq 0$; significantly though, if the D-section faces the opposite way (no afterbody), no galloping is possible. However, reattachment too may negate galloping (e.g. for $h/d = \frac{1}{4}$ or less), though not necessarily. Therefore, in summary, Table 2.1 suggests that d/h must be large enough, but not so large as to permit reattachment.

However, the conclusions above were reached strictly in terms of equation (2.17), which is a linear criterion and therefore not unequivocally reliable. For instance, the bifurcation leading to galloping may not be supercritical as in Figures 1.6(a) and 2.9, but subcritical as in Figure 1.6(b). These questions have been studied by Novak (1971, 1972) in terms of the following form of the polynomial expansion for C_{F_y}:

$$C_{F_y} = \sum_{i=1}^{m} A_i \left(\frac{\dot{y}}{V}\right)^i + \sum_{j=2}^{k} A_j \left(\frac{\dot{y}}{V}\right)^j \frac{\dot{y}}{|\dot{y}|}, \qquad (2.29)$$

where V is the wind velocity; the first sum involves odd integers (i, m), and the second even integers (j, k); note the difference in signs *vis-à-vis* (2.20).

In what follows, the following dimensionless parameters are used, some as before, some new:

$$\begin{array}{lll}
\text{dimensionless amplitude :} & Y = a/h, & \\
\text{reduced flow velocity :} & U = V/\omega h, & \\
\text{Novak's mass parameter :} & \bar{n} = \rho h^2/4m = \frac{1}{2}n, ^\dagger & (2.30) \\
\text{modified dimensionless amplitude :} & \bar{Y} = (\bar{n}/\zeta)Y, & \\
\text{"twice reduced" flow velocity :} & \bar{U} = (\bar{n}/\zeta)U. &
\end{array}$$

The notation here is different from that in Novak's papers, to conform with the rest of this chapter. Here, a is the dimensional galloping amplitude, and the height h is defined on top of Figure 2.12.

---

* I.e., without perturbation, strictly in terms of equation (2.28); as discussed later in this section.
† Note that in Novak $\bar{n} = \rho h^2/(4m) = \frac{1}{2}n$; i.e. half that in Parkinson & Smith.

Table 2.1. *The transverse force coefficient[a] for various sections in steady smooth or turbulent flow (after Blevins (1990))*

Section	h/d	$\partial C_{F_y}/\partial\alpha$ Smooth flow	Turbulent flow[b]	Reynolds number
(square, $D\times D$)	1	3.0	3.5	10^5
(rect., $\tfrac{2}{3}D$ wide, D tall)	3/2	0.	−0.7	10^5
(rect., $D/2$ wide, D tall)	2	−0.5	0.2	10^5
(rect., $D/4$ wide, D tall)	4	−0.15	0.	10^5
(rect., $\tfrac{2}{3}D$ tall, D wide)	2/3	1.3	1.2	6.6×10^4
(rect., $D/2$ tall, D wide)	1/2	2.8	−2.0	3.3×10^4
(rect., $D/4$ tall, D wide)	1/4	−10.	–	$2\times10^3 - 2\times10^4$
(Thin airfoil, width D)	–[c]	−6.3	−6.3	$>10^3$
(airfoil section, width D)	–	−6.3	−6.3	$>10^3$
(D-section, height D)	–	−0.1	0.	6.6×10^4
(half-round)	–	−0.5	2.9	5.1×10^4
(angle/chevron, height D)	–	0.66	–	7.5×10^4

[a] α is in radians; flow is left to right. $\partial C_{F_y}/\partial\alpha = -\partial C_L/\partial\alpha - C_D$, with C_{F_y} based on the dimension D, so that $\partial C_{F_y}/\partial\alpha > 0$ for galloping.
[b] Approximately 10% turbulence.
[c] Inappropriate to use h/d.

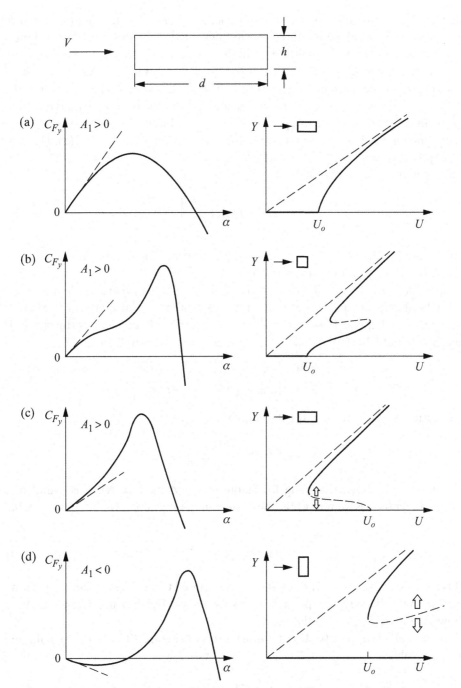

Figure 2.12. Typical lateral force coefficients and corresponding types of galloping response for prisms of different height, h, and depth, d; $Y = a/h$ and $U = V/\omega h$; from Novak (1971).

Three cases are considered, as follows, with the aid of Figure 2.12.

(i) Coefficient $A_1 > 0$. This is the case discussed in Section 2.2.2. The instability is associated with a Hopf bifurcation, beyond which the equilibrium position is no longer stable. Beyond the critical flow velocity, $U_o = \zeta/\bar{n}A_1$, the same as U_{cr} in equation (2.28), a finite flutter amplitude is obtained. The right-hand-side diagrams of

Figure 2.12 give the dimensionless amplitude, $Y = a/h$ as a function of $U = V/\omega h$. In Figure 2.12(a, b, c) we see three possibilities for $A_1 > 0$. In (a) the bifurcation is supercritical, and there is no inflection in the C_{F_y} versus α curve, and the amplitude of galloping increases continuously; in (b) there is an inflection, and hence there is a portion of the galloping response showing hysteresis. It is of interest that the hysteresis in the case of a square prism (Figure 2.11) occurring at $U > U_o$ is captured by the fifth-order polynomial representation of C_{F_y}, but it is missed if a third-order polynomial is used, and the predicted response looks like Figure 2.12(a). (If a cubic approximation is used, $C_{F_y} = A_1(\dot{y}/V) + A_3(\dot{y}/V)^3$, $A_2 = 0$, Novak (1969) shows that the limit-cycle amplitude is

$$\overline{Y} = \left[4(1 - \overline{U} A_1)\frac{\overline{U}}{3A_3}\right]^{1/2}, \tag{2.31}$$

with $\overline{Y}$ and $\overline{U}$ defined in (2.30); see also Blevins (1990)). Finally, in Figure 2.12(c) we see a case where galloping is subcritical.

   (ii) Coefficient $A_1 = 0$. This is a subclass of Figure 2.12(d) where the slope at $\alpha = 0$ in the left-hand-side diagram is zero. In this case the Den Hartog/Glauert criterion predicts no galloping. However, galloping *is* possible if sufficient perturbation is supplied (Figure 2.12(d)), yielding amplitudes $Y = a/h$ (Novak 1972)

$$Y_{1,2} = \frac{2}{3}\frac{A_2}{|A_3|}\left[\frac{8}{3\pi} \pm \sqrt{\frac{64}{9\pi^2} - 3\frac{\zeta}{\overline{n}U}\frac{|A_3|}{A_2^2}}\right] U, \tag{2.32}$$

which can only be real when the discriminant is positive, i.e. for

$$U \geq U_o = \frac{27\pi^2}{64}\frac{\zeta}{\overline{n}}\frac{|A_3|}{A_2^2}; \tag{2.33}$$

as before, $U = V/\omega h$. Checking for stability (see Appendix A), it is found that the upper branch is stable, while the lower one is unstable. The minimum stable solution is

$$Y_0 = \frac{3\pi}{4}\frac{\zeta}{\overline{n}A_2}. \tag{2.34}$$

This is a classical case of a *hard oscillator*, as opposed to a *soft oscillator* in which the oscillation is self-excited (supercritical or subcritical). The dynamics is essentially as shown in Figure 2.12(d), on the right.

   (iii) Coefficient $A_1 < 0$. This situation is shown in Figure 2.12(d). Steady galloping is still possible, for

$$U \geq U_o = \frac{\zeta}{\overline{n}}\left[\frac{64}{27\pi^2}\frac{A_2^2}{|A_3|} - |A_1|\right]^{-1}, \tag{2.35}$$

provided $A_1, A_2, A_3 \neq 0$. The threshold flow velocity is larger than in the previous case, and a large perturbation is *necessary* to precipitate galloping. Clearly, this also is a case of a hard oscillator.

   Novak & Tanaka (1974) presented numerical values for the coefficients of polynomial expressions of C_{F_y} similar but not identical to equations (2.20) and (2.29),

Table 2.2. *Aerodynamic coefficients of polynomial expansion of the transverse force coefficient, C_{F_y}, as in equation (2.28) – after Novak (1969) and Novak & Tanaka (1974); here, $C_{F_y} = \sum_i A_r(\dot{y}/V)^r, r = 1, 2, 3, \ldots$*

Coefficients	D-section	Rectangle, $h/d = \frac{3}{2}$	Square	Rectangle, $h/d = \frac{2}{3}$
A_1	-0.097431	0	2.69	1.9142
A_2	4.2554	-32.736	0	34.789
A_3	-28.835	7.4467×10^2	-1.684×10^2	-1.7097×10^2
A_4	61.072	-5.5834×10^3	0	-22.074
A_5	-48.006	1.4559×10^4	6.27×10^3	–
A_6	12.462	8.1990×10^3	0	–
A_7	–	-5.7367×10^4	-5.99×10^3	–
A_8	–	-1.2038×10^5	–	–
A_9	–	3.37363×10^5	–	–
A_{10}	–	2.0118×10^5	–	–
A_{11}	–	-6.7549×10^5	–	–

for rectangular prisms with height-to-depth ratio $h/d = \frac{3}{2}, 1$ and $\frac{2}{3}$, as well as for the D-section, based on which the critical condition and nonlinear post-critical behaviour can be assessed. Novak did this for both smooth and turbulent flow, but only the former data is given in Table 2.2, whereas discussion of the latter is deferred to Section 2.9. Thus, the D-section is stable in smooth flow but, as discussed in the foregoing, can gallop if suitably perturbed. The same is true for the $h/d = \frac{3}{2}$ rectangle. The galloping for the square and $h/d = \frac{3}{2}$ prisms is as in Figure 2.12(b) and (d), respectively.

Two sample cases of subcritical and hard-oscillator behaviour, obtained by Novak (1972), are shown in Figures 2.13 and 2.14, respectively. To obtain some of these results, because they represent nonuniform motion along the length (*span*) of the prism, and in one case nonuniform wind profile, the work in Section 2.3.2 has to be used.

In the case of Figure 2.13, $h/d = \frac{1}{2}$. According to Blevins (1990), long rectangles (small h/d) gallop with strong hysteresis, provided $h/d > 1/3$; beyond that (for smaller h/d), they are stable (Parkinson 1971; Washizu *et al.* 1978) because of reattachment of the shear layer. The $h/d = 2$ prism (Figure 2.14) also "gallops with difficulty" in the sense that it is a very hard oscillator, but in this case the reason is that h/d is large; as already stated, as h/d gets larger (shorter afterbody) galloping becomes harder in both senses.

Several important studies on various aspects of galloping of rectangular prisms were conducted, including those by Nakamura & Mizota (1975a), Nakamura & Tomonari (1977, 1981) and Washizu, Ohya, Otsuki & Fujii (1978); also, work on low-speed galloping (Section 2.4), further work on the effect of turbulence (Section 2.9), on elongated prisms (Section 2.11) and on many other aspects.

A numerical study of galloping was conducted by Robertson, Sherwin & Bearman (2003) at Re = 250, using a method developed by Li *et al.* (2002) for solving the Navier-Stokes equations in a moving frame on a fixed grid. Time-averaged values were used to calculate F_y and hence $(\partial C_{F_y}/\partial \alpha)|_{\alpha=0}$ and thereby to predict

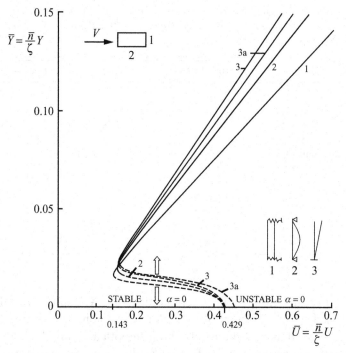

Figure 2.13. Universal galloping response of rectangular prismatic bodies with side ratio $h/d = \frac{1}{2}$ subjected to smooth flow, in terms of $\overline{Y} = (\overline{n}/\zeta)\,Y$ and $\overline{U} = (\overline{n}/\zeta)U$; curves 1, 2 and 3 were computed with wind profile exponent $\alpha = 0$; curve 3a with $\alpha = \frac{1}{6}$ (Novak 1972).

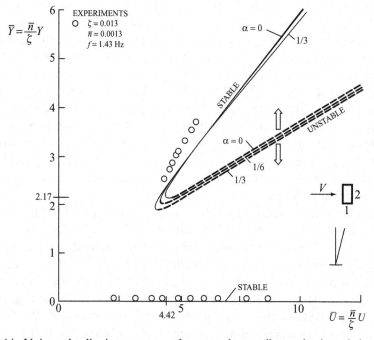

Figure 2.14. Universal galloping response of rectangular cantilevered prismatic bodies with side ratio $h/d = 2$ and aspect ratio $l/h = 3.5$ subjected to smooth flow (Novak 1972).

transverse galloping, essentially via equation (2.15). At the same time, simulations of the dynamics were conducted and galloping was observed directly. Transverse galloping was found to be possible for $d/h = 1$, 1.5 and 2, but not for $d/h \geq 3$, consistently with the quasi-steady calculations. These results are also consistent with the experimental results of Washizu *et al.* (1978) for Re $\simeq 10^5$, despite the disparity in Reynolds number.

Most of the foregoing discussion focussed on rectangular prisms, for historical and practical reasons (e.g. applications to buildings and bridge decks in wind). However, as pointed out by Luo, Yazdani, Chew & Lee (1994), there is no reason why trapezoidal prisms or indeed rectangular prisms at a non-zero mean angle of attack (tilt) should receive less attention; the wind on a building, after all, could come from any direction! Luo *et al.* (1994) studied the effects of incidence and afterbody shape on the flow structure for trapezoidal (trapezium section) and triangular (isosceles triangle section) prisms, undertaking flow visualization as well as transverse force measurements and obtaining steady and fluctuating pressure distributions. It is concluded that one cannot generally say that a particular prism shape is absolutely stable to galloping, because if it is so for one mean angle of incidence, α_0, it may be unstable for another α_0.

This work was taken farther by Luo *et al.* (1998) who conducted forced transverse oscillation experiments of square, trapezoidal and triangular prisms. For the triangular prisms, the base faced the wind (when $\alpha_0 = 0$), and similarly the thicker end of the trapezoidal prisms was windward; the tilt ranged over $0 \leq \alpha_0 \leq 32°$. The threshold of galloping predicted quasi-statically (Section 2.2.1) was compared with that involving the measured phase angles between the forced motion and the transverse force; agreement was found to be excellent. The effect of α_0 was quite important. For small α_0, cross-sectional shapes with a substantial afterbody when tilted are prone to galloping, but not so otherwise; e.g. for the triangular prism, because the sides taper away sharply (Section 2.3.4). As α_0 is increased, the prisms generally become more unstable because of increased asymmetry. In some cases, "partially unstable" galloping was found, i.e. galloping is predicted by the phase-angle criterion at low U, but stability is regained at higher U; thus, this is akin to the "low-speed galloping" discussed in Section 2.4. For large α_0, the prisms become more stable, because of reattachment of one of the shear layers to the corresponding side of the prism (Sections 2.3.4 and 2.3.5).

Further work on triangular prisms has been done, e.g. by Alonso *et al.* (2007), but with the thin edge windwards; see Figure 2.15(a). They found three zones of α_0 over which galloping is possible, as shown in Figure 2.15(b). Moreover, they found that these three zones coincide almost exactly with those predicted via the Glauert/Den Hartog criterion. Extensive work on triangular prisms, on the flow field and other aspects, was done by Buresti and associates, e.g. Iungo & Buresti (2009).

Prisms of other geometrical shapes, some to be discussed later in this book, have been studied, by Scruton (1971), Naudascher & Wang (1993), Ruscheweyh *et al.* (1996) and Deniz & Staubli (1997), for instance, and on open circular and parabolic section prisms by Weaver & Veljkovic (2005).

Some very nice results have been obtained by Slater (1969) for an angle section ("L-section") at two different orientations to the flow. The experimental results

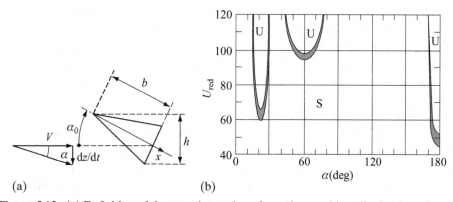

Figure 2.15. (a) Definition of the experimental configuration used in galloping experiments with isosceles triangular prisms. (b) Stability diagram for a triangular prism with $\beta = 30°$, β being the angle of the main vertex. "S" identifies a zone of stability; "U" a zone of instability; $U_{red} = V/fh$ (Alonso *et al.* 2007).

are compared with those of an analytical model equivalent to that of Section 2.2, generating the collapsed/universal amplitude-velocity curves of Figure 2.16, similar to Figure 2.11. The collapse is very successful, especially for $\alpha_0 = 90°$ when the horizontal leg is pointing downstream, over a wide range of ζ (from 0.844×10^{-3} to 1.77×10^{-3}); as expected, the agreement is less successful for $\alpha_0 = -45°$ when the L-section becomes open to the oncoming flow. Nevertheless, it must be concluded that the basic model works quite well even for sections which are rather different from the usual rectangular geometry.

Slater(1969; Appendix IV) and Novak *et al.* (1978) extended the Glauert/Den Hartog criterion for galloping to bodies at an initial angle of attack α_0, by noting that in this case

$$C_{F_y} = -(C_L + C_D \tan \alpha) \sec \alpha, \qquad (2.36)$$

where α is measured from the initial angle α_0.

Finally, in all the foregoing it was assumed that the corners of the prisms are sharp. The effect of chamfering or rounding the corners of a square-section prism on the aerodynamic forces (and hence on galloping), as well as turbulence and three-dimensionality of the flow, was studied by Tamura & Miyagi (1999). The effect can be quite important, both in smooth and turbulent flows. Similarly, the effect of cutting off a square from each of the four corners of a rectangular prism was investigated by Shiraishi *et al.* (1988).

2.3.2 Novak's "universal response curve" and continuous structures

Novak (1969, 1970, 1972) has extended the Parkinson & Smith nonlinear model to continuous elastic structures, e.g. cantilevered towers; he also proposed a "universal response curve" which would permit the prediction of galloping of a given prismatic body shape from dynamic wind-tunnel tests on a scale model, without the necessity of static force measurements.

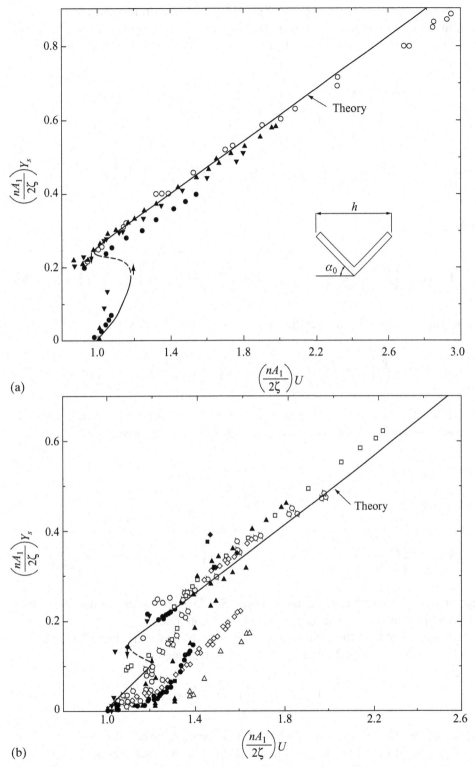

Figure 2.16. Collapsed amplitude-velocity characteristics for galloping of an angle section (a) for $\alpha_0 = 90°$ and (b) for $\alpha_0 = -45°$. In (a) the various symbols correspond to values of ζ ranging from 0.844×10^{-3} to 1.77×10^{-3}; in (b) for $\zeta = 0.742 \times 10^{-3} - 1.74 \times 10^{-3}$ and $n = 0.733 \times 10^{-3} - 6.48 \times 10^{-3}$ (Slater 1969).

Novak (1969) obtained expressions for the steady amplitudes outside and within the interval of oscillation hysteresis (see Figure 2.11). For instance, in the former range, the dimensionless amplitude $Y = a/h$, a being the dimensional amplitude, is given by

$$Y = \sqrt{2}\, U \left\{ \left(\frac{4}{35A_7} \right)^{1/3} \left[\left(A - \frac{\zeta}{\bar{n}U} + \sqrt{\left(A - \frac{\zeta}{\bar{n}U} \right)^2 - B^2} \right)^{1/3} \right. \right.$$

$$\left. \left. + \left(A - \frac{\zeta}{\bar{n}U} - \sqrt{\left(A - \frac{\zeta}{\bar{n}U} \right)^2 - B^2} \right)^{1/3} \right] + \frac{4}{21} \frac{A_5}{A_7} \right\}^{1/2}, \qquad (2.37)$$

where

$$A = A_1 + \frac{10}{35} \frac{A_5}{A_7} \left(\frac{40}{189} \frac{A_5^2}{A_7} - A_3 \right) \quad \text{and} \quad B = \sqrt{\frac{4}{35A_7} \left(\frac{20}{63} \frac{A_5^2}{A_7} - A_3 \right)^3}, \qquad (2.38)$$

in which the A_is are from a yet different expression for C_{F_y}, namely

$$C_{F_y} = A_1 \left(\frac{\dot{y}}{V} \right) - A_2 \left(\frac{\dot{y}}{V} \right)^2 \frac{\dot{y}}{|\dot{y}|} - A_3 \left(\frac{\dot{y}}{V} \right)^3 + A_5 \left(\frac{\dot{y}}{V} \right)^5 - A_7 \left(\frac{\dot{y}}{V} \right)^7; \qquad (2.39)$$

as before, $U = V/\omega h$, $\bar{n} = \rho h^2 / 4\, m$ and $\zeta = r/2ml\omega$, recalling that $\bar{n}$ here is one-half of Parkinson & Smith's n. Moreover, for high enough U the asymptote

$$Y = kU \qquad (2.40)$$

is reached, where

$$k = \sqrt{ 2 \left\{ \left(\frac{4}{35A_7} \right)^{1/3} \left[\left(A + \sqrt{A^2 - B^2} \right)^{1/3} + \left(A - \sqrt{A^2 - B^2} \right)^{1/3} \right] + \frac{4}{21} \frac{A_5}{A_7} \right\} }$$

is a function of aerodynamic constants only. Similar expressions for the three amplitudes in the interval of oscillation hysteresis are obtained.

One then obtains the "universal response curve" shown in Figure 2.17, which is similar to that of Figure 2.11, but with different axes, namely

$$\bar{U} = \frac{\bar{n}}{\zeta} U \quad \text{and} \quad \bar{Y} = \frac{\bar{n}}{\zeta} Y, \qquad (2.41)$$

cf. equations (2.30), the first of which is the "twice-reduced" flow velocity, i.e. it is a modified form of the reduced flow velocity. The key features in the galloping response curve are therefore all obtained in terms of A_1, A and B of (2.38) and k of (2.40).

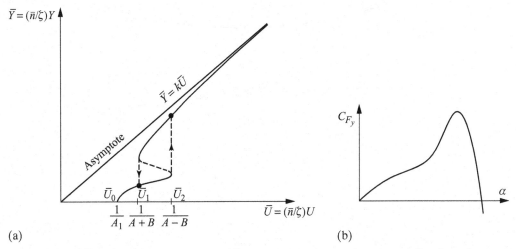

Figure 2.17. (a) Universal response curve for a square prism and corresponding lateral force coefficient, $C_{F_y} = A_1(\dot{y}/V) - A_3(\dot{y}/V)^3 + A_5(\dot{y}/V)^5 - A_7(\dot{y}/V)^7$; from Novak (1969); (b) corresponding C_{F_y} versus α curve.

The presentation in Figure 2.17 is very similar to Parkinson & Smith's (Figure 2.11), but A_1 does not appear explicitly, which is an advantage. Novak's contribution is the *explicit* realization that the galloping curve in Figure 2.17(a) is *universal* for a given cross-sectional shape of the prism. Thus, let us say that Figure 2.17(a) was obtained from wind-tunnel tests on a particular square prism. Then, this curve would be applicable to *all square prisms*, irrespective of size, mass ratio, stiffness and damping – provided, of course, that all inherent assumptions are satisfied, notably the quasi-steady one, and flow conditions are similar.

Equivalent results are obtained for other polynomial forms of C_{F_y}, e.g. in the case where the C_{Fy} versus α curve in Figure 2.17(b) has no inflection (Novak 1969).

The important conclusion is that, if plotted as in Figure 2.17(a), there is *a single curve* applicable for each prismatic shape and set of flow conditions. The galloping curves in Figures 2.13 and 2.14 have been plotted in such a fashion and are thus "universal". It is noted, nevertheless, that, if the motion is nonuniform along the body (as in curves 2 and 3 of Figure 2.13, for instance) or the flow velocity over the length of the prism is variable (curve 3a), a different galloping curve is obtained – "universal" in that sense also. These effects are considered next.

Consider a slender structure (long with respect to cross-sectional dimensions) as shown in Figure 2.18(a, b) – modelled as a pinned-pinned or cantilevered beam – subjected to cross-flow. The cross-sectional shape is presumed to be such that galloping may arise. Motions are assumed to be planar, and the kind of assumptions usually made for the linear Euler-Bernoulli beam approximation (e.g. neglect of axial motion and rotation) are presumed to hold. Even so, a rigorous solution of the problem is difficult, and Novak (1969) developed an approximate galloping theory by means of energy considerations.

Consider motions in the nth normal mode of the structure,

$$y(x, t) = y_n(x, t) = a\phi_n(x) \cos \omega_n t,$$

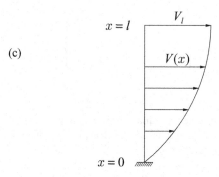

Figure 2.18. Modal forms for first-mode deformation of (a) a pinned-pinned beam, (b) a cantilevered beam; (c) a variable wind profile, $V(x)$, and definition of V_l.

where ϕ_n is the nth eigenfunction and a is the amplitude of the motion. Thus, this is a single-mode analysis. We next express the total damping force *per unit length* as

$$F(\dot{y}) = r\frac{\partial y}{\partial t} - C_{Fy}\,\tfrac{1}{2}\rho h V^2,$$

with C_{F_y} as in equation (2.39) and r being the equivalent viscous damping coefficient. The work done by this force over a period of motion T is

$$W = \int_0^T \int_0^l F(\dot{y})\dot{y}\ \mathrm{d}x\mathrm{d}t. \tag{2.42}$$

Substituting (2.20) into this equation and setting $W = 0$ for galloping yields (Novak 1969)

$$\frac{35}{64}A_7 c_7 \left(\frac{Y}{U}\right)^6 - \frac{5}{8}\,A_5\,c_5 \left(\frac{Y}{U}\right)^4 + \frac{3}{4}A_3\,c_3 \left(\frac{Y}{U}\right)^2$$

$$+ \frac{8}{3\pi}A_2\,c_2\left(\frac{Y}{U}\right) + \frac{\zeta}{\bar{n}U} - A_1\,c_1 = 0, \tag{2.43}$$

in which the A_i, $i = 1 - 7$, are as defined in equation (2.39), and

$$Y = \frac{a}{h}, \quad U = \frac{V}{\omega_n h}, \quad \zeta = \frac{r}{2m\,\omega_n}, \quad \bar{n} = \frac{\rho h^2}{4m}, \tag{2.44}$$

m being the mass per unit length, and

$$c_j = \int_0^l \mid \phi_n(x) \mid^{j+1}\ \mathrm{d}x \bigg/ \int_0^l \phi_n^2(x)\mathrm{d}x. \tag{2.45}$$

For $\phi_n(x) = 1$, corresponding to uniform lateral motion, the solution is the same as that obtained before, as it should be.

We next consider, e.g. for a vertical structure (say, a tower), that the flow velocity varies as in Figure 2.18(c); thus, expressing $V(x) = V_l v(x)$, with V_l the velocity at some reference height, and proceeding as before (but now with a variable V), we obtain for galloping the same equation (2.43), but with

$$c_j = \int_0^l v^{2-j} \mid \phi_n(x) \mid^{j+1} \Big/ \int_0^l \phi_n^2(x)\, \mathrm{d}x. \qquad (2.46)$$

Moreover, the flow (wind) profile is often expressed in terms of a power law,

$$v(x) = \left(\frac{x}{l}\right)^\alpha. \qquad (2.47)$$

The values of c_j have been tabulated by Novak (1969), as in Table 2.3, for different values of α. The effect of a nonuniform amplitude of motion on galloping response is shown in Figure 2.13. It is seen that the increase in amplitude with $\overline{U}$ is greater for flexible structures than for flexibly supported rigid ones. Similar results for a square prism are shown in Figure 2.19, where it is seen that in the hysteresis range the response is highly dependent on the mode of vibration. It is also seen that, for a nonuniform wind profile, the slope is decreased *vis-à-vis* a uniform wind.

Novak (1969) also conducted some new experiments with a cantilevered square prism, showing that the universal galloping curve works quite well, except for low $\overline{U}$ and high ζ; but still the trend is as predicted.

2.3.3 Unsteady effects and analytical models

Predictions of galloping, as presented so far, are based on quasi-steady theory, thus effectively presuming that the period of oscillation is long compared with the time taken for the flow to pass along the prism. Hémon (1999a) offers an improvement for $h/d > 1$, involving a time delay between motion and the forces generated, similar in nature to that considered in Chapter 5. Promising agreement with *ad hoc* experiments was achieved.

Grosso modo, most of the foregoing discussion was based on quasi-steady theory and hence on the statically measured lift and drag forces. As shown by Nakamura & Matsukawa (1987), however, another kind of galloping, the so-called *low-speed galloping* or *low-speed flutter*, is possible for prisms of large h/d ("short" or "shallow" prisms). This is a phenomenon that can only be detected via unsteady fluid dynamics. It is discussed in Section 2.4.

A bold attempt to dispense with the necessity of extensive C_{F_y} versus α measurements in order to describe the post-critical galloping behaviour was made by Luo and Bearman (1990). They used unsteady thin-airfoil theory which, of course, was conceived for anything but bluff bodies. For an airfoil oscillating transversely in incompressible flow, the lift force *per unit length* induced by the transverse oscillation is

$$L = -C_{mo}\,\rho\left(\tfrac{1}{4}\pi D^2\right)\ddot{y} - \frac{\partial C_L}{\partial \alpha}\rho\, V_\infty\left(\tfrac{1}{2}D\right) C(k)\dot{y}\,; \qquad (2.48)$$

Table 2.3. *Coefficients c_i with uniform and variable wind speed (Novak 1969)*

The mode of vibration	The mode $\phi_n(x)$	Case	Wind $v(x)$	c_1	c_2	c_3	c_5	c_7
	1	1a	1	1	1	1	1	1
		1b	$\left(\dfrac{x}{l}\right)^{1/6}$	$\dfrac{6}{7}$	1	$\dfrac{6}{5}$	2	6
	$\sin\dfrac{n\pi}{l}x$	2	1	1	$\dfrac{8}{3\pi}$	$\dfrac{3}{4}$	$\dfrac{5}{8}$	$\dfrac{35}{64}$
	$\dfrac{x}{l}$	3a	1	1	$\dfrac{3}{4}$	$\dfrac{3}{5}$	$\dfrac{3}{7}$	$\dfrac{1}{3}$
		3b	$\left(\dfrac{x}{l}\right)^{1/6}$	$\dfrac{18}{19}$	$\dfrac{3}{4}$	$\dfrac{18}{29}$	$\dfrac{18}{39}$	$\dfrac{18}{49}$
		3c	$\left(\dfrac{x}{l}\right)^{1/3}$	$\dfrac{9}{10}$	$\dfrac{3}{4}$	$\dfrac{9}{14}$	$\dfrac{1}{2}$	$\dfrac{9}{22}$
	$\dfrac{x^2}{l^2}$	4a	1	1	$\dfrac{5}{7}$	$\dfrac{5}{9}$	$\dfrac{5}{13}$	$\dfrac{5}{17}$
		4b	$\left(\dfrac{x}{l}\right)^{1/6}$	$\dfrac{30}{31}$	$\dfrac{5}{7}$	$\dfrac{30}{53}$	$\dfrac{10}{25}$	$\dfrac{30}{97}$
		4c	$\left(\dfrac{x}{l}\right)^{1/3}$	$\dfrac{15}{16}$	$\dfrac{5}{7}$	$\dfrac{15}{26}$	$\dfrac{5}{12}$	$\dfrac{15}{46}$

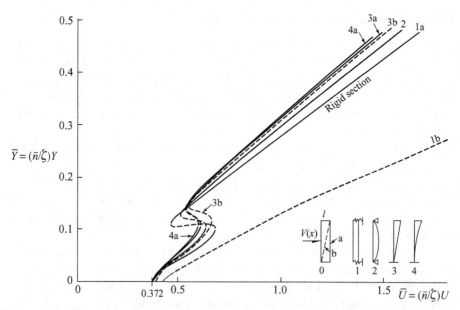

Figure 2.19. Universal galloping response of square prisms with different modal deformation and for different wind profiles, as shown. Curves 1a, 2, 3a and 4a: for $V(x) = V_l$; curves 1b and 3b: $V(x) = V_L (x/l)^{1/6}$; from Novak (1969).

C_{mo} is the potential flow inertia coefficient, and $C(k) = \mathrm{H}_1^{(2)}(k)/\{\mathrm{H}_1^{(2)}(k) + \mathrm{i}\,\mathrm{H}_0^{(2)}(k)\}$ is Theodorsen's function, with $\mathrm{H}_0^{(2)}$ and $\mathrm{H}_1^{(2)}$ being Hankel functions of the second kind and orders 0 and 1, respectively; $k = \pi/U^*$ with U^* being the reduced flow velocity, and $\partial C_L/\partial\alpha$ is the lift-curve slope. Luo & Bearman "have taken the bold step of assuming that an equation which is derived for an airfoil oscillating with small amplitude can be applied to an oscillating cylinder".

Developing the theory, the following expressions are obtained for the spectral component of the fluctuating lift coefficient $C_L'(f_N)$ at frequency f_N and the phase angle ϕ:

$$C_L'(f_N) = \left\{ \frac{1}{2}\left[\frac{C_{ml}\pi^2 \tan\alpha_{max}}{U^*} \right]^2 + \left[\frac{\partial C_L}{\partial\alpha} F(k)\tan\alpha_{max} \right]^2 \right\}^{1/2},$$

$$\phi = \tan^{-1}\left[\frac{-(\partial C_L/\partial\alpha)F(k)U^*}{C_{ml}\pi^2} \right], \tag{2.49}$$

where $C_{ml} = C_{mo} + C_{ma}$, with $C_{ma} = (1/\pi k)(\partial C_L/\partial\alpha)G(k)$ is the so-called added inertia coefficient; $F(k)$ and $G(k)$ are related to $C(k)$ by $C(k) = F(k) + \mathrm{i}\,G(k)$, and

$$\frac{\partial C_L}{\partial\alpha} = -\left(\frac{\partial C_{F_y}}{\partial\alpha} \right)\bigg|_{\alpha=0}, \qquad \alpha_{max} = \tan^{-1}\left[2\pi\frac{a}{D}\frac{1}{U^*} \right], \tag{2.50}$$

in which $U^* = V/f_N D$, a is the amplitude of oscillation of the prism, and f_N is its frequency. Thus, the theory takes into account the inertia effect and the influence of shed vorticity.

Some typical results are shown in Figure 2.20, where the abscissa $U^*/(a/D)$ was preferred instead of U^*. It is seen that at high values of $U^*/(a/D)$ both theories agree with experimental measurements (and with each other). For $25 < U^*/(a/D) < 75$ approximately, the unsteady airfoil theory provides better prediction! For $U^*/(a/D) \leq 26$, corresponding to $\alpha_{max} \simeq 13.5°$, there may be reattachment onto the sides of the prism, an effect not taken into account by the unsteady theory; as observed in Figure 2.20, the theory does less well in this range of $U^*/(a/D)$.

It ought to be stressed that this is not a wholly unsteady analytical theory, because $(\partial C_{F_y}/\partial\alpha)$ must still be provided as an empirical input. Nevertheless, it is quite remarkable how well this theory can perform, provided $U^*/(a/D)$ is not too small.

2.3.4 Some comments on the flow field

The discussion in Sections 2.3.1 and 2.3.2 has been largely phenomenological, without reference to the underlying fluid mechanics. Several attempts to remedy that have been made over the years and much insight has been gained thereby. However, the fluid mechanics is complex and depends on shape, reduced flow velocity range, Reynolds number, turbulence, and so on. Hence, few clear and universally true statements can be made. Nevertheless, some general comments are provided here, drawn mainly from the work of Bearman & Trueman (1972), Parkinson (1974,

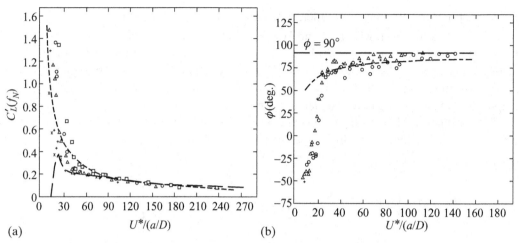

Figure 2.20. (a) The fluctuating lift coefficient $C'_L(f_N)$ as a function of $U^*/(a/D)$, a being the amplitude of galloping, for a square prism, and (b) the corresponding phase angle ϕ. Experimental data: $\square$, $a/D = 0.5$; $\circ$, $a/D = 0.675$; $\triangle$, $a/D = 1.0$; $+$, $a/D = 1.5$; $\times$, $a/D = 2.0$. $- \; -$, Quasi-steady theory prediction; $- - -$, unsteady airfoil theory prediction (Luo & Bearman 1990).

1989), Bearman *et al.* (1987) and Nakamura & Hirata (1989). The aim is to sensitize the reader regarding some of the fluid mechanics associated with galloping (or nonoccurrence thereof) and, inevitably, vortex shedding.

We first start by stressing again the importance of the afterbody shape and length (Parkinson 1974, 1989). A convenient shape for a systematic discussion is the rectangular section – see Figure 2.21. The three curves in that figure summarize a great deal of data gathered from experiments with stationary or galloping cylinders for $10^4 < \mathrm{Re} < 10^5$: for the drag coefficient $C_D = D/(\frac{1}{2}\rho V^2 h)$, the Strouhal number $S = f_v h/V$, f_v being the vortex formation frequency in one shear layer, and $Y_s = y_s/h$ the dimensionless transverse galloping amplitude; $U = V/\omega_n h$ is the reduced flow velocity.

The curves show that, as d/h increases, flow conditions change gradually, except in two ranges: $0.5 < d/h < 1.0$ and $2.5 < d/h < 3.0$. The discussion here follows closely Parkinson's (1974, 1989). For the first range, as argued by Bearman & Trueman (1971), increasing d reduces the size of the base "cavity" from which entrainment of fluid takes place by the shear layer forming a discrete vortex, and the base pressure is reduced, with an accompanying increase in C_D. As d becomes sufficiently large, however, it begins to interfere with the inward-curving shear layer, thus forcing a change in the process. Also, if d is small enough not to cause interference with the process of vortex formation, the resulting vortex at full strength will lie closer to the base of the rectangle, again contributing to a lower base pressure and a higher C_D. Thus, C_D reaches a maximum at $d/h \simeq 0.62$ (Figure 2.21). The interference between the shear layer and the downstream corner of the body has an important effect on the pressure loading on the sides of the body. As found by Brooks (1960) and Smith (1962), hard-oscillator galloping prevails for $d/h < 0.75$, whereas for $0.75 \leq d/h \leq 3.0$ prisms gallop as soft oscillators.

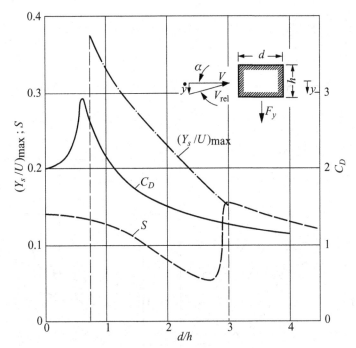

Figure 2.21. The effect of rectangular section afterbody (d/h = depth-to-height ratio) on the Strouhal number, S, drag coefficient, C_D, and amplitude of galloping, $(Y_s/U)_{max}$; from Parkinson (1974).

The other abrupt changes, for $2.5 < d/h < 3.0$, are clearly related to reattachment of the shear layers on the afterbody sides, with final separation occurring at the downstream corners. Hence, the reason why prisms with $d/h > 3.0$ are stable with regard to galloping is that the reattached shear layers provide a pressure loading on the afterbody that opposes small transverse disturbances.

The effect of chamfering or rounding the corners of a prism causes the separated shear layers to approach the sides of the prism, thus promoting reattachment (Tamura & Miyagi 1999).

As stated in the foregoing, the peak in C_D in Figure 2.21 is associated with a decrease in C_{p_b} – in effect a negative peak. Nakamura & Hirata (1989) take this further by defining the value of d/h for which this negative peak occurs as *the critical depth*. This of course varies with the shape of the body. It is also a relatively weak function of U, as shown in Figure 2.22(b). For rectangular and generalised D-sections (Figure 2.22(c)), it is seen that, for large enough U – beyond the vortex resonance zone, where in any case galloping theory needs to be modified – the critical geometry for rectangular prisms is $(d/h)_{crit} \simeq 0.6$, and $(d/h)_{crit} \simeq 0.8$ for D-sections.

Below the critical depth, i.e. for $d/h < (d/h)_{crit}$, the decrease in C_{p_b} as d/h is increased in mainly due to the increased curvature of the shear layers as they roll up, because of a reduced base cavity. Above $(d/h)_{crit}$, C_{p_b} increases with increasing d/h because the vortices are formed further downstream due to the influence of the rear corners of the body. Nakamura & Tomonari (1981) have studied this latter

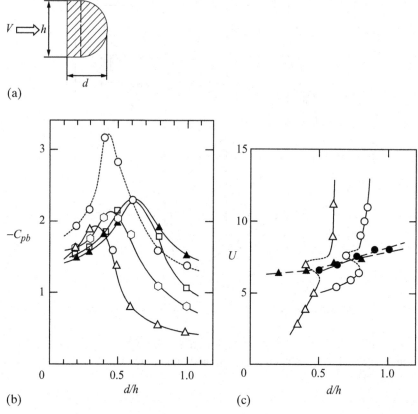

Figure 2.22. (a) Generalised D-section in cross-flow. (b) Base-suction coefficient, $-C_{p_b}$, versus d/h for oscillating rectangular prisms: ▲, stationary prism; □, $U = 8.9$; ○, U_r; "hexagon", $U = 5$; △, $U = 2.8$. Here $U = V/f_v h$, and U_r is the value at vortex-shedding resonance. (c) Critical geometry as a function of U and d/h; open symbols are for the critical depth, while filled symbols are for vortex resonance: △, ▲, rectangular prism; ○, ●, D-section prism; from Nakamura & Hirata (1989).

effect, specifically what was later called the *shear-layer/edge direct interaction* (s-l/e interaction, for short). This s-l/e interaction yields a reattachment-type pressure distribution on the side faces, characterised by a low-pressure plateau, followed by recovery to a higher pressure. Nakamura & Hirata (1989) argue that, for $d/h >$ $(d/h)_{\mathrm{crit}}$, a small angle of incidence causes a side load due to the pressure difference between the upper and lower sides of the prism which is due to this s-l/e interaction. Moreover, they argue that this is the reason why galloping can occur for $d/h >$ $(d/h)_{\mathrm{crit}}$.

Parkinson (1989) views this in terms of a secondary flow. Soft galloping is excited by an unbalanced pressure distribution on the afterbody sides when the section is subjected to a small transverse velocity. This unbalance requires a secondary flow between the two shear layers and the body. This secondary flow presumably has enough strength to trigger soft galloping only when d/h is large enough to cause this "interference".

When the prism is oscillating, the flow past the bluff body has two frequency components: the body oscillation frequency f_y and the natural vortex-shedding frequency f_{vs}. The flow associated with f_y consists of two parts: one is linked to the acceleration of the body (dominant at low U) and the other is associated with the continual variation of the angle of incidence. This latter variation produces *undulation of the wake*, which becomes progressively shorter with decreasing U. As stated by Nakamura & Hirata (1989), wake undulation can produce motion-generated vortices when the reduced velocity is small and the amplitude large; but its influence is present, even when no motion-induced vortices are generated. Motion-generated vortices are intimately connected with the so-called low-speed galloping discussed in Section 2.4.

Nakamura & Hirata (1989) showed that, whereas for rectangular prisms with $d/h = 0.3$ and 0.4 oscillating at vortex resonance, the side-face pressure distribution is reasonably uniform, for $d/h = 0.5$ and 0.6 reattachment-type pressure distributions are observed, with a significant pressure recovery, even though actual reattachment does not occur. Similar results were obtained for D-section prisms. Further insights are provided in the work of Nakamura, Hirata & Urabe (1991) aiming at understanding galloping in the presence of a splitter plate; see Section 2.5.

2.3.5 Shear-layer reattachment

As surmised from the discussion in Section 2.3.4, shear-layer reattachment onto the sides of the prism plays an important role on the occurrence, and nonoccurrence, of galloping.

Parkinson (1971) remarks that, for a square prism, one of the shear layers which has separated from one of the upstream corners of the prism reattaches to one of the downstream corners at an angle of attack $\alpha \simeq 13°$, and that this corresponds to the maximum F_y attained (see Figure 2.8). At higher α, thus at higher amplitudes of oscillation (cf. equation (2.1)), the reattachment point moves upstream and F_y becomes smaller. It is this phenomenon, related to reattachment, that is responsible for the self-limiting amplitude feature of galloping.

Parkinson (1989) provides a very useful compilation of data on the maximum amplitude and range of occurrence of soft galloping as a function of d/h, based on experimental data from Brooks (1960), Smith (1962), Laneville (1973), Novak (1974), Nakamura & Tomonari (1977) and Washizu *et al.* (1978), shown here in Figure 2.23. It is seen that, in smooth flow, the upper limit of the soft-galloping range is at $d/h \simeq 3.2$ due to diminution of the side force F_y as a result of shear-layer reattachment. As stated by Parkinson (1989), reattachment of the shear layers at the trailing edge means that "the section no longer has an afterbody, and becomes immune to transverse galloping". The decrease in $(Y_s/U)_{\max}$ with increasing d/h is also determined by reattachment effects, as may be appreciated with the aid of the inset sketch. The value of α for reattachment to occur, which determines $\dot{y}_{\max}$, decreases as d/h increases; hence, $(Y_s)_{\max}$ decreases with increasing d/h.

Turbulence causes increased entrainment of the fluid by the separated shear layers, thickening them; this promotes interference with the trailing-edge corner as d/h is increased, leading to reattachment. Therefore, as seen in Figure 2.23, with increasing turbulence intensity, Tu, the maximum d/h at which galloping is possible becomes smaller.

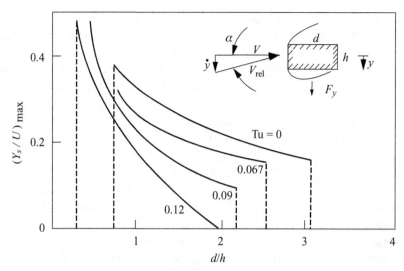

Figure 2.23. The effect of the d/h ratio for rectangular prisms on the occurrence and amplitude of galloping, for various levels of turbulence intensity, Tu; here U and Y_s are as in Section 2.2.2 (Parkinson 1989).

However, reattachment also engenders a different set of excitations on the prism, as discussed by Komatsu & Kobayashi (1980), Shiraishi & Matsumoto (1983), Nakamura & Nakashima (1986), Schewe (1989), Naudascher & Wang (1993), Deniz & Staubli (1997) and Matsumoto *et al.* (2008a), for instance, and as reviewed in Section 3.4.3. In this literature, the separated shear layer is considered to be a *leading-edge vortex*; indeed, for large enough d/h, sufficient roll-up occurs upstream of the trailing edge to justify the appellation. Also, reattachment is viewed as *impingement* of this vortex on a side of the prism. Moreover, in addition to the "normal", stationary-prism vortex shedding, there may be *motion-induced vortices*. The impingement of the leading-edge vortices on the prism, if it occurs, and possible coupling with trailing-edge vortex shedding generate a whole new set of excitations (IIE) and motion-induced excitations (MIE) in Naudascher & Rockwell's (1980) classification. Specifically, impinging leading-edge vortex (ILEV) excitation and alternate-edge vortex shedding (AEVS) may arise, as discussed in Section 3.4.3. The interested reader should also consult Rockwell & Naudascher's (1979) excellent review on impinging free shear layers.

2.4 Low-Speed Galloping

As already clarified, galloping is associated with the occurrence of negative fluid-dynamic damping. Typically, in what has been covered in the foregoing, galloping occurs at "high speeds", meaning speeds higher than for vortex-shedding lock-in, and hence in the range where the quasi-steady assumption is valid – and, incidentally, without substantial interference from vortex shedding (refer to Section 2.10). However, negative fluid-dynamic damping may also arise over a range of low reduced flow velocities, or "low speeds", below vortex-shedding lock-in. This gives rise to so-called *low-speed galloping* or *low-speed flutter*, as first discovered by Nakamura & Matsukawa (1987). More precisely, Nakamura & Hirata (1989) classify low-speed

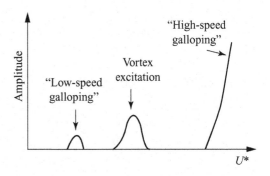

Figure 2.24. Schematic diagram showing the possible occurrence of "low-speed galloping", vortex-shedding lock-in, and "high-speed galloping", with increasing reduced flow velocity U^* [adapted from Nakamura (1990)].

galloping as one occurring *below the critical depth* (Section 2.3.4), i.e. for "short" or "shallow" prisms (prisms of small d/h); whereas high-speed galloping, i.e. the galloping discussed in all of the foregoing, occurs for $d/h > (d/h)_{crit}$.

Therefore, the generic response for prisms in cross-flow is as shown in the diagram of Figure 2.24, where it should be stressed that not all three phenomena arise *necessarily* for any given system. In addition, the low-speed galloping and vortex-excitation responses may be close enough to each other for the former to appear as a "nose" to the vortex lock-in region, as seen, for instance, in Figure 2.25(b).

In the low-speed range, where $f_y \ll f_{vs}$, variations in the angle of incidence in the course of oscillation produce an undulation of the wake and "flapping" of the separated shear layers, which can manifest itself as *motion-induced vortices*; here, f_y is the transverse oscillation frequency and f_{vs} is the vortex-shedding frequency. The development of low-speed galloping is intimately related to the undulation and "flapping" of the separated shear layers and to motion-induced vortices in the near-wake (Nakamura 1988; Nakamura & Hirata 1989; Tamura & Itoh 1999; Tamura & Dias 2003; Tamura *et al.* 2004). Therefore, this is clearly an unsteady phenomenon: to study it experimentally one must utilize either flexibly supported prisms capable of reaching a reasonably large amplitude of motion or forced oscillation experiments; its existence cannot be established from static drag and lift measurements as in the foregoing. Moreover, unlike high-speed galloping which can persist apparently without limit as the flow velocity is increased, low-speed galloping occurs over a limited range of flows.

In fact, low-speed galloping is a very complex phenomenon. Its existence was established experimentally by Nakamura & Matsukawa (1987) by direct observation in free-oscillation experiments and from an analysis of forced-oscillation data. Specifically, in the latter, the phase angle of the lift relative to displacement, ϕ, was measured over a range of $U^* = V/f_y h$.* As is well known, a large change in ϕ occurs near vortex-shedding resonance, U^*_{cr}, which in these experiments ranged between approximately 6.2 and 7 for prisms with $d/h = 0.2 - 0.6$. However, it is of interest that $\phi \simeq 60°$ to $70°$ in the range $2 < U^* < 6$ for the $d/h = 0.2$ and 0.4 prisms (see Figure 2.25(a)). Recalling that a positive ϕ indicates a positive out-of-phase component of the lift force and amplified oscillation, it is clear that we have flutter, well before vortex-shedding lock-in, i.e. low-speed galloping.

---

* Here, we use the asterisk to remind us that U^* is based on f, whereas U is based on $\omega = 2\pi f$.

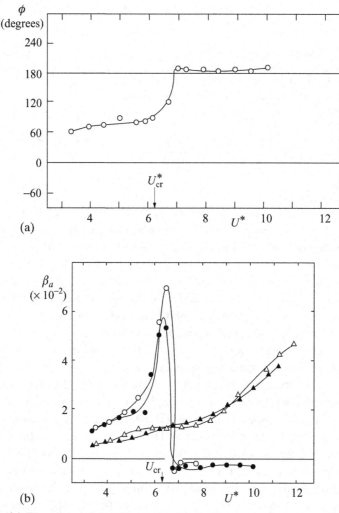

Figure 2.25. (a) The phase of the lift relative to displacement of a rectangular prism with $d/h = 0.2$ in forced oscillation, as a function of $U^* = V/f_hy$; (b) the aerodynamic growth rate, β_a, as a function of U^* for the same system: $\circ$, $\bullet$, without splitter plate; $\triangle$, $\blacktriangle$, with splitter plate; open symbols, direct measurements from free-vibration experiments; filled symbols, from forced vibration experiments (Nakamura & Matsukawa 1987).

The rate of growth of the oscillation $\beta = \beta_a - \delta_s$ has also been studied, where β_a is the aerodynamic growth (the negative aerodynamic damping) and δ_s is the logarithmic decrement in still fluid. In Figure 2.25(b), we see clearly a positive β_a in the range $3.5 < U^* < 5.5$, well before the onset of vortex-shedding lock-in.

It should be recalled here that, according to Table 2.1, prisms with $d/h = 0.25$, not too different from 0.2, should be immune to galloping, but that is for *high-speed* galloping.

Further work on this topic was done by Nakamura & Hirata (1991, 1994) and by Tamura and his associates. In addition to the phase measurements, which are along the lines of the foregoing, Nakamura & Hirata (1991) undertook mean and fluctuating pressure measurements along the sides of the prism. From these, it is clear that, for a $d/h = 0.4$ prism, the fluctuating pressures on the side faces do positive work on the

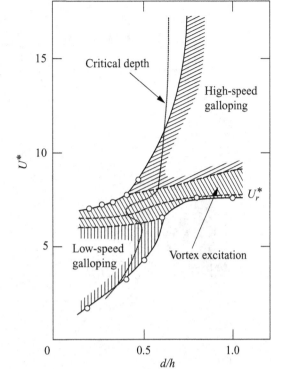

Figure 2.26. Results from free-oscillation experiments on rectangular prisms, showing regions of low- and high-speed galloping and vortex excitation; · · · ·, critical geometry; - - - , vortex-resonance velocity (Nakamura & Hirata 1991).

prism over a specific range of U^*, thus rendering low-speed galloping possible. Corresponding measurements for a $d/h = 0.6$ prism, however, display a reattachment-type pressure distribution along the side faces, as well as a decrease in suction at separation, thus clarifying why low-speed galloping does not occur in this case.

Figure 2.26 presents a map constructed by Nakamura & Hirata (1991) showing the regions of occurrence of low- and high-speed galloping and of vortex-shedding excitation for rectangular prisms. A nice summary of the foregoing and a discussion contrasting low- and high-speed galloping is provided by Nakamura & Hirata (1994).

Tamura & Itoh (1999) conducted the first numerical study of the problem, solving the continuity and Navier-Stokes equations at $\mathrm{Re} = Vh/\nu = 10^4$ for controlled transverse motions of $d/h = 0.2 - 0.8$ prisms, with no turbulence modelling. The three-dimensional computational grid involved 200 000 grid points. A number of important findings were made, in particular, that the lift and drag time histories displayed a two-state characteristic, as shown for example in Figure 2.27(a), intermittently switching from one state to the other. These were tied to different near-wake flow patterns, as shown by the vorticity contours in Figure 2.27(b, c). Where large fluctuations in lift and high mean drag occur in Figure 2.27(a), the shear layers roll up strongly and quickly (Figure 2.27(b)); whereas segments of the time histories with small fluctuating lift and lower mean drag correspond to weaker roll-up of the shear layers, farther from the base of the prism (Figure 2.27(c)).

The critical depth ratio in Tamura & Itoh's calculations, $(d/h)_{\mathrm{crit}} = 0.5$, is a little lower than the value of ∼0.6 found experimentally, no doubt because of some weakness in the numerical modelling (perhaps the absence of turbulence modelling). Because the aerodynamic forces were thereby overestimated, there were niggling

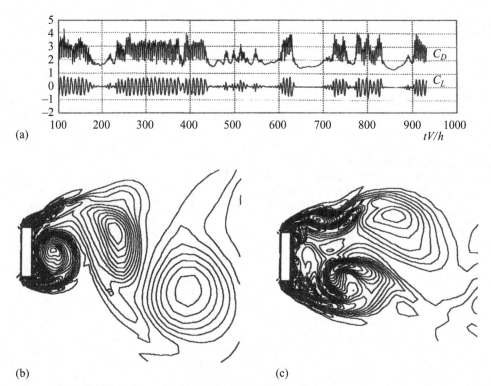

(a)

(b)   (c)

Figure 2.27. (a) Time histories of the drag and lift coefficients on a stationary rectangular prism of $d/h = 0.2$, showing switching between two different states. (b, c) Two distinct wake patterns, (b) at $tV/h = 779$, showing strong roll-up of the separated shear layers, and (c) at $tV/h = 767$, showing weaker roll-up (Tamura & Itoh 1999).

doubts concerning these results, despite the many useful insights provided. This was remedied by Tamura *et al.* (2004) in a computational fluid dynamis (CFD) study at Re $= 2 \times 10^4$ using the large eddy simulation (LES) method and a dynamic Smagorinsky model for subgrid scale turbulence. In this case, the mean drag was in excellent agreement with the measurements of Bearman & Trueman (1972) and Nakaguchi *et al.* (1967), showing a peak at $d/h \simeq 0.6$. The results obtained are in broad agreement with those in the earlier study, and will not be further elaborated here.

An experimental study was conducted by Tamura & Dias (2003), specifically on a $d/h = 0.2$ prism for $5.4 \times 10^3 <$ Re $< 32 \times 10^3$; the critical reduced velocity for vortex-shedding resonance was $U_{cr}^* = 7.1$. Amplified oscillation was observed, enveloping the lock-in range. Thus, for $\zeta = 0.10$ (Sc $= 15.5$), the oscillation range was $4 < U^* < 9$ approximately, reaching large amplitudes ($Y = y/h > 0.5$). That this was not just a wide vortex-induced vibration response was established by (i) noting that the onset value of U^* increased as ζ increased, as should be for galloping, (ii) the presence in the power spectral density (PSD) of the wake, e.g. at $U^* = 5$, of a frequency component corresponding to the oscillation frequency, and (iii) inferring the pressure loading on the side faces from the visualised streamlines, showing the existence of a clear fluctuating lift in the appropriate direction, e.g. at $U^* = 5$.

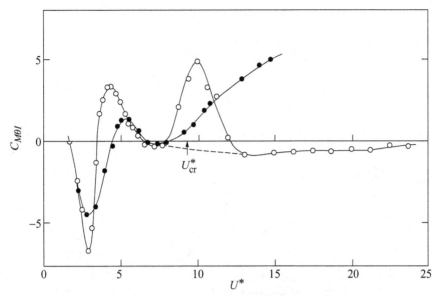

Figure 2.28. The imaginary component of the torsional moment coefficient $C_{M\theta I} = Im(C_{M\theta})$ versus $U^* = V/fh$ for a rectangular prism with depth/height $d/h = 1.5$ oscillating sinusoidally about its centre with frequency f, where $C_{M\theta} = M(t)/[\frac{1}{2}\rho V^2 h^2 \theta(t)]$: $\circ$, smooth flow; $\bullet$, turbulent flow; $---$, with vortex resonance response excised. For galloping, $C_{M\theta I} > 0$ (Nakamura & Yoshimura 1982).

Finally, in a comprehensive set of experiments Nakamura & Yoshimura (1982) studied torsional galloping (see Section 2.7) and vortex shedding of rectangular prisms, in both smooth and turbulent flows. It is of interest that, as for transverse galloping, they also observed *low-speed torsional galloping*; see, e.g., Figure 2.28.

2.5 Prisms and Cylinders with a Splitter Plate

An interesting set of experiments was conducted by Nakamura & Tomonari (1977) on the effect of mounting a long splitter plate downstream of a rectangular prism, unconnected to it and immobile. A sketch of the system is shown in the inset of Figure 2.29.

In Figure 2.29 we see the effect of the gap X between the prism and the splitter plate for a prism of variable d and $h = 10$ cm, such that $0.2 < d/h < 1.45$ and $X/h = 0.1 - 2.0$. It is seen that, in the absence of a splitter plate, prisms can gallop spontaneously (i.e. the growth rate, or negative overall aerodynamic damping, $\beta_a \equiv -\delta_a > 0$), provided that $d/h > 0.75$. With a splitter plate, however, all prisms with $d/h \geq 0.2$ can gallop spontaneously at high enough flow, for all values of X/h. Moreover, the closer the splitter plate to the prism, the higher is β_a. Nakamura & Tomonari go on to state that "What is more surprising is that not only a D-section but also a circular cylinder were susceptible to an instability of this kind". The reader is also referred to Nakamura & Matsukawa (1987) and Nakamura & Hirata (1989). Referring back to Figure 2.25, it is seen ($\triangle$, $\blacktriangle$) that the prism with a splitter plate does gallop, but it is not subject to vortex-shedding excitation.

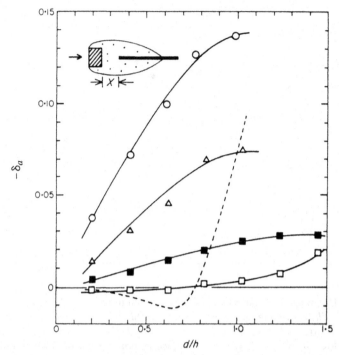

Figure 2.29. Negative aerodynamic damping, $-\delta_a$, versus depth-to-height ratio, d/h, for prisms without and with a splitter plate, at a distance X from the prism trailing edge: $- - -$, prism without a splitter plate; $\circ$, $X/h = 0.1$, $U^* = V/fh = 67$; $\triangle$, $X/h = 0.5$, $U^* = 67$; $\square$, $X/h = 2$, $U^* = 74$; $\blacksquare$, $X/h = 2$, $U^* = 114$ (Nakamura & Tomonari 1977).

It is of particular interest that similarly unexpected dynamical behaviour has been reported for shell ovalling in the presence of a splitter plate by Païdoussis & Helleur (1979), who at the time were unaware of the present work; see Chapter 6.

A very thorough study of galloping of rectangular prisms in the presence of an always unattached immobile splitter plate was conducted by Nakamura, Hirata & Urabe (1991), for systems with $d/h = 0.2 - 6.0$ and variable X/h. Both free- and forced-oscillation experiments were conducted, measuring the mean and fluctuating pressures on the prism, as well as providing flow visualization by using smoke.

Galloping can occur in virtually all cases, as seen in Figure 2.30(a): for $d/h \simeq 0$ to 5 approximately. Moreover, because of the absence of regular vortex shedding, the response is much simpler than without the splitter plate. Figure 2.30(b) is for prisms without a splitter plate, and it is given here for comparison. Two points are of interest: (i) the similarity in the results for low-speed galloping, with and without a splitter plate, and (ii) the fact that galloping vanishes for d/h between 5 and 6 with a splitter plate, whereas without one this occurs for $d/h \simeq 3$.

Figure 2.31 shows the measured aerodynamic growth rate versus U^* for a $d/h = 0.6$ prism. It is clear that the threshold values of U^* at which $\beta_a > 0$ with and without a splitter plate are quite close. Indeed, as already mentioned, the stability bounds for low-speed galloping, in particular for small d/h, are very close. At low U^*, the flow patterns with and without a splitter plate are also surprisingly similar, as may be seen in Figure 2.32.

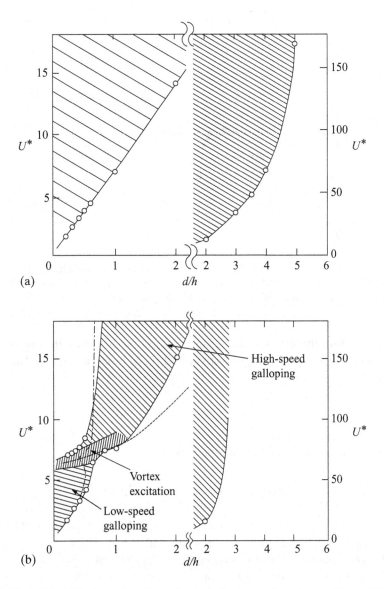

Figure 2.30. (a) Stability boundary for galloping of rectangular prisms with a splitter plate (Nakamura *et al.* 1991); (b) stability boundary without a splitter plate (Nakamura & Hirata 1994).

Here, it ought to be said that interest in the dynamics of such systems with a splitter plate derives not from an attempt to suppress galloping, which as we have seen is worse than futile, but as a means of gaining a deeper understanding of galloping *per se*, as was done by Nakamura *et al.* (1991) and is related below.

The fluctuating pressure on the prism is a function of both the displacement $y(t)$ and the angle of incidence, $\tan^{-1}[\dot{y}(t)/V]$, assuming quasi-steady fluid dynamics and so neglecting the effect of wake undulation. Thus, we can write for the quasi-steady pressure coefficient

$$C_{py}\mathrm{e}^{\mathrm{i}\phi} = -\alpha + \frac{\beta}{U^*}\mathrm{i}. \tag{2.51}$$

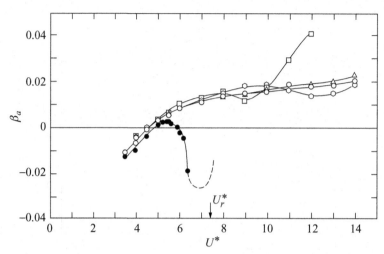

Figure 2.31. The aerodynamic growth rate β_a versus reduced flow velocity $U^* = V/f_y h$ for a $d/h = 0.6$ prism in the low-speed range. •, Without splitter plate; other symbols, with splitter plate of varying length ($6.67h$, $10h$, $15h$ and $20h$); (Nakamura et al. 1991).

For unsteady flow, the effect of wake undulation may be taken into account as a "circulation lag", $A \exp(i\gamma)$, where both A and γ are functions of U^*. Thus, (2.51) becomes

$$C_{py}e^{i\phi} = A e^{-i\gamma} \left[-\alpha + (\beta/U^*)i \right], \qquad (2.52)$$

and its imaginary part, related to galloping, is

$$\mathcal{I}m(C_{py}e^{i\phi}) = A \left[\alpha \sin \gamma + (\beta/U^*) \cos \gamma \right]. \qquad (2.53)$$

Also, as shown by Nakamura & Mizota (1975a) and Luo & Bearman (1990), for high enough U^*, the circulation lag is small and $A \simeq 1$. From measurements of C_{py} and ϕ on a square prism by Nakamura et al. (1991), it is possible to estimate $\alpha = 0.022$, $\beta/U^* = 0.07$ and $\gamma = 17°$ approximately at $U^* = 100$. Hence, $\beta/U^* \gg \alpha$

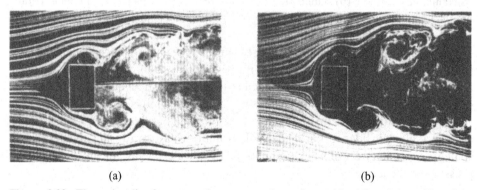

Figure 2.32. Flow visualization around a rectangular prism with $d/h = 0.6$ at $U^* = 2.8$: (a) with a splitter plate; (b) without a splitter plate (Nakamura et al. 1991).

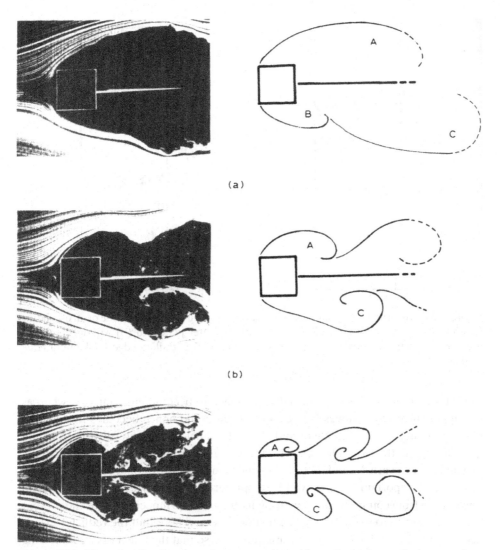

Figure 2.33. Flow visualization around a square prism with a splitter plate at the instant when the cylinder is moving downwards through the neutral position ($2\pi f_y t = 180°$): (a) $U^* = 12$; (b) $U^* = 7.0$; (c) $U^* = 5.0$ (Nakamura *et al*. 1991).

and γ is small at $U^* = 100$, and more so for $U^* < 100$, and so equation (2.53) may be written as

$$\mathcal{I}m(C_{py}\mathrm{e}^{\mathrm{i}\phi}) \simeq A\frac{\beta}{U^*}\cos\gamma,\qquad(2.54)$$

indicating that the incidence effect, arising from the second component of (2.51), is dominant.

Then, with the aid of flow visualization (Figure 2.33) and pressure distributions on the side of the square prism (Figure 2.34), Nakamura *et al.* proceed to discuss the basic mechanism of galloping associated with the incidence effect, as U^* is lowered from a high value, as related almost *verbatim* below.

At high U^*, where the shear layers are free from the direct interaction with the trailing edges (Figure 2.33(a)), the downwards cylinder motion causes the lower

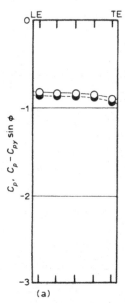
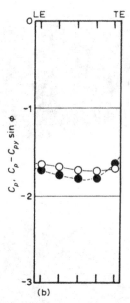

Figure 2.34. Mean and instantaneous pressure distributions on the lower side of an oscillating square prism at $U^* = 20$, $2\pi f_y t = 180°$; (a) with a splitter plate and (b) without. ○, Mean pressure; ●, instantaneous pressure. LE: leading edge; TE: trailing edge (Nakamura *et al.* 1991).

shear layer to move closer to the side and hence to become more curved, whereas the upper shear layer moves further away from it and hence becomes less curved. Correspondingly, both the mean and the instantaneous pressure distributions are quite flat along the side (Figure 2.34), and the motion-dependent pressure force is negative (downwards), tending to aid cylinder motion. We should remark that one of the most important roles played by a splitter plate is to prevent the upper and lower flows from communicating with each other. That is indeed vital for the present basic mechanism to work effectively. It should be said here that the reduced speeds involved in Figures 2.33(a) and 2.34 are quite low, so that the effect of circulation lag may be large, but not so large as to invalidate the foregoing qualitative discussion.

As U^* is lowered, the wavelength of the wake undulation is progressively shortened, and the circulation lag is correspondingly increased. There comes a stage (Figure 2.33(b)) at which the lower side impedes the shear layer from rolling up freely, and this is where the shear-layer/edge (s-l/e) direct interaction (Nakamura & Hirata 1989) occurs. As can be seen in Figure 2.33(b), the curvature of the shear layer near the trailing edge is reversed and at the same time the vortex B vanishes at this interaction. With further decrease in U^*, the process progresses steadily (Figure 2.33(c)). Thus, it is not because the circulation lag becomes more important, but mainly because of the s-l/e interaction that galloping vanishes as U^* is diminished.

As U^* is further decreased, another critical stage is reached, where the shear layer reattaches intermittently on the prism sides in the course of a cycle of oscillation. Figure 2.33(c) suggests that this critical U^* may be just below $U^* = 5.0$.

This work was extended to the case of a *circular cylinder* with an unattached splitter plate behind it. As reported by Nakamura & Tomonari (1977), the cylinder with a splitter plate can gallop; this was confirmed by later studies (Matsumoto *et al.*

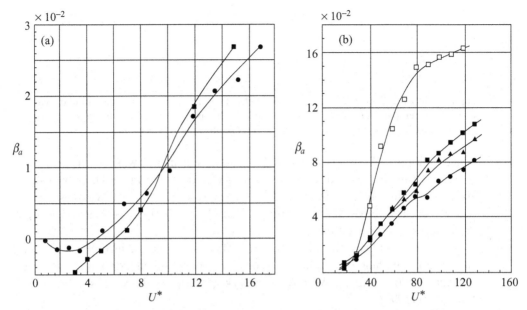

Figure 2.35. Aerodynamic growth rate of oscillation β_a versus reduced speed U^*. (a) Low-speed range, $\mu = 291$, Sc = 6.7, $L/D = 15$, L being the splitter-plate length and D the cylinder diameter; ●, free oscillation; ■, forced oscillation. (b) High-speed range, $\mu = 2160$, Sc = 93; □, $L/D = 4.2$; ●, $L/D = 10.4$; ▲, $L/D = 20.8$; ■, $L/D = 31.3$ (Nakamura, Hirata & Kashima 1994).

1990; Kawai 1990). The same system was studied more thoroughly by Nakamura, Hirata & Kashima (1994) who state epigrammatically: "It is well known that a circular cylinder can be vortex-excited but cannot gallop ... (but) a circular cylinder cannot be vortex-excited but can gallop in the presence of a long stationary splitter plate".[*]

Nakamura *et al.* (1994) measured the mean and fluctuating pressures on the cylinder and the static aerodynamic force (by tilting the incident flow relative to the cylinder-plate combination). Similar experiments as for prisms were carried out with regard to galloping and flow visualization.

Figure 2.35 shows the aerodynamic growth rate $\beta_a = \beta - \delta_s$, where β is the overall logarithmic growth rate and δ_s is the logarithmic decrement in still air, (a) at low speeds and (b) at high speeds. In these experiments, not only δ_s was varied, but also the mass ratio $\mu = m/\rho D^2$; both μ and Sc = $2\mu\delta_s$ are quoted. Figure 2.35(b) shows that the effect of varying the splitter plate length can be very significant.

The mechanism underlying galloping in this case is similar to that of rectangular prisms with a splitter plate, as may be appreciated by comparing the sketches of Figure 2.36 and Figure 2.33. In (a) the downwards cylinder motion causes the lower shear layer to move closer to the cylinder and to become more curved, while the opposite happens to the upper shear layer, thus generating a downwards force and

---

[*] The latter statement that the circular cylinder with a splitter plate cannot be vortex-excited must be qualified; if the splitter plate is *attached* to the cylinder, then "impinging leading-edge vortices", ILEV (Rockwell & Naudascher 1979) can certainly excite the system; refer to Chapter 3.

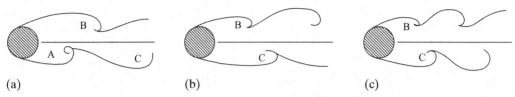

(a)   (b)   (c)

Figure 2.36. Sketches from visualization of the flow around an oscillating circular cylinder with a splitter plate ($L/D = 15$), when the cylinder is moving down through the neutral position: (a) $U^* = 10$; (b) $U^* = 5$; (c) $U^* = 3$ (Nakamura *et al.* 1994).

hence galloping. As U^* is lowered, the undulation in the wake is progressively shortened; thus, in (b) the downwards-moving cylinder now impedes the lower shear layer from rolling up freely; vortex A has now vanished. As a result, a reversal of the phase of shear-layer oscillation relative to cylinder oscillation has occurred, leading to the suppression of galloping.

The flow field around bluff bodies with splitter plates, including cylinders, and vortex shedding, or absence thereof, have been studied extensively; see, e.g., Bearman (1965), Apelt, West & Szewczyk (1973), Apelt & West (1975) and Nakamura (1996).

2.6 Wake Breathing and Streamwise Oscillation

A momentous review of flow-induced streamwise vibrations of structures is provided by Naudascher (1987).* The review deals with all kinds of instability- and movement-induced excitations resulting in *streamwise vibration* of structures: cylinders, prisms, hydraulic gates and so on. An abbreviated presentation may be found in Naudascher & Rockwell (1994, 2005).

According to Naudascher (1987), "Any bluff body in cross-flow free to vibrate in the *streamwise* direction may become excited on account of movement-induced displacements of the separating shear layers that make the near wake appear to breathe"; hence, the term "wake-breathing" related oscillation. Naudascher goes on to say that a precondition for such excitation is that upstream body movements ($\dot{x} < 0$) involve a reduction in fluid forces, whereas an enhancement of the fluid forces results from downstream movements. Two distinct types of wake breathing are recognised, and this classification is retained here.

2.6.1 Wake breathing of the first type

Wake breathing of this type may occur for bodies with a rounded shape, such that a wide or narrow wake forms behind the bluff body, depending on whether the boundary layer ahead of separation is laminar or turbulent (Naudascher & Rockwell 1994). Furthermore, Martin, Currie & Naudascher (1981) propose a mechanism related directly to the sharp change in drag in the critical Reynolds number range (sometimes referred to as a "drag crisis") as the boundary layer changes from laminar to turbulent, as shown in Figure 2.37 for a circular cylinder in a generally confined

---

* The paper, significantly, begins with the Socratic statement: *There was a time when I thought that at some distant day, after dedicating much of my professional work to a small area of Fluid Mechanics, I would be able to say: this area, at least, I understand. Today I know better.*

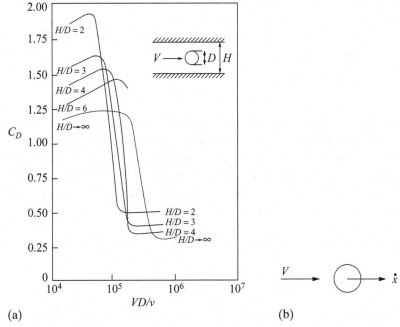

Figure 2.37. (a) Mean drag coefficient over critical Reynolds number range for a cylinder in generally confined flow (Richter & Naudascher 1976; Martin *et al*. 1981); (b) streamwise movement of the cylinder.

cross-flow. Specifically, in the half-cycle over which the cylinder is moving upstream, against the flow, the relative velocity is $V - \dot{x} > V$, where V is the free-stream flow velocity and $\dot{x}$ the cylinder velocity, with $\dot{x} < 0$ here; thus, the effective Reynolds number is higher than the mean value over the cycle of streamwise oscillation, resulting in a low C_D – in the range $0.25 < C_D < 0.50$ in the figure, depending on H/D. Similarly, when the cylinder is moving with the flow, we have $V - \dot{x} < V$ (as $\dot{x} > 0$ now) and thus a lower Reynolds number, and hence a higher C_D, $1.2 < C_D < 1.8$ in the figure. Therefore, a periodic unsteady component of drag is generated by cylinder motion, synchronous with it, resulting in net energy transfer from the fluid to the cylinder.

Clearly, therefore, the condition of instability is

$$\frac{\mathrm{d}F_D}{\mathrm{d}V} < 0, \tag{2.55}$$

where $F_D = \frac{1}{2}\rho V_{\rm rel}^2 D\, C_D$, and $V_{\rm rel} = V - \dot{x}$ is the relative velocity between the flow and the cylinder. We next write $F_D = \frac{1}{2}\rho(\nu/D)^2\, \mathrm{Re}^2\, C_D$, where $\mathrm{Re} = V_{\rm rel}D/\nu$ is the instantaneous Reynolds number. Furthermore, we approximate $\mathrm{Re} = VD/\nu$ and $\partial(\ )/\partial V = \partial(\ )/\partial V_{\rm rel}$ for small $\dot{x}$, and the above criterion leads to

$$\frac{\mathrm{d}F_D}{\mathrm{d}V} = \left(\frac{D}{\nu}\right)\frac{\mathrm{d}F_D}{\mathrm{d}\mathrm{Re}} = \frac{1}{2}\rho\left(\frac{\nu}{D}\right)\frac{\mathrm{d}(C_D\mathrm{Re}^2)}{\mathrm{d}\mathrm{Re}}\;;$$

since C_D is a function of Re, criterion (2.55) becomes

$$\frac{\mathrm{d}C_D}{\mathrm{d}\mathrm{Re}} + 2\frac{C_D}{\mathrm{Re}} < 0, \tag{2.56}$$

whereby the similarity to the Glauert/Den Hartog criterion for transverse galloping is quite obvious.

Martin *et al.* then proceed to solve the following nonlinear equation for the dynamics of a flexibly supported rigid cylinder in a cross-flow, in the critical Reynolds number range:

$$m l \ddot{x} + r \dot{x} + k x = \tfrac{1}{2} \rho V_{\text{rel}}^2 D l \, C_D', \qquad (2.57)$$

where x is the streamwise displacement of the cylinder, m its mass per unit length including the added mass, l its length and D its diameter; r is the linear mechanical damping coefficient, k the linear stiffness constant and C_D' the instantaneous drag coefficient for a fixed cylinder at the instantaneous Re – thus invoking the quasi-steady assumption. Now, V_{rel} is a function of $\dot{x}$, and furthermore C_D is a function of V_{rel}; hence, this is clearly a nonlinear equation.

Defining $\xi = x/D$, $\tau = \omega_n t$ with $\omega_n = 2\pi f_n = (k/ml)^{1/2}$, and writing $V_{\text{rel}} = V[1 - \alpha(\mathrm{d}\xi/\mathrm{d}\tau)]$ with $\alpha = \omega_n D/V$, a dimensionless form of equation (2.57) is obtained:

$$\ddot{\xi} + \varepsilon \dot{\xi} - (n/\alpha^2)\Phi'(\dot{\xi}) + \xi = 0, \qquad (2.58)$$

where the overdot now denotes $\mathrm{d}(\ )/\mathrm{d}\tau$, $\varepsilon = r/ml\omega_n$, $n = \rho D^2/2m$ is the mass ratio, and $\Phi' = \Phi - \overline{\Phi}$, with $\overline{\Phi}$ being the constant part of the drag force, not contributing to the solution. In general,

$$\Phi'(\dot{\xi}) = (1 - \alpha\dot{\xi})^2 \, C_D'(\dot{\xi}, \ \mathrm{Re}, \ S, \ H/D),$$

where $S = f_n D^2/\nu$ is the Stokes number and H/D is the relative channel confinement as in Figure 2.37. The similarity of equation (2.58) to (2.19) is evident.

This equation was solved by a perturbation scheme, finally yielding limit-cycle oscillation amplitudes of the type shown in Figure 2.38 for two values of ε/n – which is similar to the $2\zeta/n$ parameter in Section 2.2.2; e.g. in equation (2.22).

Streamwise oscillations of circular steel piles were observed during construction of a jetty at Immingham, in England, in the tidal flow of the Humber estuary. The observations made by Wootton *et al.* (1974) have been compared with the analytical amplitude predictions of Martin *et al.*, showing promising agreement. It is of interest that the predictions always satisfy criterion (2.56).

This phenomenon can only occur in the narrow range at approximately $V/fD = \mathrm{Re}^*/S$, where Re^* is the Reynolds number at which the C_D versus Re curve in Figure 2.37 has the largest negative slope; for this reason, this phenomenon is often called a *drag-crisis-induced instability*. Moreover, the limit-cycle amplitudes obtained are rather modest. Hence, wake-breathing oscillations of the first type are of limited interest, even though they have occurred in practice, as mentioned above. Nevertheless, they are of special interest in the context of this chapter because of their superficial similarity to transverse galloping.

2.6.2 Wake breathing of the second type

This type of wake breathing has nothing to do with the drag crisis, but it also leads to streamwise oscillation, as seen in Figure 2.39, where the axially sliding splitter plate was used to suppress vortex-shedding excitation.

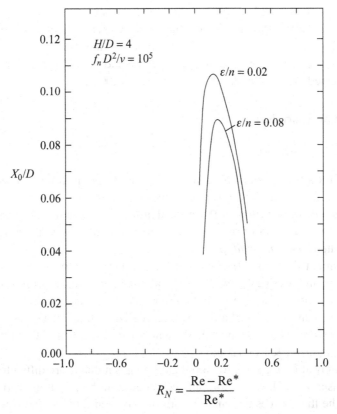

Figure 2.38. Amplitude prediction for wake-breathing streamwise oscillation in terms of a "reduced Reynolds number", R_N, where $\mathrm{Re} = VD/\nu$ is the actual Reynolds number and Re^* is a reference value depending on H/D; in this case $\mathrm{Re}^* = 1.56 \times 10^5$ (Martin $et\ al.$ 1981).

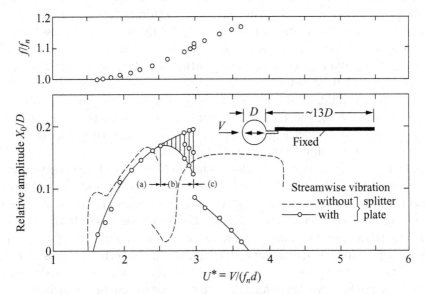

Figure 2.39. Response of a circular cylinder vibrating freely in streamwise direction; $\mathrm{Re} = 2 \times 10^3 - 9 \times 10^3$, body density $= 0.94\rho$, $\delta = 0.157$ (Naudascher & Rockwell 2005; after Aguirre 1977).

The mechanism in this case is rather intricate and appears to be related to reorganization of the streamlines and the wake, in synchronism with streamwise motions, and attendant changes in C_D. This type of wake-breathing oscillation has been studied extensively by Aguirre (1977) and Naudascher (1987). The interested reader is referred also to Naudascher & Rockwell (2005).

2.7 Torsional Galloping

2.7.1 General comments

Figure 2.4 shows the Tacoma Narrows Bridge undergoing torsional oscillations.[*] If one takes a sectional view, however, torsion means that sections of the bridge undergo *rotation*; and since most theoretical and experimental studies examine the flow around and dynamics of rigid sections, most of what follows could more accurately be labelled *rotational galloping*.

A convenient demonstration of rotational galloping is provided by mounting a triangular prism (wedge), free to rotate about its centre, in cross-flow. An isosceles triangular prism("wedge") can be induced to oscillate incessantly, whereas an equilateral one can either oscillate or rotate; see Nagashima & Hirose (1982). The mechanisms involved have been studied via flow visualization and PIV by Srigrarom (2003) and Srigrarom & Koh (2008).

Surprisingly at first glance, torsional galloping is much more difficult to analyze than the translational "heave" galloping considered in the previous sections. This is because (i) the fluid forces depend on both the angle and the angular velocity, and (ii) the relative flow velocity, V_{rel}, varies from point to point, hence giving rise to a varying angle of attack along the section. As a result of (ii) above, one cannot define an equivalent static configuration and hence cannot proceed with the use of the quasi-steady assumption in the same way as for translational galloping. Another difficulty is that the phase difference between the fluid-dynamic forces acting on the section and motion of the section changes with flow velocity. No totally satisfactory solution has been found to these difficulties, and hence analytical models of torsional galloping involve approximations going considerably farther than the quasi-steady assumption.

Indeed, as will be shown in Section 2.7.4, the quasi-steady assumption is found to be generally invalid, or at least of limited applicability, for predicting the onset of rotational galloping and the ensuing nonlinear dynamics. For this reason, the notation of *torsional flutter* is gradually supplanting "torsional galloping". In this book, however, we retain the traditional, older nomenclature.

All types of rectangular, triangular, L-shaped (angles) and H-shaped structural components are susceptible to torsional galloping. They are widely used in open civil engineering structures; for instance, structural angles are frequently used in high-voltage transmission towers, antenna masts and bridges, and bridge decks are essentially H-sections. In addition to the few celebrated disasters (e.g. the Tacoma Narrows Bridge collapse), some long, slender angle members in transmission towers have been known to experience large-amplitude oscillations when exposed to normal atmospheric winds. Advances in metallurgy have encouraged the use of lighter, more

---

[*] Actually, what is seen could be either pure torsional galloping or flutter involving both torsion and transverse motions.

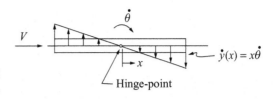

Figure 2.40. A rectangular section in rotational ("torsional") motion about a hinge-point, showing that $\alpha = \alpha(x)$, $V_{\rm rel} = V_{\rm rel}(x)$ generally.

flexible components, and hence instances of failure due to torsional galloping have been reported (Slater 1969). These are reasons why a great deal of effort has been devoted to the subject; but, as we shall see, progress has been rather tortuous.

2.7.2 Linear quasi-steady analysis

Historically, the first batch of studies on this topic attempted to adapt the analysis used for transverse galloping to torsional galloping. Figure 2.40 shows the major complication associated with torsional *vis-à-vis* translational galloping: even for this very simple system, the effective angle of attack and, hence, $V_{\rm rel}$ are functions of location along the section.

Consider a more general situation, as in Figure 2.41, where the hinge is not necessarily at the geometric centre but at the shear centre.* One approach to circumvent the problem of a variable angle of attack is to define a reference radius r_r on the rotating section (Figure 2.41), such that there is a unique value of transverse velocity $r_r\dot{\theta}$ at angle γ_r, which is supposed to be representative of motion of the whole section (e.g. Slater (1969), Blevins (1990)). One can therefore obtain a *reference angle of attack* α induced by $\dot{\theta}$ and a *reference relative velocity*, much as was done for transverse galloping. Referring to Figure 2.41(b), one obtains

$$\theta - \alpha = \tan^{-1}\left[\frac{r_r\dot{\theta}\sin\gamma_r}{V - r_r\dot{\theta}\cos\gamma_r}\right], \tag{2.59}$$

$$V_{\rm rel}^2 = \left(r_r\dot{\theta}\sin\gamma_r\right)^2 + \left(V - r_r\dot{\theta}\cos\gamma_r\right)^2, \tag{2.60}$$

which for small α (effectively small $\dot{\theta}$) may be approximated by

$$\alpha \simeq \theta - \frac{r_r\dot{\theta}\sin\gamma_r}{V} \equiv \theta - \frac{R\dot{\theta}}{V}, \tag{2.61}$$

which defines a characteristic radius $R = r_r\sin\gamma_r$, and

$$V_{\rm rel} \simeq V. \tag{2.62}$$

Even so, the choice of r_r is not obvious. Moreover, since γ_r is a function of θ, as appreciated from Figure 2.41(b), $R = R(\theta)$ and equation (2.61) is nonlinear. For rectangular sections oscillating about the geometric centre, Nakamura & Mizota

---

* The shear centre, or elastic axis, is the point on the section where an applied force produces no torsion and a moment produces no lateral motion.

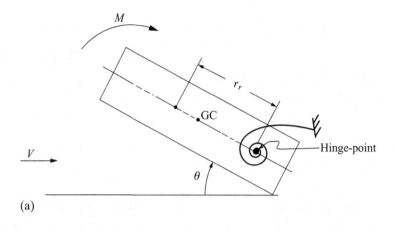

(a)

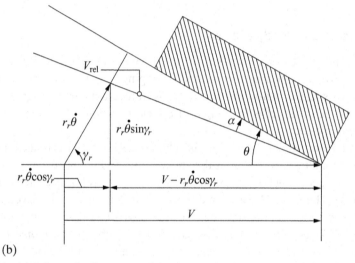

(b)

Figure 2.41. (a) Schematic of a rectangular prism section in torsional oscillation; the torsional spring of stiffness k_θ is attached to the prism at the hinge point. (b) Diagram for determining α in terms of r_r.

(1975b) chose R to be half the depth of the rectangle ($R = \frac{1}{2}d$), thus α corresponds to the instantaneous angle of attack of the leading edge. For right-angle sections facing the wind about the apex, Slater (1969) took R to be half the sum of the fore and aft lengths of the angle section.

This approach, i.e. to utilize a reference r_r, has been inspired by the use of the three-quarter chord point in airfoil flutter analysis (Bisplinghoff & Ashley 1962), although its applicability to separated flows is tenuous. This approach was also used, in a more sophisticated framework, by Modi & Slater (1974) for the angle section and by Blevins & Iwan (1974) for a two-degree-of-freedom analysis of the same problem.

Proceeding as in Blevins (1990), the equation of torsional motion is

$$J_\theta \ddot{\theta} + c_\theta \dot{\theta} + k_\theta \theta = \tfrac{1}{2}\rho V^2 h^2 C_M, \qquad (2.63)$$

Table 2.4. *Slope of torsional moment coefficient for various sections rotating about their geometric centre, extracted from Nakamura & Mizota (1975b) and Blevins (1990); α is in radians, and the flow is from the left*

Section	$\partial C_M/\partial \alpha$	Reynolds number
+ ↕ D	−0.18	$10^4 - 10^5$
2D / + 1D	−0.64	$5 \times 10^3 - 5 \times 10^5$
4D / + 1D	−18.	$2 \times 10^3 - 2 \times 10^4$
5D / + 1D	−26.	$2 \times 10^3 - 2 \times 10^4$

where J_θ is the section mass moment of inertia, c_θ and k_θ are, respectively, the torsional damping and stiffness coefficients, and h is a characteristic length; the moment M is associated with a moment coefficient C_M via $M = \frac{1}{2}\rho V^2 h^2 C_M$. As in Section 2.2.1, equation (2.12), we expand C_M in a Taylor series

$$C_M = C_M\Big|_{\alpha=0} + \left(\frac{\partial C_M}{\partial \alpha}\right)\Big|_{\alpha=0} \alpha + \cdots \tag{2.64}$$

Ignoring the static component, utilizing (2.61) for α, and defining $c_\theta/J_\theta = 2\zeta_\theta\omega_\theta$ and $k_\theta/J_\theta = \omega_\theta^2$, equation (2.63) may be written as

$$\ddot{\theta} + \left[2\zeta_\theta\omega_\theta + \frac{1}{2}\rho\,\frac{V R h^2}{J_\theta}\frac{\partial C_M}{\partial \alpha}\right]\dot{\theta} + \left[\omega_\theta^2 - \frac{1}{2}\rho\,\frac{V^2 h^2}{J_\theta}\frac{\partial C_M}{\partial \alpha}\right]\theta = 0. \tag{2.65}$$

Hence, in principle, this system is susceptible to both static torsional divergence and torsional galloping.

Divergence would arise if

$$\frac{1}{2}\rho V^2 h^2 \frac{\partial C_M}{\partial \alpha} > k_\theta, \tag{2.66}$$

with $\partial C_M/\partial \alpha > 0$; refer also to Richardson *et al.* (1965).

Galloping would arise if $\partial C_M/\partial \alpha < 0$ and the total damping vanishes, which would occur at $V_{\text{crit}} = 4J_\theta\,\zeta_\theta\,\omega_\theta/\left[\rho R h^2(\partial C_M/\partial \alpha)\right]$, or

$$\frac{V_{\text{crit}}}{f_\theta h} = \frac{-4}{[R(\partial C_M/\partial \alpha)]}\left(\frac{J_\theta\,\delta_\theta}{\rho\,h^3}\right), \tag{2.67}$$

where $f_\theta = \omega_\theta/2\pi$ and $\delta_\theta \simeq 2\pi\,\zeta_\theta$.

A compilation of values of $\partial C_M/\partial \alpha$ has been provided by Blevins (1990), extracted from the work of Nakamura & Mizota (1975b); it is given in Table 2.4 here. It is seen that the more elongated the prism, the higher is the negative value of $\partial C_M/\partial \alpha$.

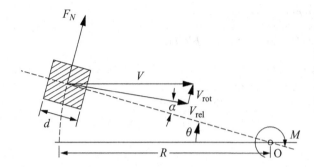

Figure 2.42. Cross-sectional view of the system considered by van Oudheusden; here the angular velocity is $R(d\theta/dt)$; in the experiments, $R/d = 7.5$ (van Oudheusden 1996).

The main problem with this type of model is the difficulty in defining an appropriate r_r and γ_r, and hence finding a suitable approximation for R. A simplified system, shown in Figure 2.42, where this difficulty does not arise (because, strictly, it is *not* a torsional system) was studied by van Oudheusden (1995, 1996). It is obvious that in this case, because the depth, d, of the body is small relative to its distance from the hinge point, the choice of r_r is obvious, and $R = r_r$. Hence, the angle of attack is simply

$$\alpha = \tan^{-1}\left(\frac{V\sin\theta - R\dot\theta}{V\cos\theta}\right) \simeq \theta - \frac{R\dot\theta}{V}. \tag{2.68}$$

In fact, this is effectively a degenerate system which could in principle be analysed in terms of translational galloping theory; refer also to Nakamura (1979).

Some recent numerical work is presented at the end of Section 2.7.4.

2.7.3 Nonlinear quasi-steady analysis

If the linear quasi-steady theory is questionable, as discussed in Section 2.7.4, nonlinear quasi-steady theory is likely more so. However, before the definite disqualification of quasi-steady theory, considerable work was done to generate a *nonlinear* quasi-steady theory for torsional galloping by Slater (1969), and was further elaborated by Modi & Slater (1974) and Blevins & Iwan (1974).

In brief, Slater begins with equation (2.63), making use of an effective r_r and γ_r, as in Section 2.7.2. The moment coefficient C_M is measured statically for an angle section and $\alpha_0 = -45°$ (see Figure 2.16), and the results are plotted as functions of θ and $\dot\theta/U$. Because the curves are approximately straight lines, a new coordinate system is introduced (ζ, ξ) for C_M, slanted to $(\theta, \dot\theta/U)$ at angle $\lambda = 35°$. In the new plot, C_M does not vary with ζ, and hence $C_M = \sum_{i=1}^{N} a_i\xi^i$ may be constructed, in which $\xi = (\dot\theta/U)\sin\lambda - \theta\cos\lambda$. Thereby, equation (2.63) takes the form of the pair (2.19) and (2.20), and thus can be and was analysed by similar methods. Results may be found in Slater (1969) and Slater & Modi (1974); in general, an unstable limit cycle is obtained, nesting within a larger stable one. Unfortunately, in the accompanying experiments, a high enough U could not be reached to observe torsional flutter of the angle section, despite the high flow capabilities ($V = 60$ m/s) of the wind tunnel.

The degenerate torsional system of Figure 2.42, however, permits simpler, less controversial analysis, even in the nonlinear domain, as obtained by van Oudheusden

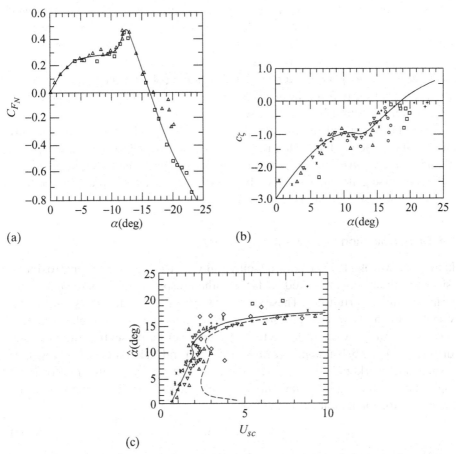

Figure 2.43. (a) Measured static aerodynamic normal force coefficient and line of best fit for the system of Figure 2.42. (b) the *negative* of the total (structural plus aerodynamic) damping coefficient. (c) Torsional limit-cycle amplitude versus a "scaled" flow velocity, $U_{sc} = (\rho d R^3/J_\theta)(U/\zeta_0)$, where $U = V/\omega R$ and ζ_0 is the viscous damping factor; from van Oudheusden (1996).

(1996). Typical theoretical and experimental results are shown in Figure 2.43: (a) the static normal force coefficient C_{F_N} versus α, (b) the negative aerodynamic damping coefficient and (c) the limit-cycle amplitude versus a scaled flow velocity, U_{sc}. Here, $U_{sc} = \mu\, U/\zeta_0$, with $\mu = \rho d R^3/J_\theta$, $U = V/\omega R$, and ζ_0 is the mechanical viscous damping factor; J_θ is the section torsional mass moment of inertia (i.e. per unit span l). The dotted line in the figure corresponds to results with an additional frictional damping modelled by van Oudheusden. It is seen that the required $\partial C_M/\partial\alpha \simeq \partial C_{F_N}/\partial\alpha < 0$ holds for $13° < \alpha < 22°$, and that negative aerodynamic damping (positive in the notation of Figure 2.43(b)) occurs at $\alpha \geq 17°$ approximately.

It is seen that theory and experiment agree fairly well, except for large amplitudes and low damping. This is discussed by van Oudheusden, and it is concluded that the validity of the quasi-steady assumption must be questioned, even though Blevins' criteria that $V/f_n d > 20$ and (transverse amplitude)/$h < 0.2$, h being the section height, are amply satisfied. This questioning of quasi-steady theory is interesting because, as was mentioned earlier, the geometry here does not provide a good test for quasi-steady torsional galloping theory. Especially since $R/h = 7.5$, torsional and transverse motions and forces are derivable from each other (Nakamura 1979; section 7); because quasi-steady theory works well for transverse galloping, so it should for torsional galloping in this case. According to Nakamura & Mizota (1975), even the criterion $\partial C_M/\partial \alpha < 0$ for torsional galloping should only apply to hinge positions "not far away from the structure"; yet, as seen in Figure 2.43, the theory does quite well, fortuitously or otherwise.

It must be concluded that, experimental support for the use of quasi-steady theory for linear and nonlinear characteristics of torsional galloping is rather sparse. The kind of *incontrovertible evidence* in support of quasi-steady theory as in the case of transverse galloping (provided by Parkinson, Novak and their co-workers) is missing in the case of torsional galloping.

2.7.4 Disqualification of quasi-steady theory

Nakamura & Mizota(1975b) state unequivocally that the use of the approximate quasi-steady theory as presented so far is rather suspect and its agreement with experiments rather fortuitous.* These statements are preambulatory to showing that torsional galloping is, in fact, an unsteady rather than quasi-steady phenomenon. In this section we provide evidence to that effect, before presenting the unsteady theory in Section 2.7.5. Nakamura & Mizota (1975b) and Nakamura (1979), together with several co-workers in other papers, consider that a major element in torsional galloping is what they call the "fluid memory" effect. To appreciate what this is, one can write the torsional moment in the form

$$M(\theta) = M_\theta \theta + M_{\dot\theta} \dot\theta, \tag{2.69}$$

where $M_\theta = \partial M/\partial \theta$ is the "steady aerodynamic torsional moment" derivative, and $M_{\dot\theta} = \partial M/\partial \dot\theta$ may be called the "quasi-steady aerodynamic damping" derivative. Assuming phase lags ϕ_θ and $\phi_{\dot\theta}$, the phase-correction factors for the quasi-steady responses to the θ and $\dot\theta$ motions are denoted by $C_\theta e^{-i\phi_\theta}$ and $C_{\dot\theta} e^{-i\phi_{\dot\theta}}$; hence we can write

$$M(\theta) = C_\theta e^{-i\phi_\theta} \overline{M}_\theta \theta + C_{\dot\theta} e^{-i\phi_{\dot\theta}} \overline{M}_{\dot\theta} \dot\theta \tag{2.70}$$

with the overbar denoting the quasi-steady value; in view of (2.69), the aerodynamic damping derivative becomes

$$-M_{\dot\theta} = -C_{\dot\theta} \cos \phi_{\dot\theta} \, \overline{M}_{\dot\theta} + (C_\theta \sin \phi_\theta/\omega) \, \overline{M}_\theta.$$

Further, assuming $C_\theta, C_{\dot\theta} \simeq 1$ and $|\phi_\theta|, |\phi_{\dot\theta}| \ll 1$, this reduces to

$$- M_{\dot\theta} = -\overline{M}_{\dot\theta} + \lim_{\omega \to 0} (\sin \phi_\theta/\omega) \, \overline{M}_\theta. \tag{2.71}$$

The important thing to notice is that a component of the stiffness is now effectively part of the aerodynamic damping. Here, the reader is referred to Section 5.2 for a similar development in connection with fluidelastic instability in cylinder arrays. Thus, the physical nature of this phase-lag-related damping can be supposed to be

---

* In fact, Nakamura & Mizota's study is a broadside on the $\partial C_M/\partial \alpha < 0$ criterion of equation (2.67) for torsional galloping, as adopted by the American Society of Civil Engineers (ASCE) (Anonymous 1961); they state that they "do not know on what rational grounds (it) was established."

similar; i.e. that it is related to the time taken for the viscous flow to adjust to the changing location/configuration of the torsionally oscillating body, or that the pressure on the body cannot adjust itself instantaneously because of the shed vorticity as the body changes position. At present, no means for predicting ϕ_θ is available.

Thus, we see that "the fluid memory effect" is really a phase-lag effect which causes a part of the displacement-dependent force to be transformed into a velocity-dependent one (the second term in (2.71)). Nakamura & Mizota state that this second component in (2.71) is not necessarily negligible and that, in fact, it can be dominant, which means that torsional galloping cannot be analysed via quasi-steady theory.

Here, we should emphasize that the "fluid-memory effect" and equivalently the "phase-lag effect" should be interpreted more broadly as evidence that unsteady effects cannot be ignored. Thus, the discussion above should *not* be interpreted as meaning that torsional galloping can be predicted via quasi-steady theory provided that it is modified to include a phase lag. The shortcomings of quasi-steady theory for torsional galloping are rather more profound; they are essentially as enumerated in Section 2.7.1. The work immediately following aims to show that the fluid-dynamic forces associated with torsional galloping are not only functions of rotational velocity but also of the angle itself (or "angle-of-attack motion").

Nakamura & Mizota (1975b) conducted experiments on rectangular prisms with h/d ratios of 1, $\frac{1}{2}$, $\frac{1}{4}$ and $\frac{1}{5}$, also varying the location of the hinge point. Both stable and unstable limit cycles were found, e.g. for the $h/d = \frac{1}{2}$ prism. In the experiments, the damping component of the unsteady aerodynamic torsional moment was measured by means of free-oscillation experiments, i.e. $M_{\dot\theta}$ in equation (2.69). From this, one can obtain a nondimensional damping derivative, k_a, defined by $k_a = -M_{\dot\theta}/[\rho f(\frac{1}{2}d)^4]$. The sign of k_a is compared with $\partial C_M/\partial \alpha$ obtained from the static measurements in Figure 2.44; both should be negative for galloping to occur. It is seen that the sign of $\partial C_M/\partial \alpha$ (quasi-steady theory) agrees with that of k_a (unsteady theory) only for small d/h and over a limited range of the hinge-point location X_p. Considering all experimental data, it is concluded that the $\partial C_M/\partial \alpha < 0$ criterion is "fairly applicable" for sections with $d/h = 1$ and 2, but not for the deeper (higher d/h) prisms – for which in any case reattachment may occur. It is also remarked that, if the hinge point is far ahead or behind the bluff body, "torsional galloping" is little different from transverse galloping.

Nakamura (1979) reconsiders the question in a different way. Pure torsion about the centre of the prism is considered as the superposition of "angle-of-attack motion", essentially transverse heaving (plunging), and "angular velocity motion" about a hinge. It is shown that the former is associated with unsteady, as opposed to quasi-steady, components in the overall aerodynamic lift and moment coefficients. Nakamura conducted experiments with prisms of depth-to-height ratios of $d/h = 1$ and 4, in either torsional or transverse translational oscillation about their centre (Figure 2.45), such that $\theta(t) = \theta_0 \cos \omega_m t$ and $y = y_0 \sin \omega_m t$. Small angular amplitudes were used, $\theta_0 = 1°$ and $2°$, and the translational ones were chosen so as to yield *the same effective angle of attack* ($\alpha = \dot y_0/V$). Lift and moment coefficients were measured in both cases. Analysis of these measurements showed that "the angle-of-attack motion" is dominant in determining both the lift and moment coefficients, in particular, for $d/h = 4$. This shows that unsteady effects are not only important, but can be

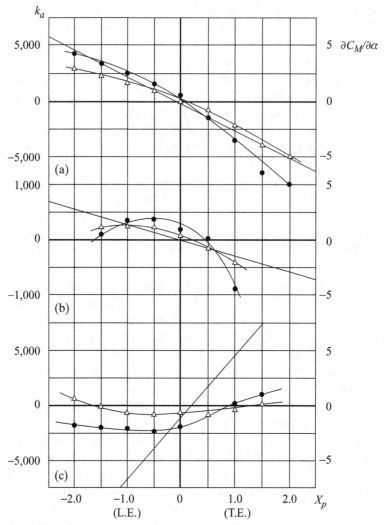

Figure 2.44. Unsteady aerodynamic damping coefficient for torsional motion, k_a, and quasi-steady moment derivative, $\partial C_M/\partial \alpha$, for various locations of the hinge point, for (a) $d/h = 1$, (b) $d/h = 2$ and (c) $d/h = 4$ rectangular prisms: $\triangle$, k_a for $U^* = 50$; $\bullet$, k_a for $U^* = 100$; $-$, $\partial C_M/\partial \alpha$. Here $X_p = x_p/d$, where x_p is location of the hinge measured from the centre of the prism (Nakamura & Mizota 1975b).

dominant. Hence, for torsional galloping one needs to use unsteady fluid-dynamic theory.

A considerable number of torsional experiments were conducted by Washizu *et al.* (1980); because of the smallness of damping, there was interference by vortex-shedding lock-in in some cases. Nevertheless, it was possible to compare Blevins' (1977) quasi-steady theory for torsional galloping with the experimental results. It was found "that the results from this theory differ from these experimental results and thus are not applicable for practical purposes"; further it is said, very diplomatically, that "unsteady aerodynamic moments ... are more complicated than those predicted by the quasi-steady aerodynamic theory".

Figure 2.45. Prismatic bar oscillating in cross-flow: (a) torsional oscillation; (b) translational "heaving" oscillation (after Nakamura (1979)).

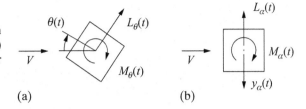

(a)   (b)

Considering all the evidence presented in the foregoing, it must be concluded that quasi-steady galloping theory is generally invalid. Still, the picture is not totally black and white. In the numerical study by Robertson *et al.* (2003), briefly discussed in Section 2.3.1, torsional galloping also was studied for rectangular sections with $d/h = 1$ to 5 at Re = 250. Torsional galloping was observed in all cases, and this was found to be consistent with results obtained by the linear quasi-steady torsional theory of Section 2.7.2 (using $R = \frac{1}{2}d$) in terms of the sign of $(\partial C_M/\partial \alpha)|_{\alpha=0}$ in Table 2.5, which in turn are consistent with those of Table 2.4. So, at least in terms of the onset of torsional galloping, quasi-steady theory *can* in some cases perform well, fortuitously or otherwise!

Table 2.5. *Variation of the moment coefficient against angle of attack in radians about a zero effective angle of attack for rectangular sections of increasing aspect ratio (Robertson, Li, Sherwin & Bearman 2003)*

d/h	1	1.5	2	3	4	5
$(\partial C_M/\partial \alpha)|_{\alpha=0}$	−0.516	−1.49	−2.46	−3.21	−9.74	−2.06

2.7.5 Unsteady theory

The unsteady theory begins by eschewing the use of a representative radius r_r and corresponding γ_r as was done in Section 2.7.2. Instead, we start with equation (2.69), by recognizing that the torsional moment $M(\theta)$ is a function of both θ and $\dot{\theta}$; thus, for small motions, $M(\theta) = M_\theta \theta + M_{\dot\theta}\dot\theta$. An outline of the theory will be given here, following Nakamura's (1979) presentation. A much more general theory for elongated and bridge-deck sections, but applicable to both transverse and torsional galloping of *any kind of sectional shape*, is given in Section 2.11.

Consider torsional motion of a rectangular prism about a point O', generally off its centre O, as shown in Figure 2.46. For small motions, so that the principle of superposition is valid, the fluid-dynamic moment $M'_\theta(t)$ per unit length about O' may be expressed as

$$M'_\theta(t) = M_\theta(t) + L_\theta(t)\,x_p + M_\alpha(t) + L_\alpha(t)\,x_p, \qquad (2.72)$$

where $L_\theta(t)$, $L_\alpha(t)$ and $M_\theta(t)$, $M_\alpha(t)$ are the fluid-dynamic lifts and moments about O, and $\alpha(t) = -\dot{\theta}(t)\,x_P/V$ is the induced incidence,* so that the α-components correspond to $\dot{\theta}$-components in (2.69).

* Note that this is not valid when $x_p \to 0$.

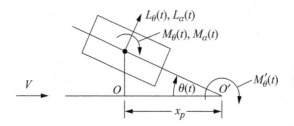

Figure 2.46. A rectangular prism oscillating in torsion about point O' at a distance x_p from its cross-sectional centre O.

Dividing by $\frac{1}{2}\rho V^2 d^2$, the moments may be converted to moment coefficients, e.g. $C_{M\theta}(t) = M_\theta/[\frac{1}{2}\rho V^2 d^2]$, and similarly the lifts by dividing by $\frac{1}{2}\rho V^2 d$. These may then be converted to transfer functions in the frequency domain, e.g. $C_{M\theta}(k)$, through $C_{M\theta}(\tau) = \mathcal{R}e[C_{M\theta}(k)\exp(ik\tau)]$, where $\tau = Vt/h$, and $k = \omega_m h/V$ is the reduced frequency, ω_m being the frequency of oscillation. Therefore, (2.72) may be rewritten in dimensionless form as

$$C'_{M\theta}(k) = C_{M\theta}(k) + C_{L\theta}(k)X_p - ik(d/h)\,C_{M\alpha}(k)X_p - ik(d/h)\,C_{L\alpha}(k)X_p^2, \quad (2.73)$$

where $X_p = x_p/d$. As suggested by Nakamura (1979), in some cases it is possible to use the following approximations: $C_{M\theta}(k) \simeq C_{M\alpha}(k)$ and $C_{L\theta}(k) \simeq C_{L\alpha}(k)$. Therefore, the imaginary component of $C'_{M\theta}$, which defines stability, may be written as

$$C'_{M\theta I} = C_{M\alpha I} + C_{L\alpha I}X_p - k(d/h)C_{M\alpha R}X_p - k(d/h)C_{L\alpha R}X_p^2, \quad (2.74)$$

where subscripts R and I stand for "real" and "imaginary". The first two terms in this equation are associated with unsteady effects, whereas the other two with quasi-steady fluid dynamics. It is noted that, if $X_p \ll 1$, then $C'_{M\theta I} \simeq C_{M\alpha I}$; hence, for small X_p (including $X_p = 0$), torsional galloping is predominantly or entirely an unsteady fluid-dynamic phenomenon, as already concluded in the foregoing. On the other hand, for $X_p \gg 1$, it is $C_{L\alpha R}$ that is dominant, and torsional galloping, if it occurs, is quite similar to transverse galloping.

Nakamura compared the theoretically predicted $C_{M\theta I}$ with the experimental values for $d/h = 1$ and 4 prisms and $-1 < X_p < 1$. The latter were obtained by free-vibration experiments measuring the rate of decay or growth. As shown in Figure 2.47, agreement between this unsteady theory and experiment is excellent; in contrast, agreement with quasi-steady theory ($C_{M\alpha I} = C_{L\alpha I} = 0$ in (2.74)) is generally rather poor.

Similarly to quasi-steady theory for transverse galloping, the unsteady theory here is semi-empirical, but clearly more elaborate measurements are needed in this case. Specifically, $C_{L\theta}$, $C_{M\theta}$, $C_{L\alpha}$ and $C_{M\alpha}$ must be measured over the desired range of U^*. For instance, in Nakamura & Mizota's (1975b) experiments, two identical models were built for each experiment: one was placed within the test-section of a wind tunnel, whereas the other, used to correct for inertial effects, was positioned outside; they were oscillated simultaneously by a mechanical shaker. In the first part of the experiment, the fluid-dynamic lift $L_\theta(t)$ and moment $M_\theta(t)$ were measured, for $\theta(t) = \theta_0 \cos \omega_m t$, with $\theta_0 = 1°$ and $2°$. The frequencies used were $f_m = \omega_m/2\pi = 0.9$–$3.3$ Hz and $V = 6$–8 m/s, so that $U^* = 20$–60, approximately; $Re \simeq 10^5$. In the second part of the experiment, $L_\alpha(t)$ and $M_\alpha(t)$ were measured for $y_\alpha(t) = y_0 \sin \omega_m t$, with y_0 chosen so that the effective angle of attack $\alpha(t) = \dot{y}_\alpha(t)/V$ would be equal to $\theta(t)$, even though, in general, this need not be so. The signals were then processed

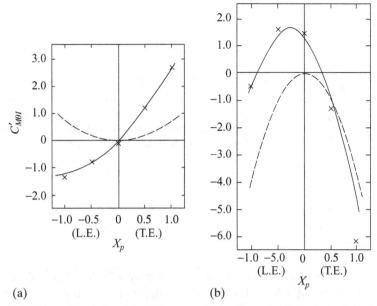

Figure 2.47. Effect of pivotal position $X_p = x_p/d$ (see Figure 2.46) on the out-of-phase component of the aerodynamic moment coefficient for rectangular prisms in torsion: (a) $d/h = 1$; (b) $d/h = 4$. ×, Experiment; — , unsteady theory; – – – , quasi-steady theory (Nakamura 1979).

so as to yield magnitude and phase for each of the four quantities measured, or equivalently the real and imaginary components as in (2.73).

It is obvious that, to obtain the required input data for this theory, a major experimental effort is necessitated. The situation is therefore similar to that of the unsteady models for fluidelastic instability in cylinder arrays (Section 5.3.3), and the quandary is also the same: might it not be easier to measure the torsional critical flow velocities directly in model experiments, rather than the L_θ, M_θ, L_α and M_α data (or at least the last two sets) required as input for the unsteady theory?

Furthermore, this unsteady model probably breaks down for values of U^* substantially above the galloping threshold, because, with increasing amplitudes, the principle of superposition no longer holds true; if tried by the authors concerned, the results were not published.

Finally, it should be mentioned that torsional galloping is sensitive to the level of turbulence intensity in the free stream; the reader is referred to Section 2.9.

2.8 Multi-Degree-of-Freedom Galloping

2.8.1 Quasi-steady models

Some early work on two- and three-degree-of-freedom galloping was conducted by Richardson, Martucelli & Price (1965), both theoretically and experimentally, related to overhead transmission lines, involving horizontal ("pendular"), vertical ("radial") and rotational motions of generally iced conductors. In the limit, the criteria obtained reduce to the Den Hartog/Glauert criterion. Examples of more

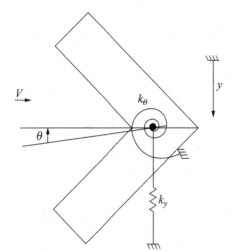

Figure 2.48. Elastically supported two-degree-of-freedom angle section, with the springs attached at the elastic axis; translational and rotational dashpots c_y and c_θ, not shown, are also attached there (Blevins 1990).

recent work on the subject may be found in Yu *et al.* (1993a, b), Yu, Popplewell & Shah (1995a, b) and Macdonald & Larose (2006, 2008a, b) and the references cited therein. This work is very particular to overhead transmission lines, and will not be discussed in detail here.

Systematic studies on two-degree-of-freedom combined torsional and transverse galloping were conducted by Slater (1969) and Blevins & Iwan (1974); see also Modi & Slater (1974, 1983). In both, quasi-steady theory was used for the torsional component also, which has since been found to be generally inappropriate (Section 2.7.4); for this reason, the presentation of this work in what follows is rather brief.

The prism considered is an angle section, as shown in Figure 2.48. The supporting springs are attached on the elastic axis (or shear centre; refer to Section 2.7.2) of the prism. Since the elastic axis does not coincide with the inertial axis (going through the centre of mass), the sectional (per unit length) equations of motion in the two degrees of freedom, y and θ, are coupled inertially:

$$m\ddot{y} + 2m\zeta_y\omega_y\dot{y} + s_x\ddot{\theta} + k_y y = F_y = \tfrac{1}{2}\rho\, V^2 h\, C_y,$$

$$J_\theta\ddot{\theta} + 2J_\theta\zeta_\theta\omega_\theta\dot{\theta} + s_x\ddot{y} + k_\theta\theta = F_M = \tfrac{1}{2}\rho\, V^2 h^2\, C_M, \qquad (2.75)$$

where $s_x = \int_S \xi\rho_p\, d\xi\, d\eta$, in which ρ_p is the density of the section material and ξ and η are orthogonal coordinates in the section, with ξ going through both the elastic and inertial axes. The other symbols are self-evident from the foregoing in this chapter; it is nevertheless recalled that h is the maximum height of the section facing the free stream.

In accordance with quasi-steady theory, for small motions

$$\alpha = \theta - \frac{r_r\dot{\theta}}{V} + \frac{\dot{y}}{V}, \qquad (2.76)$$

where, following Slater (1969), r_r is half the sum of the streamwise fore and aft lengths of the angle section from the support ("hinge") point. It is therefore clear that, in addition to inertial coupling, there is aerodynamic coupling in the two degrees of freedom through (2.76).

A linear analysis of the coupled equations (2.75) yields

$$
V = \text{minimum} \begin{cases} \dfrac{-4m\zeta_y\omega_y - 4b_2^2 J_\theta \zeta_\theta \omega_\theta}{\rho h(r_r b_2 + 1)(b_2 h \partial C_M/\partial\alpha - \partial C_y/\partial\alpha)}, \\[4mm] \dfrac{-4b_1^2 m\zeta_y\omega_y - 4J_\theta \zeta_\theta \omega_\theta}{\rho h(r_r + b_1) - (-b_1 \partial C_y/\partial\alpha + h \partial C_M/\partial\alpha)}, \end{cases}
\tag{2.77}
$$

where $\omega_y^2 = k_y/m$ and $\omega_\theta^2 = k_\theta/J_\theta$, and

$$
b_1 = \frac{s_x}{m}\frac{\omega_1}{\omega_y^2 - \omega_1^2}, \qquad b_2 = \frac{s_x}{J_\theta}\frac{\omega_2^2}{\omega_\theta^2 - \omega_2^2},
$$

$$
\omega_{1,2}^2 = \frac{\omega_y^2 + \omega_\theta^2 \pm \left\{(\omega_y^2 + \omega_\theta^2)^2 - 4\omega_y^2\omega_\theta^2(1 - s_x^2/J_\theta m)\right\}^{1/2}}{2(1 - s_x^2/J_\theta m)}.
\tag{2.78}
$$

If $s_x = 0$ and hence $b_1 = b_2 = 0$, which is *not* the case for the angle section, equations (2.77) yield the flow velocities for galloping in the individual degrees of freedom. For the angle section, fully coupled behaviour was expected, involving both degrees of freedom, and with the oscillation frequency not coinciding with either ω_θ or ω_y.

In Slater's (1969) experiments with angle sections, despite $s_x \neq 0$, the above expectations did not materialize. Coupling was weak – basically because ms_x^2/J_0 is small, J_0 being the moment of inertia about the inertial axis – and the observed coupled motions occurred at the natural frequencies of the individual degrees of freedom, ω_y and ω_θ, as illustrated in Figure 2.49. It is recalled that predominantly torsional galloping could not be obtained in Slater's experiments.

An important experimental observation was that transverse and torsional displacements of the same frequency were in phase for translational resonant (synchronised or locked-in) oscillations, whereas for torsional resonant oscillations they were 180° out of phase.* This suggests the existence of two distinct virtual centres of rotation, or virtual "hinge" points – cf. Kosko (1968) and Wardlaw (1967). Therefore, based on the observation that the oscillations occur essentially in one or the other degree of freedom, in Slater's analytical models transverse and torsional galloping were considered separately, the former in the manner of Parkinson & Smith (1964) and the latter as in Section 2.7.3.

Blevins & Iwan (1974) neglected inertial coupling from the start, and they approximated $C_y \simeq 0.656\alpha - 7.83\alpha^3$ and $C_M \simeq -0.105\alpha + 9.34\alpha^3$ – purposely considering a low-order polynomial to allow nonlinear analysis to go farther. Taking the ratio of the magnitude of the structural damping and aerodynamic forces to that of the inertial and spring forces to be small and characterised by the parameter ϵ, the equations of motion were recast as asymptotic expansions to the first order of ϵ. Then, assuming amplitudes and phases to be slowly varying functions, a sophisticated analysis was undertaken by a method similar to the method of multiple scales (Appendix A), but here with two variables, y/h and θ. Two cases were examined:

---

* An interesting aspect of resonant vortex-shedding response is this: in transverse synchronised oscillations, the vortex-shedding frequency is controlled by prism motions, but in torsional lock-in it is the vortex shedding that controls the oscillation (Modi & Slater 1988).

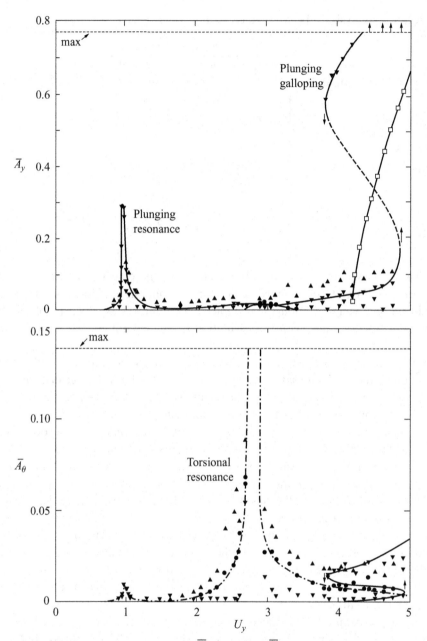

Figure 2.49. Displacement measurements $\overline{A}_y = y/h$ and $\overline{A}_\theta$ versus $U_y = V/\omega_y h$ for a 3-inch angle section at $\alpha_0 = -45°$ with $\omega_\theta/\omega_y = 2.92$, $n_Y = \rho h^2 l/2m = 2.98 \times 10^{-3}$, $n_\theta = \rho h^4 l/2J_\theta = 19.52 \times 10^{-3}$, $\zeta_y = c_y/2m\omega_y = 4.12 \times 10^{-3}$, $\zeta_\theta = c_\theta/2J_\theta\omega_\theta = 5.13 \times 10^{-3}$: ▲, ▼, motion with frequency ω_y; ●, motion with ω_θ (Slater 1969). The results from the Blevins & Iwan (1974) model in the vicinity of $U_y = 4.5$ have been added ($-\square-\square-$).

one where there is no internal resonance, and the other when there is, i.e. when $\omega_\theta/\omega_y \simeq \frac{1}{3}$, 1 or 3.

In the case of no internal resonance, the solutions are characterised by either a steady transverse motion and sensibly no torsion, or the inverse. In the case of predominantly transverse galloping, the predicted amplitude is compared with Slater's experimental data in Figure 2.49. In Slater's experiments, $\omega_\theta/\omega_y = 2.92$, which might be thought to be sufficiently close to 3 for internal resonance to occur. However, the difference is greater than the half-power bandwidth of a lightly damped oscillator, $\Delta\omega = 2\zeta\omega_n$, with either ζ_θ and ω_θ or ζ_θ or ω_y; and so the results for no internal resonance should apply. The onset of galloping is really well predicted in Figure 2.49 (squares), but the S-shape in the experimental results could not be captured, probably because of the low-order polynomial approximations of the aerodynamic coefficients. In addition, the amplitude for $U_y = V/\omega_y h > 4.5$ diverges substantially from the experimental values; h is the maximum width of the section, normal to the free stream.

The case of internal resonance has also been studied by Blevins & Iwan, necessitating more nonlinear analysis. Some theoretical results in the vicinity of $\omega_\theta/\omega_y = 1$ and $U_y = 5.0$ are obtained (see Blevins (1990)). In the region $\omega_\theta/\omega_y \simeq 1$, there is strong coupling between transverse and torsional motions; outside that range, the motion is essentially purely transverse and the amplitude agrees quite well with that predicted by the uncoupled solution (i.e. with $\omega_\theta/\omega_y \neq 1$).

Unfortunately, no comparison could be made with experiments, either for predominantly torsional galloping or for strongly coupled transverse galloping, and hence the general validity of this model could not be tested. However, in view of the general disqualification of the torsional quasi-steady component of the model, the overall model (Section 2.7.4) is suspect.

2.8.2 Unsteady models

There has been very strong impetus for developing unsteady models for combined transverse-torsional galloping: the galloping of H-section prisms, representing bridge sections. The famous collapse of the first Tacoma Narrows Bridge served as a powerful incentive. In due course, reliable unsteady models of galloping were developed. They are presented in Section 2.11.

2.9 Turbulence and Shear Effects

We have already come across the effect of turbulence on galloping in Table 2.1 – at least on soft, self-excited galloping, i.e. according to the Den Hartog/Glauert criterion: some sections will gallop in smooth flow but not in turbulent flow, and *vice versa*. This is not surprising, since turbulence changes both the lift and drag of the bluff body, the details of separation, as well as reattachment, if any. In short, turbulence changes the pressure distribution on the afterbody which, as we have seen (Section 2.3.4), controls the occurrence or otherwise of galloping.

Considerable work on the effect of turbulence on galloping was done in the 1970s and 1980s by Novak & Davenport (1970), Laneville & Parkinson (1971), Novak (1972, 1974), Otsuki *et al.* (1974), Novak & Tanaka (1974), Nakamura & Tomonari (1977), Kwok & Melbourne (1980), Nakamura & Yoshimura (1982), Bokaian & Geoola (1982) and Bearman *et al.* (1987). In addition, a great deal of work was done on the effect of turbulence specifically on the flow field around bluff bodies; refer, e.g., to Bearman (1972), Bearman & Morel (1983), Nakamura & Ohya (1984), Bearman *et al.* (1987). Some of the salient findings are that small-scale turbulence increases the growth rate of the separated shear layers via increased mixing, whereas

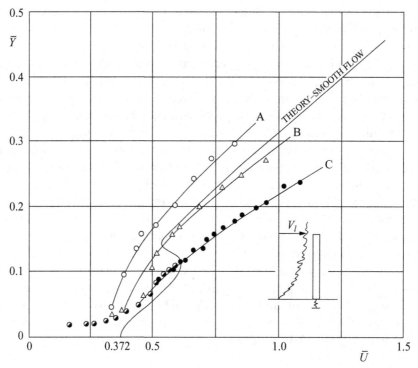

Figure 2.50. Universal galloping response of a pivoted rigid (flexibly mounted rigid-cantilever) square prism in shear flow, showing the effect of turbulence: ○, smooth flow; △, Tu = 4.7%, representative of "open country"; ●, Tu = 8.6%, representative of "urban area"; all with $\zeta = 0.0040$–0.0053; half-filled ○, Tu = 8.6%, $\zeta = 0.0084$–0.0110 (Novak & Davenport 1970). Here and in the figures that follow, $\overline{U}$, $\overline{Y}$ and $\bar{n}$ are as defined in equation (2.30).

large-scale turbulence weakens regular vortex shedding by reducing spanwise correlation; both depend on d/h of the prism. In addition, the mean base pressure can vary significantly with both the scale and intensity of turbulence.

Novak & Davenport (1970) were the first to study the effect of turbulence on galloping. They developed an analytical model based on the work of Parkinson & Smith (1964) and Novak (1969), in which the wind velocity is taken to comprise a mean component and a time-varying, turbulence-related component. The resulting nonlinear differential equation involves coefficients which are random in time and is therefore difficult to solve. A more pragmatic, experimental approach was then pursued. Some typical results for a pivoted rigid (i.e. a flexibly supported rigid-cantilever) square prism in flow with different levels of turbulence are shown in Figure 2.50. It is seen that in this case turbulence has an appreciable stabilizing effect on galloping: the effective critical flow velocity is raised and the galloping amplitude is diminished. Moreover, it is seen that below the galloping threshold there is significant turbulence-induced "buffeting" response of the prism, in contrast to, say, Figure 2.11.

Some very-high-quality measurements by Bearman *et al.* (1987) confirm the stabilizing effect of turbulence for square prisms in uniform flow and uniform transverse motion of the prism. The results for $A_1 = (\partial C_{F_y}/\partial \alpha)\,|_{\alpha=0}$ were: $A_1 = 5.4$ for Tu = 0.05%; $A_1 = 3.9$ for Tu $\simeq$ 6.5–7.0% and $A_1 = 3.4$ for Tu = 10.5–11.5%;

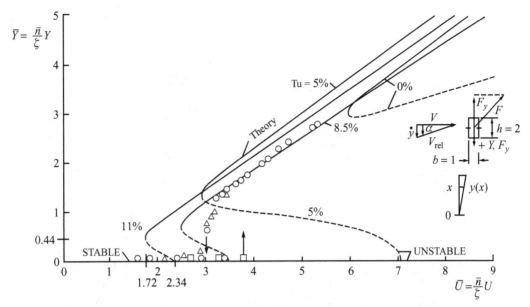

Figure 2.51. Universal galloping curves of a pivoted rigid rectangular prism ($h/d = 2$) show-ing the effect of turbulence, with Tu $= 0, 5\%, 8.5\%$ and 11%. In the experiments, Tu $= 11\%$: $\circ$, $\zeta = 0.006$; $\triangle$, $\zeta = 0.012$; $\square$, $\zeta = 0.014$ (Novak 1971).

refer to equation (2.28). In contrast, a recent numerical (LES) study by Tamura & Ono (2003) finds virtually no effect on the threshold of galloping, but, in common with the results of Figure 2.50, predicts the disappearance of the S-shaped curve for Tu $= 12\%$; refer also to Lidner (1992).

More extensive studies, along the same lines, were conducted by Novak (1971, 1972) and Laneville & Parkinson (1971). Figure 2.51 shows the results for a $h/d = 2$ prism. It is seen that at turbulence intensity Tu $\simeq 0\%$ we have hard galloping. With increasing Tu, the galloping becomes softer. Thus, (i) the critical flow velocity $\overline{U}_{cr}$ is reduced, (ii) the minimum amplitude to trigger galloping $\overline{Y}_{cr}$ is diminished and (iii) the maximum amplitude is higher. For example, at Tu $= 0\%$ we have $\overline{U}_{cr} \simeq 6$ and $\overline{Y}_{cr} \simeq 3$, whereas at Tu $= 11\%$ we have $\overline{U}_{cr} = 1.72$, a dramatic reduction, with $\overline{Y}_{cr} \simeq 0.44$; moreover, at $\overline{U} = 6$, for Tu $= 11\%$ we have $\overline{Y} = 3.5$, higher than for smooth flow. Here, $\overline{U}$ and $\overline{Y}$ are as defined in (2.30). In other words, in this case the effect of turbulence is *destabilizing*. The reason why the system becomes softer with increasing Tu is appreciated from Figure 2.52: the slope of C_{F_y} at the origin, i.e. $A_1 \equiv (\partial C_{F_y}/\partial \alpha)|_{\alpha=0}$ changes from negative to positive, and becomes increasingly positive with increasing Tu; see also Novak (1972; figure 10).

Laneville & Parkinson also show that the longitudinal turbulence scale, L_{Tu} (for L_{Tu}/h in the range of 1.58 to 5.0) has no appreciable effect on C_{F_y} and galloping for $h/d = \frac{1}{2}$ and 1 prisms.

Similar results for a $h/d = \frac{1}{2}$ are shown in Figure 2.53, displaying the opposite trend: the effect of turbulence is *stabilizing*. Indeed, for a sufficiently high level of Tu, namely Tu $= 8.5\%$, galloping is prevented altogether, and one obtains turbulent buffeting alone. For a shear profile $\alpha = 1/6$, turbulence (Tu $= 5\%$) changes the response from subcritical to supercritical.

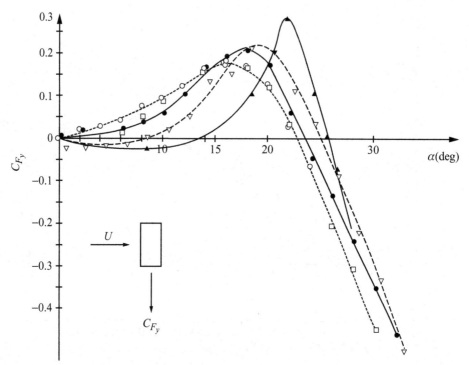

Figure 2.52. Measurements of C_{F_y} versus α for a $h/d = 2$ prism: ▲, Tu = 0.07%; ▽, Tu = 6.7%; ●, Tu = 9.0%; ○, □, Tu = 12% for different scales, obtained with small and large grid turbulence, respectively (Laneville & Parkinson 1971).

In a more recent study by Hémon *et al.* (2001), this stabilizing effect for the $h/d = \frac{1}{2}$ prism is found to be nonmonotonic: i.e. increasing Tu up to 5% decreases the critical velocity $U^* = V/fh$ from 65 to 42. Then, for higher, Tu the effect is stabilizing, and ceases at Tu = 7.5% in accordance with previous findings. This latter behaviour is reproduced by the LES study of Tamura & Ono (2003).

Further studies of this type were conducted by Novak & Tanaka (1974), for D-sections, $h/d = \frac{2}{3}$ and $\frac{3}{2}$ prisms, and cruciform sections. For a pendularly mounted D-section with variable mean angle of attack due to "blow-back", it was found that, for low $\overline{U}$ (small blow-back angle $\overline{\theta}$), turbulence has a destabilizing effect (cf. Table 2.1). It is interesting that as $\overline{\theta}$ increases, galloping eventually ceases, at high $\overline{U}$.

Nakamura & Tomonari (1977) conducted similar studies with rectangular prisms with $d/h = 0.2$–1.0 (i.e. $h/d = 1$–5), focussing attention on self-excited, or soft, galloping and the effect of turbulence on the critical depth (Section 2.3.4); soft galloping commences around the critical depth. It was found that turbulence, by influencing the position and shape of the shear layers relative to the trailing-edge corners, can decrease the value of the critical depth, thus affecting the manifestation of shear-layer/edge direct (s-l/e) interaction, and hence hasten the onset of soft galloping.

Indeed, as shown by Laneville, Gartshore & Parkinson (1975), Nakamura & Tomonari (1976, 1981) and Courchesne & Laneville (1982), turbulence can significantly reduce the value of the critical depth of both rectangular prisms and D-sections.

The effect of turbulence on low-speed torsional galloping (see Section 2.4) can be even more pronounced (Nakamura & Yoshimura 1982; Tamura *et al.* 2004).

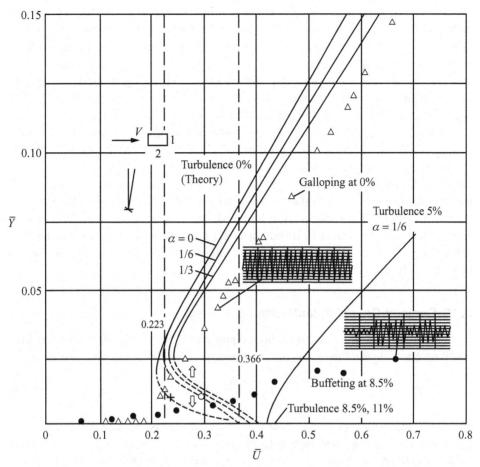

Figure 2.53. Universal galloping curves for a pivoted rigid rectangular prism ($h/d = \frac{1}{2}$) showing the effect of turbulence and shear for $n = 3.4 \times 10^{-4}$, $\zeta = 0.007$: $\triangle$, Tu = 0%; $\bullet$, Tu = 8.5% (Novak 1974).

In Figure 2.28 we see that the onset of low-speed galloping is delayed and the maximum amplitude diminished for a $d/h = 1.5$ prism. The effect becomes stronger for $d/h = 2$, and there is a total eclipse of low-speed torsional galloping for $d/h = 3, 4$ and 5, although for $d/h = 3$ and 4 high-speed galloping persists (Nakamura & Yoshimura 1982).

In summary, then, increasing turbulence intensity causes hard galloping (requiring an initial perturbation) to become soft (self-excited), and soft galloping to become weaker and eventually vanish. Moreover, this statement holds also for torsional galloping in the high-speed range, whereas turbulence always weakens low-speed torsional galloping, eventually suppressing it altogether. Here, we refer back to Figure 2.23, the data compiled by Parkinson (1989). It is seen that the range over which soft galloping is possible shifts to lower values of d/h as turbulence intensity is increased – both the low d/h and the high d/h ends of the range. Moreover, for a given d/h, the maximum galloping amplitude is diminished as Tu is increased.

The effect of shear, i.e. of a sheared incident flow profile (see the paragraph including equations (2.46) and (2.47)), is as one might guess: because most of the prism is subjected to an average flow velocity smaller than that for a uniform flow profile, the effect is stabilizing. In Figure 2.19, for uniform transverse motion of a square prism (curves 1a and 1b), the effect is very strong; for pivoted (rigid-cantilever) motion (curves 3a and 3b), the effect is considerably weaker, presumably because the motion near the support is small in any case.

Similar results are shown in Figure 2.53 for a $h/d = \frac{1}{2}$ prism: as α is increased from 0 to 1/6 and 1/3, the system is progressively stabilised, requiring a higher flow velocity for galloping and resulting in lower galloping amplitudes.

Nevertheless, shear can also generate galloping-type instability in situations where the system would be stable in uniform flow. This has been shown to occur for a circular cylinder in a shear layer at high reduced flow velocities (Yu, Yu & Chen 2004). Modelling the effect of turbulence as a parametric stochastic perturbation, Zhu *et al.* (2008/9) show via sophisticated analysis that the system can be stabilised by sufficiently strong turbulence.

2.10 Conjoint Galloping and Vortex Shedding

As outlined in Section 2.2 of this chapter, equation (2.28), the threshold of galloping, based on quasi-steady theory, occurs at a nondimensional velocity $(V/\omega_n h)$ given by

$$U_g = \frac{2\zeta}{nA_1}, \tag{2.79}$$

where A_1 is the slope of C_y as a function of α, ζ is the viscous damping factor and $n = (\rho h^2)/(2m)$ is the mass parameter. Furthermore, the velocity at which vortex shedding in the wake of a stationary cylinder is equal to the cylinder natural frequency (see Chapter 3 of this book), resulting in a vortex-shedding-induced resonance, may be expressed as

$$U_v = \frac{1}{2\pi S}, \tag{2.80}$$

where S is the Strouhal number. As discussed by Parkinson (1989), for aeroelastic systems U_g will usually be greater than U_v; however, if either ζ is small (low structural damping) or n is large (as obtained with low-density structures) then U_g may be close to or even less than U_v; furthermore, for realistic structural parameters in hydroelastic systems, where ρ is of the order of 1000 times greater than in airflow, U_g will almost always be less than U_v. Hence, there are a significant number of cases where the potential exists for interaction between galloping- and vortex-shedding-induced resonance, and they cannot be considered as separate entities. The complexities induced by the simultaneous, or near-simultaneous, occurrence of galloping- and vortex-shedding-induced resonance is the subject of this section.

Possibly the earliest example of an aeroelastic response due to combined vortex-shedding and galloping was reported by Scruton (1960) who considered the translational response of a square-section prism in cross-flow. He presented results showing the nondimensional velocity, U, required to initiate instability as a function of

$m\delta/\rho h^2$. For $m\delta/\rho h^2 \geq 8$, approximately, the vibrational response was of a classical nature, consisting of a resonance with vortex shedding at low values of U, an instability due to galloping at a higher U and a stable region between the galloping and vortex-shedding responses. However, for $m\delta/\rho h^2 \leq 8$ there were no longer distinct vortex-shedding and galloping responses as U was increased; instead, there was one common instability boundary. Novak (1971) also reports some examples of where there appears to be an aeroelastic response due to combined galloping and vortex-shedding for a rectangular section with $d/h = 0.5$ and low values of mass damping ($m/\rho h^2 = 130$ and damping between 0.37 and 2.12% critical).

Parkinson and co-workers (Parkinson & Sullivan 1979; Parkinson & Wawzonek 1981) also reported examples of aeroelastic responses which appeared to be due to combined vortex shedding and galloping. Considering three-dimensional models of tall towers with square cross-sections in a boundary layer tunnel, Parkinson & Sullivan (1979) showed that, for very lightly damped models, galloping occurred at a velocity lower than that predicted by quasi-steady theory. It was concluded that this was due to the interaction between the vortex-shedding and galloping responses. Also, if the velocity required to initiate vortex shedding, U_v, was close to and lower than that where galloping would occur, U_g, then there was no recovery in the intermediate velocity range. In addition, when $U_v \simeq U_g$ it was observed that the vibrational response was greater than that predicted by either mechanism alone, again indicating significant interaction between the vortex-shedding and galloping mechanisms. Parkinson & Wawzonek (1981) and Bearman et al. (1987) confirmed that the behaviour reported by Parkinson & Sullivan (1979) also occurred for two-dimensional square sections, and demonstrated that, although U_g could be accurately predicted using quasi-steady theory for high values of damping, for low damping it could not. Additional results attributed to Wawzonek (1979) by Bearman & Luo (1988) suggest that when U_g is greater than U_v two distinct peaks occur (as expected), but when U_g is smaller than U_v no vibration occurs until $U = U_v$.

Experiments on two-dimensional square sections, with the corners of the squares being either sharp or rounded with two different radii, mounted such that they could move in the transverse direction only, in water flow were conducted by Bokaian & Geoola (1984c). Several different sets of experiments were completed covering a wide range of mass-damping, with the damping being in the range $0.008 \leq \zeta \leq 0.288$ and the mass parameter varying from $0.0317 \leq n \leq 0.0847$ (corresponding to $5.9 \leq m/\rho h^2 \leq 15.8$). For the sharp-edged square sections (the results obtained with the rounded corners were similar and will not be discussed here), it was shown that the dynamic response as a function of nondimensional velocity, $U = V/\omega_n h$, could be categorised by one of three different types of behaviour depending on the value of the mass-damping parameter. (In fact, Bokaian & Geoola (1984c) categorised the response in terms of the so-called *response parameter*, $K_s = \zeta/(nU_v^2)$, where for their experiments on the sharp-edged squares $U_v = 1.27$.)

For all three categories of vibrational response, as U was increased the vibrational response started at velocities just below the vortex resonance speed, U_v; the difference in behaviour occurring for velocities greater than U_v. For the lowest range of mass-damping, $\zeta/n \leq 1.069$ ($K_s \leq 0.66$), as U was increased beyond U_v, there was a continual increase in vibrational amplitude with increasing U, and the authors noted that "the vortex resonance amplitude is inseparable from galloping";

for the intermediate range of mass-damping, $1.183 \leq \zeta/n \leq 2.27$ ($0.73 \leq K_s \leq 1.4$), there was an initial build-up in amplitude as U was increased beyond U_v, followed by a small decrease prior to the galloping response, and the authors referred to this as a "build up-drop off behaviour"; whereas for the highest range of mass-damping, $\zeta/n \geq 2.66$ ($K_s \geq 1.64$), a complete separation between the vortex-shedding and galloping responses was obtained.

A detailed comparison of the results presented in the preceding paragraphs is given by Parkinson (1989), but possibly the most significant conclusion coming from this work (as succinctly expressed by Luo & Bearman (1990)) is that for systems with only moderate mass (high n) and damping (low ζ), where U_g is predicted to be only slightly higher or even lower than U_v, experimental observations show that galloping always commences at $U \simeq U_v$.

Motivated by the work of Parkinson and co-workers, Bearman & Luo (1988) conducted a detailed experimental investigation of the unsteady fluid force acting on a square-section prism forced to oscillate transverse to the flow. Using a scotch-yoke mechanism, the amplitude of cylinder oscillation was varied over the range $0.25 \leq a/h \leq 2.0$. The oscillation frequency was held at a constant value of $f_n = 8\,\mathrm{Hz}$; however, varying the wind-tunnel velocity enabled values of U up to a maximum of 180 to be obtained, with the Reynolds number being in the range $10^4 \leq \mathrm{Re} \leq 8 \times 10^4$. A pressure-averaging technique was used to obtain the fluctuating lift force, C_L'.

Typical results for C_L' and ϕ (the phase angle by which lift leads the displacement) are presented in Figure 2.54; note that negative ϕ implies that the fluid damping is positive. For low values of U there is a high value of C_L' caused by lock-in of the vortex shedding with the body oscillation; however, as U is increased, C_L' decreases until it reaches a minimum value; then it steadily increases until attaining the value measured on a stationary cylinder of 1.45. However, it should be noted that when C_L' is large, the fluid damping is positive and so galloping is not possible; hence, the aeroelastic response, resulting from a forced oscillation due to the fluctuating lift force, will be relatively small. More importantly, at these low values of U where the flow is controlled by the vortex shedding, the quasi-steady theory (which predicts a galloping instability for low damping) is not valid. Based on flow visualization by Luo (1985), it is suggested that the minimum in C_L' is associated with momentary reattachment of the flow on the side face. Bearman & Luo (1988) suggest that the flow may be divided into three distinct regions of U: (i) for U below that where minimum C_L' is obtained, where there are multiple lock-in regions between the vortex shedding and body oscillation; (ii) between the U for minimum C_L' and that when C_L' reaches the quasi-steady value of 1.45; and (iii) for U above this. The quasi-steady theory will definitely work in region (iii) and will not work in region (i), but is expected to work in region (ii). Interestingly, conclusions similar to these had previously been reached by Nakamura & Mizota (1975a) for rectangular sections with $d/h = 1, 2$ and 4 oscillating transverse to the flow.

Despite the conclusions previously obtained by Bearman & Luo (1988), Corless & Parkinson (1988) attempted an analysis of combined vortex shedding and galloping for the transverse motion of a square section in cross-flow using a quasi-steady model. They employed the Parkinson & Smith (1964) model of galloping (presented in Section 2.2.2) and the Hartlen & Currie (1970) model for vortex shedding (this is described in some detail in Section 3.3.4 of this book).

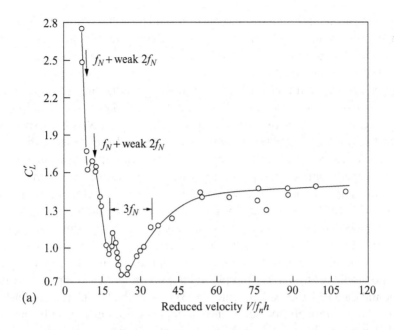

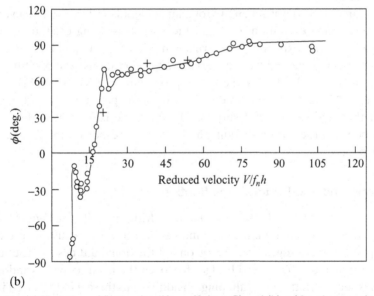

Figure 2.54. (a) Variation of fluctuating lift coefficient C'_L and (b) ϕ (the phase angle by which lift leads the displacement) as a function of reduced velocity $U^* = V/f_n h$ for a square-section prism being forced to oscillate transverse to the flow with $a/h = 0.675$ (reproduced from Bearman & Luo (1988)).

The equation of motion of the section is given by

$$\ddot{Y} + Y = nU^2 \left(\overline{C}_{F_y} + \overline{C}_v\right), \qquad (2.81)$$

where Y, τ and n are as given by equation (2.18); $\overline{C}_{F_y}$ is the same as the force coefficient used in the standard Parkinson & Smith (1964) analysis, equation (2.20), with the addition of the structural damping term, and $\overline{C}_v$ is the solution to the standard Hartlen & Currie wake-oscillator model, except that an additional inertial coupling term is included to account for the effects of the cylinder acceleration on the wake vortices. The resulting equations were solved using the method of multiple scales, which gives a series solution for the amplitude of conductor motion (full details are given by Corless & Parkinson). Unfortunately, the theoretical solutions given are for individual terms in the series, which are compared with the experimental results of Bearman *et al.* (1987), but the final sum of all the terms in the series is not presented. Hence, although there appears to be qualitative agreement between theory and experiment, it is not possible to assess quantitative agreement.

In a later paper (Corless & Parkinson 1993) the same authors present an improved solution to the same equations – still using the method of multiple scales. They concluded that the analysis agrees qualitatively with experimental results; for example, small vibrational amplitudes are predicted when U is less than U_v, even if U is greater than U_g; non-zero amplitudes are predicted in the resonance region, with the possibility of hysteresis and phase jumps. However, a comparison with the experimental data of Brika & Laneville (1993) is described by Corless & Parkinson as being only qualitatively correct. Hence, it appears, not unexpectedly, that quasi-steady theory is not able to predict the vibrational response due to combined vortex shedding and galloping.

A similar approach was taken by Bokaian & Geoola (1984c), where in equation (2.81) the galloping excitation, $\overline{C}_{F_y}$, and the vortex-shedding excitation, $\overline{C}_v$, were represented by series expressions in terms of $\alpha = \tan^{-1}(\dot{y}/V)$, with the coefficients in the series being obtained from experimental observations. The resulting nonlinear equations were solved using the first approximation of Krylov and Bogoliubov. The results obtained were compared with their own experimental results, discussed earlier in this section, and, not surprisingly, for those cases where there was interaction between the vortex-shedding and galloping responses the agreement was relatively poor.

2.11 Elongated and Bridge-Deck Sections

The catastrophic failure of the Tacoma Narrows Bridge in 1940 (see Figure 2.4) indicated very clearly the potential for wind-induced vibrations to cause severe damage to long-span flexible bridges. Since then, one of the dominant design criteria for all new suspension and cable-stayed bridges has been the need to avoid wind-induced instabilities such as flutter and galloping. In addition to these wind-induced instabilities, bridge decks are also susceptible to buffeting and vortex-induced vibrations, which will not be discussed in this section. It should be emphasised that the aim of this section is not to give a definitive description of the phenomena and sequence of events leading to the ultimate failure of the Tacoma Narrows Bridge; this is something which has been studied extensively, and thus, the relevant literature is extremely large. However, the reader is cautioned that although the literature does contain some simple explanations for this failure (such as stating that the vibrations were due to vortex shedding or galloping) the precise mechanism appears to be much more complex than this, and is still a subject of considerable controversy; see, for example Matsumoto *et al.* (2003a). On the contrary, the main aim of this section is to give some insight into the procedures required to analyze bridge decks in cross-flow.

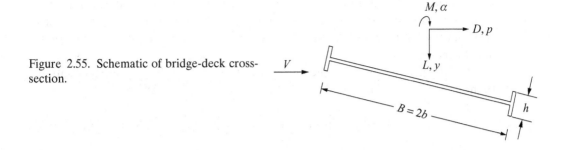

Figure 2.55. Schematic of bridge-deck cross-section.

Bridge decks have cross-sectional shapes which may often be approximated as rectangular or H-sections, although the ratio of the deck chord, B, to depth, h (see Figure 2.55),[*] is typically large; hence, when exposed to wind normal to their longitudinal axes these sections are potentially susceptible to both translational and torsional galloping, discussed in the earlier sections of this chapter. However, in addition to these two types of galloping, which occur predominantly in one degree of freedom, and are a direct result of negative fluid damping in that degree of freedom (referred to as damping-controlled instabilities in this book), there is a second type of instability that is due to a coupled-mode flutter. As Scanlan & Tomko (1971), pointed out, this phenomenon is very similar to flutter of aircraft wings, and the analytical methods employed to investigate it follow very closely those used in classical aircraft aeroelasticity; see, for example, Fung (1955) and Bisplinghoff, Ashley & Halfman (1955).

Before attempting an analysis of bridge-deck instabilities, it is important to appreciate that even though V/fB for bridge decks, where f is the structural natural frequency of the mode of interest, is often quite large and may be greater than 10,[†] which is usually taken as the limiting value of when quasi-steady aerodynamics is valid,[‡] there are other limiting factors which preclude the use of quasi-steady aerodynamics for the analysis of bridge-deck sections. As outlined by van Oudheusden (1995), a second prerequisite for the use of quasi-steady aerodynamics is the necessity to be able to "define a steady situation (in which the structure is in rest with regard to some suitably chosen reference frame) which is aerodynamically equivalent to the unsteady situation". Although this is easily achieved for translational galloping (see Figure 2.1), it is apparent from Figure 2.40 that this is not possible for torsional galloping; hence, the use of quasi-steady aerodynamics is not valid in the analysis of torsional galloping, and by extension, coupled-mode flutter, of bridge-deck sections.

The analysis presented in the following follows very closely the work of Scanlan and co-workers (Scanlan & Tomko 1971; Scanlan, Beliveau & Budlong 1974; Scanlan 1978, 1990; Kumarasena, Scanlan & Ehsan 1992) and considers the flutter of a

---

[*] Previously in this chapter the width ("depth" in the foregoing) of rectangular sections has been denoted by d (see Figure 2.12); however, in order to maintain conformity with the majority of the literature in this area, in this section the width of the bridge deck, or deck chord, is represented by the symbol B. Similarly, what we call depth (h) here was previously referred to as "height".

[†] For example, the natural frequencies of bridge decks are typically small, on the order of 0.5 Hz, and even though B may be of the order of 20 m, then, using these values, $V/fB \geq 10$ is satisfied when $V \geq 20$ m/s.

[‡] The rationale for requiring $V/fB \geq 10$ for quasi-steady aerodynamics to be applicable is discussed in Section 4.2.1 of this book.

two-dimensional section of a bridge deck subject to a steady wind acting normal to its longitudinal axis, as shown in Figure 2.55. Assuming the deck to be much stiffer in the in-plane direction compared with the vertical or torsional directions, and furthermore assuming the centre of mass to be roughly coincident with the elastic axis (taken as being at the mid-point of the deck width), then the equations of motion, per unit length, in the vertical and torsional directions are

$$m\left[\ddot{y} + 2\zeta_y\,\omega_y\,\dot{y} + \omega_y^2\,y\right] = L, \tag{2.82a}$$

$$I\left[\ddot{\alpha} + 2\zeta_\alpha\,\omega_\alpha\,\dot{\alpha} + \omega_\alpha^2\,\alpha\right] = M, \tag{2.82b}$$

where m and I are the sectional mass and moment of inertia per unit span, y and α are the vertical and torsional displacements of the section, ω_y and ω_α are the vertical and torsional natural structural frequencies, ζ_y and ζ_α are the structural viscous damping factors and L and M are the aerodynamic "lift" force (note that L is positive in the downwards direction, the reverse of that normally employed for a lift force) and torsional moment per unit span acting on the section. If the centre of mass and elastic axis are not coincident, as is typically the case for aircraft wings, there are additional structural coupling terms on the left-hand side of equations (2.82a, b) which have a significant effect on whether or not flutter will occur and, if it does occur, on the velocity at which it will do so.

Given that quasi-steady aerodynamics is not applicable, then clearly L and M will depend on the frequency of oscillation of the bridge deck. This, of course, is exactly the same situation as for oscillating airfoils for which a theoretical solution (Theodorsen 1935) exists for L and M as functions of the nondimensional frequency of oscillation, $\omega c/V$, where c is the chord of the airfoil. Given the existence of this theoretical solution for L and M, and that there is a passing similarity between the seemingly streamlined cross-sectional shapes of bridge decks and airfoils, it is not surprising that attempts have been made to use this theoretical solution to analyze the stability of bridge decks. However, one important difference between the aerodynamics of bridge decks and airfoils is that the flow will typically separate from the leading edge of a bridge deck, whereas Theodorsen's theoretical solution for an airfoil assumes attached flow. This difference has a significant effect on the unsteady aerodynamics, and hence, even though Theodorsen aerodynamics may be a useful tool to investigate certain phenomena, as pointed out by Scanlan (1990) and Ge & Tanaka (2000), it cannot be used to obtain a reliable estimate of when instability will occur for bridge decks. Indeed, as will be seen later in this section, the aeroelastic response of bridge decks is very sensitive to their particular cross-sectional shape; thus, all new bridge decks have to be investigated individually.

Realizing that the aerodynamic forces on the bridge deck will vary with the frequency of oscillation, a convenient way of expressing L and M is as follows (Kumarasena, Scanlan & Ehsan 1992):

$$L = \frac{1}{2}\rho V^2(2B)\left[KH_1^*(K)\frac{\dot{y}}{V} + KH_2^*(K)\frac{B\dot{\alpha}}{V} + K^2H_3^*(K)\,\alpha + K^2H_4^*(K)\frac{y}{B}\right], \tag{2.83a}$$

$$M = \frac{1}{2}\rho V^2(2B^2)\left[KA_1^*(K)\frac{\dot{y}}{V} + KA_2^*(K)\frac{B\dot{\alpha}}{V} + K^2A_3^*(K)\,\alpha + K^2A_4^*(K)\frac{y}{B}\right], \tag{2.83b}$$

where B is the width of the bridge deck, $K = \omega B/V$ is the nondimensional frequency of oscillation, and $H_i^*(K)$, $A_i^*(K)$, $i = 1 - 4$, which are functions of K are referred to as either the *flutter* or *aerodynamic* derivatives. In earlier work (Scanlan & Tomko 1971; Scanlan, Beliveau & Budlong 1974; Scanlan 1978) the H_4^* and A_4^* terms were ignored; although written as being multiplied by y they could equally be interpreted as inertia terms, multiplied by $\ddot{y}$, and so they are dominated by the added mass terms and are typically very small. In some references (for example, Matsumoto *et al.* (2008a)) and some of the earlier papers by Scanlan and co-workers, the expressions for L and M given in equations (2.83a, b) are written in a slightly different form, namely with $b = B/2$ replacing B and $k = \omega b/V$ replacing K; however, the form used in equations (2.83a, b) seems to be the preferred notation. It should be appreciated that using b instead of B changes the definition and numerical values of the aerodynamic derivatives; indeed, if the derivatives based on b and k are written as $\tilde{H}_i^*(K)$, $\tilde{A}_i^*(K)$, $i = 1$–4, then it is apparent that $\tilde{H}_1^*(k) = 4H_1^*(K)$, $\tilde{H}_2^*(k) = 8H_2^*(K)$, $\tilde{H}_3^*(k) = 8H_3^*(K)$, $\tilde{H}_4^*(k) = 4H_4^*(K)$, $\tilde{A}_1^*(k) = 8A_1^*(K)$, $\tilde{A}_2^*(k) = 16A_2^*(K)$, $\tilde{A}_3^*(k) = 16A_3^*(K)$ and $\tilde{A}_4^*(k) = 8A_4^*(K)$. Unfortunately, it is not always clearly indicated in the literature whether the aerodynamic derivatives are based on B and K or b and k, and thus, there is considerable need for caution when using this data. Expressions for L and M can be written using complex notation as opposed to purely real numbers; this was done by Nakamura (1978). Some of these different notations are discussed by Zasso (1996).

From equations (2.83a, b) it is readily apparent that H_1^*, H_2^*, A_1^* and A_2^* are velocity-dependent damping terms, with H_1^* and A_2^* being the *direct-derivative terms*, and A_1^* and H_2^* being the *cross-derivative*, or coupling terms. Equally H_3^*, H_4^*, A_3^* and A_4^* are stiffness terms. A necessary condition for pure torsional flutter (usually referred to as torsional galloping) to occur is that A_2^* must be positive, and equally for pure vertical galloping H_1^* must be positive.

It should be stressed that the aerodynamic derivatives contain *all possible wind-induced excitations*, including all vortex-shedding-related phenomena. That is why the methods described in this section supersede, in a sense, everything else discussed in this chapter and also some of the things discussed in Chapter 3. This generality is at once a strength and a weakness. As in CFD analyses of a system, the simulation is, or can be, very close to reality, almost mimicking the natural behaviour; on the other side of the coin, the physical mechanisms underlying the physical phenomena involved are not necessarily elucidated thereby.

As previously mentioned, one of the complications associated with bridge-deck aerodynamics is that the flow will typically separate from the leading edge; hence, analytical methods, such as those presented by Theodorsen (1935) for the flow over airfoils, cannot be employed to determine the aerodynamic derivatives for bridge decks. Although numerical methods (briefly discussed later in this section) are now being proposed, at present, resort must be made to experimental techniques to obtain accurate estimates of the aerodynamic derivatives.

Two distinct experimental techniques are discussed in the literature: (i) the *direct* method, involving measuring forces on a sectional model undergoing forced harmonic oscillation, and (ii) the *indirect* method, where the bridge-deck motion, subject to wind-induced effects, is measured and the aerodynamic terms are inferred from this measured motion. Scanlan & Tomko (1971) employed the indirect method using wind-tunnel models; this involved: (i) measuring the vertical motion of the bridge deck while the torsional motion was prohibited; (ii) measuring the torsional motion with no vertical motion; and (iii) measuring the unrestricted torsional and vertical motion simultaneously. In all cases the aeroelastic frequencies and damping levels were determined as a function of velocity. From (i) and (ii) the direct aerodynamic derivatives could be obtained, and from (iii) the coupling terms were inferred. Full details of this approach are given by Scanlan & Tomko (1971) and Scanlan, Beliveau & Budlong (1974). To verify their experimental technique, it was first applied to a NACA 0012 airfoil and the results obtained were compared with Theodorsen's analytical expressions. Results are also presented (for $H_i^*(K)$, $A_i^*(K)$, $i = 1 - 3$) for a number of different bridge decks including the original Tacoma Narrows and different-size H-sections. Some of these results, including those for the original Tacoma Narrows Bridge, are reproduced in Figure 2.56. It is apparent that for the Tacoma Narrows Bridge deck, along with truss-stiffened bridge decks 2 and 3, A_2^* becomes positive as V/fB is increased, indicating the likelihood of pure torsional flutter (for the airfoil, pure torsional flutter is not possible), the change in sign of A_2^* with increasing V/fB also demonstrates very clearly that quasi-steady aerodynamics is not applicable: hence, any models of torsional flutter using quasi-steady aerodynamics, such as those described in Section 2.7.2, are not applicable for these sections.

The data of Figure 2.56 also indicates the potential for pure vertical galloping, as indicated by positive H_2^*, for the original Tacoma Narrows Bridge deck. Scanlan & Tomko (1971) suggest that the peak in H_2^*, clearly evident for the Tacoma Narrows section (along with all the other H-sections tested by Scanlan & Tomko), is due to vortex shedding. Hence, it is apparent that there are a number of distinctly different potential instability mechanisms affecting the Tacoma Narrows Bridge, and thus, the controversy over exactly what was the instability mechanism leading to its destruction; see, for example, Matsumoto et al. (2003a).

Since Scanlan's original work, a number of other researchers have developed alternative indirect methods of obtaining the aerodynamic derivatives; a discussion of these different indirect methods is given by Brownjohn & Bogunovic Jakobsen (2001) and Chen, He & Xiang (2002). An alternative approach is the *direct* method, where the bridge-deck section is forced to oscillate harmonically, either in torsional or plunging motion, and the aerodynamic forces and moments, and their phase with respect to the forced oscillation, measured. The aerodynamic forces can be measured via force balances, although it is then necessary to subtract the structural inertial force or moment from the value measured in wind, or obtained from measured pressure distributions. Examples of using the direct approach are given by Matsumoto and co-workers. Matsumoto et al. (2005) obtained the aerodynamic derivatives for rectangular sections using pressure measurements with B/h ranging from 5 to 20, while Matsumoto et al. (2008a) used force measurements to obtain the aerodynamic derivatives for H-sections with B/h ranging from 2 to 20. Examples of the aerodynamic derivatives for H-sections measured by Matsumoto et al. (2008a) are shown in Figure 2.57 (note that these aerodynamic derivatives are based on b and k, whereas those of Figure 2.56 are based on B and K), where they are compared with the analytical solution for an airfoil (thin plate) obtained using Theodorsen's analysis. The results presented in Figure 2.57 show that, depending on B/h, there can be considerable difference between the aerodynamic derivatives for H-sections

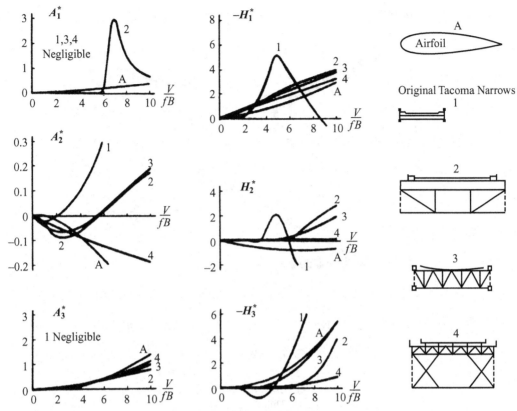

Figure 2.56. Measured aerodynamic derivatives for the original Tacoma Narrows Bridge and three other truss-stiffened bridge decks, and the theoretical results for an airfoil. (Reproduced from Scanlan & Tomko 1971), the symbols A and 1-4 refer to the cross-sections shown on the right-hand side of the figure.

and the Theodorsen solution. It is also apparent that changing B/h can have a significant effect on the magnitude of the aerodynamic derivatives; see, for example, the $\tilde{H}_1^*$ term. These results, along with those of Figure 2.56, illustrate very clearly the inapplicability of using Theodorsen aerodynamics to analyze the stability of bridge decks, and the need to determine the aerodynamic derivatives for the particular cross-sectional shape of the bridge deck.

Sarkar *et al.* (2009) recently initiated a research program to compare the various ways of determining the aerodynamic derivatives. They compared their own experimental results, based on pressure measurements using the direct method, with other results from both the direct and indirect approaches. This was done for five different cross-sections, including three streamlined sections and two rectangular sections with $B/h = 2$ and 5. Their results show that values of the damping terms A_2^* and H_2^* obtained from measurements on forced oscillation are extremely sensitive to errors in the phase measurement, while the stiffness terms are not. In general, it was found that, provided the amplitudes of oscillation are not too different, there is reasonable agreement between the different approaches; however, their results suggest that changing the amplitude of oscillation could have a big effect on the estimates of the aerodynamic derivatives, suggesting considerable nonlinear behaviour.

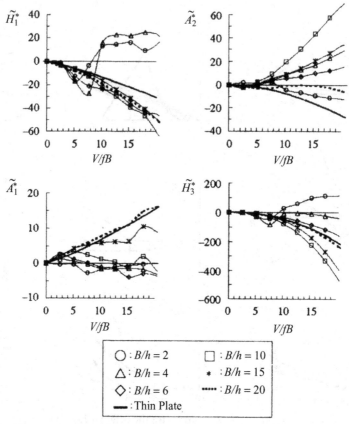

Figure 2.57. Measured values of sample aerodynamic derivatives for H-shaped bridge-deck sections with different B/h compared with the theoretical solution for a thin airfoil obtained from Theodorsen (1935) (reproduced from Matsumoto *et al.* 2008a).

Although the data of Figures 2.56 and 2.57 indicate the possibility of single-degree-of-freedom damping-controlled instabilities in either the torsional or vertical directions, there is also the possibility of yet another type of instability involving both the vertical and torsional modes. This is a coupled-mode flutter very similar to that obtained on aircraft wings. As discussed by Scanlan (1990) this instability is governed mainly by stiffness effects as opposed to damping terms; other examples of stiffness-controlled instabilities occurring on overhead transmission lines and in heat-exchanger tube arrays are discussed in more detail in Chapters 4 and 5 of this book. The complication associated with this analysis is that the aerodynamic derivatives are functions of K, but K cannot be determined until the frequency of oscillation, which is part of the solution, is known; hence, some sort of iterative solution is required. The method originally proposed by Scanlan and co-workers (see Simiu & Scanlan (1978)) is to ignore the H_4^* and A_4^* terms; then, expressing equations (2.82a, b) and (2.83a, b) in nondimensional form, the following are obtained:

$$\eta'' + 2\zeta_y \, K_y \, \eta' + K_y^2 \eta = \frac{\rho B^2}{m} \left[K H_1^* \, \eta' + K H_2^* \, \alpha' + K^2 H_3^* \alpha \right], \qquad (2.84a)$$

$$\alpha'' + 2\zeta_\alpha \, K_\alpha \, \alpha' + K_\alpha^2 \alpha = \frac{\rho B^4}{I} \left[K A_1^* \, \eta' + K A_2^* \, \alpha' + K^2 A_3^* \alpha \right], \qquad (2.84b)$$

where the prime indicates differentiation with respect to nondimensional time $s = Vt/B$, $\eta = y/B$ is the nondimensional vertical displacement, $K_y = \omega_y B/V$, and $K_\alpha = \omega_\alpha B/V$.

Defining a stability boundary as the point where α and η are purely harmonic, neither growing nor decaying with time, then it is apparent that on a stability boundary the solutions to equations (2.84a, b) will be of the form $\eta = \eta_o \exp(iKs)$ and $\alpha = \alpha_o \exp(iKs)$, where K is the nondimensional frequency of oscillation. Thus, on a stability boundary

$$\left[-K^2 + 2i\zeta_y K_y K + K_y^2 - \frac{\rho B^2}{m} iK^2 H_1^* \right] \eta_o - \frac{\rho B^2}{m} K^2 \left[iH_2^* + H_3^* \right] \alpha_o = 0, \qquad (2.85a)$$

$$\left[-\frac{\rho B^4}{I} iK^2 A_1^* \right] \eta_o + \left[-K^2 + 2i\zeta_\alpha K_\alpha K + K_\alpha^2 - \frac{\rho B^4}{I} K^2 \left(iA_2^* + A_3^* \right) \right] \alpha_o = 0, \quad (2.85b)$$

For a nontrivial solution of equations (2.85a, b) the determinant of the matrix equation must be equal to zero, yielding an equation which may be expressed in its real and imaginary parts as a function of X, where $X = \omega/\omega_\alpha$ (see Simiu & Scanlan (1978) for a full derivation of these equations). Each of these equations is then solved in terms of X for different values of K, and the value of K at which the two solutions are consistent is the true solution at the stability boundary.

An alternative approach is to express the equations in matrix form, for example, as

$$[M]\ddot{p} + [B]\dot{p} + [E]p = 0, \qquad (2.86)$$

where

$$[M] = \begin{bmatrix} m & 0 \\ 0 & I \end{bmatrix},$$

$$[B] = \begin{bmatrix} 2m\zeta_y\omega_y - \rho VBKH_1^* & -\rho VB^2 KH_2^* \\ -\rho VB^2 KA_1^* & 2I\zeta_\alpha\omega_\alpha - \rho UB^3 KA_2^* \end{bmatrix},$$

$$[E] = \begin{bmatrix} m\omega_y^2 & -\rho V^2 BK^2 H_3^* \\ 0 & I\omega_\alpha^2 - \rho V^2 B^2 K^2 A_3^* \end{bmatrix}$$

are, respectively, the inertia matrix, the total damping matrix, including both structural and aerodynamic damping terms, and the total stiffness matrix, including both structural and aerodynamic stiffness terms (again, the H_4^* and A_4^* terms have been ignored).

These equations can then be solved using any of the standard iterative techniques employed in classical aeroelasticity, for example the p-k method given by Hodges & Pierce (2002). This method has the advantage that the H_4^* and A_4^* terms can easily be incorporated; they merely add additional terms to the total stiffness matrix; in addition, it can be extended to account for the effect of swaying motion, as discussed later in this section.

An example of using the eigenvalue-based approach is given by Matsumoto et al. (2008a) who, using the aerodynamic derivatives given in Figure 2.57, obtained results

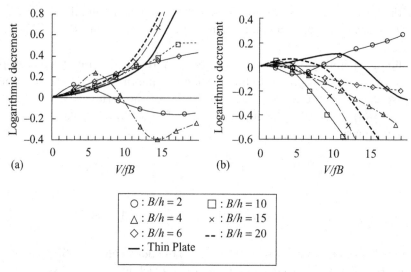

Figure 2.58. Variation of aeroelastic logarithmic decrement, δ, (a negative δ implies an unstable system) with nondimensional velocity obtained using the data of Figure 2.57 ($B = 1.5$ m, $f_y = 4.5$ Hz, $f_\alpha = 6.0$ Hz, $m = 1.96$ kg/m, $I = 4.91 \times 10^{-3}$ kg m²/m): (a) vertical mode, (b) torsional mode (Matsumoto et al. 2008a).

for the aeroelastic logarithmic decrement of the vertical and torsional modes as a function of nondimensional velocity; a negative aeroelastic logarithmic decrement indicating instability. These results, reproduced in Figure 2.58, show very clearly that the velocity at which instability occurs, and even whether it is the vertical or torsional mode which becomes unstable, depends very strongly on the particular value of B/h.

The two methods described in the previous paragraphs have the disadvantage that they are iterative. Although this is not a problem if one is interested only in the stability of the system, if the combined effects of wind-induced vibrations and other external forcing (for example, turbulent buffeting, or even control of the bridge vibrations) are desired, then the iterative nature of these approaches becomes extremely problematic. One method of eliminating the need for iterative techniques is to represent the unsteady aerodynamic forces, which are given at discrete values of K, by so-called *approximate rational functions* of the aerodynamic terms in the Laplace domain. The equations can then be solved without recourse to iteration, and thus, the effects of additional forcing functions can easily be accounted for. This technique has been used extensively in the aeronautical field to analyze flutter of aircraft wings, and many different methods are available to determine appropriate rational functions for the aerodynamic forces; see, for example, Tiffany & Adams (1988). This technique has also recently been employed to analyze the stability of bridge decks by Chen, Kareem & Matsumoto (2001) who considered the effect of buffeting, and Kobayashi & Nagaoka (1992), Wilde & Fujino (1998), Wilde, Fujino & Kawakami (1999) and Fujino (2002) who attempted to use either active or passive control to alleviate the effects of wind-induced forces. Although a detailed description of this method is beyond the scope of this book (the interested reader is referred to Tiffany & Adams (1988)), it should be appreciated that removal of the iterative step in the methodology comes at the cost of increasing the size of the resulting matrices. Indeed, new eigenvalues appear as part of the solution – these being associated with the functions describing the aerodynamic forces – and thus, it is necessary to separate out the eigenvalues which are true characteristics of the system from those which are associated with the new representation of the aerodynamics.

No matter which method is used to solve the equations, if the complete form of the aerodynamics given by equations (2.83a, b) is employed, there are eight aerodynamic derivatives which must be determined experimentally, and given that these are functions of the frequency of oscillation, it is apparent that considerable experimental data is required prior to any theoretical analysis. However, Bartoli & Mannini (2008) have shown that for structural parameters typical of real bridge decks the instability is dominated by the aerodynamic derivatives H_1^*, A_2^* and A_3^*, and in some cases by H_1^* and A_2^* only. These results are significant in two ways: first, the amount of experimental data required is reduced considerably; second, these results suggest that the instability is dominated by the direct aerodynamic derivatives and that the coupling terms are less important. It is worth noting that it is much more difficult to make accurate measurements of the coupling terms than the direct terms.

The initial work of Scanlan considered the bridge to be sufficiently rigid for in-plane or swaying motion, compared with the vertical and torsional directions, such that it could be ignored; this, of course, is exactly the same as is typically done for the analysis of wing sections. However, as bridges have become more flexible, it has become apparent that, in some cases, the in-plane motion may not be negligible and can have an effect on the stability of the bridge. An example showing the importance of in-plane motion is given by Miyata (2003); some results are reproduced in Figure 2.59 showing considerable in-plane motion of an aeroelastic wind-tunnel model of the Akashi Kaikyo Bridge. To account for this, an in-plane equation must be added to equations (2.82a, b); thus, the complete set of equations is now given by

$$m \left[\ddot{y} + 2\zeta_y \, \omega_y \, \dot{y} + \omega_y^2 \, y \right] = L, \tag{2.87a}$$

$$I \left[\ddot{\alpha} + 2\zeta_\alpha \, \omega_\alpha \, \dot{\alpha} + \omega_\alpha^2 \alpha \right] = M, \tag{2.87b}$$

$$m \left[\ddot{p} + 2\zeta_p \, \omega_p \, \dot{p} + \omega_p^2 \, p \right] = D, \tag{2.87c}$$

where p is the in-plane displacement of the section, ω_p and ζ_p are the natural structural frequency and the structural viscous damping factor in the in-plane direction, D is the drag force on the section and all other terms are as defined with respect to equations (2.82a, b). The additional complications when considering the unsteady aerodynamic forces on the right-hand side of these equations are: first, that it is necessary to account for the drag, D, and second, that the lift, L, and moment, M,

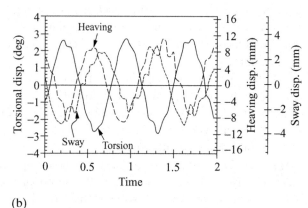

Figure 2.59. Motion of the 1/100 scale aeroelastic model of the Akashi Kaikyo Bridge: (a) motion of bridge at onset of flutter; (b) time history response at mid-point of centre span (Miyata 2003).

will be functions of the in-plane motion. Hence, in their most complete form the aerodynamic forces become

$$L = \frac{1}{2}\rho V^2 B \left[KH_1^* \frac{\dot{y}}{V} + KH_2^* \frac{B\dot{\alpha}}{V} + K^2 H_3^* \alpha + K^2 H_4^* \frac{y}{B} + KH_5^* \frac{\dot{p}}{V} + K^2 H_6^* \frac{p}{B} \right],$$

(2.88a)

$$M = \frac{1}{2}\rho V^2 B^2 \left[KA_1^* \frac{\dot{y}}{V} + KA_2^* \frac{B\dot{\alpha}}{V} + K^2 A_3^* \alpha + K^2 A_4^* \frac{y}{B} + KA_5^* \frac{\dot{p}}{V} + K^2 A_6^* \frac{p}{B} \right],$$

(2.88b)

$$D = \frac{1}{2}\rho V^2 B \left[KP_1^* \frac{\dot{p}}{V} + KP_2^* \frac{B\dot{\alpha}}{V} + K^2 P_3^* \alpha + K^2 P_4^* \frac{p}{B} + KP_5^* \frac{\dot{y}}{V} + K^2 H_6^* \frac{y}{B} \right].$$

(2.88c)

The lift and moment equations contain two new aerodynamic derivatives each, and there are six aerodynamic derivatives for the drag equation; hence, in total, there are now eighteen aerodynamic derivatives which are required for a complete description of the unsteady aerodynamic forces. (It should be noted that equations (2.88a, b, c) are not consistent with (2.83a, b), the difference being the factor of 2 in the expressions for L and M. Hence, the aerodynamic derivatives of equations (2.88a, b, c) are not the same as either $H_i^*(K)$, $A_i^*(K)$, $i = 1$–4 used in equation (2.83a, b) or $\tilde{H}_i^*(K)$, $\tilde{A}_i(K)$, $i = 1$–4 as discussed below (2.83a, b); this again indicates the need for extreme caution when using data from the literature.)

Given this vast amount of required data, it is not surprising that some authors have used physical reasoning to conclude that some of the aerodynamic derivatives will be very small and can be neglected. For example, Miyata (2003) considers a

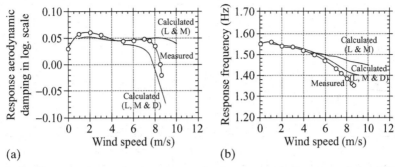

Figure 2.60. Change of aeroelastic response in the torsional direction of the 1/100 scale aeroelastic model of the Akashi Kaikyo Bridge with wind-tunnel speed: (a) aeroelastic logarithmic decrement, δ, (negative δ implying an unstable system); (b) aeroelastic frequency (Miyata 2003).

simplified set of equations for the aerodynamic forces, where the effect of in-plane motion on L and M and the effect of vertical motion on D are ignored; hence,

$$L = \frac{1}{2}\rho V^2 B\left[KH_1^*\frac{\dot{y}}{V} + KH_2^*\frac{B\dot{\alpha}}{V} + K^2H_3^*\alpha + K^2H_4^*\frac{y}{B}\right], \qquad (2.89a)$$

$$M = \frac{1}{2}\rho V^2 B^2\left[KA_1^*\frac{\dot{y}}{V} + KA_2^*\frac{B\dot{\alpha}}{V} + K^2A_3^*\alpha + K^2A_4^*\frac{y}{B}\right], \qquad (2.89b)$$

$$D = \frac{1}{2}\rho V^2 B\left[KP_1^*\frac{\dot{p}}{V} + KP_2^*\frac{B\dot{\alpha}}{V} + K^2P_3^*\alpha + K^2P_4^*\frac{p}{B}\right]. \qquad (2.89c)$$

Using expressions of this form for the aerodynamics, some results obtained by Miyata for the torsional response of a wind-tunnel model of the Akashi Kaikyo Bridge are reproduced in Figure 2.60. These results, in particular the torsional damping presented in Figure 2.60(a), show that as wind speed is increased there is a significant difference between the experimental and theoretical results when only the lift and moment aerodynamics are accounted for, and that this difference can be reduced considerably via the addition of the drag terms.

Some recent effort has gone into obtaining the aerodynamic derivatives using numerical methods. For example, Larsen (1998) uses a discrete vortex method to obtain the aerodynamic derivatives for two different cross-sections. A comparison between his results and the experimental data of Scanlan & Tomko (1971) shows the numerical data to be reasonably good, but probably not good enough to obtain accurate stability predictions. The coupled fluid-structural problem around a simplified bridge deck was also considered by Frandsen (2004), assuming the flow to be laminar.

A number of authors have considered the possibility of using either active or passive control to raise the flutter speed of bridge sections. For example, Kobayashi & Nagaoka (1992) considered active control using flat-plate airfoils, with chord lengths of 10% of the bridge deck, placed close to the leading and trailing edges of the bridge. Using aerodynamic derivatives obtained via Theodorsen's analysis (assuming the flow to remain attached) they predicted that flutter could be eliminated with the appropriate active control; wind-tunnel experiments using a two-dimensional model

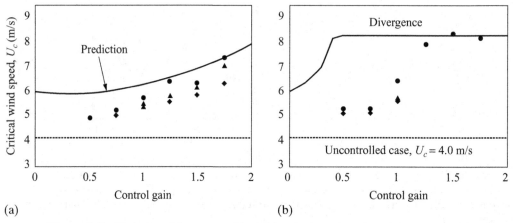

Figure 2.61. Variation of flutter speed for bridge-deck as a function of control gain: (a) control pattern 1, (b) control pattern 2 (Fujino 2002).

suggested that, although flutter could not be totally eliminated, the flutter speed could be doubled compared with when there was no active control. Wilde & Fujino (1998) also considered a very similar configuration and suggested that active control is capable of raising the flutter speed to any desired value.

Wilde, Fujino & Kawakami (1999) and Fujino (2002) considered the possibility of using passive control to raise the flutter speed. Their control system consisted of a pendulum attached to the centre of gravity of the bridge deck connected to control surfaces positioned at the leading and trailing edges of the deck. With the aid of aerodynamic derivatives based on Theodorsen theory, an appropriate phase difference between the pendular motion and the control surface displacement was determined. Two different control patterns were considered, in the first (control pattern 1) the direction of rotation of the two control surfaces was the same, whereas in the second (control pattern 2) the two control surfaces rotated in opposite directions. Some results obtained by Fujino are shown in Figure 2.61, where they are compared with experimental results; the theoretical results for control pattern 2 indicate that, when the control gain is greater than approximately 0.5, the instability switches from being dynamic to static (divergence). Although there is some disagreement between the experimental and theoretical results, it is apparent that, depending on the control gain, a significant increase in the flutter speed of the bridge can be achieved.

2.12 Concluding Remarks

As is often the case, different disciplines of engineering owe their existence to either persistent problems defying solution via conventional means or to spectacular failures that could not be ignored. The latter provided the impetus for the genesis of Wind Engineering. Specifically, one can identify the destruction of the old Tacoma Narrows Bridge in 1940 and the collapse of the cooling towers at Ferrybridge in 1965 (CEGB 1966) as the two major events triggering the systematic study of wind effects on structures, which eventually blossomed into Wind Engineering.

In fact, again, as is often the case, such mishaps do not come without warning. A precursor to the Tacoma Narrows Bridge disaster may be considered to be the

Brighton Chain Pier collapse in 1833 and again in 1836 (Russell 1841; Scott 2001). The collapse of this bridge also occurred because of large-amplitude wind-induced torsional oscillation. This bridge too collapsed in broad daylight, and "Lt. Col. William Reid recorded its last moments in meticulous detail, but the full significance of his sketches and notes would not be appreciated for more than a century" (Scott 2001). One of the sketches shows a side-elevation view of the bridge with the top of the deck showing over half the span and the underside over the other half, as reproduced in Russell (1841) and Simiu & Scanlan (1996)! The collapse of Ellet's Wheeling Bridge in 1854, in a similar torsional oscillation, and other occurrences are described in Scott (2001).

Some discussion on the Tacoma Narrows Bridge – the *Galloping Gertie*, as she was nicknamed by the locals – has already been given in Sections 2.1 and 2.11. It would have been appropriate to provide somewhere in this chapter an account of what exactly happened to the bridge on 7 November 1940. However, as mentioned already in Section 2.11, things are not clear. What seems fairly definite from the film which recorded the catastrophe,* is that the bridge first vibrated transversely at 0.62 Hz with relatively modest amplitude; then switched to a large-amplitude torsional mode at 0.23 Hz, in which it failed (Blevins 1990; Scott 2001). The "modest" transverse amplitude was, in fact, ~ 0.6 m, large enough to necessitate closure of the bridge. Such transverse oscillations, though worrying, occurred several times; they were responsible for the Galloping Gertie *sobriquet*, and they were the delight of adventurous motorists and pedestrians. It was the $\sim \pm 35°$ torsional oscillation that was unprecedented, and it was the one that destroyed the bridge.

Tests on a model of the bridge (ASCE 1961) give two ranges of transverse vibration with vortex-shedding lock-in, before the torsional mode is excited at higher wind velocity, again with lock-in; see Blevins (1990; figure 4-15). However, even today the precise mechanism of excitation and the sequence of oscillatory behaviour culminating in the torsional oscillation (clearly visible in Figure 2.4) leading to collapse are uncertain, even though different plausible scenaria exist. One of the problems is the uncertainty in some of the pertinent data, notably the wind speed and direction. Thus, as recently as 2003, a study on why torsional flutter occurred when it did and not earlier, as it should have done, appeared (Matsumoto *et al.* 2003a), suggesting this was due to "interference" between transverse and torsional motions. So, we shall content ourselves with citing some of the numerous pertinent references, namely Farquharson (1949–1954), Smith & Vincent (1950), ASCE (1955, 1961), Steinman & Watson (1957), Scruton (1981), Scanlan (1980, 1990), Billah & Scanlan (1991), Kubo, Hirata & Mikawa (1992), Larsen (2000a, b), Scott (2001), Miyata (2003) and Matsumoto *et al.* (2003a).

It is of interest that, as in the case of fluidelastic instability of cylinder arrays (Section 5.1) and ovalling oscillation of chimney stacks (Chapter 6), early on, vortex shedding was considered to be the culprit for the Tacoma Narrows Bridge disaster (von Karman & Dunn 1952; Beckett 1969). With the present knowledge on ILEV and AEVS (Komatsu & Kobayashi 1980; Naudascher & Wang 1993; Deniz & Staubli 1997) and the observation of vortex-shedding lock-in in the ASCE model

---

* In fact, the bridge was under constant observation by transit and ciné-film camera since its opening on 1 July 1940, over the whole four months of its life (Farquharson 1949–1954, part I, chapter III).

experiments, the thinking on the subject may have come full circle: vortex shedding may have played an essential role after all (Larsen 2000a, b).

In this chapter we have presented a variety of aeroelastic and hydroelastic phenomena which can conveniently be grouped under the umbrella of galloping. Yet, several topics have been left out, even though related to galloping, e.g. coupled galloping conductors and the possibility of chaotic motions (Cook & Simiu 1990).

We shall close the chapter by listing some of the methods proposed for suppressing galloping oscillations. Methods based on modifying the fluid mechanics include the use of fairings (Wardlaw 1971, 1980, 1990; Nagao *et al.* 1993), side plates (Naudascher *et al.* 1981) and oscillating flaps (Mizota & Okajima 1992). Methods based on increased damping include the use of vibration absorbers (Irwin, Cooper & Wardlaw 1976), tuned mass dampers (Simiu & Scanlan 1996) and nutation dampers (Modi, Welt & Seto 1995). Alleviation of wind-induced vibration, in general, is discussed, e.g. in Blevins (1990), Naudascher & Rockwell (1994), and Wardlaw (1990) specifically on bridges. See also discussion in Section 2.11 on active- and passive-control methods for bridges.

Vortex-Induced Vibrations

The presence of a chapter on vortex-induced vibrations (VIV) in a book on cross-flow-induced instabilities may seem surprising. It fact, it is often emphasized in books on flow-induced vibrations that VIV fundamentally differs from fluid-elastic instabilities (Chen 1987; Blevins 1990; Dowell 1995; Simiu & Scanlan 1996). Still, most reviews have discussed the effect of structural motion on the fluid forces inducing VIV (Griffin 1985; Parkinson 1989; Pantazopoulos 1994; Hori *et al.* 1997; Sarpkaya 2004; Williamson & Govardhan 2004; Gabbai & Benaroya 2005); this brings the topic closer to fluid-elasticity, that is discussed throughout this book. Moreover, some recent work has explicitly treated VIV as an instability (Cossu & Morino 2000; de Langre 2006). Literature on vortex-induced vibrations is vast and continuously growing, both on fundamental issues and on methods for their prediction in engineering, where applications are numerous. Here, we shall focus on the essential ideas underlying the phenomenology and models of VIV, particularly in relation to fluidelastic instabilities.

Notation used in Chapter 3	
Cylinder diameter :	D
Free-stream velocity :	U
Reduced flow velocity :	$\mathrm{Ur} = U/fD, f$ being the frequency in Hz
Lateral displacement of cylinder :	y
Mass ratio :	$m^* = m_S/(\rho\pi D^2/4)$, defined in equation (3.23), m_S being the cylinder mass per unit length and ρ the fluid density

3.1 Elementary Case

To illustrate first the physics of VIV we consider the experiment described in Khalak & Williamson (1996). A rigid cylinder is subjected to a uniform cross-flow of water. It is allowed to vibrate in the cross-flow direction only, through springs, as shown Figure 3.1.

The characteristics of the cylinder are its diameter D, length L and mass M. In water but without flow, the frequency of oscillation of the cylinder is f_s^0. As the flow

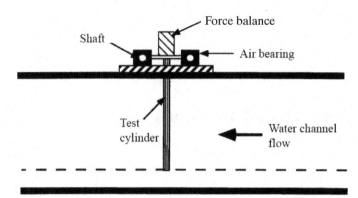

Figure 3.1. Elastically supported rigid cylinder in cross-flow (Khalak & Williamson 1996).

velocity is varied, the motion of the cylinder is analysed in terms of its amplitude A and the main frequencies in its spectrum.

Two essential dynamical features are observed, as seen in Figure 3.2: a significant oscillatory motion, of about a diameter in magnitude, in a limited range of velocities, and simultaneously, a significant change in the frequency. This motion is therefore not the result of a galloping instability as described in Chapter 2, because of the limited range of velocities where it occurs. A more detailed analysis shows that one of the frequencies present in the spectrum is actually that of flow oscillations in the wake, even in the range of velocities where the cylinder barely moves.

To investigate this phenomenon we may initially use a simple model described in de Langre (2006). The cylinder motion in the direction transverse to the flow is assumed to follow that of an oscillator:

$$M\ddot{y} + Ky = 0, \tag{3.1}$$

where K is the stiffness of the spring and M is the inertial mass of the system such that $M = K/(2\pi f_s^0)^2$. The lift forces acting on a stationary cylinder have been measured extensively in the literature. Following Norberg (2003), we model it as

$$F_L(t) = \frac{1}{2}\rho U^2 DLC_L(t), \tag{3.2}$$

where the dimensionless lift coefficient varies, as a first approximation, as $C_L(t) = C_L^0 \sin(2\pi f_{vs}t)$ with $C_L^0 = 0.3$ and $f_{vs} = \mathrm{St}U/D$, where $\mathrm{St} = 0.2$ is the Strouhal number. Note that this oscillating lift coefficient satisfies an oscillator equation

$$\ddot{C}_L + (2\pi f_{vs})^2 C_L = 0. \tag{3.3}$$

This is the central idea of the *wake oscillator model*, refer for instance to Hartlen & Currie (1970), discussed at the end of this chapter. The equation of motion for the cylinder therefore reads

$$\frac{K}{(2\pi f_s^0)^2}\ddot{y} + Ky = \frac{1}{2}\rho U^2 DLC_L(t). \tag{3.4}$$

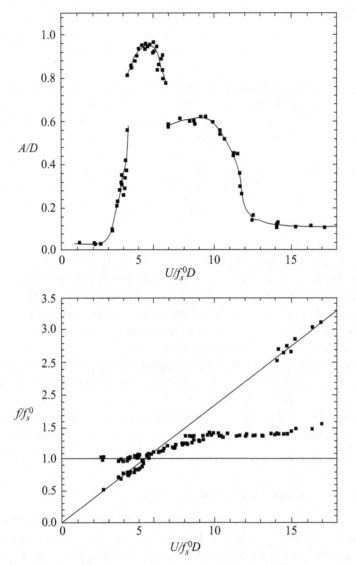

Figure 3.2. Amplitude and frequency of response of an elastically supported cylinder in cross-flow (Khalak & Williamson 1996).

Because the motion of the cylinder in Figure 3.2 is of the order of a diameter, it can be expected to significantly modify the vorticity dynamics leading to lift. This may be taken into account by forcing the equation of lift, equation (3.3), as follows (Facchinetti *et al.* 2004a):

$$\ddot{C}_L + (2\pi f_{vs})^2 C_L = \alpha \frac{\ddot{y}}{D}. \tag{3.5}$$

Here, the coefficient α may be derived by analysing the case where the motion of the cylinder is prescribed as $y(t) = y_0 \sin(2\pi f_f t)$. Using equation (3.5), the resulting amplitude of lift oscillation is then

$$C_L^F = \frac{\alpha y_0 / D}{1 - (f_f / f_{vs})^2}. \tag{3.6}$$

This amplitude becomes larger than that of the existing free lift when $C_L^F >$ C_L^0. This defines a range of forcing frequency f_f which may be compared with experimental data, for instance those in Blevins (1990), where the lift frequency is found to be driven by the forcing frequency. Typical experimental results show that, for $y_0/D = 0.1$, the range of frequency where this happens is approximately $0.8 < f/f_{vs} < 1.2$. Using equation (3.6) this leads to $\alpha \simeq 4C_L^0$.

Using (3.1), (3.2) and (3.3) the set of equations representing the coupled dynamics of the oscillator and of the lift variable read

$$\frac{K}{(2\pi f_s^0)^2}\ddot{y} + Ky = \frac{1}{2}\rho U^2 DLC_L(t), \tag{3.7}$$

$$\ddot{C}_L + (2\pi f_{vs})^2 C_L = \alpha\frac{\ddot{y}}{D}. \tag{3.8}$$

A straightforward modal analysis of this set of linear equations leads to the frequencies, shown in Figure 3.3(a), where they are compared with the experimental data. It is found that in a range near $f_{vs} = f_s^0$ only one frequency exists in the system, as in the experiments. In this same range, as seen in Figure 3.3(c), there exists a mode with a negative damping, i.e. a mode which is unstable. This instability is identical in form to coupled-mode flutter. This can be related to the sudden change of amplitude found in the experiments; see Figure 3.3(b).

The elementary case presented here suggests that vortex-induced vibrations are a true case of *flow-induced instability*, resulting from a coupling between the motion of the structure and the lift dynamics.

Moreover, these vibrations can be modelled using experimental data obtained from the lift dynamics on a stationary cylinder and the lift dynamics on a cylinder in forced oscillation.

3.2 Two-Dimensional VIV Phenomenology

Vortex shedding is the result of a complex three-dimensional fluid-dynamics process; see for instance Williamson (1996). When the body of interest is slender (Figure 3.4), considering a cross-section and the two-dimensional fluid dynamics in this section is a first step in the understanding of vortex shedding and of the coupling with the motion of the body. This two-dimensional configuration has been analysed by many authors, as in the previous chapter for galloping. Numerical simulations are often based on two-dimensional assumptions, for reasons of simplicity; see for instance Anagnostopoulos (2002), Al Jamal & Dalton (2005) and Mittal & Singh (2005). Similarly, experimental data, three-dimensional by nature, are often analysed through a two-dimensional approach, both in terms of flow visualisation and load analysis. Finally, the two-dimensional approximation is presently the basis of most models of vortex-induced vibrations. It is therefore a necessary step in the presentation of VIV.

The fundamental issues of vortex shedding from two-dimensional bluff bodies, in general, and circular cylinders, in particular, are presented in Sections 3.2.1 and 3.2.2, respectively. The effects of motion of the body on the wake are detailed in Section 3.2.3. In Section 3.2.4 we show how an elastic body interacts with vortex shedding.

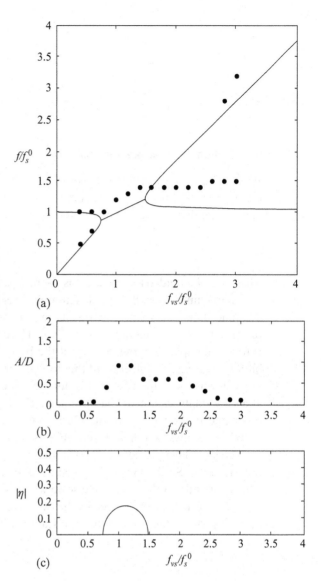

Figure 3.3. Elementary case of vortex-induced vibration. Comparison between experimental results (Khalak & Williamson 1997) and the model involving a lift oscillator, equations (3.7) and (3.8); see also de Langre (2006). Effect of the flow velocity on several parameters: (a) frequency, (b) amplitude in the experiments, (c) negative damping in the model.

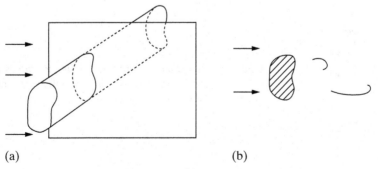

Figure 3.4. (a) The three-dimensional vortex shedding process; (b) two-dimensional approximation.

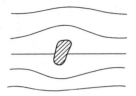

Figure 3.5. Flow past a bluff body at low Reynolds number: creeping flow.

3.2.1 Bluff-body wake instability

We consider a flow of uniform velocity U on a bluff body of arbitrary shape (Figure 3.4(b)). Because of the existence of solid boundaries in the flow, viscous forces play a role; they are scaled by the Reynolds number

$$\text{Re} = \frac{U\ell}{\nu}, \tag{3.9}$$

where ν is the fluid viscosity and ℓ is a length scale still to be defined.

In the limit of small Reynolds numbers, viscous effects are dominant in the flow dynamics near the solid. This results in streamlines that adapt to the body shape, referred to as creeping flow (Figure 3.5). Here, vorticity produced at the boundary is totally dissipated in the vicinity of the body. The flow profile downstream of the bluff body is again uniform and the flow pattern does not vary with time.

Conversely, at high Reynolds numbers the effect of viscosity is dominant only at a very small scale near the boundary. The high level of produced vorticity is not dissipated in the vicinity of the bluff body and the flow profile downstream of the bluff body shows two shear layers (Figure 3.6).

A shear layer is naturally unstable, via a purely inviscid mechanism (Ho & Huerre 1984). Such an instability results in the production of vortices, the spacing of which depends on the transverse scale where shear exists. Here, two shear layers of opposite vorticity are produced by the two edges of the bluff body. These two unstable shear layers interact in a coupled instability. This instability has been extensively

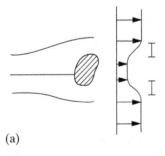

(a)

Figure 3.6. Flow past a bluff body at high Reynolds number: (a) two shear layers over the flow profile; resulting in (b) vortex shedding.

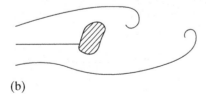

(b)

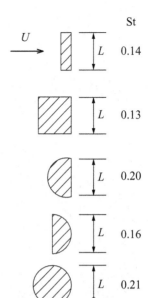

Figure 3.7. Typical values of the Strouhal number St, equation
(3.10), for some sections at large Reynolds number.

analysed theoretically, experimentally and numerically. It results in large alternating
vortices being created, usually referred to as *vortex shedding*. The first observations
and analysis of this phenomenon were made by Bénard (1908) and von Kármán
(1911), respectively. This is a *global* instability in the sense that the whole wake is
affected. It is very robust: its source is the vorticity continuously produced by the bluff
body, but detailed analysis of the dynamics of the instability shows that upstream
perturbations have but a small effect (Huerre & Monkewitz 1990). The alternating
vortices define a self-sustained oscillation with a very well defined dominant vortex-
shedding frequency, f_{vs}. If this frequency is assumed to depend on the flow velocity U
and on a typical dimension of the bluff body, say ℓ, elementary dimensional analysis
leads to

$$f_{vs} = \text{St}\frac{U}{\ell},$$ (3.10)

where St is called the *Strouhal number*. In the limit of large Reynolds numbers (here
10^3), typical values of this dimensionless number are given in Figure 3.7 for several
section shapes. These values are naturally related to the particular choice of the
reference length scale ℓ which is generally taken as the cross-flow frontal dimension
of the body. The order of magnitude of the Strouhal number based on this cross-flow
dimension is St $\simeq 0.2$.

As emphasized above, vortex shedding is not related to the details of the flow
near the bluff body itself, but to the flow profile created by the bluff body: computations (Pier & Huerre 2001) and even experiments (Afanasyev & Korabel 2006) have
shown that vortex shedding with all its characteristics exists without the presence
of a bluff body, provided an appropriate flow profile is considered! The relation to
the shear-layer instability helps explain why vortex shedding, and therefore vortex-
induced vibrations, can be observed on any shape of bluff body. This also gives
grounds for the simple idea that the shedding frequency should be defined using

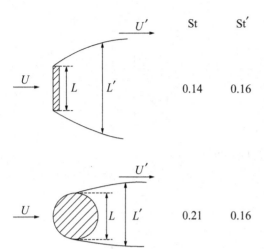

	St	St$'$
	0.14	0.16
	0.21	0.16

Figure 3.8. Universal Strouhal number St$'$, equation (3.11), based on the characteristics of the shear layer emanating from the bluff body (Roshko 1955).

characteristics of this flow: an alternative definition of the Strouhal number would be

$$f_{vs} = \text{St}' \frac{U'}{L'}, \tag{3.11}$$

where U' and L' are a velocity and a dimension, both referring to the flow profile dowstream of the bluff body (Roshko 1955; Anderson & Szewczyk 1996; Zdravkovich 2003). By doing so, the shedding frequencies of various sections can be related by means of a universal Strouhal number, St$'$ (see Figure 3.8).

We have here briefly described two limit cases: that of small Reynolds numbers and that of large Reynolds numbers where vortex shedding takes place. The transition from one regime to the other depends on the body shape, which affects the Reynolds number above which vortex shedding exists. Similarly, the effect of Reynolds number on the characteristics of the global wake instability, such as the Strouhal number, is complex and section-dependent. We shall from now on consider the particular case of a circular cross-section, which is of utmost practical and theoretical interest.

3.2.2 Wake instability of a fixed cylinder

The much simplified two-dimensional case of uniform flow on a fixed cylinder has been extensively studied since the work of Bénard (1908) and von Kármán (1911). Literature on experimental data, numerical studies and theoretical analysis is more than abundant; it can be found in books and reviews (see for instance Williamson (1996), Chen (1987), Zdravkovich (2003), Sumer & Fredsøe (1997) and Norberg (2003)). As noted in the previous section, the dimensionless number that controls most of the aspects of the wake dynamics is the Reynolds number. We now summarize the main features of the phenomenology of vortex shedding from a stationary cylinder, with the aim of understanding the coupling with motions of the cylinder, as will be analysed in the forthcoming sections.

Figure 3.9. Fluctuating lift resulting from vortex shedding on a fixed cylinder in steady uniform flow.

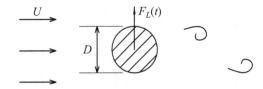

In this perspective, the load on the cylinder resulting from vortex shedding needs to be quantified. As fluctuations of vorticity develop in the unstable wake, fluctuating forces and moments result on the cylinder. We consider here only the component of the force per unit length transverse to the direction of the flow, which is the lift force $F_L(t)$, as shown in Figure 3.9. We may define its dimensionless form, which is the time-varying lift coefficient,

$$C_L(t) = \frac{F_L(t)}{\frac{1}{2}\rho U^2 D}, \tag{3.12}$$

where ρ is the fluid density and D is the diameter of the cylinder. How to define a two-dimensional fluctuating load $F(t)$ from a three-dimensional experiment or simulation of vortex shedding is an important issue in relation to correlation effects. Here, again, the fictitious configuration of two-dimensional vortex shedding is presented first for the sake of clarity.

The fluctuating lift coefficient $C_L(t)$ is the load image of the fluctuating vorticity in the wake. Because a dominant frequency exists there, which is the shedding frequency, it can be expected to be observed also in load fluctuations. Following Norberg (2003), we therefore characterize here the lift coefficient by its dominant frequency f_L and its root mean square (r.m.s.) value C_L^{rms}. Note that it need not be assumed here that lift is a harmonic function of time, nor that the lift frequency is identical to the shedding frequency, f_{vs}; see for instance the power spectra of lift oscillation depending on the Reynolds number in Sumer & Fredsøe (1997). In this section we use the dimensionless Strouhal number defined as $\mathrm{St} = f_L D / U$. The effect of the Reynolds number on the topology of vortex shedding, on the Strouhal number and on the r.m.s. lift coefficient is schematically summarized in Figure 3.10. Several domains may be identified as follows.

(i) For very small Reynolds numbers, $\mathrm{Re} < 40$, no vortex shedding occurs. The flow pattern ranges from creeping flow to flow with a recirculation bubble. No Strouhal number can be defined, nor a fluctuating lift.

(ii) Above the critical Reynolds value, when $40 < \mathrm{Re} < 300$, a laminar state of vortex shedding develops, with a well-defined von Kármán street of alternating vortices. The resulting Strouhal number increases with Reynolds number (Williamson 1996), as does the lift coefficient.

(iii) In a very wide range, $300 < \mathrm{Re} < 10^6$, vortex shedding exists but undergoes several transitions of topology and dynamics; refer for instance to Zdravkovich (2003). This results in a complex and not so well-defined evolution of the Strouhal number and of the lift coefficient.

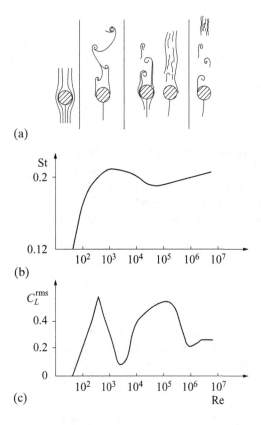

(a)

(b)

(c)

Figure 3.10. Effect of the Reynolds number on vortex shedding from a fixed cylinder in uniform flow. (a) Flow topology; (b) corresponding Strouhal number; (c) r.m.s. value of the fluctuating lift coefficient; adapted from Norberg (2003) and Blevins (1990).

(iv) Finally, when Re $> 10^6$, a regime of fully turbulent vortex shedding sets in, with a well-defined Strouhal number and lift coefficient.

These four regimes make a very crude approximation of the succession of two-dimensional and three-dimensional dynamics. The terminology is not uniformly accepted among authors; see for instance the summary given in Chen (1987). Other dimensionless parameters, that scale the effect of the surface roughness of the cylinder or of the turbulence intensity of the upstream flow, influence the properties of the wake and therefore that of the lift (Sumer & Fredsøe 1997). Suffice it to say here that vortex shedding, from its onset to very large Reynolds numbers, is not a uniform process; consequently, the characteristics of fluctuating lift, St and C_L^{rms}, are strongly Reynolds number dependent.

The effect of vortex shedding is also noticeable on the drag force, in the direction of the flow, defined by the drag coefficient

$$C_D(t) = \frac{F_D(t)}{\frac{1}{2}\rho U^2 D}. \tag{3.13}$$

Both the time-averaged value $\overline{C_D}$ and the r.m.s value C_D^{rms} depend on the Reynolds number (Blevins 1990), and the dominant frequency of the fluctuating drag is typically $f_D \simeq 2f_s$. This will be discussed briefly in Section 3.4.2.

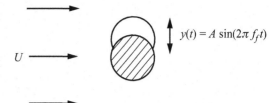

Figure 3.11. Cylinder forced to oscillate transversely to a uniform flow.

3.2.3 Wake of a cylinder forced to move

Considering that displacements in vortex-induced vibrations may be of the order of one cylinder diameter, the effect of cylinder motion on vortex shedding itself needs to be assessed. The natural step in that direction is to force the cylinder to move, following a prescribed law, and to measure the resulting evolution of shedding and resultant loads (see Figure 3.11). If the prescribed motion is harmonic, say

$$y(t) = A \sin 2\pi f_f t, \tag{3.14}$$

then two dimensionless numbers can be defined: the reduced amplitude A/D and the reduced frequency f_f/f_{vs}^0, where f_{vs}^0 is the shedding frequency that would be observed if there were no motion of the cylinder. Because the Reynolds number also influences vortex shedding, the presentation of results becomes much more complex. The frequency of shedding is expected to vary with all three parameters, as will the lift force, in magnitude and phase.

3.2.3 (a) Pattern of vortex shedding

The dynamics of vortex shedding from a fixed cylinder is further complicated by the motion of the cylinder. For instance, in the range of $300 < \mathrm{Re} < 1000$ a detailed analysis (Williamson & Roshko 1988; Blackburn & Henderson 1999) reveals the following features, as shown in Figure 3.12.

(i) In a limited range of forcing frequency and amplitude, the shedding pattern is qualitatively not modified: two vortices of opposite sign are shed in a period of motion. This is referred to as the "2S" mode (for 2 "single" vortices).

(ii) For higher frequencies and amplitudes, a well-defined distinct mode is observed, whereby *two pairs* of vortices are shed per period. Because they form in pairs of opposite vorticity on each side, this is referred to as the "2P" mode (for 2 "pairs" of vortices) (see in Figure 3.12).

(iii) Outside the range of amplitude and frequencies corresponding to these two regimes, the interaction is much more complex, with for instance nonsymmetric "P+S" modes; see Williamson & Roshko (1988).

These features certainly depend on the Reynolds number, but they show that the vortex-shedding process may be significantly altered by the motion of the cylinder, when the amplitude of motion is of the order of a diameter, and the frequency close to the original shedding frequency. This is expected to play a role in the coupling between the wake and a cylinder free to move.

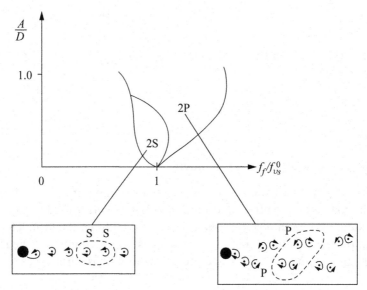

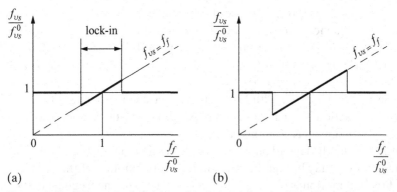

Figure 3.12. Patterns of vortex shedding depending on the amplitude and frequency of the forced cylinder motion, adapted from Williamson & Roshko (1988).

3.2.3 (b) Shedding frequency

Even when the pattern of shedding is not modified, the shedding frequency may be influenced by the motion of the cylinder. At a given Reynolds number, Figure 3.13 shows that this shedding frequency, f_{vs}, may deviate from its nominal value, f_{vs}^0, when the forcing frequency, f_f, approaches f_{vs}^0 (Sarpkaya 2004). Not only does it leave its original value, but it also follows closely the frequency of the external forcing. This phenomenon is referred to as *lock-in* or *wake capture*. Note that these terms are also used in the case of a cylinder free to vibrate, which differs in many aspects, involving for instance a varying flow velocity instead of varying forcing frequency. The range of lock-in varies with the amplitude of forcing (see Figure 3.13(a, b)). A summary of the effect of both dimensionless parameters is shown in Figure 3.14.

Figure 3.13. Schematic definition of lock-in of the shedding frequency, f_{vs}, on the external forcing frequency, f_f. (a) $A/D = 0.05$, (b) $A/D = 0.25$ and Re = 1500; adapted from Sarpkaya (2004).

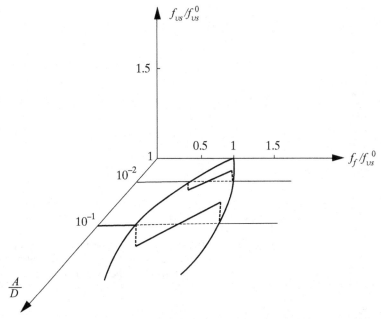

Figure 3.14. Schematic view of the effect of the forcing frequency f_f and amplitude A on the shedding frequency f_{vs}.

3.2.3 (c) Lift force

Knowledge of the lift force acting on a cylinder forced to oscillate in flow is essential for the prediction of the motion of a freely oscillating cylinder. A large number of data is available. The two dimensionless numbers A/D and f_f/f_{vs}^0, as well as the Reynolds number, affect this force. Because this data is commonly used for the prediction of the effect of flow on a cylinder with a given free frequency, the reduced velocity is also used as an alternative dimensionless parameter, in place of f_f/f_{vs}^0,

$$\mathrm{Ur} = \frac{U}{f_f D} = \frac{1}{\mathrm{St}} \left(\frac{f_f}{f_{vs}^0} \right)^{-1}. \tag{3.15}$$

Note that here the frequency is that of a forced motion, not that of a free motion, as in most applications in which the reduced velocity is utilized. For a given displacement of the cylinder, $y = A \sin(2\pi f_f t)$, the resulting lift force is not necessarily harmonic. Still, it is possible to define its amplitude F_L^0 and phase φ at that same frequency by standard Fourier analysis, so that it is approximated as

$$F_L(t) = F_L^0 \sin(2\pi f_f t + \varphi). \tag{3.16}$$

Both amplitude and phase depend on the dimensionless numbers defined above. There exist in the literature several ways of presenting these dependencies, which are now summarized.

(i) The simplest presentation refers to the *modulus and phase of the lift coefficient* $C_L(t)$, as defined in equation (3.2). By stating

$$C_L(t) = \overline{C}_L \sin(2\pi f_f t + \varphi), \tag{3.17}$$

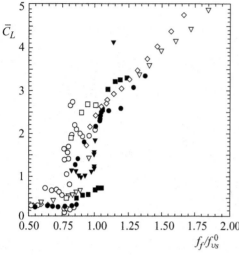

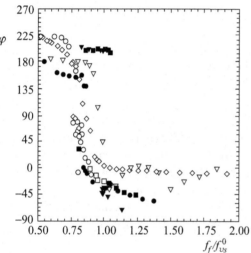

Figure 3.15. Effect of the forcing frequency on amplitude and phase of the fluctuating lift coefficient; from Carberry *et al.* (2005).

both coefficients $\overline{C}_L$ and φ are expected to depend on the dimensionless parameters, A/D and f_f/f_{vs}^0 (or equivalently Ur). The main features of this dependence are shown on Figure 3.15, adapted from Carberry *et al.* (2005). For a given amplitude of motion, say $A/D = 0.5$, the magnitude of the lift force $\overline{C}_L$ in Figure 3.15(a) changes abruptly when the frequency of motion approaches the frequency of natural shedding. Simultaneously, its phase φ jumps from out-of-phase to in-phase (Figure 3.15(b)). Both features show a profound change in the nature of the force, which will appear more clearly in other presentations in what follows. When considering a given frequency of motion equal to that of shedding, $f_f/f_{vs}^0 = 1$, the effect of the amplitude of motion on the amplitude of lift is also strong: Figure 3.16 shows that the fluctuating lift is first enhanced by cylinder motion, up to approximately $A/D = 0.5$; moreover, the fluctuating lift almost disappears near $A/D = 1.5$.

(ii) Alternatively, the data on lift may be presented in terms of *phased lift coefficients*. This formulation, most commonly used, is of interest in understanding the

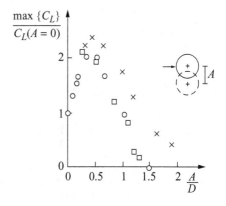

Figure 3.16. Effect of the amplitude of motion on the amplitude of the fluctuating lift coefficient; from Sumer & Fredsøe (1997).

energy balance that controls vortex-induced vibrations. Here, the same lift coefficient is written as

$$C_L(t) = (\overline{C}_L \cos\varphi)\sin(2\pi f_f t) + (\overline{C}_L \sin\varphi)\cos(2\pi f_f t). \qquad (3.18)$$

The phased coefficients $(\overline{C}_L \cos\varphi)$ and $(\overline{C}_L \sin\varphi)$ multiply terms that are, respectively, in phase with the displacement y (or equivalently the inertia $-\ddot{y}$) and with the velocity $\dot{y}$. The first one is therefore related to conservative fluid forces and the second one to nonconservative fluid forces that may result in a positive or negative work over a cycle of motion. The phased coefficients $(\overline{C}_L \cos\varphi)$ and $(\overline{C}_L \sin\varphi)$ have been used with various sign definitions and notations. They correspond, for instance, respectively, to C_a and $-C_{dh}$ in Sarpkaya (1979), to C_{mh} and C_D in Sarpkaya (2004), to $-C_{la}$ and $-C_{lv}$ in Hover et al. (2000) and to C_{Lm} and C_{LD} in Molin (2002). In Figure 3.17, from Sarpkaya (1979) at $A/D = 0.5$, it is shown that most of the evolution of the amplitude of the lift coefficient comes from the lift in phase with cylinder displacement. The phase shift near $f_f/f_{vs}^0 = 1$ is also seen to originate from the change of sign of the velocity-phased lift. When the velocity-phased lift coefficient is positive, the work of lift over the cycle is positive. The same data may also be

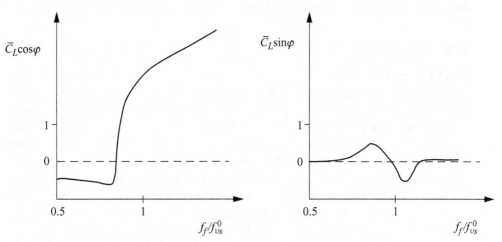

Figure 3.17. Effect of the frequency of forcing on the phased lift coefficients, equation (3.18); adapted from Sarpkaya (2004).

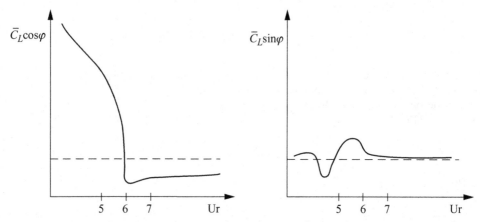

Figure 3.18. Same data as in Figure 3.17, but using $Ur = f_{us}^0/(St f_f)$ on the abscissa.

plotted as a function of the reduced velocity Ur (Figure 3.18). The abrupt changes are now found near $Ur = 1/St \simeq 5$. The effect of the amplitude of motion may be analysed from level plots of the phased coefficients, as in Figure 3.19. Whereas the conservative effect is not significantly affected by the amplitude of motion, it is the nonconservative part which will allow a limit cycle to be established.

(iii) Finally, the data may be presented in terms of *inertia* and *drag coefficients*, as in Figure 3.20. Here, the force is expressed in terms of that acting in still fluid by stating (Sarpkaya 1979; Molin 2002; Sumer & Fredsøe 1997)

$$C_L(t) = -\rho \left(\pi \frac{D^2}{4} \right) C_M \ddot{y} - \frac{1}{2} \rho D C_D \dot{y} |\dot{y}|. \tag{3.19}$$

By inserting $y(t)$ in the above equation we have (Sarpkaya 1979)

$$C_M = (\overline{C_L} \cos \varphi) \frac{Ur^2}{2\pi^3 (y_0/D)}. \tag{3.20}$$

As expected, the limit for small reduced velocity in Figure 3.20(a) is that in still fluid, namely $C_M = 1$. This is discussed in more detail in Section 3.4.1 on added mass.

3.2.4 Cylinder free to move

3.2.4 (a) Parameters

Instead of forcing the motion of the cylinder, we now consider the case where the motion is a result of its interaction with the wake. The most common case is when the dynamical system can be modelled as a linear oscillator throughout the range of amplitude of interest. It should be noted that this is a constraint made on laboratory set-ups to make the analysis of data and comparison with models simpler. In practice, the range of amplitude of motion that may arise from VIV often requires nonlinear effects to be accounted for in the structural dynamics of the bluff body. Moreover, as discussed in the last section of this chapter, considering a single degree of freedom is certainly restrictive for three-dimensional systems where large sets of modes can contribute to the response. Notwithstanding this, the model where the cylinder dynamics can be modelled via the system of Figure 3.21 is the basis of

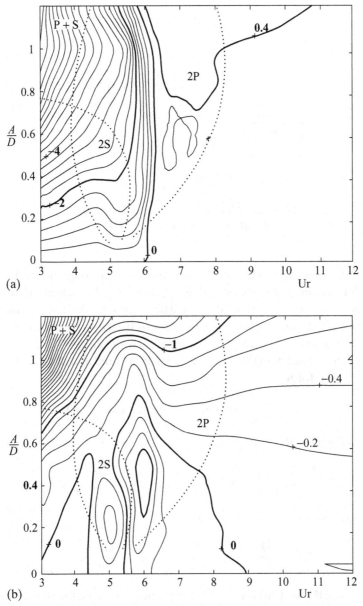

Figure 3.19. Level of phased lift coefficients, equation (3.18), as a function of both reduced frequency and reduced amplitude of motion. (a) Coefficient in phase with the displacement, $\overline{C}_L \cos \varphi$; (b) coefficient in phase with the velocity, $\overline{C}_L \sin \varphi$. Adapted from Hover *et al.* (1998).

the understanding of the phenomenology of coupled motion of a cylinder and its oscillating wake since the early work of Feng (1968).

The free dynamics of the cylinder is now assumed to be defined by the cylinder mass m_S, natural frequency in still fluid, f_s^0, and damping in the absence of fluid η_S. Alternative sets can be used, such as with the stiffness k_S instead of the frequency, or the frequency of free motion in the absence of fluid. The motion of the cylinder, $y(t)$, that results from the coupling with the oscillating wake is here analysed in its

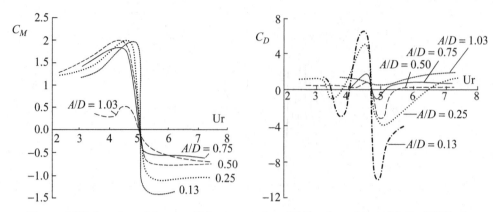

Figure 3.20. Inertia and drag coefficients, equation (3.19), adapted from Molin (2002), showing a noticeable effect of the amplitude of forced motion, A/D.

simplest characteristics: the amplitude A and its dominant frequency f_s. In addition, the dominant frequency of vortex shedding, f_{vs}, which can be measured from velocity fluctuations in the wake, is also discussed, particularly in relation to the cases treated in the previous sections. At this stage, because of the large number of parameters involved, some dimensional analysis is necessary. The results, A, f_s, f_{vs}, are functions of the parameters describing the characteristics of the fluid and of the structure. This reads, in a general form

$$[A, f_s, f_{vs}] = \text{function}\,(U, \rho, \nu; D, m_S, f_s^0, \eta_S). \tag{3.21}$$

Using Buckingham's π-theorem, a dimensionless form of this same relation reads

$$\left[\frac{A}{D}, \frac{f_s}{f_s^0}, \frac{f_{vs}}{f_s^0}\right] = \text{function}\left(\frac{UD}{\nu}, \frac{U}{f_s^0 D}, \frac{m_S}{\rho\pi D^2/4}, \eta_S\right). \tag{3.22}$$

The dimensionless groups on the right-hand side

$$\text{Re} = \frac{UD}{\nu}, \quad \text{Ur} = \frac{U}{f_s^0 D}, \quad m^* = \frac{m_S}{\rho\pi D^2/4}, \quad \eta_S, \tag{3.23}$$

are, respectively, the Reynolds number, the reduced velocity, the reduced mass and the damping coefficient in the absence of fluid. Because seven dimensionless groups are involved here, many alternative choices may be, and have been, used in the literature for comparison with models or in comparison between experiments. This

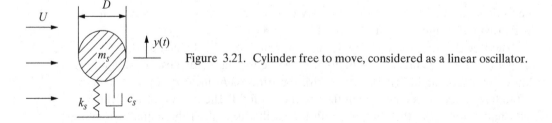

Figure 3.21. Cylinder free to move, considered as a linear oscillator.

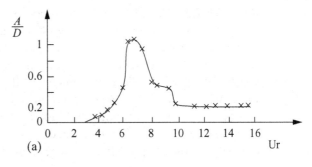

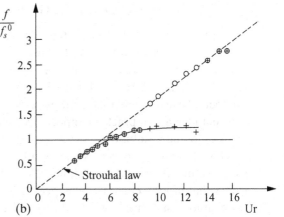

Figure 3.22. Effect of the flow velocity on the motion of an elastically supported cylinder (Sumer & Fredsøe 1997): (a) amplitude and (b) frequency of motion. Also shown is the frequency of vortex shedding (circles).

results in a large set of possible graphical representations where one parameter is presented as a function of another. Here, we seek to point out the main features of the phenomenology by relating these graphs.

3.2.4 (b) Dynamics of a given mechanical system

For a given cylinder with dimensional caracteristics D, m_S, f_s^0, η_S in a given fluid with properties ρ, v, equation (3.22) shows that the effect of the flow velocity can be analysed using the Reynolds number and the reduced velocity only. Disregarding first the effect of Reynolds number, a typical evolution of the dimensionless amplitude with the reduced velocity is shown in Figure 3.22(a).

The most important feature is the large amplitude of motion in the range of reduced velocity of approximately 5 to 10. This must be considered with the simultaneous evolution of the frequency of motion and of shedding (Figure 3.22(b)) A reference line is shown as the frequency of free shedding from a fixed cylinder, defined by the Strouhal law $f_{vs}^0 = \mathrm{St}U/D$, so that $f_{vs}^0/f_s^0 = \mathrm{St}\mathrm{Ur}$. Another line shows the frequency of free motion, $f_s/f_s^0 = 1$. Near $\mathrm{St}\mathrm{Ur} = 1$ the frequency of (free) shedding equals the frequency of (free) motion. The results on shedding and motion frequencies show that, outside the range 5 to 10 mentioned above, they both follow the Strouhal law, while inside the range 5 to 10 they deviate from this law, and their common dynamics seems to be "captured" by the frequency of free motion. This is akin to lock-in, as defined in the previous section in the case of a forced wake. The effect of the mass ratio m^* is essential and will be discussed in the next section in relation to specific models.

3.2.4 (c) Effect of the cylinder parameters on the main features of VIV

From the results of Figure 3.22, two key quantities can be extracted in terms of practical application: the maximum amplitude of motion A_{max}, presumably reached near Ur = 1/St, and the range of reduced velocity where lock-in occurs, also in the vicinity of Ur = 1/St. From the dimensional analysis presented in Section 3.2.4(a), these are expected to depend on Re, m^*, η_S. Disregarding the effect of Reynolds number again, the maximum amplitude is often related to both mass and damping simultaneously by using a combined parameter such as the Scruton number or equivalently the Skop-Griffin number, defined, respectively, as

$$Sc = \pi(1 + m^*)\eta_S/2; \; SG = 2\pi^3 St^2(1 + m^*)\eta_S = 4\pi^2 St^2 Sc. \qquad (3.24)$$

In Figure 3.23(a) it appears that the maximum amplitude does not exceed one or two diameters, even for very low mass-damping. Moreover, for heavier or more heavily damped systems the motion is significantly reduced. Simultaneously, the range of lock-in shown in Figure 3.23(b) decreases with increasing Skop-Griffin number. These two essential features of *self-limited* oscillation amplitude and *limited* lock-in range are of practical importance, and they show that VIV is clearly distinct from galloping instabilities.

Finally, Figure 3.24 shows that Figures 3.22(a) and 3.23(a) and (b) can be understood as the projection of a unique VIV surface on orthogonal planes. This surface shows that motion can only be expected in a limited region of the set of parameters characterising the flow and the cylinder, namely the reduced velocity and the Skop-Griffin number.

The effect of the Reynolds number on these results is complex, as can be expected from the dependence of the wake dynamics on Reynolds numbers (Figure 3.10). For instance, the maximum amplitude at low Skop-Griffin number, Figure 3.23(a), is actually dependent on the Reynolds number (Klamo *et al.* 2005; Govardhan & Williamson 2006).

3.3 Modelling Vortex-Induced Vibrations

3.3.1 A classification of models

From the phenomenology of vortex shedding and vortex-induced vibrations summarized in the previous section one can expect a large variety of models to exist. In fact, because of the practical and theoretical importance of VIV, models have been developed and used since the 1960s. Reviews show not only a large number of them, but also significant differences in the fundamental aspects of their formulations (Dowell 1995; Hori *et al.* 1997; Gabbai & Benaroya 2005).

We shall therefore focus here on a classification of these models, with particular attention to their relation with the formulations of fluidelastic effects presented in other chapters of this book. Following the previous sections, we shall only consider the simple case of two-dimensional VIV in the cross-flow direction, the motion of the solid being defined by the displacement $y(t)$. Note that in this section the dynamics of the solid need not necessarily be modelled as an elastic oscillator, although we shall only consider this simple case in the following.

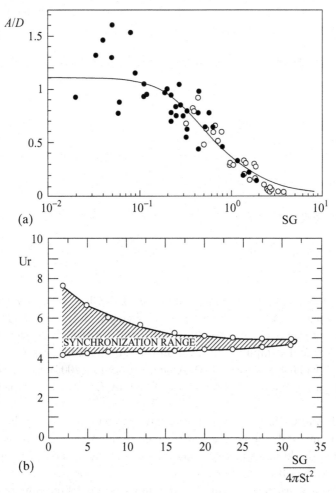

Figure 3.23. Effect of the mass damping parameter SG on (a) the maximum amplitude of motion at lock-in (Williamson & Govardhan 2004), (b) the range of lock-in in terms of reduced velocity (Chen 1987).

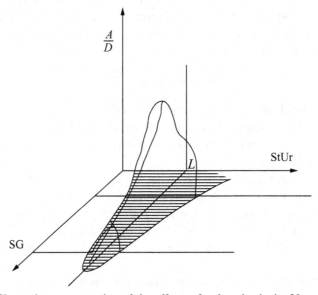

Figure 3.24. Shematic representation of the effects of reduced velocity Ur and Skop-Griffin number SG on the amplitude of motion due to VIV.

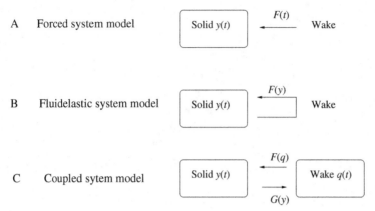

Figure 3.25. A classification of VIV models.

We shall not discuss numerical simulations of the flow, that only use primitive information on the flow (incoming velocity, viscosity and possibly parameters in models of turbulence). The possibility of this latter approach is growing fast, as can be seen from recent computations, even in three-dimensional cases. The reader is referred to Al Jamal & Dalton (2005), Anagnostopoulos (2002), Lucor *et al.* (2005), Mittal & Singh (2005) and Guilmineau & Queutey (2002) as examples of recent work.

All models to be presented therefore have in common that (i) they are based on some experimental results such as described in the previous sections, essentially data on vortex shedding from a stationary or oscillating cylinder, and (ii) they aim at predicting some other experimental results, namely vortex-induced vibration of free cylinders, and for more complicated configurations (three-dimensional, etc.). Hence, they are all empirical, or *phenomenological*, by nature, although these words are commonly used only for some of them.

We propose now a simple classification of models, as illustrated on Figure 3.25. It is based on the formulation of the fluid force applied on the solid in the direction of lift. More precisely, we consider the part of the fluid force that exceeds the conventional inviscid added mass effect, following Williamson & Govardhan (2004),

$$F_{\text{Fluid}} = -m_A \ddot{y} + F, \tag{3.25}$$

where $m_A = \rho \pi D^2 / 4$ (this force decomposition is discussed in detail in Section 3.4.1 on added mass). Hence the proposed classification is as follows.

Type A: *Forced system models*, where F is independent of y, and therefore only depends on time, $F(t)$.

Type B: *Fluidelastic system models*, where F depends on y, denoted as $F[y(t), t]$. The dependence in y may include all time derivatives, integrals and even time delays.

Type C: *Coupled system models*, where F depends on another variable related to the wake dynamics, say q, the evolution of which depends on y. Here, we have $F[q(t), t]$ and the effect of y on the evolution of q is taken in the most general form $G[y(t), t]$.

All type A models will therefore be based only on data from experiments with a fixed cylinder, because the displacement cannot be accounted for. Conversely, models of types B and C can incorporate data from experiments with an oscillating cylinder. Type A is a particularly simple case of types B and C, where the dependence on y is ignored. Type B is also a particular case of type C, where q is just y.

In the appraisal of the models in the forthcoming sections, several criteria will be considered: the simplicity of the formulation, its relation to the physical mechanisms at work, its accuracy in predictions, its generality in predictions, its inherent ability to be generalized to complex cases, and finally its simplicity of use. The particular weight to give to each of these criteria is left to the reader, depending on the particular need for a model.

3.3.2 Type A: Forced system models

Models of this type are based on data from measurements of forces on a stationary cylinder. In Section 3.2.2 it was shown that the force could be quantified by the amplitude of the fluctuating lift coefficient C_L^{rms} and the dimensionless quantity characterising the main frequency of fluctuation, St. A natural model for the force is therefore

$$F(t) = \frac{1}{2}\rho U^2 D C_L \sin\left(2\pi \mathrm{St}\frac{U}{D}t\right), \tag{3.26}$$

where $C_L = \sqrt{2}C_L^{\mathrm{rms}}$. The motion of the cylinder in the lift direction is then defined by

$$(m_S + m_A)\ddot{y} + 2\eta_S(m_S + m_A)(2\pi f_s^0)\dot{y} + (m_S + m_A)(2\pi f_s^0)^2 y = F(t), \tag{3.27}$$

where f_s^0 is the frequency of oscillation of the cylinder in still fluid, η_S is the corresponding damping coefficient, $m_A = \rho\pi D^2/4$ is the added mass, and m_S is the mass of the cylinder. The solution is found from classical linear vibration theory and, using dimensionless variables, reads

$$\frac{y(t)}{D} = \frac{1}{2\pi^3}\frac{C_L \mathrm{Ur}^2}{(1+m^*)[(1-\mathrm{Ur}^2\mathrm{St}^2)^2 + 4\eta_S^2\mathrm{Ur}^2\mathrm{St}^2]^{1/2}} \sin\left(2\pi\mathrm{St}\frac{U}{D}t + \varphi\right), \tag{3.28}$$

where the phase φ is defined by

$$\tan\varphi = \frac{2\eta_S\mathrm{St}\mathrm{Ur}}{1 - \mathrm{St}^2\mathrm{Ur}^2}; \tag{3.29}$$

the dimensionless numbers are

$$m^* = \frac{m_S}{m_A}, \quad \mathrm{Ur} = \frac{U}{f_s^0 D}. \tag{3.30}$$

Figure 3.26 shows the evolution of the amplitude and phase, as well as frequency of the motion of the cylinder, with the flow velocity, according to this model. A resonance occurs at $\mathrm{StUr} = 1$, corresponding to the frequency of shedding being equal to that of the structure, as

$$\mathrm{StUr} = \mathrm{St}\frac{U}{f_s^0 D} = \frac{\mathrm{St}U/D}{f_s^0} = \frac{f_{vs}}{f_s^0}. \tag{3.31}$$

The amplitude of motion increases up to a maximum which depends on the damping η_S, as discussed below. The phase between the displacement $y(t)$ and the force $F(t)$

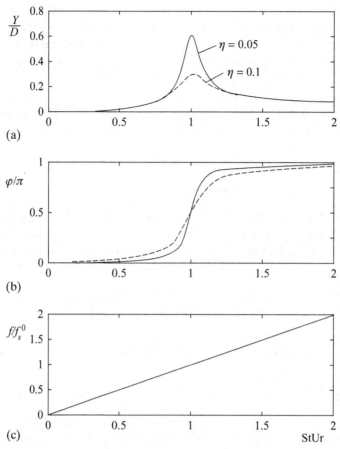

Figure 3.26. Response curves of model of Type A, equation (3.28): (a) amplitude, (b) phase, (c) frequency of motion.

rapidly changes from 0 to π near the resonance. Throughout all the variation of the flow velocity, the frequency of motion is identical to that of the force, given by the Strouhal law, $f_s = f_{vs} = \text{St}U/D$, so that $f/f_s^0 = \text{StUr}$. The maximum A of the resonance curve, which occurs near $\text{StUr} = 1$ reads

$$\frac{A}{D} = \frac{C_L}{4\pi^3(1+m^*)\text{St}^2\eta_S}. \tag{3.32}$$

In the evolution of this maximum, the mass and damping parameters of the fluid-structure system are seen to be grouped in a single dimensionless parameter, which is in fact the Skop-Griffin number defined previously, equation (3.24). Then, equation (3.32) reads simply

$$\frac{A}{D} = \frac{C_L}{2\text{SG}}. \tag{3.33}$$

This result, using $C_L = 0.35$, is compared in Figure 3.27 with the experimental data shown previously. The decrease of amplitude is well decribed for high values of SG, but the limitation of amplitude at low mass-damping is not predicted. This could be expected, as the model does not include any influence of the motion on the fluid

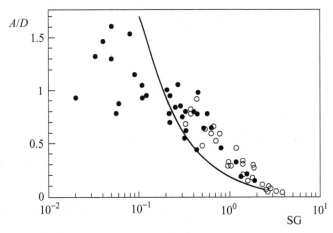

Figure 3.27. Type A model : amplitude of response at resonance, equation (3.33), and experiments from Williamson & Govardhan (2004).

force. In conclusion, it appears that the simplest type of model where the lift force is presumed to be independent of cylinder motion is not able to predict essential aspects such as the lock-in of frequency or an adequate limitation of amplitude. Still, because it overpredicts the motion in the range of low mass-damping parameters, it may be used as a first, quite overconservative estimate.

3.3.3 Type B: Fluidelastic system models

This type of model gathers several kinds and even families of models with the common feature that the fluid force acting on the cylinder is somehow made dependent on its motion. In that sense, the formulation here bears many similarities to the modelling of fluidelastic effects analysed in other chapters, such as in galloping of sections or instabilities in arrays of tubes. The motion of the cylinder, $y(t)$, may be taken into account in the force by several means. These may be global characteristics such as the average amplitude of motion, A, and its dominant frequency, f, or the value of the instantaneous displacement $y(t)$ and its derivatives $\dot{y}(t), \ddot{y}(t)$.

3.3.3 (a) Modified forcing model

Here, the basic formulation of models of type A is kept, whereby a sinusoidally varying force is used to simulate the fluid loading, but the characteristics of the motion of the cylinder are taken into account in the parameters of the forcing. Following Blevins (1990), the simplest idea to account for the self-limited motion observed in experiments is to make the amplitude of the force, namely the fluctuating lift coefficient, dependent on the amplitude $C_L(A/D)$. The force then reads

$$F(A, t) = \frac{1}{2}\rho U^2 D \, C_L\left(\frac{A}{D}\right) \, \sin\left(2\pi \mathrm{St}\frac{U}{D}t\right). \tag{3.34}$$

In the data from experiments on forced oscillations at coincidence, such as in Figure 3.16, the lift is found to first increase with the displacement, as the wake is organised and enhanced by the oscillation; for larger amplitudes, the cylinder

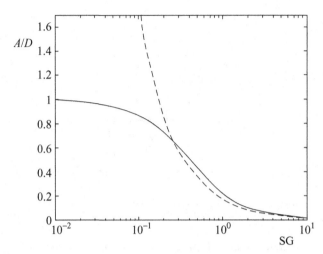

Figure 3.28. Type B modified forcing model: effect of the SG number on the amplitude of the limit cycle, in comparison with a model of Type A. - - -, Type A model ; —, type B modified forcing model.

displacement has a destructive effect on the wake, up to the point where the actual lift coefficient decreases. A straightforward polynomial fit reads (Blevins 1990)

$$C_L\left(\frac{A}{D}\right) = C_L^0 + \alpha\frac{A}{D} + \beta\left(\frac{A}{D}\right)^2, \tag{3.35}$$

where $C_L^0 = 0.35$, $\alpha = 0.60$ and $\beta = -0.93$. The effect of a force defined by equation (3.34) on the maximum amplitude of motion can easily be derived by using equations (3.35) and (3.34):

$$\frac{A}{D} = \frac{C_L(A/D)}{2\text{SG}}; \tag{3.36}$$

so that the amplitude is defined by the polynomial equation

$$C_L^0 + (\alpha - 2\text{SG})\frac{A}{D} + \beta\left(\frac{A}{D}\right)^2 = 0. \tag{3.37}$$

Figure 3.28 shows on a Griffin plot the comparison between equation (3.37) and that of a model of type A, equation (3.33).

Though the modified forcing model still does not simulate the effect of lock-in of frequencies at all, it avoids one the main limitations of type A models: large amplitudes at low SG numbers.

3.3.3 (b) Advanced forcing model

The approach in the preceding model may be considerably improved by taking into account *both* the amplitude and frequency of the motion, in *both* the amplitude and the phase of the force. The most commonly used formulation is based on the phased lift coefficients previously defined. Experimental data now exists to support such models, where the phased lift coefficients have been obtained with good consistency by many authors (see for instance Sarpkaya (2004), Gopalkrishnan (1993) and Hover *et al.* (1998)). Then, the force reads

$$F\left(\frac{A}{D}, \mathrm{Ur}, t\right) = \frac{1}{2}\rho U^2 D \left[C_{my}\left(\frac{A}{D}, \mathrm{Ur}\right)\sin\omega t - C_{dy}\left(\frac{A}{D}, \mathrm{Ur}\right)\cos\omega t\right], \quad (3.38)$$

in which we have used Sarpkaya's notation for the displacement- and velocity-phased coefficients:

$$C_{my} = \overline{C}_L \cos\varphi, \quad C_{dy} = -\overline{C}_L \sin\varphi, \quad (3.39)$$

where $\overline{C}_L$ and φ have been defined in equation (3.17). Note that these coefficients were built to express the evolution of the force created by a given harmonic motion. To use them in a case where the motion of the cylinder is free requires that the phase of the displacement be arbitrarily fixed by stating that $y = A\sin\omega t$, and that the reduced velocity Ur be defined using the frequency of free motion,

$$\mathrm{Ur} = \frac{U}{fD}, \quad f = \omega/2\pi. \quad (3.40)$$

Note also that the force F in most of these models is assumed to account for all the flow-induced forces, so that no specific added mass or added damping should be taken into account in the equation of the oscillator. The equation of motion now reads

$$m_S\ddot{y} + 2\eta_S m_S(2\pi f)\dot{y} + m_S(2\pi f)^2 y = F\left(\frac{A}{D}, \mathrm{Ur}, t\right), \quad (3.41)$$

where f is here the natural frequency without added mass. Although equation (3.38) is also applicable outside the coincidence condition, it may be used (Gopalkrishnan 1993) to estimate the maximum amplitude of motion: this is expected to occur when the frequency of forcing becomes identical to the frequency of free motion with added mass, and when the damping is exactly balanced by the velocity-phased fluid forces. This is achieved at

$$\frac{A}{D} = \frac{C_{dy}(A/D, \mathrm{Ur} = 1/\mathrm{St})}{2\mathrm{SG}}, \quad (3.42)$$

which implicitly defines A/D. This is identical in form to equation (3.36) obtained for the much simpler model, but several comments need to be made. First, equation (3.42) allows the accurate computation of the motion at any flow velocity because it is based on data not only at coincidence. Second, the particular form of the dependence of C_{dy} on amplitude needs not be parabolic. Finally, this allows the relation between SG and A/D to be defined through simple analysis of experimental data, using a graphical intersection between $C_{dy}(A/D, \mathrm{Ur} = 1/\mathrm{St})$ and $2\mathrm{SG}A/D$; see Gopalkrishnan (1993).

3.3.3 (c) Time domain fluidelastic models

An even more general formulation of a force dependent on the motion of the cylinder requires that the time-dependent displacement function $y(t)$ be explicitly used, and not only its amplitude or frequency. This is done in the model of Chen *et al.* (1995a), where the force reads formally

$$F(y, \dot{y}, \ddot{y}) = \left[-m\left(\frac{A}{D}, \mathrm{Ur}\right)\ddot{y} - \rho U c\left(\frac{A}{D}, \mathrm{Ur}\right)\dot{y} - \rho U^2 k\left(\frac{A}{D}, \mathrm{Ur}\right)y \right]$$
$$+ \frac{1}{2}\rho U^2 C_L \sin \omega_{vs} t. \tag{3.43}$$

Here, the force is assumed to be composed of a classical fluidelastic force, in brackets, that includes added inertia, added damping and added stiffness as in other sections of this book, and of a forcing term at the shedding frequency given by the Strouhal law. The contribution of the fluidelastic force is to modify the frequency (inducing lock-in) and to induce the initial instability that leads to growth. A normalized form of the added stiffness and added damping coefficients measured in experiments is presented in Figure 3.29, as a function of the reduced velocity and of the amplitude of motion. Note the similarity with phased lift coefficients, Figure 3.18. Because the coefficient still depends explicitly on the amplitude and frequency of motion, through the reduced velocity, the force may not be fully expressed in terms of y and its time derivative.

A similar but more complex model is proposed by Simiu & Scanlan (1996), where the force reads

$$F(y, \dot{y}, \ddot{y}, t) = \frac{1}{2}\rho U^2 D\left[c(\mathrm{Ur})\left(1 - \varepsilon\frac{y^2}{D^2}\right)\frac{\dot{y}}{U} + k(\mathrm{Ur})\frac{y}{D} + C_L(\mathrm{Ur})\sin \omega_{vs} t \right],$$
$$\tag{3.44}$$

the reduced velocity being defined with the frequency of motion. In this formulation the amplitude effect is accounted for by a nonlinear term $y^2\dot{y}$; thus, harmonic motion is not assumed. This is more general in form than equation (3.43).

3.3.4 Type C: Coupled system models

3.3.4 (a) Limitations of type A and B models
In the two previous types of model, A and B, the fluid force is assumed to be either a given function of time or to be directly related to the displacement, or a combination of both. This allows accounting for the simple observation that in vortex-induced vibrations the forcing mechanism is perturbed by the response of the system it excites. Numerous experimental data therefore have to be included in the model, such as the dependence of coefficients on the reduced velocity or the amplitude of motion. This approach bears some conceptual limits which may be summarized as follows.

(i) By their nature these models are limited to harmonic motion of the cylinder, because coefficients were measured only for such motions. Generalisation to more complex motion in time gives rise to difficulties, exactly as in galloping or instabilities of tube arrays: a way to define the equivalent frequency needs to be chosen, and multifrequency motion may not be taken into account easily. This is not a major limitation, because vortex-induced vibrations often result in quasi-harmonic motion, but it complicates the practical use of such models.

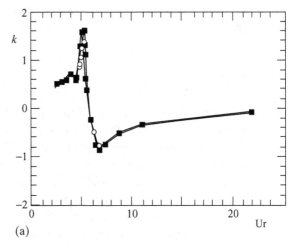

Figure 3.29. Force coefficients from Chen *et al.* (1995a), equation (3.43), as a function of the reduced velocity: (a) stiffness coefficient, (b) damping coefficient.

(a)

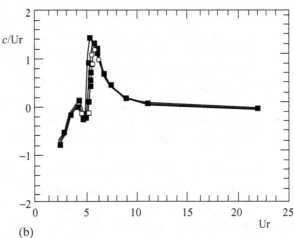

(b)

(ii) More fundamentally, the formulation of these models bears no relation to the physics of wakes as global modes (Chomaz *et al.* 1988; Huerre & Monkewitz 1990; Zielinska & Wesfreid 1995; Chomaz 2005): forces in vortex-induced vibrations are the result of wake dynamics that follow specific rules. Hence, all the data used in models of types A and B are actually the image of a more profound evolution.

3.3.4 (b) General features of Type C models

All these limitations, as well as physical observations of the wake dynamics, have led to the development of models whereby the fluid force is the result of the wake dynamics, itself influenced by the cylinder motion. All models of type C therefore consider VIV as resulting from the coupling of two systems: the cylinder and the wake. The corresponding formulation may be represented by two equations (Dowell 1995): one for the cylinder variable $y(t)$,

$$m\ddot{y} + \cdots = F(q, ...),\qquad (3.45)$$

where F defines the effect of the wake on the cylinder, and another for a wake variable $q(t)$,

$$W[q(t)] = G(y, ...),\qquad(3.46)$$

where W defines the dynamics of the free wake and G defines the effect of the cylinder on the wake. The original idea of using a wake oscillator equation was proposed by Bishop & Hassan (1964). Numerous models have been proposed since then, starting with the pioneering work of Hartlen & Currie (1970). They differ in the formulations of F, W and G and the meaning of the variable q. In all cases the parameters of the model are built on basic experimental data on (i) the wake dynamics with a fixed cylinder, to derive W; (ii) the corresponding lift, to derive F; (iii) the wake dynamics with a cylinder forced to move, to derive G. Depending on the experiments chosen for this and on the degree of accuracy required by the authors in matching this experimental data by their model, these functions may be quite complex. This has led to some doubts; refer for instance to the reviews by Sarpkaya (1979, 2004) on the soundness of this approach: if adding new terms or new coefficients allows any new experiment to be accounted for, where is the limit? Moreover, why should a particular mathematical form have any relation to the wake dynamics?

These questions can only be answered in general by stating that wake oscillator models are no more empirical than models of other types: essentially they have been designed to understand what may cause the fluid forces that other models take for granted.

3.3.4 (c) The Hartlen and Currie model

The essential concepts used in type C models were introduced in the seminal paper by Hartlen & Currie (1970). The first aspect is the choice of the wake variable as being the lift coefficient, $q(t) = C_L(t)$. Therefore, the function F in equation (3.45) simply reads $F(t) = \rho U^2 D q(t)/2$.

Second is the form of the equation governing the dynamics of a free wake, namely W in equation (3.46). Two essential features of the wake as a global mode need to be represented: (i) the oscillating wake results from a self-sustained flow instability, as a *linearly unstable* global mode, and (ii) this instability is self-limited as a *nonlinear* global mode. A Rayleigh equation captures these two features, as

$$W(q) = \ddot{q} - a\dot{q} + b\dot{q}^3 + \omega^2 q = 0,\qquad(3.47)$$

where a and b are positive coefficients. Here, the term $-a\dot{q}$ allows for self-sustained oscillation at the frequency ω, whereas the cubic term limits the amplitude of oscillations (Nayfeh & Mook 1979). Note that using a van der Pol equation, as has been done by most authors since then, is equivalent to using a Rayleigh equation.

Finally, the effect of the cylinder motion on the wake dynamics, G in equation (3.46), needs to be defined. As stated by Hartlen and Currie, this was chosen "rather arbitrarily" as being proportionnal to the cylinder velocity, $G = c\dot{y}$. This particular choice has been much discussed since then; refer for instance to Facchinetti *et al.* (2004a).

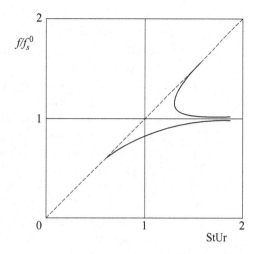

Figure 3.30. Lock-in of frequency using the original wake oscillator model of Hartlen & Currie (1970).

By analysing the dynamics of this coupled model in various cases, Hartlen and Currie showed that lock-in of frequencies and self-limited amplitude of motion could be reproduced qualitatively by this very simple model (see Figure 3.30).

3.3.4 (d) The model of Facchinetti, de Langre and Biolley

To underline some issues related to all wake oscillator models we summarize some features of the model discussed in Facchinetti *et al.* (2004a), with a van der Pol oscillator, and used in Mathelin & de Langre (2005) and Violette *et al.* (2007). It includes most of the evolutions of wake oscillator models since the work of Hartlen & Currie (1970); see for instance Skop & Griffin (1973), Skop & Balasubramanian (1997) and the summary in Gabbai & Benaroya (2005), while keeping a simple form. The equations of motion for the two variables read

$$\left(m_S + \rho D^2 \frac{\pi}{4}\right)\ddot{y} + \left(c_s + \frac{1}{2}\rho U\, D\mathrm{St}C_D\right)\dot{y} + k_S y = \frac{1}{4}\rho U^2 D C_L^0 q, \qquad (3.48)$$

$$\ddot{q} + 2\pi\mathrm{St}\frac{U}{D}\varepsilon(q^2 - 1)\dot{q} + \left(2\pi\mathrm{St}\frac{U}{D}\right)^2 q = \frac{A}{D}\ddot{y}. \qquad (3.49)$$

Here, the fluid variable $q(t)$ is interpreted as a dimensionless reduced lift coefficient $q(t) = 2C_L(t)/C_L^0$. In equation (3.48) the fluid force is seen to be composed of three terms: (i) an inertia force, assumed to be obtainable from an inviscid added mass model; (ii) a flow-induced damping resulting from drag (Skop & Balasubramanian 1997); (iii) a force related to vortex shedding through the variation of the lift coefficient variable $q(t)$. In parallel, the equation pertaining to the dynamics of the wake, equation (3.49), shows that the wake variable is controlled by a van der Pol equation. It produces a self-limited oscillation of amplitude 2 at the frequency of vortex shedding defined by the Strouhal law, when the forcing term is absent. Finally, the effect of cylinder motion on the wake dynamics has been derived by a systematic analysis of possible first-order forms, y, $\dot{y}$, $\ddot{y}$: the coupling function G was found to be proportionnal to $\ddot{y}$ rather than to $\dot{y}$.

The coefficients of these equations that need to be derived from elementary experiments are of three kinds. First, those that can be derived from knowledge of flow forces on a fixed cylinder: the drag coefficient, C_D, and the reference fluctuating lift coefficient, C_L^0. Abundant litterature exists on the values of these coefficients and their dependence on the Reynolds number. Typical values are $C_D = 1.2$ and $C_L^0 = 0.3$. Second are parameters related to the dynamics of a wake on a fixed cylinder: the Strouhal number St, and the growth rate coefficient ε, which are here used to define the frequency and damping of the wake variable q. Whereas the Strouhal number is abundantly documented, the determination of ε is still needed. Finally, the parameter A which scales the coupling of the cylinder motion to the wake dynamics also needs to be determined from experiments. A simple method is proposed in Facchinetti *et al.* (2004a) to derive A and ε, based on experimental results with forced cylinder motion. First, the ratio A/ε is adjusted on the amplitude of the fluid force as a function of the amplitude of the displacement at resonance. Then, the range of lock-in as a function of the same parameter is used to adjust ε, leading to $\varepsilon = 0.3$ and $A = 12$.

The coupled model may now be used to simulate vortex-induced vibrations. In a Griffin plot, Figure 3.31(a), the model is found to underestimate the structure oscillation amplitude, when using the values of parameters proposed above. Yet, the qualitative influence of the Skop-Griffin parameter is recovered. In particular, the asymptotic self-limited response amplitude at low SG is assured by the fluid damping $C_D/(4\pi St)$, using here $C_D = 1$.

The effect of the mass ratio at low SG is also shown in Figure 3.31(b) compared with the experimental results from Williamson & Govardhan (2004). As m^* tends to zero, the widening of the lock-in domain is clearly unbounded and significant structural oscillations persist at high Ur. This latter phenomenon is quite consistent with experimental data, where it was coined as "resonance for ever" (Govardhan & Williamson 2002).

3.3.4 (e) Lock-in as a case of linear coupled-mode flutter

The ability of type C models to represent lock-in of frequencies was recognized in the original paper by Hartlen & Currie (1970). For the purpose of understanding the relation between VIV and fluidelastic effects, which are discussed in other chapters of this book, we explore here the elementary mechanism that enables these models to simulate this lock-in effect. The approach follows that given in de Langre (2006), of which some preliminary elements may be found in Nakamura (1969).

The set of coupled wake and cylinder equations (3.48) and (3.49) are now written in a much simpler form by using dimensionless time, and by neglecting all velocity-dependent terms and all nonlinear terms. The latter two simplifications are made with the aim of pointing out the mechanism responsible for the evolution of frequencies which does not depend much on damping effects or nonlinear effects. The new system of equations reads

$$\ddot{y} + y = DM(\text{StUr})^2 q, \tag{3.50}$$

$$\ddot{q} + (\text{StUr})^2 q = (A/D)\ddot{y}. \tag{3.51}$$

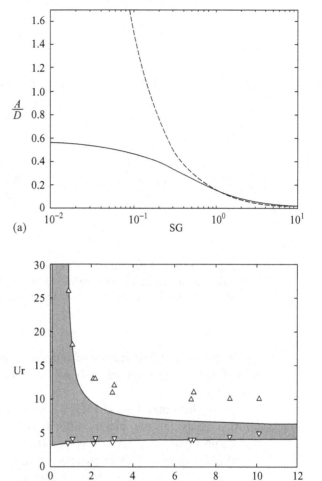

Figure 3.31. Results of a Type C model (Facchinetti *et al.* 2004a). (a) Griffin plot, showing self-limitation of amplitude: —, wake oscillator model; - -, type A model. (b) Effect of the mass ratio m^* on the range of lock-in at low SG: in grey, lock-in predicted by a wake-oscillator; symbols show the limit of lock-in found in experiments by Govardhan & Williamson (2002).

where $M = C_L^0/(4\pi^3 \mathrm{St}^2(1+m^*))$ and $\mathrm{Ur} = U/f_s^0 D$. The dynamics corresponding to this simple system of coupled linear oscillators is fully determined by standard modal analysis, setting $(y, q) = (y_0, q_0)e^{i\omega t}$. Provided that the product AM is less than 1, which is the case of practical interest, the evolution of modal frequencies, ω_1 and ω_2, corresponding to the two modes of the system, can easily be tracked when the reduced velocity is varied. Here, the frequency may be real (neutrally stable mode), or complex (damped or unstable mode). The exact solution of this modal problem is given in de Langre (2006), but suffice it to say that two distinct cases exist in terms of the reduced velocity, as follows.

(i) In the ranges $\mathrm{StUr} < (1 + \sqrt{AM})^{-1}$ or $\mathrm{StUr} > (1 - \sqrt{AM})^{-1}$ two modes are found, with distinct real frequencies, denoted as ω_W and ω_S. They may be attributed, respectively, to the wake dynamics, mode "W" in Figure 3.32(a), and to the solid dynamics, mode "S", depending on the corresponding modal components (y_0, q_0). In each mode, both components y_0 and q_0 exist, although one is clearly dominant. The frequency of the solid mode remains close to that of the free solid ($\omega_S \simeq 1$), whereas the frequency of the wake follows that of the free wake

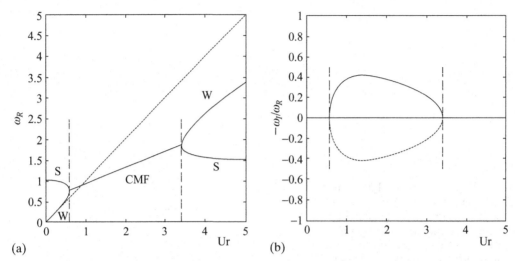

Figure 3.32. Effect of the reduced velocity on a coupled linear system representing the cylinder and the wake, equations (3.50) and (3.51). (a) Evolution of frequencies and (b) growth rates of the two modes of the system (de Langre 2006).

($\omega_W \simeq StUr$). In these ranges of reduced velocity the coupling between the two oscillators therefore induces only a small alteration of the frequencies and modal components.

(ii) In the range $(1 + \sqrt{AM})^{-1} < StUr < (1 - \sqrt{AM})^{-1}$ a completely different solution exists: the two modes differ only by their imaginary part, so that one of them is unstable, $\omega_I < 0$, whereas the other is damped (see Figure 3.32(b)).

This evolution, via which two neutral modes with distinct real frequencies eventually merge to bring about two modes, one stable and one damped, is generic of coupled-mode flutter or binary flutter (Blevins 1990; Dowell 1995; Païdoussis 1998; Schmid & de Langre 2003). This range of coupled-mode flutter combining the wake dynamics and the cylinder dynamics (CMF in Figure 3.32) may be associated with lock-in via several features. First, the frequency of oscillation of the system deviates from the Strouhal law, as can be seen in Figure 3.32(a). Second, the existence of a new source of instability, distinct from the wake instability is expected to result in larger motion of the solid. Finally, the phase between the lift, q, and the displacement, y, shifts progressively from 0 to π (see de Langre (2006) for details).

This linear model does not directly provide predictions of the amplitude of motion of the cylinder. Yet, the high growth rate that results from coupled-mode flutter is clearly the energy source needed to cause high-amplitude motion in this range. The amplification of motion at lock-in, observed in experiments, can therefore be understood as the consequence of an amplification of energy input caused by coupled-mode flutter. This eventually results in a higher saturated amplitude when nonlinear effects are introduced. This interpretation is consistent with the analysis of the evolution of phase and modal components in the lock-in range. The linear coupled-mode flutter model therefore provides an interpretation for both the frequency evolution and for the amplification of cylinder motion in the range of lock-in.

In fact, considering lock-in in VIV as a linear instability is quite natural, when considering the exponential growth of amplitude from an initial condition, as observed in experiments by Brika & Laneville (1993). A numerical analysis of the linear stability of coupled modes involving the flow and the cylinder may be found in Cossu & Morino (2000).

3.4 Advanced Aspects

3.4.1 The issue of added mass

Added mass in vortex-induced vibrations has been the subject of several controversies, resulting in some confusion on its physical meaning, its value and even its relevance. We focus now on some of the essential technical aspects behind these controversies to show similarities and differences between approaches, and hopefully, to clarify the situation. Several distinct issues have appeared when considering added mass in relation to VIV. Can added mass still be defined when VIV occurs? If so, is there any relation to its value in inviscid still fluid? What is the meaning of a negative added mass? Finally, what must be done in practice when computing the response of systems to VIV?

3.4.1 (a) The Lighthill decomposition controversy

In Leonard & Roshko (2001) and Sarpkaya (2001), the fundamental issue of the forces acting on a moving body in flow were discussed. A summary of the discussion may be found in Williamson & Govardhan (2004). To emphasize the main argument, let us consider the two-dimensional forced motion of a circular cylinder, as in the previous sections, with $y(t)$ being its cross-flow displacement and $F_F(t)$ the resulting fluctuating lift. Using the inviscid added mass $m_A^{\text{inv}} = \rho \pi D^2 / 4$, one may define the force

$$F_V = F_F + m_A^{\text{inv}} \ddot{y}, \tag{3.52}$$

often named the "vortex lift". The central point of the controversy is whether this force, F_V, is then only dependent on $\dot{y}$, as stated by Leonard & Roshko (2001) based on a general result by Lighthill (1986), or whether it still contains inertia terms, as stated by Sarpkaya (2001) based on the solution for a moving sphere by Stokes (1851) and abundant experimental evidence. The decomposition of the lift force F_F as the sum of an inertia term $-m_A^{\text{inv}} \ddot{y}$ and a remaining force named F_V, equation (3.52), is certainly acceptable, because this is nothing but a change of variables. The discussions on amplitude and phases of F_V compared with those of F_F are therefore legitimate; see for instance Williamson & Govardhan (2004) who consider the case of jumps between branches of lock-in.

As to whether or not the remaining force F_V is only related to the cylinder velocity, there is certainly experimental evidence that F_V cannot be expressed only as a term proportional to cylinder velocity. Still it must be emphasized that F_F has only been measured for harmonic motions, of finite amplitude, so that the general form of the dependence of F_V on the displacement and its derivatives has not been established. Hence, the uncertainties that remain *vis-à-vis* the precise dependence of F_V on displacement, should not raise any doubt as to the legitimacy of using a decomposition of forces as in equation (3.52).

One should also note that this decomposition is distinct from the numerical technique whereby an inviscid added mass is subtracted from both sides of an equilibrium equation in computational fluid-structure interaction (Bélanger *et al.* 1995); this latter is a simple technique to avoid numerical instabilities of explicit numerical schemes when the added mass is larger than the physical mass of the immersed system.

3.4.1 (b) Experimental results for added mass

As discussed in Section 3.2.3, there are several ways to decompose the lift force on a cylinder forced to oscillate in flow. The lift coefficient in phase with the displacement, $\overline{C}_L \cos \varphi$, may be used to define an added mass coefficient

$$C_M = (\overline{C}_L \cos \varphi) \frac{\mathrm{Ur}^2}{2\pi^3 (y_0/D)}. \tag{3.53}$$

Figure 3.33(a) illustrates the values of C_M as a function of the reduced velocity $\mathrm{Ur} = U/f_f D$, from experiments by Gopalkrishnan (1993) in the case of a cylinder forced to move with $y(t) = y_0 \sin(2\pi f_f t)$. We use for this case the terminology of *forced* added mass coefficient. These results show that the added mass coefficient varies significantly with the reduced velocity. As expected, it starts from its value in still fluid $C_M = 1$ near $\mathrm{Ur} = 0$, where the influence of flow is negligible. It then strongly increases, before suffering a sudden change in sign near $\mathrm{Ur} = 6$, and then settles at a nearly constant value of about $C_M = -0.5$. Many other experiments confirm this evolution, displaying a noticeable effect of the amplitude of motion y_0/D which we shall not discuss here; see Sarpkaya (2004). The existence of a negative inertia coefficient is indubitable.

In the case of a cylinder free to move, a *free* added mass coefficient may also be defined from the lift force and it has been measured (Vikestad *et al.* 2000). This free added mass coefficient is shown in Figure 3.33(b) as a function of the reduced frequency defined here as $\mathrm{Ur} = U/f_s^0 D$, where f_s^0 is the frequency of oscillation in still fluid. Here again the added mass coefficient changes sign, but this time near $\mathrm{Ur} = 7$. The value of free added mass at low reduced velocity does not come close to unity, as one would expect, and the value for high velocities tends to $C_M = -1$ instead of $C_M = -0.5$.

These two sets of data may first be compared by rescaling the axis of the reduced velocity. If the velocity for both sets of data is referred to the frequency of *actual* motion, which varies with Ur, a good overlap is observed in the range of velocities near the change in sign in Figure 3.34(a). Alternatively, referring both sets of data to a *constant* frequency, independent of Ur, can be made by using now the data of Figure 3.33(a) to estimate the variation of frequency in the free motion case: the reduced velocity must be rescaled by a factor $[(1 + m^*)/(m^* + C_M)]^{1/2}$, using here $m^* = 1.65$ (Vikestad *et al.* 2000). In this alternative rescaling, the two sets of data are also found to be consistent (see Figure 3.34(b)).

The consistency between results from free or forced vibration in the range of velocities near $\mathrm{Ur} = 1/\mathrm{St}$ shows that the change in sign of the added mass coefficient is a central feature of VIV. Forces in phase with acceleration do change sign, such

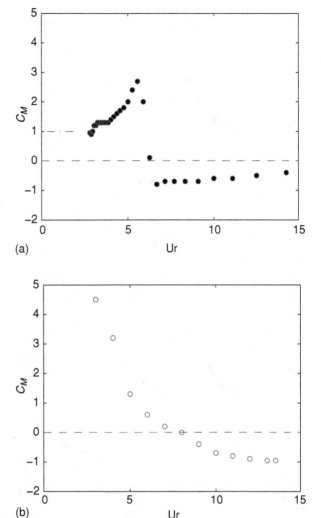

Figure 3.33. Effect of the reduced velocity on the added mass coefficient. (a) Forced added mass coefficient, from Gopalkrishnan (1993), for a cylinder under forced motion; (b) free added mass coefficient from Vikestad *et al.* (2000) for a cylinder in free motion, $m^* = 1.65$.

that they act in the same direction as the acceleration. Note that this puzzling effect is a true change of sign of the coefficient of inertia: it is not a fictitious negative added mass as is found in fluidelastic formulations when a stiffness force, proportionnal to the displacement, is expressed in terms of the acceleration in harmonic motion; in that case, a force $F = -k_A y$ becomes $F = -(-k_A/\omega^2)\ddot{y}$, so that $m_A = -k_A/\omega^2$. Here, the fluid force does not depend on y because of the invariance of the system with translation in the cross-flow direction.

As often emphasized, added mass in *still* fluid is not the mass of a physical system. It is nevertheless always positive because it is the scaling factor of the kinetic energy of the fluid in relation with the solid motion (see for instance de Langre (2002)). Here, in the case of a *moving fluid* with its own vorticity dynamics there is no *a priori* reason that the force in phase with the acceleration should be opposite or in the same direction as the acceleration. In fact, as discussed below, the change in sign in added mass can be seen as the result of the dynamics of a wake oscillator.

Considering the evolution of frequencies in VIV (Figure 3.22(b)), it is not surprising that the apparent added mass varies significantly, whether positive or negative.

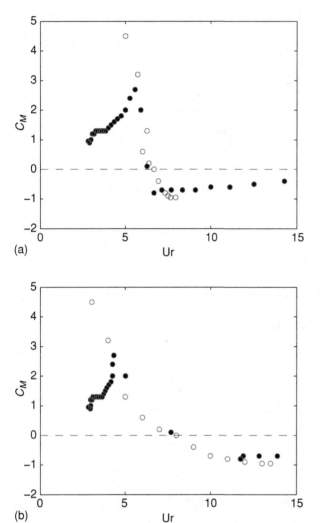

Figure 3.34. Comparison of added mass coefficients derived from free and forced motion. Experimental data from Figure 3.33. (a) Comparison using a reduced velocity based on the frequency of actual motion, as in Figure 3.33(a). (b) Comparison of both sets of data using a reduced velocity based on a constant frequency, as in Figure 3.33(b).

For instance, an increase in the frequency of motion of the solid, as observed in the lock-in process, necessarily corresponds to a decrease in added mass, because no fluid-stiffness term exists.

At this stage, added mass, free or forced, appears to have a rather complex evolution with the reduced velocity. We now examine how models of VIV can help in understanding this.

3.4.1 (c) Modelling forced added mass

In type A models, the force exerted by the fluid is composed of an inviscid inertia term and a forcing term, independent of the cylinder motion. Forced added mass, which is the part of the fluid force related to the acceleration when the latter is prescribed, is therefore exactly $C_M = C_M^{inv} = 1$. It is independent of the reduced velocity. This type of model fails to reproduce an essential feature of added mass in VIV, which is the change in sign of C_M (Figure 3.35).

For type B models, the fluid force depends explicitly on the motion of the cylinder. Experimental data on added mass, or phased lift coefficients, is precisely

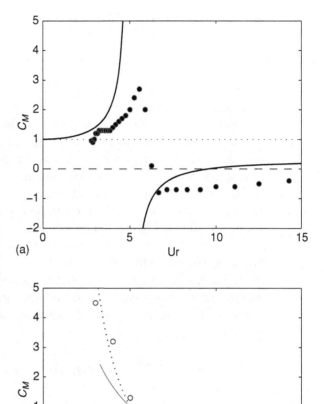

Figure 3.35. Modelling added-mass coefficients in vortex induced vibration. (a) Forced added mass coefficient: ..., type A model; −, type C model (wake oscillator), equation (3.57). (b) Free added mass coefficient: ..., type A model, equation (3.60); − , type C model (wake oscillator), equation (3.61). Experimental data as in Figure 3.33.

the input to the models, so that the comparison of models and experiments on added mass for type B models is pointless.

In type C models, where the fluid force results from the wake dynamics, forced added mass may easily be computed from the response of the wake oscillator to external forcing. We illustrate this using the much simplified linearized model of equation (3.51),

$$\ddot{q} + (\text{StUr})^2 q = (A/D)\ddot{y}. \tag{3.54}$$

Under a forced motion $y = y_0 \sin t$ (here, the timescale has been chosen the same as that of the forcing), the response of the wake oscillator is

$$q(t) = -\frac{A/D}{1 - (\text{StUr})^2}\ddot{y}. \tag{3.55}$$

Using the formulation for the total fluid force,

$$F_F(t) = -m_A^{\text{inv}}\ddot{y} + \frac{1}{4}\rho U^2 DC_L^0 q, \tag{3.56}$$

the forced added mass coefficient reads

$$C_M = 1 + \frac{A C_L^0}{4\pi^3} \frac{\mathrm{Ur}^2}{1 - (\mathrm{StUr})^2}. \tag{3.57}$$

Using the same values as before for the parameters, $A = 12$, $C_L^0 = 0.3$ and $\mathrm{St} = 0.2$, the evolution of the forced added mass coefficient is shown in Figure 3.35(a). An important new feature emerges: the change in sign of the added mass coefficient is reproduced, and clearly results from the resonance curve of the wake oscillator, in its simplest form. Hence, the sole ingredient that is needed to explain the existence of a negative added mass is a linear wake oscillator. Further refinements allow the evolution of C_M with Ur and y_0/D to be captured (see Violette (2009)), but the essential feature lies in the oscillating nature of the wake.

3.4.1 (d) Modelling free added mass
In the framework of a type A model, as given in equation (3.26), the response of the cylinder to the forcing is also straightforward. Using again a very simple case of negligible damping for the sake of clarity, we have

$$y(t) = \frac{\frac{1}{2}\rho U^2 D C_L^0}{k_S - (2\pi \mathrm{St} U/D)^2 (m_S + m_A^{\mathrm{inv}})} \sin(2\pi \mathrm{St} U t/D). \tag{3.58}$$

Considering that the fluid force is

$$F_F(t) = -m_A^{\mathrm{inv}} \ddot{y} + \frac{1}{2}\rho U^2 D C_L^0 \sin(2\pi \mathrm{St} U t/D), \tag{3.59}$$

one may derive the added mass coefficient by writing $F_F(t) = -\rho\pi D^2 C_M \ddot{y}/4$. This yields simply

$$C_M = 1 + (1 + m^*) \frac{1 - \mathrm{St}^2 \mathrm{Ur}^2}{\mathrm{St}^2 \mathrm{Ur}^2}. \tag{3.60}$$

The variation of this free added mass coefficient with Ur in type A models is shown in Figure 3.35(b), in comparison with the data of Vikestad et al. (2000). Note that the frequency of motion of the cylinder is, by the nature of the model, set to follow that given by the Strouhal law. Hence, the deviation from its original value is not, strictly speaking, due to a change in added mass but to the forcing by the Strouhal law.

Finally, free added mass may be easily computed from the linearized version of the type C model, presented in Section 3.3.4. In the range of coupled mode flutter (i.e. lock-in) the added mass coefficient may be derived from the evolution of the real part of the complex frequency,

$$C_M = 1 + (1 + m^*) \frac{1 - \omega_R^2}{\mathrm{St}^2 \mathrm{Ur}^2}. \tag{3.61}$$

Figure 3.36. Relation between free and forced added mass. Experimental data correpond to Figure 3.33(a) and (b): full circles come from forced motion experiments and open circles from free motion experiments. Model using a linearized wake oscillator, equation (3.61): — (thin line), unstable mode, which models free added mass; - -, cylinder mode which models forced added mass; — (thick line), common mode in the range of lock-in. The reduced velocity is based on a constant frequency, as in Figure 3.34(b).

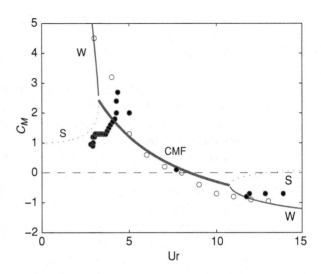

Figure 3.35(b) shows that the experimental data from Vikestad *et al.* (2000) is well represented by this. It may therefore be stated that the evolution of free added mass may be seen as resulting from the coupled dynamics of a wake oscillator and a solid oscillator.

3.4.1 (e) Modelling the differences between forced and free added mass in VIV

As shown in Figure 3.34(b), free and forced added mass may be compared when the reduced velocity is referred to the same frequency. Even though the values of added mass near $Ur = 1/St$ coincide, a significant difference is found for lower reduced velocities. For higher reduced velocities there is a difference between the data here also, although it is less apparent.

This may be understood by using again the wake oscillator approach, in its simplest linear form (Section 3.3.4 (e)). Outside the range of lock-in, modal analysis yields two modes with two distinct frequencies, ω_S and ω_W. Two values of the added mass may therefore be defined, corresponding to these two values of frequencies (see Figure 3.36). One of them corresponds to the wake mode, marked W, which will dominate in the *free* motion, as discussed by de Langre (2006). The other one is associated with the solid mode, marked S, which only contributes to the motion when the solid is *forced*. This explains why forced added mass tends to the inviscid value of $C_M = 1$ for low reduced velocities, whereas the free added mass does not. Free and forced added mass are therefore both included in the dynamics of the coupled wake and cylinder oscillators.

3.4.1 (f) Concluding remarks on added mass

As seen in the foregoing, using the added mass concept in the context of vortex-induced vibration certainly encompasses some ambiguities. Even in simple configurations, the added mass then depends on the reduced velocity and on the type of motion, free or forced.

The dependence on reduced velocity bears some relation to that of fluidelastic stiffness or damping coefficients in several other chapters of this book. It results from the existence of two distinct timescales, convection and oscillation, the ratio of which defines the reduced velocity. Because vortex shedding occurs at a timescale proportional to convection, through the Strouhal number, the dependence on reduced velocity is also found (actually the dependence is on StUr, as found in all models). It may be understood through the dynamics of a wake oscillator, which has its own frequency and transfer function.

The dependence on the type of motion found here is even more specific: forced or free motions are associated with different modes of the coupled system, and therefore different relations between cylinder and wake oscillation.

Therefore, whereas use of the concept of added mass is certainly legitimate in *representing* fluid forces in vortex-induced vibrations, it may not be appropriate for *understanding* the dynamics involved in vortex-induced vibrations: behind its rather complex evolution is hidden the dynamics of a global wake mode, which can be represented, in some aspects, by a simple wake oscillator.

3.4.2 From sectional to three-dimensional VIV

From the start of the present chapter we have focussed on the much idealized case of sectional VIV with cross-flow motion only. The phenomenology of the wake dynamics and of its coupling with motion of the section, as well as models thereof, have been developed in the literature using that framework. Yet, vortex shedding is essentially a three-dimensional phenomenon. In practice, structures in industrial applications are of many types of geometry and the flow is not always uniform; refer for instance to Furnes (2000) and Le Cunff *et al.* (2002). Some extensions of the concepts developed in the two-dimensional sectional case to more realistic configurations are certainly needed.

In Figure 3.37(a) the idealized sectional model is recalled, where motion is limited to the cross-flow direction. Still, in the two-dimensional sectional approximation, Figure 3.37(b) shows the case where the motion of the cylinder is unrestricted. The reader is referred to Jauvtis & Williamson (2003) and to references therein for a thorough discussion of the similarities and differences with the case of restricted motion; refer also to Kim & Perkins (2002) and Chaplin *et al.* (2005) for pratical applications. We emphasize here that two-dimensional unrestricted VIV (which is often referred to as XY-VIV) is a mechanism totally different from cylinder wake galloping discussed previously in Chapter 2. This is evident in both the wake phenomenology and the motion of the cylinder.

Figure 3.37(c) illustrates the case of the three-dimensional VIV of compact bluff bodies, here a sphere. In that case the wake dynamics itself is multidirectional, and the motion of the sphere in the plane transverse to the flow is much richer that for the two-dimensional case (Thompson *et al.* 2001; Provansal *et al.* 2004; Govardhan & Williamson 2005).

The most important case of a slender flexible body is illustrated in Figures 3.37(d) and (e). In the first, all characteristics of the impinging flow and of the flexible structure are supposed to be independent of the z-axis, which is that of the structure.

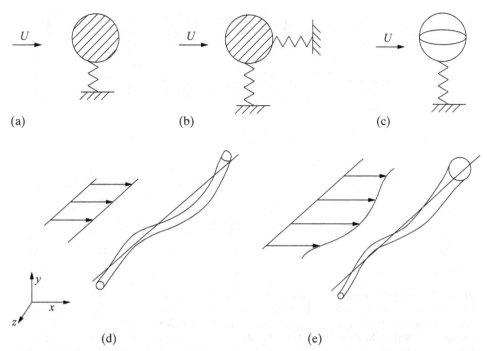

Figure 3.37. Two-dimensional and three-dimensional VIV. (a) Idealized 2-D sectional case. (b) 2-D unrestricted VIV, with motion in the lift and drag directions. (c) 3-D VIV of a compact bluff body. (d) 3-D VIV of a uniform slender body in uniform flow. (e) 3-D VIV of a nonuniform slender body in nonuniform flow.

Although this seems to be a pure extension of the sectional case, several new aspects must be taken into account. First, the wake dynamics with uniform flow on a rigid uniform cylinder is not uniform: secondary instabilities of the flow result in phase shifts and decreasing correlation of the lift force along the z-axis (see Williamson (1996)). Second, this z-dependence of the wake dynamics is modified by the motion of the cylinder. Finally, the flexible cylinder is a continuous system and therefore its dynamics cannot be reduced to a few degrees of freedom. Depending on the boundary conditions that apply at the end of the cylinder, this dynamics is better represented in a modal framework or in a wave framework.

The application of the models presented in the previous sections to this configuration of a slender flexible body raises a few questions. First, for all models the issue of spatial correlation needs somehow to be taken into account. Hopefully, when significant cylinder motion occurs, correlation is essentially the result of cylinder motion, not of the wake dynamics itself, and decorrelating the loading along the structure is not needed. Second, when several modes contribute to the motion of the cylinder, the interplay of the frequency of motion and of the wake frequency in the lock-in process becomes even more complex: two modal frequencies may exist in the range of lock-in (Willden & Graham 2004; Huera Huarte et al. 2006). In that case, defining a reduced frequency to use in type B models raises practical difficulties. These difficulties, and possible corresponding solutions, are conceptually indentical to those encountered in the use of fluidelastic coefficients when

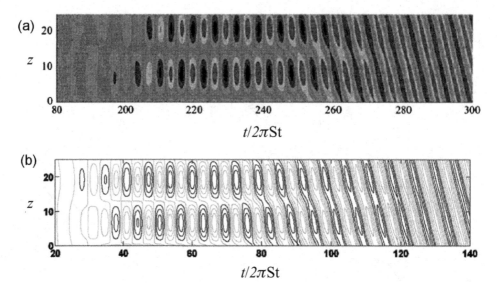

Figure 3.38. Stationary and propagating vortex-induced waves along an infinite tensionned cable in uniform flow. Evolution with time of the displacement along the cable. A transition is observed from stationary to propagating waves. (a) Motion derived from Direct Numerical Simulation (DNS) of the flow coupled with the dynamics of the cable (Newman & Karniadakis 1997). (b) Motion derived using a series of a wake oscillators coupled with the dynamics of the cable (Violette *et al.* 2007).

predicting the post-instability motion of tubes with nonlinear supports (Hadj-Sadok *et al.* 1997). Finally, when the flexible system is long (see the discussion in Vandiver (1993) for the meaning of "long") propagating waves may dominate instead of stationary waves. Computational results (Newman & Karniadakis 1997) and experimental results (Alexander 1981; Facchinetti *et al.* 2004b) have shown the existence of such vortex-induded waves (VIW). Type C models have been used to model this (Facchinetti *et al.* 2004b; Mathelin & de Langre 2005; ,Violette *et al.* 2010). Figure 3.38 shows the comparison between a full DNS computation by Newman & Karniadakis (1997) and a wake oscillator approach in the case of a uniform flow across an infinitely long flexible cable. The cable motion is seen to progressively switch from stationary to propagating waves. Here again, a wake oscillator approach clearly captures the essential features of the vortex-induced waves.

The most general case of nonuniform flow over a nonuniform flexible structure is illustrated Figure 3.37(e). Note that the case of uniform flow over a cylinder of a nonuniform section, such as a cone, falls in that category. Then, the wake dynamics itself is again further complicated, so that even the local frequency of vortex shedding from a fixed cylinder may significantly differ from that given by the Strouhal law (Griffin 1985; Papangelou 1992; Piccirillo & Van Atta 1993). This results from complex diffusion processes between sections of the wake; see for instance the models in Noack *et al.* (1991), Balasubramanian & Skop (1996) and Facchinetti *et al.* (2002). The resulting pattern of motion is certainly complex. In the case of systems with fixed boundary conditions where modes are well defined, a systematic comparison of computational methods with a set of experimental results may be found in Chaplin *et al.* (2005). A case where stationary and nonstationary waves

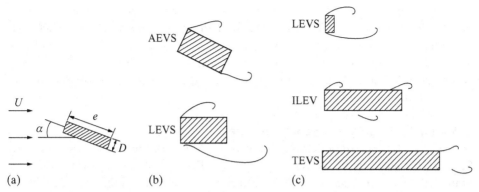

Figure 3.39. (a) Inclined rectangular cross-section; (b) vortex dynamics patterns when the angle of attack is varied; (c) vortex dynamics patterns when the aspect ratio is varied. LEVS: leading-edge vortex shedding; ILEV: impinging leading-edge vortices; TEVS : trailing-edge vortex shedding; AEVS: alternate-edge vortex shedding (Naudascher & Wang 1993; Naudascher & Rockwell 1994; Deniz & Staubli 1997).

coexist in shear flow has been computed by Lucor *et al.* (2006) and modelled by Violette *et al.* (2007, 2010).

3.4.3 VIV of noncircular cross-sections

We have limited the presentation so far to the case of circular cross-sections, because of its simplicity and its wide range of applications. Yet, as noted in Section 3.2, vortex shedding may occur in the wake of any bluff body, as it results from the instability of the flow profile created by the presence of the body itself. Hence vortex-induced vibrations can be expected in flexible slender bodies of any cross-section, and more generally for a deformable bluff body of any shape.

We only give here some of the essential differences between VIV of circular cross-sections and VIV of noncircular cross-sections. This is illustrated for the case of the rectangular cross-section, Figure 3.39(a), defined by its aspect ratio e/d and the angle of attack of the free-stream flow, α. The presentation here is based on the approach adopted by Naudascher & Wang (1993), Naudascher & Rockwell (1994) and Deniz & Staubli (1997) where a complete description of the issue is presented. References therein are not cited here for the sake of brevity, but one should recall that the work in Naudascher & Wang (1993), Naudascher & Rockwell (1994) and Deniz & Staubli (1997) is itself based on a large body of anterior work, by these authors and others.

Before exploring the issue of vortex shedding, one must keep in mind that noncircular cross-sections are prone to other types of cross-flow-induced instabilities, such as galloping described in Chapter 2. In the case of the rectangular cross-section, the aspect ratio and the angle of attack have an essential role on the possibility of instability. Even below the instability threshold, damping induced by these fluidelastic effects may be negative so that the response of the structure to vortex-induced forces may be indirectly enhanced. We shall not explore coupling between VIV and galloping in more detail: this was done in Section 2.10.

From a purely descriptive point of view, the vorticity dynamics created by a noncircular cross-section is more complex than the simple alternate vortex shedding resulting in the Bénard-Kármán vortex street as described Section 3.2. From observations, four types of vorticity dynamics have been identified (Figure 3.39(b, c)); see Naudascher & Wang (1993). They are now described in relation to their influence on the Strouhal number.

3.4.3 (a) Evolution of the Strouhal number

Let us first consider the effect of the angle of attack α, on the Strouhal number $St = fD/U$, here for the aspect ratio $e/D = 2$, Figure 3.40(a), following Deniz & Staubli (1997). For a small angle of attack, the vorticity dynamics is classically referred to as leading-edge vortex shedding (LEVS) as the vortices are shed in fact from the leading edge, Figure 3.39(b). The Strouhal number is nearly constant with α.

A change in regime is found in terms of the Strouhal number, near $\alpha = 6°$, when separation occurs at the upper edge. A new pattern of vortex shedding is observed, referred to as alternate-edge vortex shedding, or AEVS (Figure 3.39(b)). Actually, AEVS belongs to the same class of mechanism as LEVS, directly related to the wake instability. It can therefore be described in the same framework, provided the correct scaling parameters are used. For instance, in AEVS, the dependence of the Strouhal number on the angle of attack may be easily accounted for by considering that the scale of the transverse flow gradient which controls the wake instability is no longer D, but is now the cross-flow distance between the two shedding edges, $H = D \cos \alpha + e \sin \alpha$: the Strouhal number in AEVS in Figure 3.40(a) does decrease as $1/[\cos \alpha + (e/D) \sin \alpha]$.

Figure 3.40(b) shows a simplified view of the effect of the aspect ratio e/D on the Strouhal number. Several domains can be defined. First, at low aspect ratio values, $e/D < 3$, the Strouhal number is close to 0.2, consistently with what was described at the beginning of this chapter. Leading-edge vortex shedding (LEVS) is observed. At high aspect ratios, $e/D > 10$, the value is also close to 0.2, and weakly dependent on the aspect ratio. The vortices are now shed from the trailing edge, so that this is referred to as trailing-edge vortex shedding (TEVS). In these two limiting cases, the pattern of vortex shedding is essentially the same: alternate vortices resulting from the wake instability. As the wake flow profile is scaled by the transverse dimension, D, the frequency is properly described by the Strouhal number based on this dimension, fD/U, as expected. Its dependence on e/D is weak and mainly related to the corresponding modifications of the width of the wake; refer to the discussion on the universal Strouhal number in Section 3.2.1. We shall not describe these cases of LEVS, AEVS and TEVS in more detail, because they belong to the same class of vortex-shedding and VIV as the circular cross-section. Of course, the data on the fluctuating lift coefficients, fluctuating moment coefficients, limit amplitude and other parameters relevant to modelling VIV differ from the circular case; refer for instance to Deniz & Staubli (1997).

3.4.3 (b) Impinging leading-edge vortices

For intermediate values of the aspect ratio, the evolution of the Strouhal number is much more complex (see Figure 3.40(b)): discontinuities appear and even multiple

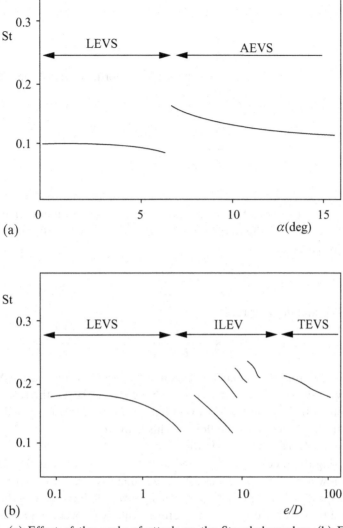

Figure 3.40. (a) Effect of the angle of attack on the Strouhal number. (b) Effect of the aspect ratio on the Strouhal number. LEVS, AEVS, ILEV and TEVS are the vorticity modes described in Figure 3.39.

values of St can be found for some aspect ratios. This suggests that a new kind of vorticity dynamics is involved, which differs from classical Bénard-Kármán vortex shedding. The vorticity dynamics is then referred to as impinging leading-edge vortices (ILEV). This may be described qualitatively as interactions between vortices created by the shear layer at the leading edge and the downstream part of the bluff body: this has led to the use of the term "impinging". It is actually a particular case of amplification of shear-layer instability through a fluid-dynamic feedback loop; see Naudascher & Rockwell (1994). This is similar in principle to the instability of a jet impacting on a wedge. The unstable hydrodynamic mode selected by this feedback loop is such that there must be an integer ratio between the distance from emission (the leading edge) to reception (the trailing edge), here e, and the wavelength of

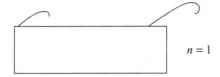

Figure 3.41. Vorticity dynamics in ILEV modes.

the mode, λ, so that $e = n\lambda$. This is illustrated Figure 3.41. Assuming that the phase velocity of the mode is U, the frequency of oscillation is $f = U\lambda$. By combining these two equations we have

$$f = \frac{Ue}{n},$$ (3.62)

so that the Strouhal number reads

$$St = \frac{fD}{U} = \frac{n}{e/D}.$$ (3.63)

In Figure 3.40(b) this behaviour corresponds to curves of decreasing St for $n = 1, 2, \ldots$ Only a portion of these materialize. Note that in ILEV the relevant length scale is no longer that of transverse flow gradients, D, but that of a specific feedback mechanism in the direction of the flow, e. This leads to a strong dependence of the Strouhal number fD/U on the e/D ratio.

Considering the mechanism involved in ILEV, it is now clear that (a) it is not expected at low aspect ratio, because it would then involve very high frequencies, (b) at large aspect ratio the number of wavelengths n involved renders the feedback loop sensitive to pertubations and is therefore unlikely to occur in practice, and (c) because the shedding of vortices is not decorrelated between the two leading edges, an alternating load can be expected. Note that ILEV is not affected by the presence of a splitter plate (Naudascher & Wang 1993), because it does not involve the wake instability, contrary to LEVS, TEVS and AEVS.

In terms of vortex-induced vibrations, the fluctuating pressure field associated with the travelling vortices on the two sides of the bluff body results in a fluctuating lift and fluctuating moment. Transverse and torsional degrees of freedom may therefore be excited. Yet, as the motion of the rectangular section affects the shedding, a more detailed analysis shows that the resulting phase between moment and angular velocity renders torsional VIV more unlikely to happen (Naudascher & Wang 1993).

3.4.3 (c) Possibilities of lock-in

Considering all these mechanisms, a chart may be drawn concerning the risk of lock-in (Figure 3.42). We only give it here for transverse vibrations. We consider, for the sake of simplicity, that lock-in occurs when the condition $StUr = 1$ is satisfied. Of course, as emphasized throughout this chapter, lock-in is not a resonance, and

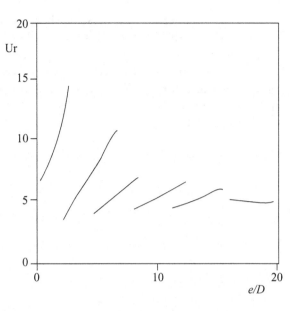

Figure 3.42. Possibilities of lock-in on a rectangular cylinder, as a function of the reduced velocity Ur = U/fD and the aspect ratio e/D; Adapted from Naudascher & Wang (1993).

significant amplitude can occur outside this strict condition. For a given aspect ratio, several successive lock-ins may clearly occur as the flow velocity is increased, in the range of e/d where ILEV is possible.

More generally, LEVS, TEVS and AEVS are true Bénard-Kármán vortex-shedding patterns and can be modelled in the same framework as for circular cylinders: all models presented in this chapter could in some way be used to predict VIV, provided the correct coefficients are used. Conversely, ILEV does not belong to the same group of vorticity dynamics. It involves the dynamics of two shear layers, each of them being influenced by the motion of a leading and a trailing edge.

3.4.4 Summary and concluding remarks

The physics of vortex-induced vibrations is rich in complexity, and modelling of VIV is not a trivial task. Yet, engineering applications, as well as fundamental issues, sustain continuing work in this field.

Whereas more and more experimental data are produced, with detailed flow measurements, more information is now also gained from computations. This continuous increase of available information on *what* happens in vortex-induced vibrations (i.e. the phenomenology) allows and requires a corresponding increase in depth of the models addressing *why* all this happens! Simply incorporating new data in models through new coefficients or new fitted functions is an endless task. As an example, one may consider the effect of the Reynolds number on the extent of lock-in: if a database of force coefficients is used, such as in type B fluidelastic models, this can only be taken into account by incorporating data at various levels of Reynolds numbers. Conversely, if the extent of lock-in can be related to more general parameters where the Reynolds number effect is known, such as the Strouhal number or the lift coefficient, as was done in Section 3.3.4, the consistency of data is ensured, as well as simplicity.

Similar issues are, in fact, present when discussing models of other mechanisms presented in this book, for instance in the case of fluidelastic instabilities of tube arrays in cross-flow: models such as those of Price & Païdoussis (1984a) and Lever & Weaver (1986) are based on some physics of the fluid dynamics that generate the pertinent forces, and are therefore more general in application than those only based on interpolating data.

4 Wake-Induced Instabilities of Pairs and Small Groups of Cylinders

4.1 The Mechanisms

The mechanisms of wake-induced instabilities of pairs and small groups of cylinders, on the one hand, and cylinder arrays involving perhaps hundreds of cylinders, on the other, are fundamentally similar. Therefore, in principle, one could conceive of a common introduction for both Chapters 4 and 5; in practice, however, there are sufficient peculiarities in both, in particular with regard to sign conventions, to make it desirable to have different introductory sections in the two chapters, despite the repetitiousness in terms of the basic ideas involved. Wake-induced flutter is discussed in Section 4.2.

As in the introductory sections for other chapters, what is presented here is highly simplified, leaving most of the complexities and refinements to the sections that follow. Specifically, following Païdoussis & Price (1988), in Section 4.1.1, a modified form of quasi-steady theory is presented, suitable for the subject matter of Chapters 4 and 5. In Section 4.1.2, the damping-controlled instability mechanism is revisited in the context of this modified quasi-steady theory.

In terms of systems to which this work can be related, the main ones are overhead transmission lines subjected to wind and clustered offshore oil and gas production risers subjected to ocean currents.

Notation used in Chapter 4	
Cylinder diameter:	D
Free-stream flow velocity:	U
Relative flow velocity:	U_r
Local wake velocity:	$\overline{U}$, defined in equation (4.2)
Reduced flow velocity:	$V = U/\omega D$
Streamwise displacement	x (positive upstream), see Figure 4.1
Transverse displacement:	y (positive downwards), see Figure 4.1
Streamwise separation between cylinders:	X, see Figure 4.5
Transverse separation between cylinders:	Y, see Figure 4.5
Mass per unit length:	m
Flow-retardation parameter	μ, see equation (4.5)

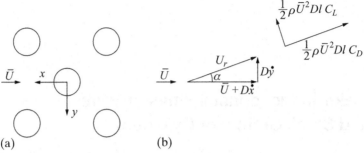

Figure 4.1. (a) Cross-sectional view of a small group of cylinders in cross-flow; (b) velocity vector diagram.

4.1.1 Modified quasi-steady theory

Consider a small group of cylinders subjected to cross-flow, as shown in Figure 4.1. For simplicity, consider for the moment that all the cylinders are rigid, except the central one, which is flexibly mounted; its motions are defined by the nondimensional displacements x and y, as shown ($x = x^*/D$, where x^* is the dimensional displacement and D the conductor diameter, and similarly for y). Conventionally, y is taken positive downwards, as in Chapter 2, and x positive in the countercurrent direction, but the fluid forces on the cylinder are positive upwards and to the right.

Hence, assuming no mechanical coupling between x- and y-direction motions, the equation of motion of the cylinder in the y-direction may be written as

$$(ml\ddot{y} + c\dot{y} + ky)\,D = F_y, \tag{4.1}$$

where F_y is the fluid-dynamic force, l is the length of the cylinder, and c and k are the effective mechanical damping and stiffness of the cylinder, respectively.

According to quasi-steady theory, the forces acting on the oscillating cylinder are approximately the same as the static forces at each point of the cylinder undergoing oscillation, provided that the approach velocity is properly adjusted to take into account the velocity of the cylinder, in the manner shown in Figure 4.1(b), where $\overline{U}$ is the local wake velocity. Thus, F_y may be written as

$$F_y = -\tfrac{1}{2}\rho U_r^2\, lD\,(C_L \cos\alpha + C_D \sin\alpha), \tag{4.2}$$

where U_r and α are defined in Figure 4.1(b) and given by

$$U_r = [(\overline{U} + D\dot{x})^2 + (D\dot{y})^2]^{1/2}, \quad \alpha = \sin^{-1}(D\dot{y}/U_r);$$

C_L and C_D are the static lift and drag coefficients, respectively, which for small motions about the equilibrium position may be expressed in linearised form as

$$C_L = C_{L_0} + (\partial C_L/\partial x)x + (\partial C_L/\partial y)y,$$

and similarly for C_D. Then, equation (4.2) may be linearised to give

$$F_y = -\tfrac{1}{2}\rho\,\overline{U}^2\,lD\left[C_{L_0} + 2C_{L_0}(D\dot{x}/\overline{U}) + (\partial C_L/\partial x)x + (\partial C_L/\partial y)y + C_{D_0}(D\dot{y}/\overline{U})\right]. \tag{4.3}$$

For a regular symmetric geometrical pattern of the cylinders, as shown in Figure 4.1(a), $C_{L_o} = 0$ and $\partial C_L/\partial x = 0$. Hence, equation (4.3) simplifies to

$$F_y = -\tfrac{1}{2}\rho\,\overline{U}^2\,lD\left[(\partial C_L/\partial y)y + C_{D_0}(D\dot{y}/\overline{U})\right]. \tag{4.4}$$

This simplification is, indeed, the main reason for choosing the symmetric system of Figure 4.1(a) in this discussion. Nevertheless, as will be seen in the sections that follow, this simplification is not generally possible, e.g. for a staggered pair of cylinders.

The discussion so far has been in terms of traditional quasi-steady fluid dynamics. It is known, however, that there is a time lag between cylinder displacement and the fluid-dynamic forces generated thereby. This may be thought to be related to the delay in the two fluid streams on either side of the cylinder readjusting to the changing configuration as the cylinder oscillates (Lever & Weaver 1982); alternatively, it may be thought to be associated with the retardation that the fluid experiences as it nears the cylinder, notably in the vicinity of a stagnation point, in conjunction with intercylinder positions having meanwhile changed as a result of cylinder motions (Simpson & Flower 1977; Price & Païdoussis 1984a, 1986a, b). As a first approximation, this time delay may be expressed as (Price & Païdoussis 1984a)

$$\tau = \mu D/\overline{U}, \tag{4.5}$$

where $\mu \sim \mathcal{O}(1)$ and μ should not be confused with the dynamic viscosity. Perhaps this time lag may most easily be conceived as a delay in the viscous wake adjusting continuously to the changing conditions imposed by the vibrating cylinder (Païdoussis *et al.* 1984; Granger & Païdoussis 1996); thus, it is associated with local viscous effects related to cylinder motion and their convection downstream, related to the time $D/\overline{U}$. Hence, taking this effect into account, and assuming harmonic motions, such that $y = y_0 \exp(i\omega t)$, equation (4.1) may be written as

$$F_y = -\tfrac{1}{2}\rho\,\overline{U}^2 lD\left[e^{-i\omega\tau}(\partial C_L/\partial y)y + C_{D_0}(D\dot{y}/\overline{U})\right]. \tag{4.6}$$

4.1.2 The damping-controlled mechanism*

Equations (4.1) and (4.6) may be combined and written in the form

$$\ddot{y} + \left[\left(\frac{\delta}{\pi}\right)\omega_n + \frac{1}{2}\left(\frac{\rho\overline{U}D}{m}\right)C_{D_0}\right]\dot{y} + \left[\omega_n^2 + \frac{1}{2}\left(\frac{\rho\overline{U}^2}{m}\right)\left(\frac{\partial C_L}{\partial y}\right)e^{-i\omega\tau}\right]y = 0, \tag{4.7}$$

where ω_n is the *in vacuo* radian natural frequency of the mechanical system and δ the logarithmic decrement. For harmonic motions, $y \propto \exp(i\omega t)$, utilizing (4.5) the damping term is found to be

$$\left[\left(\frac{\delta}{\pi}\right)\omega_n\omega + \frac{1}{2}\left(\frac{\rho\overline{U}D}{m}\right)C_{D_0}\omega - \frac{1}{2}\left(\frac{\rho\overline{U}^2}{m}\right)\left(\frac{\partial C_L}{\partial y}\right)\sin(\mu\omega D/\overline{U})\right]. \tag{4.8}$$

When this damping term becomes negative, amplified oscillatory motion will ensue, i.e. an oscillatory instability (in the linear sense) will arise. At the threshold of

---

* Originally (Païdoussis & Price 1988), this was referred to as *the negative damping mechanism* to indicate that it was a velocity-dependent, damping-controlled mechanism, in contrast to the stiffness-controlled one. However, the present description is more unambiguous.

instability, $\overline{U} = \overline{U}_c$, the damping will be zero. Hence, presuming that $\mu \omega D/\overline{U}$ is sufficiently small for $\sin(\mu \omega D/\overline{U}) \simeq \mu \omega D/\overline{U}$, one obtains

$$\frac{\overline{U}_c}{\omega_n D} = \left\{ \frac{2/\pi}{-C_{D_0} + \mu(\partial C_L/\partial y)} \right\} \frac{m\delta}{\rho D^2}, \tag{4.9}$$

which means that the critical reduced flow velocity is linearly dependent on the mass-damping parameter, $m\delta/\rho D^2$, or half the Scruton number.

Clearly, flutter can only arise if the quantity in brackets is positive, i.e. if

$$-C_{D_0} + \mu(\partial C_L/\partial y) > 0, \tag{4.10}$$

i.e. if $\partial C_L/\partial y$ is positive (noting that ΔC_L and Δy are in opposite directions) and sufficiently large. Thus, an expression similar to the criterion for galloping in Section 2.2, namely the inequality (2.9) of Glauert and Den Hartog, is obtained. In fact, it can be shown that the two are identical, as follows. Starting with (2.9),

$$C_{D_0} + \frac{\partial C_L}{\partial \alpha} < 0,$$

we first modify it for the presence of a time delay into

$$C_{D_0} + \mathcal{R}e\left[e^{-i\omega\tau}(\partial C_L/\partial \alpha)\right] < 0. \tag{4.11}$$

Next, from the definition of α, we can express y as a function of α for harmonic motions as $y = (U/i\omega D)\alpha$; hence, we can write $\partial C_L/\partial \alpha = (U/i\omega D)(\partial C_L/\partial y)$. Assuming $\omega\tau$ to be small and making use of (4.5), inequality (4.11) may be written as

$$C_{D_0} - \mu(\partial C_L/\partial y) < 0, \tag{4.12}$$

which is identical to (4.10). Hence, the negative damping instability found here is indeed a form of galloping.

The important point should be made that if there were no time delay, i.e. if $\mu = 0$, no oscillatory instability could arise; as $C_D > 0$ generally, inequality (4.10) could never be satisfied. However, there is another mechanism via which flutter can arise, which does not depend on either a time delay or negative damping, namely the wake-flutter mechanism which is discussed next.

4.1.3 The wake-flutter mechanism

This mechanism has been studied extensively in connection with bundles of overhead transmission lines, typically involving two or four conductors ("cylinders") subjected to crosswind, often referred to as *sub-span oscillations*; refer to Simpson (1971b), Price (1975a, b), Simpson & Flower (1977) and many other references cited in the sections that follow.

In the spirit of obtaining a simple physical explanation for wake-induced flutter, a mechanism leading to instability is proposed by use of a simple two-conductor model; furthermore, the windward conductor is taken to be fixed and only the leeward conductor is free to oscillate (as shown later in Section 4.2, it is the leeward of the two conductors which extracts energy from the flow). As shown schematically

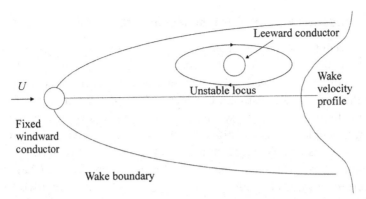

Figure 4.2. Wake flutter of a leeward conductor in the wake of another, showing a typical trajectory during flutter.

in Figure 4.2, the velocity distribution in the wake of the windward conductor is approximately parabolic, with a minimum on the wake centreline.

Considering the equilibrium position of the leeward conductor to be above the wake centreline* and supposing that it is given a small clockwise elliptical perturbation, then, as shown in Figure 4.2, the leeward conductor moves downstream in a region of flow with a greater velocity than it opposes when it moves back upstream. Assuming that the flow pattern around the leeward conductor remains roughly constant as it traverses this elliptical path implies that the conductor's drag coefficient will also remain constant, and so the variation in the leeward conductor's drag force will be governed by the wake-velocity distribution. Hence, the leeward conductor experiences a higher drag force when it moves downstream than it opposes when it moves back upstream, and so it extracts energy from the flow, causing the motion to increase in amplitude – i.e. the conductor motion is unstable. It is also apparent that, again considering the leeward conductor to be positioned above the wake centreline, an initial *anticlockwise* perturbation results in energy being dissipated by the conductor, resulting in a stable motion. So, the direction of the perturbation plays a critical role in determining whether or not instability occurs. Following the same argument, if the leeward conductor is positioned below the wake centreline, then an anticlockwise motion is required to produce instability. Finally, the wake centreline represents a position where instability cannot occur. All of these conclusions are consistent with both experimental investigations and more sophisticated analytical models; however, what is not explained by this simple model is the origin of the elliptical perturbation – this requires a more complete analysis which is presented in the next section. The most important conclusion from this simple analysis, however, is that the phenomenon producing this wake-induced instability, in its simplest form, is a *displacement-dependent* or *stiffness-controlled* mechanism, in contrast to that presented in Section 4.1.2 which is a velocity-dependent (damping-controlled) mechanism.

---

* Ideally, for a full-scale transmission line the two conductors are aligned in the horizontal plane, with the leeward conductor positioned on the wake centreline. However, in practice, the individual conductors have slightly different lengths (this may occur either in the installation of the transmission line, or due to differential creep of the conductors) and the leeward conductor may be at any transverse position in the wake of the windward conductor.

4.2 Wake-Induced Flutter of Transmission Lines

Wake-induced flutter, or sub-span oscillation as it is more usually referred to among power-line engineers, is an aeroelastic instability which occurs on "bundles" of over-head power conductors; specifically, when one or more of the conductors are positioned in the wake of an upstream conductor. In principle this type of instability, and variants thereof, can occur whenever a bluff body is positioned in the wake of another. Examples of this phenomenon producing severe vibration problems include flexible risers of offshore structures (as will be discussed in Section 4.3) and heat-exchanger tube arrays in coolant cross-flow to be discussed in Chapter 5.

The terminology "wake-induced galloping" is also sometimes used to describe this instability; however, this nomenclature is confusing, because the physical mechanisms leading to galloping and wake-induced flutter are very different. Galloping in its simplest form requires only one degree of freedom (Den Hartog 1932), and is the result of negative fluid damping; whereas, as will be shown later, wake-induced flutter requires at least two degrees of freedom and can occur despite the aerodynamic damping being positive.

Before embarking on a description of wake-induced flutter, it is worth noting that this is but one of at least four different aeroelastic phenomena which can cause vibration problems on overhead transmission lines. A summary of these phenomena is given by Simpson (1983), but briefly their characteristics are as follows. *Aeolian, or vortex-induced, vibration* is due to a frequency resonance between the oscillatory aerodynamic force, arising via the alternate shedding of vortices from the conductors, with one of the conductor structural natural frequencies; it can occur on single conductors. (A comprehensive review of vortex shedding is given in Chapter 3.) Because transmission lines have a large number of natural frequencies and harmonics, given approximately by their stretched-string equivalents, a resonance will exist for almost the entire practical range of wind speeds, and thus, this vibration is observed over a very wide range of frequencies – typically from 8 to 60 Hz. The maximum amplitude of vibration for Aeolian vibration is relatively small, on the order of one conductor diameter, but its omnipresent nature often leads to rapid fatigue problems for conductors. A related problem is *buffeting*, where a conductor is subject to periodic force excitation via vortices shed from an upstream conductor; although the amplitude of oscillation due to buffeting is typically smaller than for vortex-induced oscillations, the frequency range over which it can occur is greater, 0.1 to 60 Hz, and once again it can lead to fatigue problems. A far more dramatic aeroelastically induced vibration problem is *galloping*, which occurs on ice-coated conductors. This instability usually occurs in the fundamental mode of vibration of the transmission line, in the frequency range 0.1 to 0.5 Hz, and can have extremely large amplitudes of up to 10 m peak-to-peak – leading to clashing and "trip-outs" between individual conductors. In its most severe form, galloping is so violent that it can cause structural damage to the conductor towers. The mechanism leading to this instability was first identified by Den Hartog (1932) who showed that the ice coating on the conductors converts the symmetric circular cross-section, which does not produce a steady lift force, into an asymmetric cross-section which does. Den Hartog demonstrated that for vertical conductor motion the steady lift force induces a negative fluid-damping leading to the instability. For further information, the reader is referred to Section 2.2.

Unlike vortex-induced vibrations or galloping, *sub-span oscillation (wake-induced flutter)* can only occur on bundles of conductors, and specifically when one conductor lies in the wake of another (a sub-span is the length of conductor between either two spacers, used to physically separate the conductors, or a spacer and a tower end point). In the United Kingdom, "twin-conductor" bundles were first introduced in the early 1950s when transmission line voltages began to exceed 200 kV. At these voltages, power losses due to Corona discharge, and the resulting radio interference, become sufficiently severe to merit their reduction by the use of two or more conductors rather than one larger diameter conductor; further increases in transmission line voltage saw the advent of "quad" conductor bundles in the early 1960s. The first recorded incidents of sub-span oscillation were observed in the United Kingdom very soon after the introduction of bundled conductors (Rissone *et al.* 1968), and there are reports of sub-span oscillation first being observed in the U.S.S.R. as early as the early 1950s (Liberman 1974). In terms of vibrational amplitude and range of frequency over which it occurs, sub-span oscillation is intermediate to galloping and vortex-induced vibrations. It usually occurs in the fundamental mode, or first harmonic, of one of the sub-spans, typically 0.5 to 2.5 Hz; hence, the terminology sub-span oscillation. In its most common form, the oscillation consists of near-horizontal pairs of conductors performing antiphase motion, the locus of each conductor describing a shallow ellipse with the major axis, on the order of 10 conductor diameters peak-to-peak, inclined at a small angle to the horizontal; and so impacting between conductors is possible (Rowbottom & Aldam-Hughes 1972). Although the vibrational amplitude is not as large as for galloping, it is sufficient to cause structural damage to either the conductors themselves or to the spacers separating the conductors; in addition, although not as omnipresent as vortex-induced oscillation, sub-span oscillation can occur for sufficiently long periods to be responsible for fatigue problems on conductor bundles.

When attempting to provide a comprehensive and accurate model of wake-induced flutter there are two significant complexities which must be accounted for. The first is the unsteady nature of the aerodynamic forces acting on the individual conductors making up the bundle, the unsteadiness arising from the oscillation of the conductors. Here, it should be realised that although conductor bundles may consist of groups of two, three, four or more individual conductors, as far as the aerodynamics is concerned all bundles may be broken down into more basic configurations of single conductors, aerodynamically isolated from other conductors in the bundle, and pairs of conductors with the leeward conductor positioned in the wake of a windward one. Furthermore, the spacing between the conductors is sufficient to ensure that, except when undergoing the most severe vibrations, the windward conductor is aerodynamically isolated from the motion of the leeward one. Hence, the most basic configuration required to obtain an understanding of wake-induced flutter is a twin-conductor bundle with one of the conductors in the wake of the other (see Figure 4.2). Furthermore, although the stranding of the conductors affects their aerodynamics, as a first approximation the conductors may be considered as being "rough" circular cylinders.

The aerodynamic forces acting on the leeward of the two conductors are influenced by its own motion as well as that of the windward conductor. Although the aerodynamic forces acting on the windward conductor are not influenced by the motion of the leeward conductor, the windward conductor is coupled structurally to the leeward one via both the spacers separating the conductors and the towers supporting the conductor bundles. Hence, the windward conductor oscillates with an amplitude of motion approximately equal to the leeward one. However, it is the leeward of the two conductors which initiates the instability and extracts energy from the airflow. Consequently, as a first step towards understanding the mechanism leading to wake-induced flutter it is reasonable to consider the windward of the two conductors as fixed; this is the scenario analysed in Section 4.2.1. The complications arising from the motion of the windward conductor are discussed in Section 4.2.2.

A second significant complication in the analysis of wake-induced flutter of transmission line bundles is the three-dimensionality of the structural modes. Again, in the spirit of attempting to obtain a physical understanding of the mechanism leading to wake-induced flutter, the complexities of the three-dimensional structural dynamics can be ignored, and a two-dimensional analysis considered. In this two-dimensional analysis the two conductors are considered as rigid elements flexibly mounted so that each conductor is free to move independently in both the in-flow and transverse directions. This is the assumption made in Sections 4.2.1 and 4.2.2, and the complexities associated with the three-dimensional structural dynamics are discussed in Section 4.2.3.

4.2.1 Analysis for a fixed windward conductor

The definitive work for the analysis of wake-induced flutter of two conductors with the windward conductor fixed is due to Simpson (1970, 1971a, b). A summary of the aerodynamic aspects of this work is given first, and the structural components are reviewed later.

4.2.1 (a) Aerodynamic model for an oscillating conductor in the wake of a fixed conductor

Using a quasi-steady analysis, Simpson (1971a) obtained expressions for the aerodynamic forces acting on an oscillating conductor in the wake of a fixed windward one. The essence of Simpson's aerodynamic analysis can be seen from Figure 4.1 where, in this case, the velocity diagram is applied to the leeward conductor of Figure 4.2. The leeward conductor, located at a position where the local wake velocity (assumed to be in the same direction as the free-stream velocity) is $\overline{U}$, is given a small perturbation. The main thrust of the stability analysis is to determine whether this perturbation grows or decays, indicating instability or stability, respectively.

Expressing the conductor motion in nondimensional form, then $x = x^*/D$ and $y = y^*/D$, where x^* and y^* are the dimensional perturbation displacements in the in-flow and transverse directions, respectively, and D is the conductor diameter; the perturbation velocities of the conductor are $\dot{x}^* = \dot{x}D$ and $\dot{y}^* = \dot{y}D$. Accounting for its own motion, the resultant velocity acting on the oscillating leeward conductor is U_r, inclined at angle α to wake velocity $\overline{U}$ (see Figure 4.1(b)), where

$$U_r = [(\overline{U} + D\dot{x})^2 + (D\dot{y})^2]^{1/2}, \tag{4.13}$$

and

$$\alpha = \sin^{-1}(D\dot{y}/U_r) = \cos^{-1}((\overline{U} + D\dot{x})/U_r). \tag{4.14}$$

The expression for U_r is nonlinear in $\dot{x}$ and $\dot{y}$; however, realizing that the initial perturbation can be made very small, equation (4.13) can be linearised by neglecting second-order terms in $\dot{x}$ and $\dot{y}$, giving

$$U_r = \overline{U}(1 + D\dot{x}/\overline{U}). \tag{4.15}$$

At this point the quasi-steady assumption is invoked. In its most general form this states that the lift and drag coefficients acting on an oscillating body are unaffected by its motion, and consequently the magnitude and direction of the aerodynamic forces may be calculated using the resultant velocity vector, accounting for the motion of the body, and force coefficients measured on a static body. As discussed by both Fung (1955) and Blevins (1977b), this assumption is reasonable provided the motion of the body is not too violent; using different arguments, they both suggest that quasi-steady aerodynamics is valid provided

$$U/fD > 10, \tag{4.16}$$

where f is the frequency of oscillation, D is a characteristic length of the body (in this case, the conductor diameter) and U is the local velocity. For sub-span oscillation, where typical frequencies of oscillation are quite low, this is always easily satisfied.

Accounting for its motion, and making use of the quasi-steady assumption, it is shown that the lift and drag forces acting on the leeward conductor are rotated through angle α with respect to their alignment when it is stationary (see Figure 4.1). An additional assumption used here is that the lift force is always perpendicular to the resultant velocity vector; as will be discussed later, this may not necessarily be correct and the orientation of the lift force can change. This has a small effect on the final expressions for the aerodynamic forces acting on the oscillating leeward conductor. However, maintaining the assumption of the lift force being normal to the resultant velocity vector, and using the force and velocity diagram shown in Figure 4.1(b), the two-dimensional aerodynamic forces acting on a length l of the oscillating leeward conductor are given by

$$F_x = \tfrac{1}{2}\rho U_r^2 Dl\,(\overline{C}_L \sin\alpha - \overline{C}_D \cos a), \tag{4.17a}$$

$$F_y = \tfrac{1}{2}\rho U_r^2 Dl\,(-\overline{C}_L \cos\alpha - \overline{C}_D \sin\alpha), \tag{4.17b}$$

where ρ is the air density, and $\overline{C}_L$ and $\overline{C}_D$ are the lift and drag coefficients based on the local wake velocity $\overline{U}$ (*not* the free-stream velocity U). Making use of the expressions for $\sin\alpha$, $\cos\alpha$ and U_r given previously, the force expressions may be rewritten as

$$\begin{Bmatrix} F_x \\ F_y \end{Bmatrix} = \frac{1}{2}\rho Dl\overline{U}^2(1 + D\dot{x}/\overline{U})\begin{Bmatrix} \overline{C}_L D\dot{y}/\overline{U} - \overline{C}_D(1 + D\dot{x}/\overline{U}) \\ -\overline{C}_L(1 + D\dot{x}/\overline{U}) - \overline{C}_D D\dot{y}/\overline{U} \end{Bmatrix}. \tag{4.18}$$

Before equations (4.18) can be utilised in a stability analysis, a number of problems must be overcome. First, the aerodynamic force coefficients $\overline{C}_L$ and $\overline{C}_D$ are not known. Although possibly these could now be calculated via CFD calculations, the more usual practice, and certainly the only possibility in the early 1970s, is to obtain these coefficients experimentally; see for example Counihan (1963), Ko (1973), Cooper (1974), Price (1975a, b) and Wardlaw *et al.* (1975).* In all of these experimental investigations the force coefficients are given in terms of the free-stream velocity, U, and not the local-wake velocity $\overline{U}$ as required. However, considering for example the lift force, it can be written as either $\frac{1}{2}\rho Dl \overline{U}^2 \overline{C}_L$ or $\frac{1}{2}\rho Dl U^2 C_L$, where C_L is the lift coefficient based on the free-stream velocity U, and $\overline{C}_L$ is based on the local wake velocity $\overline{U}$; then, it is apparent that

$$\overline{U}^2 \overline{C}_L = U^2 C_L, \tag{4.19}$$

and similarly for the drag coefficient.

The second problem is that, as the leeward conductor moves, the relative separation between conductors changes, and C_L and C_D are strong functions of conductor separation. In fact, C_L and C_D vary in a nonlinear manner with conductor separation; but, realizing that the ultimate objective is to perform a linearised stability analysis, the conductor motion can be restricted to small displacements, and the expressions for C_L and C_D linearised, giving

$$\begin{aligned} C_L &= C_{L0} + xC_{Lx} + yC_{Ly}, \\ C_D &= C_{D0} + xC_{Dx} + yC_{Dy}, \end{aligned} \tag{4.20}$$

where C_{L0} and C_{D0} are the equilibrium values of C_L and C_D, and $C_{Lx} = \partial C_L / \partial x$ represents the variation of C_L with x, and similarly for C_{Ly}, C_{Dx} and C_{Dy}. The terms C_{Lx}, C_{Ly}, C_{Dx} and C_{Dy} are aerodynamic stiffness coefficients; they indicate that a static displacement of the conductor induces changes in the static forces acting on it, and consequently they are responsible for changing the total stiffness, and hence natural frequency, of the conductor as the wind speed is increased.

Making use of equations (4.19) and (4.20), the aerodynamic force expressions given by equation (4.18) can be rewritten as

$$\begin{Bmatrix} F_x \\ F_y \end{Bmatrix} = -\frac{1}{2}\rho Dl U^2 \left(\begin{Bmatrix} C_{D0} \\ C_{L0} \end{Bmatrix} + \frac{D}{\overline{U}} \begin{bmatrix} 2C_{D0} & -C_{L0} \\ 2C_{L0} & C_{D0} \end{bmatrix} \begin{Bmatrix} \dot{x} \\ \dot{y} \end{Bmatrix} + \begin{bmatrix} C_{Dx} & C_{Dy} \\ C_{Lx} & C_{Ly} \end{bmatrix} \begin{Bmatrix} x \\ y \end{Bmatrix} \right). \tag{4.21}$$

One final complication is that the value of $\overline{U}$, the local wake velocity, is generally not known. One way of estimating $\overline{U}$ is to assume that the static drag coefficient acting on the leeward conductor is constant for all positions throughout the wake (this effectively implies that the flow pattern around the conductor does not change with wake position), and that the variation of the leeward conductor's static drag is purely due to the variation in local velocity; then

$$\overline{U}^2 C_{D\infty} = U^2 C_{D0}, \tag{4.22}$$

where $C_{D\infty}$ is the leeward conductor's drag coefficient measured outside of the wake. For convenience this is written as

$$\overline{U} = bU, \tag{4.23}$$

---

* Recent attempts by Wu *et al.* (1999, 2002) using a free-streamline model to predict the aerodynamic forces for a smooth circular cylinder in the wake of another show considerable promise.

where $b = (C_{D0}/C_{D\infty})^{1/2}$. Finally, the aerodynamic forces acting on an oscillating leeward conductor may be written as

$$\begin{Bmatrix} F_x \\ F_y \end{Bmatrix} = -\frac{1}{2}\rho Dl U^2 \left(\begin{Bmatrix} C_{D0} \\ C_{L0} \end{Bmatrix} + \frac{D}{bU}\begin{bmatrix} 2C_{D0} & -C_{L0} \\ 2C_{L0} & C_{D0} \end{bmatrix}\begin{Bmatrix} \dot{x} \\ \dot{y} \end{Bmatrix} + \begin{bmatrix} C_{Dx} & C_{Dy} \\ C_{Lx} & C_{Ly} \end{bmatrix}\begin{Bmatrix} x \\ y \end{Bmatrix} \right).$$

(4.24)

It should be emphasised that three significant assumptions were employed in the derivation of equation (4.24). First, quasi-steady aerodynamics were employed; second, small conductor motion was assumed so that the expressions could be linearised; and finally, it was assumed that the static lift force acting on the leeward conductor is always normal to the resultant velocity vector.

The terms $-\frac{1}{2}\rho Dl U^2\{C_{D0}, C_{L0}\}^{\mathrm{T}}$ in equation (4.24) represent the static aerodynamic forces acting on the conductor, and they are responsible for establishing its static equilibrium position. The matrix terms proportional to $\dot{x}$ and $\dot{y}$ are aerodynamic damping terms, which change the overall damping of the system and are linearly proportional to the free-stream velocity U. The matrix terms proportional to x and y are aerodynamic stiffness terms, which change the overall stiffness and hence natural frequencies of the system; they are proportional to U^2. Neither the static aerodynamic forces nor the aerodynamic stiffness terms depend on b, hence, only the damping terms require that the local wake velocity, $\overline{U}$, be known.

The aerodynamic forces given in equation (4.24) can be used in a full stability analysis of twin-conductor bundles, and this is done later in this section; however, useful information concerning possible instability mechanisms can be obtained by examining the damping matrix. In particular, if the damping matrix is positive definite, then a damping-initiated instability is not possible. To determine whether or not the damping matrix is positive definite it is first decomposed into its symmetric and skew-symmetric components as shown below:

$$\begin{bmatrix} 2C_{D0} & -C_{L0} \\ 2C_{L0} & C_{D0} \end{bmatrix} = \begin{bmatrix} 2C_{D0} & C_{L0}/2 \\ C_{L0}/2 & C_{D0} \end{bmatrix} + \begin{bmatrix} 0 & -3C_{L0}/2 \\ 3C_{L0}/2 & 0 \end{bmatrix}.$$

(4.25)

The skew-symmetric components play no role in determining whether or not the matrix is positive definite; hence, they can be ignored. The test for positive definiteness is that all minor determinants of the symmetric components should be positive. Hence, the damping matrix is positive definite provided that:

$$C_{D0} > 0 \quad \text{and} \quad 2C_{D0}^2 - C_{L0}^2/4 > 0.$$

(4.26)

These conditions are satisfied for all practical spacings of twin-conductor bundles, indicating that a damping-controlled instability is not possible. (However, as shown by Zdravkovich (1977), Price & Païdoussis (1984b) and Bokaian & Geoola (1984a), for very small values of in-line separation between pairs of circular cylinders, typically less than approximately 3 cylinder diameters, C_{D0} can be very small and even negative and C_{L0} can be very large; hence, a damping-controlled instability is possible. But the small value of in-line separation required to produce this is well below what is normally encountered in overhead transmission lines.) Hence, if instability is to occur it must be from sources other than aerodynamic damping. Using a slightly different explanation, exactly the same conclusion was obtained by Rawlins (1976).

Origin of the lift force and its effect on the damping matrix. The aerodynamic forces acting on a leeward conductor oscillating in the wake of a fixed windward conductor, as given by equation (4.24), rely on the assumption that the static lift force always acts perpendicular to the resultant velocity vector. In fact, even for a fixed leeward conductor the origin of its static lift is uncertain, and a number of different mechanisms have been proposed.

Maekawa (1964) was the first to observe that the lift force on the leeward of two cylinders acts towards the wake centreline. As pointed out by Maekawa, this is the opposite of what is expected from a simple application of inviscid flow theory and the velocity distribution across the wake. Maekawa suggested that the lift was due to "buoyancy" resulting from the variation of static pressure across the wake. Using Taylor's (1947) analysis for a body in a shear flow, Savkar (1970) developed an expression for the lift force on a cylinder in the wake of another, but, similarly to the inviscid analysis of Maekawa, it acted in the wrong direction, i.e. away from the wake centreline. Mair & Maull (1971) suggested that the lift force was in fact nothing more than a resolved component of the drag force, which they concluded was directed towards the centreline. Rawlins (1974a) resorted to the use of inviscid flow theory but, accounting for the different vorticity shed by the two boundary layers separating from the leeward cylinder, obtained an expression which was exactly the negative of Savkar's – so acting in the right direction. Price (1975b, 1976) conducted a series of experiments to evaluate the relative contributions from these different mechanisms; he found that, although all of them had some contribution, none of them could be said to be dominant.

The uncertainty in specifying exactly what is the mechanism leading to the lift force does have an effect on the expressions for the aerodynamic forces acting on an oscillating leeward conductor. The expressions given in equation (4.24) are based on the assumption that the static lift force always acts perpendicular to the resultant velocity vector. Simpson & Price (1974) and Price (1975b) derived alternative expressions for these aerodynamic forces based on other hypotheses. If it is assumed that the lift force is purely due to buoyancy and, hence, always normal to the wake centreline, the following expressions are obtained:

$$\begin{Bmatrix} F_x \\ F_y \end{Bmatrix} = -\frac{1}{2}\rho D l U^2 \left(\begin{Bmatrix} C_{D0} \\ C_{L0} \end{Bmatrix} + \frac{D}{bU} \begin{bmatrix} 2C_{D0} & 0 \\ 0 & C_{D0} \end{bmatrix} \begin{Bmatrix} \dot{x} \\ \dot{y} \end{Bmatrix} + \begin{bmatrix} C_{Dx} & C_{Dy} \\ C_{Lx} & C_{Ly} \end{bmatrix} \begin{Bmatrix} x \\ y \end{Bmatrix} \right).$$

(4.27)

The only difference between these new expressions for the aerodynamic forces and those given by equation (4.24) is in the off-diagonal terms of the damping matrix. Simpson & Price (1974) and Price (1975b) also developed the following expressions for the aerodynamic forces based on the assumption that the lift force is purely due to resolved drag:

$$\begin{Bmatrix} F_x \\ F_y \end{Bmatrix} = -\frac{1}{2}\rho D l U^2 \left(\begin{Bmatrix} C_{D0} \\ C_{L0} \end{Bmatrix} + \frac{(C_D C_{D\infty})^{1/2} D}{(1+r^2)^{3/4} U} \begin{bmatrix} 2+r^2 & r \\ r & 1+2r^2 \end{bmatrix} \begin{Bmatrix} \dot{x} \\ \dot{y} \end{Bmatrix} \right.$$

$$\left. + \begin{bmatrix} C_{Dx} & C_{Dy} \\ C_{Lx} & C_{Ly} \end{bmatrix} \begin{Bmatrix} x \\ y \end{Bmatrix} \right),$$

(4.28)

where $r = C_{L0}/C_{D0}$ and $C_{D\infty}$ is the free-stream drag coefficient. There is considerable modification of the aerodynamic damping matrix *vis-à-vis* that given by either equation (4.24) or (4.27), but once again there is no difference in the stiffness matrix.

Although the uncertainty in the origin of the lift force results in different forms for the aerodynamic damping matrix, the instability is not due to negative aerodynamic damping. As will be shown later, it is governed by the aerodynamic stiffness matrix, which is unaffected by the choice of mechanism leading to the lift force, and so this uncertainty in the origin of the lift force is not particularly serious. (Provided C_{D0} is positive, both the resolved drag and buoyancy assumptions lead to damping matrices which are positive definite.)

4.2.1 (b) Structural model

In addition to the aerodynamic forces, as derived in Section 4.2.1(a), a structural model is required for the leeward conductor before a stability analysis can be performed. The structural dynamics of transmission lines, accounting for the inertial and stiffness properties of the conductors and spacers making up the bundles, plus possibly the stiffness properties of the transmission line towers, is an extremely complex problem, which is discussed in Section 4.2.3. In the analysis presented here, and in the papers cited in the remainder of this section, the objective is not to develop an analytical model which accurately predicts the natural frequencies and stability of transmission lines, but instead to develop a structural model which mimics the relevant parameters which are important *vis-à-vis* wake-induced flutter. As shown previously, an elliptical motion is required for the leeward conductor to extract energy from the wake. However, this elliptical motion is not a natural structural mode for a transmission line; it occurs as a result of an aeroelastically induced coalescence of two modes.

The simplest case of wake-induced flutter occurs on horizontal twin-conductor bundles where the two structural modes which coalesce to produce the unstable mode are so-called in-plane and out-of-plane modes, which oscillate in or out of the plane, respectively, of the catenary in which the transmission line hangs (see Figure 4.3). Because of the static drag force acting on it as it hangs as a catenary, the transmission line is swept back and inclined to the vertical by an amount known as the "blow-back" angle. This is important because the in-plane and out-of-plane modes are now no longer in the vertical and horizontal directions, respectively (see Figure 4.3).

In a full-span mode the conductor bundle oscillates with no relative motion between conductors (Figure 4.4(a)). To a first-order approximation its natural frequencies are given by the stretched string frequencies of the conductor length between end supports. As there is no relative motion between conductors, the effect of spacer stiffness and damping on this type of mode is negligible. In addition, although a spacer may move in a full-span oscillation, its mass is typically less than 1% of the total conductor mass, and so it may also be neglected. Thus, it can be said that spacers have little or no effect on the natural frequencies, mode shapes or damping levels of full-span modes.

For a sub-span mode the conductors oscillate out of phase with each other (Figure 4.4(b)). To a first-order approximation, the natural frequency is given by the stretched string frequency of the sub-span length between conductor spacers. If

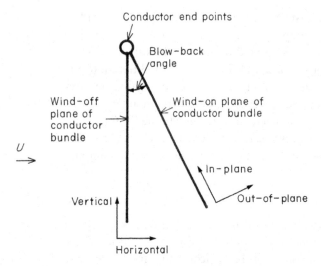

Figure 4.3. Cross-sectional view of the in-plane and out-of-plane directions of oscillation of a transmission line when subject to a static drag force.

the spacers are rigid, then no conductor motion is propagated to adjacent sub-span lengths, and the spacers effectively force nodes at their locations. Flexible spacers, however, allow conductor motion to permeate over a number of sub-spans and their locations no longer dictate the nodal positions. In addition, if the spacers are constructed using dissipative elements, then they may introduce mechanical damping into the sub-span modes. Thus, the construction and positioning of spacers has a very strong effect on the frequency, damping and shape of sub-span modes.

For quad- (i.e. four-conductor) or other multiconductor bundles the situation is more complex. Because of the larger number of conductors, there is an increase in the number of sub-span modes and, because of the spacer geometry, both in-plane and out-of-plane modes may be sub-span modes. Thus, there are a far greater number of modes which may coalesce to become unstable and the effect of spacers is not as straightforward as for twin-conductor bundles.

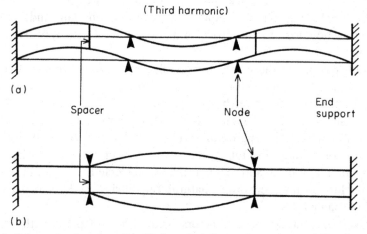

Figure 4.4. Schematic of modes of oscillation of a twin-conductor transmission line with three sub-spans: (a) a full-span mode; (b) a sub-span mode.

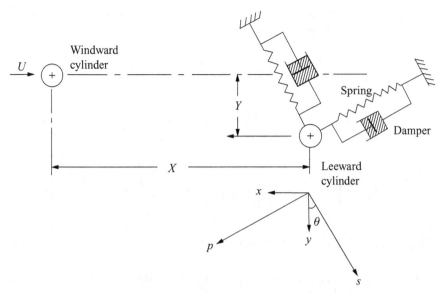

Figure 4.5. Schematic for the mechanical supports of the leeward conductor.

One thing which is common to both twin and quad bundles is that the two structural modes which coalesce to become unstable will almost certainly have different natural frequencies, and their coalescence is a direct result of the aerodynamic stiffness terms acting on the leeward conductor given in equations (4.24), (4.27) or (4.28). The manner in which these modes coalesce is very similar to that of the wing bending and torsional mode coalescence in classical aircraft wing flutter (Fung 1955); thus, the use of the terminology *wake-induced flutter*.

The first derivation of a simple but representative two-dimensional structural model for wake-induced flutter was given by Simpson (1971b) who did not account for mechanical damping. For completeness, the equations developed here do include the mechanical damping terms. The analytical model used to represent the structural support of the leeward conductor is shown schematically in Figure 4.5. The conductor is mounted via an orthogonal spring-damper system, with principal axes representing the in- and out-of-plane directions; furthermore, the mounting system may be inclined at any angle θ to the vertical, representing the conductor blow-back.

If f_p and δ_p are, respectively, the cylinder natural frequency and logarithmic decrement in the p-direction, then the equation of motion in that direction may be written as

$$(ml\ddot{p} + c_p\dot{p} + k_p p)D = F_p, \tag{4.29}$$

where m is the conductor mass per unit length, $c_p = 2\delta_p[k_p ml/(4\pi^2 + \delta_p^2)]^{1/2}$ is the viscous damping coefficient[*], $k_p = 4\pi^2 ml f_p^2$ is the effective stiffness of the conductor-mounting system, F_p is the external force in the p-direction arising from the aerodynamics and p is the nondimensional displacement of the conductor. A similar expression may also be obtained in the s-direction.

---

[*] Recall that $\delta_p = \zeta_p(4\pi^2 + \delta_p^2)^{1/2}$, where $2\zeta_p \omega_n = c_p/ml$ and $\omega_n = (k_p/ml)^{1/2}$.

The p- and s-equations may easily be transformed into the (x, y)-coordinate system by use of a transformation matrix $[T]$, where

$$\{p, s\}^{\mathrm{T}} = [T]\{x, y\}^{\mathrm{T}},$$

and

$$[T] = \begin{bmatrix} \cos\theta & \sin\theta \\ -\sin\theta & \cos\theta \end{bmatrix}, \tag{4.30a}$$

giving

$$\left\{ [M]\begin{Bmatrix} \ddot{x} \\ \ddot{y} \end{Bmatrix} + [C]\begin{Bmatrix} \dot{x} \\ \dot{y} \end{Bmatrix} + [K]\begin{Bmatrix} x \\ y \end{Bmatrix} \right\} D = \begin{Bmatrix} F_x \\ F_y \end{Bmatrix}, \tag{4.30b}$$

where $[M] = ml\begin{bmatrix} 1 & 0 \\ 0 & 1 \end{bmatrix}$ is the structural mass matrix,

$[C] = [T]^{-1}\begin{bmatrix} c_p & 0 \\ 0 & c_s \end{bmatrix}[T]$ is the structural damping matrix,

$[K] = [T]^{-1}\begin{bmatrix} k_p & 0 \\ 0 & k_s \end{bmatrix}[T]$ is the structural stiffness matrix, and

$\{F_x, F_y\}^{\mathrm{T}}$ is the vector of external aerodynamic forces given by equations (4.24), (4.27) or (4.28).

It should be stressed that this structural simulation is strictly limited to situations where there is no structural coupling between the in-plane and out-of-plane modes; this is a reasonable assumption for twin-conductor bundles, but is unlikely to be correct for quad bundles. If structural coupling does exist between these modes, then the form of the structural equations will be significantly more complex than those presented in equation (4.30b).

Simpson (1971b) used an alternative, and more general, method to determine the structural stiffness matrix (he did not consider structural damping). If ω_1 and ω_2 are the two natural frequencies of the structural system with mode shapes given by $\{1, g\}$ and $\{-g, 1\}^*$, respectively, then as shown by Simpson (1971b) the stiffness matrix is

$$\begin{bmatrix} K_{xx} & K_{xy} \\ K_{yx} & K_{yy} \end{bmatrix},$$

where

$$K_{xx} = \frac{ml\left(\omega_1^2 + g^2\omega_2^2\right)}{1 + g^2}, \quad \frac{K_{yy}}{K_{xx}} = k^2 = \frac{\omega_2^2 + g^2\omega_1^2}{\omega_1^2 + g^2\omega_2^2} \quad \text{and} \quad \frac{K_{xy}}{K_{xx}} = \frac{K_{yx}}{K_{xx}} = \varepsilon = \frac{g\left(\omega_1^2 - \omega_2^2\right)}{\omega_1^2 + g^2\omega_2^2}. \tag{4.31}$$

The advantage of this method is that it enables the structural stiffness matrix to be formulated no matter what the source of coupling between the x- and y-directions, provided g, which is effectively a measure of this structural coupling, is known. If

---

* Hence, modes ω_1 and ω_2 are sensibly the x- and y-direction modes.

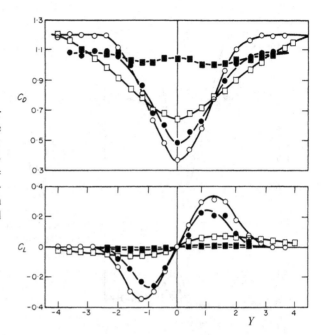

Figure 4.6. Static force coefficients for a smooth circular cylinder in the wake of another with free-stream Re $= 2.97 \times 10^4$ (Price 1975a): $\square$, $X = 19.5$ and 1.5% free-stream turbulence intensity; $\blacksquare$, $X = 19.5$ and 11% free-stream turbulence intensity; $\circ$, $X = 5.8$ and 1.5% free-stream turbulence intensity; $\bullet$, $X = 5.8$ and 11% free-stream turbulence intensity.

we restrict the analysis to those cases where the in-plane and out-of-plane modes are uncoupled – hence, coupling between the x- and y-directions is purely due to conductor blow-back – then it is possible to relate g to θ, with positive θ corresponding to positive g (and negative θ corresponding to negative g). The reader is reminded that physically only a positive θ can exist.

4.2.1 (c) Typical results

A number of authors have presented analytical and numerical solutions for the dynamic stability of one flexibly mounted conductor in the wake of a fixed windward one. An essential prerequisite for all of these solutions is that extensive aerodynamic data be available, in particular, the static lift and drag coefficients for the leeward conductor and how they vary with the conductor's wake position. This type of data has been measured by a number of different researchers, for example Counihan (1963), Ko (1973), Cooper (1974), Price (1975a, b) and Wardlaw *et al.* (1975).

Price (1975a, b) measured the aerodynamic data for a number of different cross-sectional shapes, including smooth circular cylinders (at Re $= 3.6 \times 10^4$ with free-stream turbulence intensities of 1.5% and 10%) and for two standard CEGB (the U.K. Central Electricity Generating Board) "conductors" with outer layers of either 24 or 18 strands. (The conductors were constructed from rigid cores to maintain the straightness of the two-dimensional models, with strands wrapped around the cores to mimic real conductors.) Some sample data from Price (1975a, b) are shown in Figures 4.6 and 4.7. In all cases there is a significant variation of C_L and C_D with Y, with the C_D curves being symmetric about the wake centreline and the C_L curves being antisymmetric. Using the complete data-set given by Price (1975a, b) the static aerodynamic force coefficients (C_{L0} and C_{D0}) and the aerodynamic stiffness terms

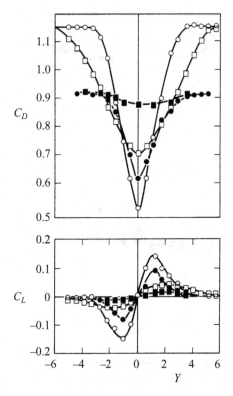

Figure 4.7. Static force coefficients for an 18-strand conductor in the wake of a similar conductor (Price 1975a): ●, $X = 8.4$, Re $= 2.1 \times 10^4$ and 10% free-stream turbulence intensity; ■, $X = 25.4$, Re $= 2.1 \times 10^4$ and 10% free-stream turbulence intensity; ○, $X = 8.4$, Re $= 2.5 \times 10^4$ and 2% free-stream turbulence intensity; □, $X = 25.4$, Re $= 2.5 \times 10^4$ and 2% free-stream turbulence intensity.

(C_{Lx}, C_{Ly}, C_{Dx} and C_{Dy}) can be obtained at any position in the wake; hence, the aerodynamic forces acting on a leeward conductor, given by equations (4.24), (4.27) or (4.28), can be evaluated.

Price (1975a) used two different methods of solution to evaluate the dynamic stability of the leeward conductor; in both of these methods, the structural damping was ignored. (Unless external damping devices are added to a transmission line its structural damping is very small, and so this is a reasonable approximation.) Hence, using the form of the aerodynamics given by equation (4.24), equation (4.30b) may be rewritten as

$$[I] \left\{ \begin{matrix} \ddot{x} \\ \ddot{y} \end{matrix} \right\} + \frac{\frac{1}{2}\rho DU}{mb} \left[\begin{matrix} 2C_{D0} & -C_{L0} \\ 2C_{L0} & C_{D0} \end{matrix} \right] \left\{ \begin{matrix} \dot{x} \\ \dot{y} \end{matrix} \right\} + \frac{1}{ml} \left[\begin{matrix} K_{xx} & K_{xy} \\ K_{yx} & K_{yy} \end{matrix} \right] \left\{ \begin{matrix} x \\ y \end{matrix} \right\}$$

$$+ \frac{\frac{1}{2}\rho U^2}{m} \left[\begin{matrix} C_{Dx} & C_{Dy} \\ C_{Lx} & C_{Ly} \end{matrix} \right] \left\{ \begin{matrix} x \\ y \end{matrix} \right\} = -\frac{\frac{1}{2}\rho U^2}{m} \left\{ \begin{matrix} C_{D0} \\ C_{L0} \end{matrix} \right\}. \qquad (4.32)$$

Considering only the homogeneous part of the equations, which determines the stability of the system, and assuming solutions of the form

$$x = x_o \exp(\lambda Ut/D) \quad \text{and} \quad y = y_o \exp(\lambda Ut/D), \qquad (4.33)$$

the following expressions are obtained:

$$\left[\begin{matrix} \lambda^2 + 2\lambda n C_{D0}/b + \chi + n C_{Dx} & -\lambda n C_{L0}/b + \varepsilon\chi + n C_{Dy} \\ 2\lambda n C_{L0}/b + \varepsilon\chi + n C_{Lx} & \lambda^2 + \lambda n C_{D0}/b + k^2\chi + n C_{Ly} \end{matrix} \right] \left\{ \begin{matrix} x_o \\ y_o \end{matrix} \right\} = 0, \qquad (4.34)$$

where

$$n = \frac{\rho D^2}{2m}, \quad \chi = K_{xx}\frac{D^2}{U^2 ml}, \quad k^2 = \frac{K_{yy}}{K_{xx}} \quad \text{and} \quad \varepsilon = \frac{K_{xy}}{K_{xx}} = \frac{K_{yx}}{K_{xx}}.$$

For a nontrivial solution of equation (4.34) the determinant of the matrix must be zero, which gives the following characteristic equation for the eigenvalues λ:

$$\lambda^4 + h_3\lambda^3 + h_2\lambda^2 + h_1\lambda + h_0 = 0, \tag{4.35}$$

where

$h_3 = 3nC_{Do}/b,$

$h_2 = \chi(1 + k^2) + n(C_{Ly} + C_{Dx}) + 2n^2\left(C_{L0}^2 + C_{D0}^2\right)/b^2,$

$h_1 = nC_{D0}(2k^2\chi + 2nC_{Ly} + \chi + nC_{Dx})/b + nC_{L0}(nC_{Lx} - 2nC_{Dy} - \varepsilon\chi)/b,$

$h_0 = (\chi + nC_{Dx})(k^2\chi + nC_{Ly}) - (\varepsilon\chi + nC_{Dy})(\varepsilon\chi + nC_{Lx}).$

The stability of the system may now be determined using Routh's stability criteria, which indicate that dynamic instability will occur if

$$T_3 = h_1 h_2 h_3 - h_1^2 - h_0 h_3^2 \leq 0. \tag{4.36}$$

Although equation (4.36) can easily be solved numerically, an alternative method of solution proposed by Simpson (1971a) was to ignore the aerodynamic damping terms (the structural damping had already been ignored). If this is done, because *all* the damping terms have now been set to zero, the Routh test determinant T_3 is automatically zero and, hence, it no longer gives a useful measure of stability. An alternative approach is to realise that a stability boundary will occur when there is a frequency coalescence of the solutions of the revised characteristic equation

$$\lambda^4 + \bar{h}_2\lambda^2 + h_0 = 0, \tag{4.37}$$

where

$$\bar{h}_2 = \chi(1 + k^2) + n(C_{Ly} + C_{Dx}),$$

$$h_0 = (\chi + nC_{Dx})(k^2\chi + nC_{Ly}) - (\varepsilon\chi + nC_{Dy})(\varepsilon\chi + nC_{Lx}).$$

Frequency coalescence is obtained when

$$(\bar{h}_2)^2 - 4h_0 = 0, \tag{4.38}$$

which gives the following quadratic in $\chi = K_{xx}[D^2/U^2ml]$:

$$\chi^2((1 - k^2)^2 + 4\varepsilon^2) + 2\chi n((1 - k^2)(C_{Dx} - C_{Ly}) + 2\varepsilon(C_{Lx} + C_{Dy}))$$

$$+ n^2(C_{Dx} - C_{Ly})^2 + 4n^2C_{Lx}C_{Dy} = 0. \tag{4.39}$$

Equation (4.39) gives two values of χ, and hence airspeed, corresponding to the stability boundaries; the system being stable for all airspeeds below the lower boundary and above the upper boundary, and unstable for airspeeds between these two values.

There are two advantages to evaluating the stability of the system by use of equation (4.39) rather than equation (4.36). First, equation (4.39) is a simple quadratic in χ, and so the stability boundaries may be obtained very easily without recourse to computationally intensive methods. However, the most significant advantage is

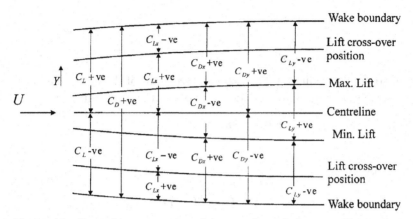

Figure 4.8. General representation of the aerodynamic force coefficients for one cylinder in the wake of another.

realised when it is recognised that physically meaningful solutions to equation (4.39) require that its roots be real and positive. This leads to the condition that instability will only occur provided the following condition is satisfied:

$$\varepsilon^2((C_{Lx} - C_{Dy})^2 - (C_{Dx} - C_{Ly})^2) + (1 - k^2)(\varepsilon(C_{Dx} - C_{Ly})(C_{Lx} + C_{Dy})$$
$$- (1 - k^2)C_{Lx}C_{Dy}) \geq 0. \tag{4.40}$$

Making use of equation (4.40), it is possible to draw a number of conclusions concerning the requirements for flutter to occur on different mechanical systems. Price (1975a, b) presents three examples, two of which are discussed in the following. To help in the interpretation of these results it should be realised that, as indicated in Figure 4.8, the wake may be divided into six distinct regions, each of which has different combinations of sign for the aerodynamic stiffness coefficients C_{Lx}, C_{Ly}, C_{Dx} and C_{Dy}*.

(i) First example: an uncoupled mechanical system. If the in-plane and out-of-plane modes are structurally uncoupled, as would be obtained with zero blow-back on a twin-conductor bundle, then $g = 0$ and from equation (4.31) $\varepsilon = 0$ and $k^2 = \omega_y^2/\omega_x^2$. Hence, equation (4.40) reduces to

$$C_{Lx}C_{Dy} \leq 0.$$

Furthermore, the solution to equation (4.39) is given by

$$\chi = n \left[\frac{(C_{Ly} - C_{Dx}) \pm 2\sqrt{-C_{Lx}C_{Dy}}}{(1 - k^2)} \right]. \tag{4.41}$$

In the positive Y half of the wake, C_{Dy} is positive everywhere (see Figure 4.8), and so flutter can only occur when the conductor is positioned in the outer region of the wake, beyond the lift cross-over point where C_{Lx} is negative and, in general, of

---

* When considering Figure 4.8 it should be remembered that a forward movement of the leeward conductor, +ve x, results in a decrease in separation between conductors, −ve ΔX; whereas a downward movement , +ve y, results in a +ve ΔY.

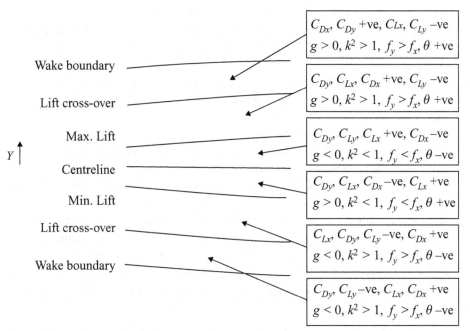

C_{Dx}, C_{Dy} +ve, C_{Lx}, C_{Ly} −ve
$g > 0, k^2 > 1, f_y > f_x, \theta$ +ve

C_{Dy}, C_{Lx}, C_{Dx} +ve, C_{Ly} −ve
$g > 0, k^2 > 1, f_y > f_x, \theta$ +ve

C_{Dy}, C_{Ly}, C_{Lx} +ve, C_{Dx} −ve
$g < 0, k^2 < 1, f_y < f_x, \theta$ −ve

C_{Dy}, C_{Lx}, C_{Dx} −ve, C_{Lx} +ve
$g > 0, k^2 < 1, f_y < f_x, \theta$ +ve

C_{Lx}, C_{Dy}, C_{Ly} −ve, C_{Dx} +ve
$g < 0, k^2 > 1, f_y > f_x, \theta$ −ve

C_{Dy}, C_{Ly} −ve, C_{Lx}, C_{Dx} +ve
$g < 0, k^2 > 1, f_y > f_x, \theta$ −ve

Wake boundary

Lift cross-over

Max. Lift

Y

Centreline

Min. Lift

Lift cross-over

Wake boundary

Figure 4.9. Undamped flutter prediction for a typical transmission line simulation.

small magnitude; furthermore, in this region C_{Dx} is positive and C_{Ly} negative, and so flutter is only possible if $k^2 > 1$ or $\omega_y > \omega_x$. For the negative Y half of the wake using similar arguments it can be shown that flutter is again possible only beyond the lift cross-over point, but in this case the requirement is that $\omega_y < \omega_x$.

(ii) Second example: transmission line simulation. Assuming that the blow-back angle θ is small, which implies that g is also small, then equation (4.31) may be written to order g as

$$K_{xx} \approx ml\omega_1^2, \quad k^2 = \frac{K_{yy}}{K_{xx}} \approx \frac{\omega_2^2}{\omega_1^2} \quad \text{and} \quad \varepsilon = \frac{K_{xy}}{K_{xx}} = \frac{K_{yx}}{K_{xx}} \approx g(1 - k^2); \quad (4.42)$$

hence, the ω_1 mode is predominantly in the x-direction, and ω_2 is in the y-direction. The condition for flutter, equation (4.40), then reduces to

$$g\,(C_{Dx} - C_{Ly})\,(C_{Lx} + C_{Dy}) - C_{Lx}C_{Dy} \geq 0, \quad (4.43)$$

and the critical flutter speeds are given by

$$\chi = \frac{n}{1-k^2}\left[C_{Ly} - C_{Dx} - 2g\,(C_{Lx} + C_{Dy}) \pm 2\,(g\,(C_{Dx} - C_{Ly})\,(C_{Lx} + C_{Dy}) - C_{Lx}C_{Dy})^{1/2}\right].$$
$$(4.44)$$

To clarify the different possible solutions available from equation (4.44) the wake is divided into seven different regions (see Figure 4.9), which are discussed in the following.

a. On the wake centreline $C_{Dy} = C_{Lx} = 0$, and so there is no possibility of flutter.
b. Between the centreline and the lift peak on the positive Y side of the wake, C_{Dy}, C_{Ly} and C_{Lx} are positive, whereas C_{Dx} is negative. Hence, equation (4.43)

can be satisfied only if g is negative; switching to the negative Y side of the wake results in a change of sign for C_{Lx} and C_{Dy}, and so flutter is possible only when g is positive. The greater the magnitude of g (greater blow-back, or coupling between the x- and y-directions), the more likely it is that the magnitude of $g(C_{Dx} - C_{Ly})(C_{Lx} + C_{Dy})$ will be greater than $C_{Lx}C_{Dy}$ (which is negative); hence, the greater the mechanical coupling between the modes, the more likely it is that flutter will occur.

c. Between the lift peak and the lift cross-over point on the positive Y side of the wake, C_{Dy}, C_{Dx} and C_{Lx} are positive and C_{Ly} is negative and small. Hence, equation (4.43) can only be satisfied if $g > 0$. Changing sides of the wake results in a change in sign for C_{Dy} and C_{Lx}, so now $g < 0$ is required for flutter to occur.

d. Between the lift cross-over point and the wake boundary the x-aerodynamic stiffness terms, C_{Lx} and C_{Dx}, are very small, and so the dominating term in equation (4.43) is $-gC_{Ly}C_{Dy}$. With positive Y, C_{Ly} is negative and C_{Dy} positive, and so flutter is possible only when $g > 0$. Switching to the negative Y side of the wake causes C_{Dy} to become negative, and so flutter is now possible only when $g < 0$.

Recalling that for all examples in case (*ii*) g is small, and taking into account typical values of the aerodynamic stiffness terms, it is evident that equation (4.44) is dominated by the term

$$\chi = \frac{nC_{Ly}}{1 - k^2}. \tag{4.45}$$

Hence, in addition to specifying whether positive or negative g is required for flutter to occur it is also possible to specify whether $k^2 = (\omega_y/\omega_x)^2$ must be greater or less than one. Thus, the frequency ratios required for flutter to occur can be obtained, and are also indicated on Figure 4.9, as is the required sign of θ when the structural coupling is purely due to blow-back.

The discussion presented in the preceding paragraph has shown that, with the exception of the region very close to the wake centreline, wake-induced flutter can occur for all transverse positions of the leeward conductor provided the correct combination exists for the coupling, g (or blow-back angle, θ) and frequency ratio, $k^2 = (\omega_2/\omega_1)^2$, between the two structural modes. Essentially the same conclusions were also obtained by Tsui (1977, 1978), who rederived Simpson's (1971a, b) analytical model. However, Tsui retained the aerodynamic damping terms and obtained stability boundaries via the Routh stability criteria in essentially the same manner as that given by equation (4.36).

Making use of the measured force coefficient data, solutions to equation (4.39) were obtained by Price (1975a, b) for a variety of different structural parameters, and the solutions compared with experimental results obtained using a specially designed apparatus where the leeward of the two conductors was supported on a pantograph spring system. By altering the initial lengths of the springs, different natural frequencies could be obtained; in addition, by rotating the pantograph, different values of blow-back angle could be simulated. A very similar experimental

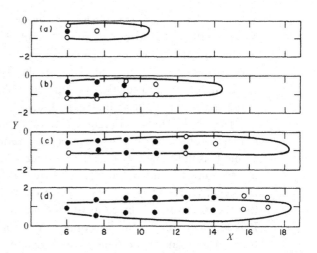

Figure 4.10. Flutter boundaries for a smooth circular cylinder with $\omega_1 = 26.4$ rad/s and $\omega_2 = 25.3$ rad/s (Price 1975a): ●, self-starting experimental boundary; ○, hard-oscillating experimental boundary; —, theoretical boundary. (a) $U = 20$ m/s, $g = 0.268$; (b) $U = 25$ m/s, $g = 0.268$; (c) $U = 30$ m/s, $g = 0.268$; (d) $U = 30$ m/s, $g = -0.268$.

facility was developed, and an experimental research program pursued, by a team of researchers at the National Research Council of Canada (Watts & Ko 1973; Wardlaw *et al.* 1975).

Some typical comparisons between experimental and theoretical results are shown in Figures 4.10 and 4.11, where for a specific set of structural parameters the unstable wake regions are given for fixed airspeeds. Two sets of experimental results are presented in each figure: one for self-starting flutter where instability was initiated without any external disturbance, whereas the second, "hard-oscillation" boundary was obtained when the conductor was given a relatively large initial perturbation of at least 3 conductor diameters. It is apparent that for all the results presented in Figures 4.10 and 4.11 the areas of the wake in which hard-oscillation flutter occurred

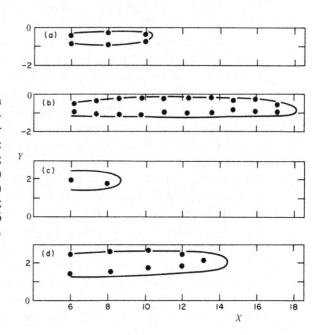

Figure 4.11. Flutter boundaries for a smooth circular cylinder with two intensities of free-stream turbulence for $U = 20$ m/s and $g = 0.268$ (Price 1975a): ●, self-starting experimental boundary; —, theoretical boundary. (a) and (b) $\omega_1 = 16.6$ rad/s, $\omega_2 = 16.0$ rad/s; (a) 11% turbulence; (b) 2% turbulence; (c) and (d) $\omega_1 = 15.9$ rad/s, $\omega_2 = 16.6$ rad/s; (c) 11% turbulence; (d) 2% turbulence.

are larger than for self-starting flutter; this was the case for all of the results presented by Price (1975a, b). Although not excellent, there is reasonable agreement between the theoretical and experimental results, with the agreement being better for the hard-oscillation experimental values. It should be remembered that the theoretical values are obtained using the "undamped" solutions and so no account was taken of either structural or aerodynamic damping.

The results shown in Figures 4.10 (a–c) are all for positive g and $\omega_1 > \omega_2$ (the reader is reminded that the ω_1 mode is predominantly in the x-direction and the ω_2 mode is in the y-direction). In agreement with the argument summarised in Figure 4.9, flutter occurs in the region between the wake centreline and the lift peak for the negative Y side of the wake. Comparing Figures 4.10 (a–c) shows very clearly that increasing the airspeed causes an increase in the area where instability occurs. It is also apparent from a comparison of Figures 4.10 (c) and (d) that changing the sign of g causes the region where instability occurs to switch to the other side of the wake, and that the instability regions obtained with positive and negative g are approximately mirror images (the theoretical curves are exactly mirror images), which is also in agreement with the arguments summarised in Figure 4.9.

The results presented in Figure 4.11 show that changing the frequency ratio of the structural modes from $\omega_1 > \omega_2$ to $\omega_2 > \omega_1$, while maintaining a positive g, causes the unstable region of the wake to switch from the negative Y to the positive Y side of the wake and also moves the unstable region outside the lift peak; again, this is in agreement with the results indicated in Figure 4.9.

One final observation from the results of Figure 4.11 is the stabilizing effect of increased free-stream turbulence; this can be seen by comparing the unstable areas in Figures 4.11(a) and (b) which are for 11% and 2% turbulence intensities, respectively; similarly for Figures 4.11(c) and (d). As shown in Figures 4.6 and 4.7, the effect of increased turbulence, for a fixed value of X, is to reduce the maximum C_L and to increase the minimum C_D; hence, there is a general reduction in magnitude of the aerodynamic stiffness terms, leading to an overall stabilizing effect. These results are consistent with field trials on transmission lines reported by Rawlins (1974b) and Wardlaw et al. (1975), who both independently noted that wake-induced flutter was mostly a problem for transmission lines situated in flat open terrain (where the turbulence intensities are typically low) and that its severity was reduced by an increase in atmospheric turbulence intensity.

The theoretical results presented in Figures 4.10 and 4.11 were obtained using undamped theory, where the effects of both structural and aerodynamic damping are ignored. Although this gives good agreement with experimental results for the majority of cases, there are some examples where the agreement is not good. The reason for this can be illustrated via the theoretical results shown in Figure 4.12. Here, the range of airspeed where instability will occur is given as a function of frequency ratio for one particular wake position; two sets of results are presented, one where there is no damping and the second where the aerodynamic damping is accounted for. Over most of the frequency range there is virtually no difference between the two solutions. However, one substantial difference is that the damped theory predicts a minimum frequency ratio below which instability will not occur, whereas the undamped theory suggests that instability will occur

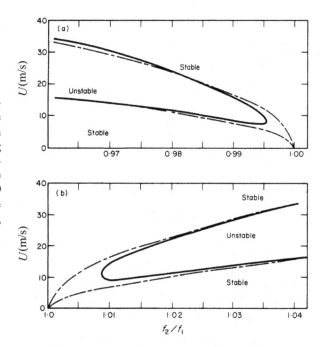

Figure 4.12. Variation of critical velocity range with frequency ratio for the damped and undamped solutions with $g = 0.268$ (Price 1975a): ——, accounting for the aerodynamic damping; - -, neglecting the aerodynamic damping. (a) $X = 6.0$, $Y = -0.8$, $\omega_1 = 26.4 \to 25.9$ rad/s, $\omega_2 = 25.3 \to 25.9$ rad/s; (b) $X = 6.0$, $Y = 2.0$, $\omega_1 = 25.2 \to 25.9$ rad/s, $\omega_2 = 26.4 \to 25.9$ rad/s.

at zero velocity when $\omega_2 = \omega_1$. Obviously the undamped results for $\omega_2 \approx \omega_1$ are physically unreasonable, and this represents a severe weakness of the undamped theory.

Accounting for both the aerodynamic and structural damping, Price & Piperni (1988) showed that the main effect of increasing the structural damping is to move the critical frequency ratio below which instability will not occur away from unity. An example of this is shown in Figure 4.13 for four different levels of structural damping.

A similar validation of theoretical results for wake-induced flutter, accounting for the aerodynamic damping terms, but not the structural damping terms, is also given by Tsui (1977, 1978) who used aerodynamic data from Ko (1973) and compared the theoretical stability boundaries with the experimental results of Watts & Ko (1973); the agreement was generally very good.

4.2.1 (d) Nonlinear analysis

The analyses and results summarised in the previous section are based on linear theory, and consequently they give no indication of the system behaviour once instability has occurred; to do so requires that nonlinear effects be accounted for. Because transmission lines are so flexible, and typical amplitudes of oscillation for sub-span oscillation are small compared with the length of a transmission line, it is reasonable to consider the structural terms as being linear. However, as can be seen from Figures 4.6 and 4.7, the aerodynamic lift and drag coefficients acting on the leeward conductor are strongly nonlinear with respect to its displacement; hence, nonlinear analyses of sub-span oscillation have typically concentrated on the aerodynamic nonlinearities.

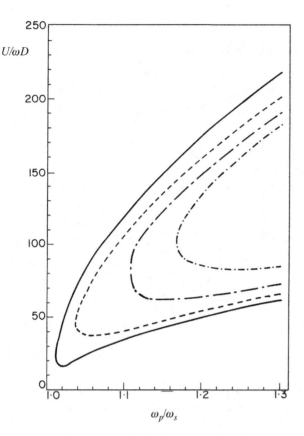

Figure 4.13. Variation of critical velocity range with frequency ratio for different levels of mechanical damping, with equal damping in the two modes (Price & Piperni 1988): $X = 10.7$, $Y = -0.8$, $m/\rho D^2 = 1706$, $\theta = 20°$; ——, $\delta_p = \delta_s = 0.0$; – –, $\delta_p = \delta_s = 0.1$; —·—, $\delta_p = \delta_s = 0.3$; ----, $\delta_p = \delta_s = 0.5$.

An initial attempt at a nonlinear analysis of wake-induced flutter was undertaken by Simpson (1971a) who employed a Runge-Kutta method to predict the amplitude of oscillation for the leeward conductor; the aerodynamic coefficients were approximated using simple analytic functions and no account was taken of the aerodynamic damping. Some preliminary calculations including the effect of both mechanical and aerodynamic damping were performed by Allnutt et al. (1980). The methodology employed by Allnutt was to consider a position in the wake where instability was predicted using linearised theory, and then assume that a stable limit cycle existed and that it could be represented by sinusoidal motion in both the in-plane and out-of-plane directions; in common with the analysis presented by Simpson (1971a), the approximation of the nonlinear aerodynamic terms was very simplistic. Oliveira & Mansour (1983) used simple analytical approximations for the aerodynamic coefficients measured by Cooper (1974) and solved the resulting equations via a Krylov & Bogoliubov method of averaging. However, no account was taken of the structural coupling between the in- and out-of-plane modes (interestingly, linearised analyses indicate that this case should be stable, but no discussion of this apparent contradiction is given by Oliveira & Mansour) and structural damping was also ignored.

A more complete analysis accounting for structural damping as well as the non-linearities associated with the aerodynamic terms is given by Price & Piperni (1988). The complete nonlinear form of the resultant wake velocity acting on the leeward conductor, as given by equation (4.13), was employed, and so the aerodynamic forces

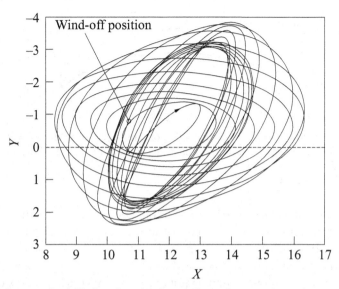

Figure 4.14. Locus of oscillation for the leeward conductor when given an initial displacement in the y-direction of one conductor diameter (Price & Piperni 1988): $X = 10.7$, $Y = -0.8$, $m/\rho D^2 = 1706$, $\theta = 20°$, $\omega_p/\omega_s = 1.1$, $U/\omega_s D = 83.5$, $\delta_p = \delta_s = 0.05$.

acting on the oscillating leeward conductor are simply

$$F_x = \tfrac{1}{2}\rho U_r^2 Dl \left(C_L \sin\alpha - C_D \cos\alpha\right), \tag{4.46a}$$

$$F_y = \tfrac{1}{2}\rho U_r^2 Dl \left(-C_L \cos\alpha - C_D \sin\alpha\right). \tag{4.46b}$$

To account for the nonlinear variation of C_L and C_D with wake position, the previously measured force coefficient data was interpolated in polynomial form. The resulting equations were solved using a standard Runge-Kutta numerical integration procedure; the leeward conductor was given a small initial displacement, and the numerical integration was allowed to proceed until either the conductor converged back on to its equilibrium position, indicating stability, or achieved a steady-state limit cycle. It should be appreciated that this procedure was extremely computationally intensive, and typically it was necessary to allow the integration to run for over 50 cycles of oscillation before a steady limit cycle was obtained; a typical example of this is shown in Figure 4.14.

Using simulations similar to that shown in Figure 4.14, the effect of structural damping on the amplitude of the final limit-cycle oscillation was investigated. An example of the results obtained is shown in Figure 4.15, where the half peak-to-peak magnitudes of the major and minor axes of the limit-cycle ellipses are presented. For structural logarithmic decrements of less than approximately 0.18, increasing the structural damping has very little effect on the amplitude of the major axis; what is even more surprising is that the amplitude of the minor axis increases. It is this increase in the amplitude of the minor axis which suggests why increasing structural damping is so ineffective in reducing the overall amplitude of oscillation. If the oscillation were a simple one-degree-of-freedom motion, then increasing the structural damping would cause a reduction in the vibrational amplitude. However, because this is a coupled two-degree-of-freedom oscillation, the effect of an increase

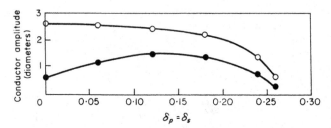

Figure 4.15. Variation of the leeward conductor limit-cycle amplitudes with mechanical damping when there is equal mechanical damping in both modes of vibration (Price & Piperni 1988): $X = 10.7$, $Y = -0.8$, $m/\rho D^2 = 1706$, $\theta = 20°$, $\omega_p/\omega_s = 1.1$, $U/\omega_s D = 64.0$; $\circ$, major axis; $\bullet$, minor axis.

in structural damping is more complex. Not only does additional damping increase the energy absorption of the conductor, it also causes a change in the conductor modal pattern. In particular, the ellipse is now much "broader", and thus, because of the increased differential in drag across the ellipse, more energy is absorbed by the conductor. Hence, additional damping not only causes increased energy dissipation, but also, albeit indirectly, an increase in energy absorption. This process continues until the trajectory becomes approximately circular, at which point the maximum possible energy is absorbed, and it is only then that increasing the structural damping has a significant effect on the amplitude of the major axis.

The results presented in Figure 4.15 are for equal structural damping in the two wind-off modes of vibration. Although this may be achievable for quad bundles, where the two modes of oscillation may be sub-span modes, it is unlikely to be so for twin bundles where the in-plane mode will be a full-span mode which will be lightly damped. To investigate the efficacy of mechanical damping for twin bundles, increasing the mechanical damping only in the out-of-plane mode was investigated, and some typical results are presented in Figure 4.16. For a logarithmic decrement of 0.7, which is far in excess of that obtainable on transmission lines, the conductor is still unstable and the amplitude of the major axis is reduced by 35% only *vis-à-vis* when the structural damping is zero, whereas a 100% increase in the minor axis amplitude is obtained. Thus, for twin-conductor bundles, or for any situation where only the out-of-plane mode is damped, damping is very ineffective in reducing the amplitude of oscillation.

Experimental support for the above conclusion concerning mechanical damping comes from the work of Claren, Diana & Nicoline (1974), Hardy & Bourdon (1979)

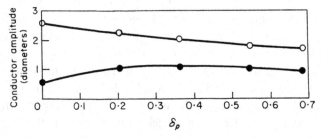

Figure 4.16. Variation of the leeward conductor limit-cycle amplitudes with mechanical damping when there mechanical damping in the out-of-plane mode only (Price & Piperni 1988): $X = 10.7$, $Y = -0.8$, $m/\rho D^2 = 1706$, $\theta = 20°$, $\omega_p/\omega_s = 1.1$, $U/\omega_s D = 64.0$; $\circ$, major axis; $\bullet$, minor axis.

and Hardy & Van Dyke (1995). Claren *et al.* conducted experiments on a triple bundle fitted with spacer dampers with a hysteretic loss coefficient of 0.15; they observed very little reduction in the amplitude of oscillation of the conductors *vis-à-vis* the same bundle fitted with rigid spacers. However, increasing the hysteretic loss coefficient to 0.35 produced a reduction in conductor amplitude of approximately one third. Unfortunately, the modal logarithmic decrements were not measured in these experiments. Hardy & Bourdon (1979) reported that, for transmission lines fitted with spacer dampers having a hysteretic loss coefficient of 0.28, no reduction in conductor amplitude could be observed; once again, no values of modal logarithmic decrements were reported for these experiments. The ineffectiveness of increased structural damping to alleviate the effects of wake-induced flutter has also been observed in field tests of full-scale quad transmission lines (Hardy & Van Dyke 1995).

One of the major disadvantages of the analysis proposed by Price & Piperni (1988) is that the computational time required to obtain a solution is significant. Price & Maciel (1990) showed that the same results could effectively be obtained using the Krylov & Bogoliubov method of averaging instead of the Runge-Kutta numerical integration, but with considerable saving in computational effort.

4.2.2 Analysis for a moving windward conductor

The analysis given in the previous section is strictly for a flexibly mounted leeward conductor in the wake of a fixed windward conductor. In a transmission line bundle, although it is the leeward conductor, or conductors, which extract energy from the airflow and are primarily responsible for the instability, the structural coupling between conductors causes the windward conductors to oscillate. Hence, although analyses based on a fixed windward conductor are useful in explaining the physics of wake-induced flutter, a complete analysis must account for the motion of the windward conductor – in particular, if an attempt is being made to calculate the amplitude of oscillation of the conductors.

Allowing the windward conductor to move introduces several complexities in the modelling of the aerodynamic forces. First, the aerodynamic forces acting on a leeward conductor are strongly dependent on its position in the wake, and this wake position is affected by the motion of both conductors. Consequently, motion of the windward conductor has a direct and significant influence on the aerodynamic forces acting on the leeward conductor; in contrast, motion of the leeward conductor has no direct effect on the aerodynamic forces acting on the windward conductor.

The simplest approach accounting for the windward conductor motion is probably that employed by Tsui & Tsui (1980). They considered the effective wake displacement of the leeward conductor to be the instantaneous sum of the displacements of the two conductors, and thus, the linearised expressions for the force coefficients on the leeward conductor are

$$C_{L_l} = C_{L0_l} + (x_w - x_l)C_{Lx_l} + (y_w - y_l)C_{Ly_l},$$
$$C_{D_l} = C_{D0_l} + (x_w - x_l)C_{Dx_l} + (y_w - y_l)C_{Dy_l},$$
(4.47)

where the subscripts l and w indicate the leeward and windward conductors, respectively. Using essentially the same procedure as that employed for the fixed windward conductor, and realizing that the lift coefficient for the windward conductor is zero, the aerodynamic forces acting on the two conductors may then be written as

$$
\begin{Bmatrix} F_{xw} \\ F_{zw} \\ F_{xl} \\ F_{zl} \end{Bmatrix} = -\frac{1}{2}\rho D l U^2 \left(\begin{Bmatrix} C_{D0_w} \\ 0 \\ C_{D0_l} \\ C_{L0_l} \end{Bmatrix} \right.
$$

$$
+ \frac{D}{\overline{U}} \begin{bmatrix} 2\left(\overline{U}/U\right) C_{D0_w} & 0 & 0 & 0 \\ 0 & \left(\overline{U}/U\right) C_{D0_w} & 0 & 0 \\ 0 & 0 & 2C_{D0_l} & -C_{L0_l} \\ 0 & 0 & 2C_{L0_l} & C_{D0_l} \end{bmatrix} \begin{Bmatrix} \dot{x}_w \\ \dot{y}_w \\ \dot{x}_l \\ \dot{y}_l \end{Bmatrix}
$$

$$
\left. + \begin{bmatrix} 0 & 0 & 0 & 0 \\ 0 & 0 & 0 & 0 \\ -C_{Dx_l} & -C_{Dy_l} & C_{Dx_l} & C_{Dy_l} \\ -C_{Lx_l} & -C_{Ly_l} & C_{Lx_l} & C_{Ly_l} \end{bmatrix} \begin{Bmatrix} x_w \\ y_w \\ x_l \\ y_l \end{Bmatrix} \right). \tag{4.48}
$$

Unfortunately, despite its very attractive simplicity, this analysis has a number of significant weaknesses. First, there is a time delay between when the windward conductor moves and when its effect is transmitted via the wake to the leeward conductor. Second, and more importantly, the Tsui & Tsui model takes no account of the windward conductor's vibrational velocity on the formation of its wake. As demonstrated by Simpson & Flower (1977) these effects have a profound influence on the aerodynamic forces acting on the leeward conductor. It is interesting to note that Tsui & Tsui (1980) were aware of the earlier work by Simpson & Flower (1977), but incorrectly judged that the added complexities in the aerodynamic modelling would not lead to any significant change in the stability of the conductors.

The most comprehensive treatment of the wake dynamics and aerodynamic modelling for a moving windward conductor is that of Simpson & Flower (1977), and this is the analysis which is summarised in the following paragraphs. However, prior to this, an interesting development was proposed by Rawlins (1976), who did account for the windward conductor vibrational velocity. In particular, he showed that this results in an inclination of the wake velocity *vis-à-vis* the free-stream direction, as well as a change in its magnitude. These effects resulted in a modification of the aerodynamic damping matrix compared with that shown in equation (4.48), although the stiffness matrix was unchanged. Rawlins, however, did not account for the time delay between the motion of the windward conductor and its effect being felt by the leeward conductor.

The complete analysis presented by Simpson & Flower (1977) is too detailed to be reproduced here, but some of the more important salient points are outlined in the following. One significant aspect of the aerodynamic modelling introduced by Simpson & Flower was to consider the effect of the time delay between motion of the windward conductor and its effect being felt by the leeward one. Considering, for example, purely the x-motion of the conductors, then if τ is the time taken for the wake to convect from the windward conductor at time $t - \tau$ to the leeward conductor at time t, it is apparent that

$$U\tau/D = X + x_{w\tau} - x_l, \qquad (4.49)$$

where $x_{w\tau}$ indicates the value of x_w at time $t - \tau$; it is referred to as a "retarded variable". Had the windward conductor retained its velocity $\dot{x}_{w\tau}$ throughout the time interval τ then the apparent nondimensional separation between conductors, as viewed from the leeward conductor, is

$$\xi = U\tau/D + \tau\dot{x}_{w\tau} \approx X + x_{w\tau} - x_l + (XD/U)\dot{x}_{w\tau}. \qquad (4.50)$$

The main effect of the transverse motion of the windward conductor is that the wake is shed at an inclination to the free-stream velocity, which to first order may be approximated as

$$\alpha_\tau = D\dot{y}_{w\tau}/U. \qquad (4.51)$$

And so the streamwise and transverse separations between conductors become (again neglecting all second-order terms)

$$\xi = U\tau/D + \tau\dot{x}_{w\tau} \approx X + x_{w\tau} - x_l + (XD/U)\dot{x}_{w\tau} - (YD/U)\dot{y}_{w\tau}, \\ \eta = Y - y_{w\tau} + y_l. \qquad (4.52)$$

Unfortunately, accounting for motion of the windward conductor in the manner described above takes no account of the windward conductor's acceleration. The acceleration in the transverse direction has a relatively minor effect. However, Simpson & Flower (1977) show that the oscillatory streamwise acceleration can result in a successive accumulation and deficit of fluid in the wake, causing the wake to "bunch" and so change its width as a function of time. Hence, streamwise acceleration of the windward conductor results in a change in the "effective" transverse wake position of the leeward conductor.

One further complication is that the wake position of the leeward conductor, given by equation (4.52), contains retarded variables for the windward conductor; these may be accounted for via the following procedure. Considering, for example, the x-motion of the windward conductor, which can be written as

$$x_w(t) = A\exp(kt)\sin(\omega t), \qquad (4.53)$$

where A is the amplitude of the motion, ω is the frequency of oscillation and k is the imaginary part of the eigenvalue. Then, the position of the windward conductor at time $t - \tau$ may be written as

$$x_w(t - \tau) = x(t)f - \tau g\dot{x}(t), \qquad (4.54)$$

where

$$f = \exp(-k\tau)\cos(\omega\tau) + k\tau g \quad \text{and} \quad g = \exp(-k\tau)\sin(\omega\tau)/\omega\tau.$$

Similar expressions can be obtained for the other retarded variables. Some indication of the complexity that this introduces may be gained by rewriting the expression for streamwise separation between conductors, given by equation (4.52), which becomes

$$\xi = X + fx_w - x_l + \{XD((f - g)/U)\}\dot{x}_w \\ - (YDf/U)\dot{y}_w - g(XD/U)^2\ddot{x}_w + gXYD^2/U^2\ddot{y}_w. \qquad (4.55)$$

The wake position and, consequently, the aerodynamic forces acting on the leeward conductor are now frequency-dependent (via the f and g terms) and are functions of both the velocity and acceleration of the windward conductor.

One final aspect of the aerodynamic modelling for the leeward conductor aero-dynamics introduced by Simpson & Flower (1977) relates to the retardation of flow as it approaches the stagnation region of a bluff body. Because the flow is slowed down, it arrives at a time Δt later than it would have arrived if the flow velocity had remained constant; Simpson & Flower (1977) showed that Δt could be expressed approximately as $\mu D/bU$, where μ is of order 1 (experiments reported by Simpson & Flower indicated that μ appeared to be a function of both wake position and Reynolds number). This introduces an effective displacement (with respect to the position evaluated based on a constant approach velocity) of the body in the y-direction of $-\mu D/(bU)\dot{y}$. Although this has no effect on the windward conductor, it is very significant for the leeward conductor where the aerodynamic force coefficients strongly depend on its y-position.

This retardation effect is not restricted to when the windward conductor is oscil-lating; it applies equally to a fixed windward conductor. To illustrate how important this can be, we will revisit the case of a fixed windward conductor, where the aero-dynamic forces on the leeward conductor, as given by equation (4.24), are

$$\left\{ \begin{matrix} F_x \\ F_y \end{matrix} \right\} = -\frac{1}{2}\rho D l U^2 \left(\left\{ \begin{matrix} C_{D0} \\ C_{L0} \end{matrix} \right\} + \frac{D}{bU} \begin{bmatrix} 2C_{D0} & -C_{L0} \\ 2C_{L0} & C_{D0} \end{bmatrix} \left\{ \begin{matrix} \dot{x} \\ \dot{y} \end{matrix} \right\} + \begin{bmatrix} C_{Dx} & C_{Dy} \\ C_{Lx} & C_{Ly} \end{bmatrix} \left\{ \begin{matrix} x \\ y \end{matrix} \right\} \right).$$

Accounting for the flow retardation introduces new terms into the final column of the damping matrix, which now becomes

$$\frac{D}{bU} \begin{bmatrix} 2C_{D0} & -C_{L0} - \mu C_{Dy} \\ 2C_{L0} & C_{D0} - \mu C_{Ly} \end{bmatrix} \left\{ \begin{matrix} \dot{x} \\ \dot{y} \end{matrix} \right\}. \tag{4.56}$$

If C_{Ly} is positive and large, then the damping term $C_{D0} - \mu C_{Ly}$ may be small, or even negative – indicating the possibility of a damping-controlled instability. For example, the data presented in Figure 4.6 shows that at $X = 5.8$ and $Y = 0.5$, $C_{D0} = 0.4$ and $C_{Ly} = 0.3$; hence, a value of $\mu = 1.33$ would lead to zero aerodynamic damping in the y-direction. As shown in Chapter 5, this flow retardation can easily lead to a damping-controlled instability in heat-exchanger tube arrays subject to coolant cross-flow (Price & Païdoussis 1986b; Païdoussis & Price 1988).

Accounting for all of the effects described in the preceding paragraphs, Simpson & Flower (1977) developed expressions for the aerodynamic forces on a pair of conductors. These expressions, given in terms of aerodynamic stiffness, damping and inertia matrices, are extremely long and are not reproduced here. Part of the reason for this complexity is that the lift force was divided into contributions from circulation, buoyancy and resolved drag; hence, the damping matrix contains all of the different terms shown in equations (4.24), (4.27) and (4.28). It is worth noting, however, that the aerodynamic forces are frequency dependent, as exempli-fied via equation (4.55), and so solution of the resulting stability problem requires an iterative approach of the type found in many aeroelastic problems.

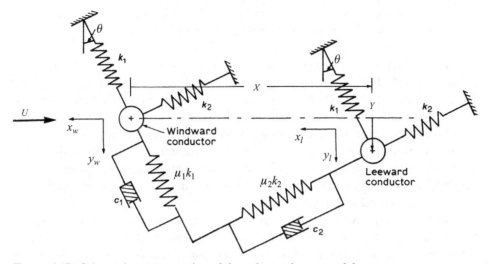

Figure 4.17. Schematic representation of the twin-conductor model.

It is unfortunate that, despite the *tour de force* by Simpson & Flower (1977) in deriving the expressions for the aerodynamic forces, they give very few examples of their application, and the limited examples they do give are for a complete three-dimensional transmission line. Simpson (1977) did employ the Simpson & Flower (1977) theoretical model to show that both static and dynamic instabilities are possible for pure in-flow motion of a conductor pair. The dynamic instability was a coupled-mode flutter and was dominated by the C_{Dx} term of the leeward conductor, which must be negative for instability to occur. The instability mode, at the lower critical velocity, was such that both conductors had approximately the same magnitude of oscillation and moved in-phase with each other; Simpson (1977) concluded that this instability could possibly explain horizontal "snaking" motions observed on some multiconductor power lines.

Price & Abdallah (1990), however, did conduct a comprehensive investigation of system parameters on the stability of a flexibly mounted pair of conductors. In fact, they extended the analysis of Simpson & Flower to include the nonlinear aerodynamic terms, allowing them to investigate the effect of system parameters on the conductor's amplitude of oscillation. In order to perform the nonlinear analysis it was necessary to remove the frequency dependence of the aerodynamics – and so terms f and g were set to 1 and 0, respectively. In addition, Price & Abdallah (1990) considered the lift force as being purely due to circulation, neglecting any contributions due to buoyancy or resolved drag. All of the other effects introduced by Simpson & Flower were retained in both the linear and nonlinear forms of the aerodynamics.

The structural model employed by Price & Abdallah (1990) is shown schematically in Figure 4.17. Both the windward and leeward conductors are now mounted via orthogonal spring systems, but, in addition, to simulate the spacers the two conductors are coupled in the in-plane and out-of-plane directions via springs, with stiffnesses $\mu_1 k_1$ and $\mu_2 k_2$, respectively, as well as linear mechanical dampers with damping coefficients c_1 and c_2. Price & Abdallah (1990) found that this tandem model exhibited an elliptical motion similar to that observed on full-scale transmission lines when the structural coupling terms between the two conductors were set with $\mu_1 = \mu_2 = 0.1$, and thus, these values were used for all the results they presented.

A typical set of linearised stability results for $X = 11.0$ and $Y = -0.5$, is presented in Figure 4.18. For any particular frequency ratio above a certain critical minimum there is a range of velocity where the system is unstable and outside of which it is stable. The results also show that instability is confined to frequency ratios greater than 1.0, in agreement with the fixed-windward-conductor analysis.

For the wake position $X = 11.0$ and $Y = 0.5$ (the mirror image about the wake centreline of the position discussed in the previous paragraph), no instability was obtained for frequency ratios between 0.7 and 1.3. This also is in agreement with the fixed-windward-conductor analysis.

For the two other wake positions investigated by Price & Abdallah (1990) ($X = 11.0$ and $Y = 1.4$ or -1.4), however, there was a significant difference between their results and those obtained assuming a fixed windward conductor; this is illustrated in Figure 4.19. What is so surprising about these results is that instability is possible over the complete range of frequency ratios from 0.7 to 1.3, not just for $f_p/f_s < 1$ as predicted by the fixed-windward-conductor analysis. This somewhat puzzling behaviour was explained by an examination of the root locus plots of the system eigenvalues. These showed that for the cases presented in Figure 4.18 the instabilities were due to a frequency coalescence of the structural modes, as predicted for a fixed windward conductor. However, for some of the results presented in Figure 4.19 this was not so, and the instability was confined to one mode and appeared to be damping-controlled (as opposed to the stiffness-controlled binary flutter obtained in all previous examples discussed in this section). Price & Abdallah (1990) showed that this new wake-induced instability is caused by a combination of movement of the windward conductor wake relative to the leeward conductor (hence, inducing a changing force field on the leeward conductor) and the time delay in this force field being felt by the leeward conductor. The time delay is essential: it produces a phase difference between the leeward conductor displacement and the changing force field, so enabling a finite amount of energy to be extracted by the leeward conductor (or dissipated) per cycle of oscillation.

Price & Abdallah (1990) also employed their nonlinear analysis to investigate the effect of system parameters on the amplitude of oscillation. A typical set of results is presented in Figure 4.20, showing the variation of the limit-cycle major axis with increasing wind speed. For low values of wind speed the amplitude of the major axis is zero, indicating that the system is stable. At a certain critical wind speed the amplitude ceases to be zero, indicating instability. As wind speed is further increased, the limit-cycle amplitudes also increase until a nondimensional wind speed of approximately 80, where the amplitudes suddenly drop to zero, indicating that the system has restabilised – in agreement with the results obtained from the linear analysis shown in Figure 4.18.

The effect of mechanical damping on the system stability can also be observed from the results of Figures 4.18 and 4.19. The results presented in Figure 4.18, representing a frequency coalescence instability, show that although increased damping can raise the minimum wind speed for frequency ratios close to 1, as soon as the frequency ratio moves away from 1 the effect of increased damping becomes negligible.

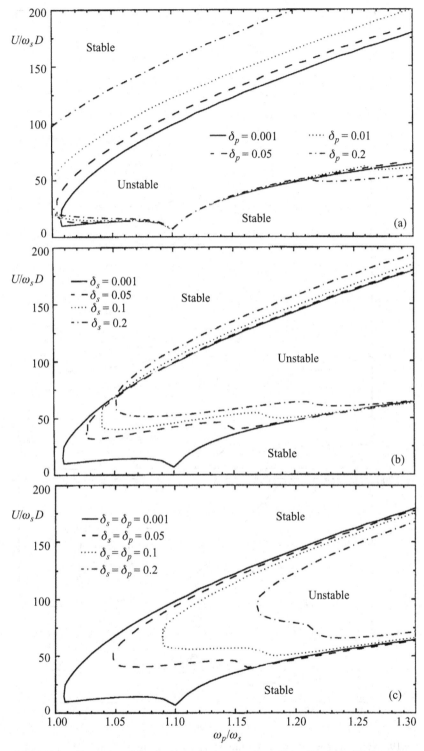

Figure 4.18. Nondimensional flutter boundaries as a function of "wind-off" frequency ratio for different levels of mechanical damping (Price & Abdallah 1990): $X = 11.0$, $Y = -0.5$, $\theta = 20°$, $\mu_1 = \mu_2 = 0.1$. (a) $\delta_s = 0.001$; (b) $\delta_p = 0.001$; and (c) $\delta_s = \delta_p$; other values of in-plane and out-of-plane logarithmic decrements, δ_s, and δ_p, respectively, are given in the figure.

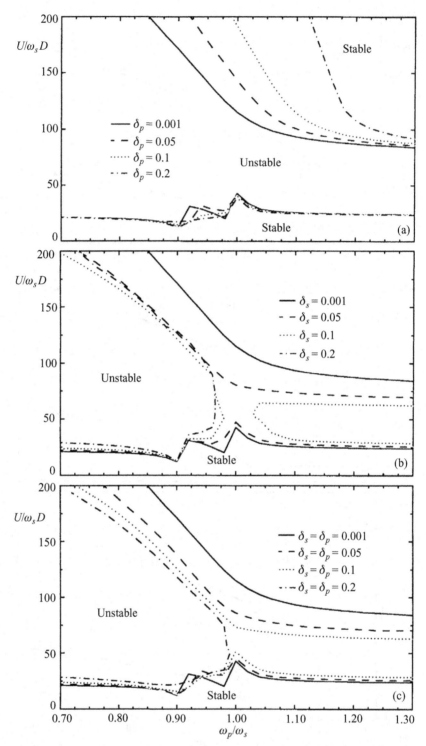

Figure 4.19. Nondimensional flutter boundaries as a function of "wind-off" frequency ratio for different levels of mechanical damping (Price & Abdallah 1990): $X = 11.0$, $Y = 1.4$, $\theta = 20°$, $\mu_1 = \mu_2 = 0.1$. (a) $\delta_s = 0.001$; (b) $\delta_p = 0.001$; and (c) $\delta_s = \delta_p$; other values of in-plane and out-of-plane logarithmic decrements, δ_s, and δ_p, respectively, are given in the figure.

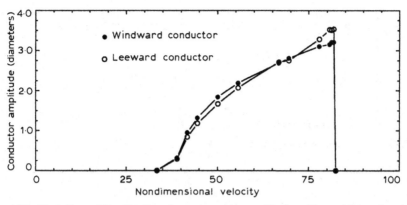

Figure 4.20. Variation of the limit-cycle major axis amplitude with nondimensional velocity for two flexibly mounted conductors (Price & Abdallah 1990): $X = 11.0$, $Y = -0.5$, $\theta = 20°$, $\mu_1 = \mu_2 = 0.1$, $\delta_s = \delta_p = 0.05$, $\omega_p/\omega_s = 1.09$.

This is in agreement with the results obtained when the windward conductor was held fixed. What is even more surprising is that, even when the instability is damping-controlled, as shown in Figure 4.19, increasing the damping still has very little effect on the minimum wind speed for instability to occur. Increasing the mechanical damping can reduce the amplitude of oscillation once the conductors have become unstable, but Price & Abdallah (1990) demonstrated that this is most effective when the damping is introduced in both the in-plane and out-of-plane modes.

Possibly one of the most interesting results to emerge from the analysis conducted with a flexibly mounted windward conductor is related to the effect of frequency detuning on the instability. The analysis based on a fixed windward conductor shows that detuning the modes causes an increase in the wind speed required to initiate instability. However, the results obtained with a flexibly mounted windward conductor suggest that this is only partially correct. Not surprisingly, for the damping-controlled instability mechanism at $X = 11.0$ and $Y = 1.4$ (see Figure 4.19), detuning the modes is ineffective in increasing the velocity threshold of instability. Also, as can be observed from the results of Figure 4.21, when the instability is damping-controlled, frequency detuning also has a minimal effect on the amplitude of oscillation. In fact, the minimum limit-cycle amplitudes are obtained in the approximate range $0.98 \leq \omega_2/\omega_1 \leq 1.02$, although there is a very rapid increase in limit-cycle amplitude as the frequency ratio moves outside this range.

The damping-controlled instability obtained by freeing the windward of a pair of conductors is due to aerodynamic forces resulting from wake effects, and not from the cross-sectional shape of the conductor as is the case for classical galloping discussed in Chapter 2. However, galloping instabilities can also be induced by aerodynamic forces acting on the complete bundle, as opposed to the individual conductors. This has been demonstrated by a number of authors (Brzozowski & Hawks 1976; Hoover & Hawks 1977; Gawronski & Hawks 1977; Nakamura 1980; Zhang, Popplewell & Shah 2000) who have shown that the full-span modes of a conductor bundle (where there is no relative motion between individual conductors) have the potential to develop classical Den Hartog galloping, but where the negative $C_{L\alpha}$ is a consequence of the "bulk-bundle" aerodynamics and does not require the presence of icing.

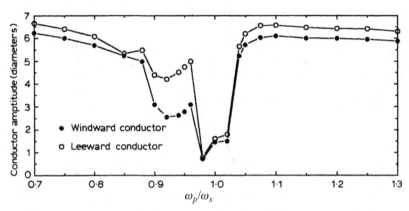

Figure 4.21. Variation of the limit-cycle major axis amplitude with wind-off frequency ratio for two flexibly mounted conductors (Price & Abdallah 1990): $X = 11.0$, $Y = 1.4$, $\theta = 20°$, $\mu_1 = \mu_2 = 0.1$, $\delta_s = \delta_p = 0.001$, $U/(D\omega_s) = 47.3$.

4.2.3 Three-dimensional effects and application to real transmission lines

The analytical models discussed in the preceding sections are so-called semi-rigid models, where the conductors are rigid but flexibly mounted. These models are strictly two-dimensional, and do not account for any of the three-dimensional effects of real transmission lines.

The first stage in the analysis of real transmission lines is to accurately obtain the natural frequencies and mode shapes of the conductor bundle. This in itself is a difficult problem which must account for effects such as: the conductor weight causing bundles to hang in a catenary shape, the inertia and stiffness properties of the spacers used to separate individual conductors in a bundle, the tower and insulator flexibility at the ends of each conductor and the possible structural coupling between adjacent spans (a span being the length of conductors between two towers) – the degree of this coupling strongly depends on the manner in which the conductors are supported at the towers. Different methods of solution are available in the literature for the structural analysis of transmission lines: these include transfer matrix (Simpson 1966, 1972; Claren et al. 1971, 1974) and finite-element methods (Hrudey, Cowper & Lindberg 1973; Allnutt, Price & Tunstall 1980), and some evidence exists showing that accurate estimates of the structural values can be obtained in the absence of any wind effects (Price 1982).

Once accurate structural information is available, coupling the aerodynamic and structural elements is reasonably straightforward, and thus, in principle, estimates of the wind speed at which a transmission line will become unstable can be obtained. However, in reality the magnitude and direction of the local atmospheric wind and turbulence conditions vary significantly along a transmission line (with lengths sometimes exceeding 500 m). In addition, because of the differential creep of conductors, even the "wind-off" position of conductors relative to each other in a bundle is unknown. Hence, it is not surprising that no detailed comparisons between theoretical and experimental stability boundaries for transmission lines are available in the literature; consequently, the type of verification of theoretical analyses that one would normally expect to see does not exist*. A number of three-dimensional analyses, however, do exist, as do results from field trials on real transmission lines. Thus, it is possible to obtain at least a qualitative appreciation of the application of analytical methods to real transmission lines.

One of the earliest attempts to analyse the stability of a three-dimensional transmission line was by Rawlins (1976, 1977) who used a transfer matrix approach. Considering a twin-conductor bundle with fixed ends, Rawlins showed that four different propagation modes exist for the travelling waves in a bundle. Three of these modes are dissipative, with energy being transferred from the conductors back to the airflow, and only one of the modes extracts energy from the wind leading to the instability. Considering, for example, the equilibrium position of the leeward conductor to be above the wake centreline, this "unstable" mode consists of clockwise elliptical motion of the leeward conductor (as exemplified in Figure 4.2), with no motion of the windward conductor. It is these three dissipative modes which induce the windward conductor motion. Rawlins also demonstrates that one of the important functions of the spacers separating the conductors is to transfer energy from the unstable mode to the dissipative modes. Furthermore, Rawlins concluded that the use of unequal sub-span lengths (the length of conductors between two spacers) will enhance this energy transfer and so suppress the occurrence of wake-induced flutter in conductor bundles. This represents the first suggestion in the literature of the beneficial effect of using unequal sub-span lengths, a theme that will be revisited later in this section. The initial analysis presented by Rawlins (1976) was for a simple twin bundle with no catenarity and rigidly supported at its ends; Rawlins (1977) extended this to account for a number of effects including tower mobility at the ends of the conductors, catenary effects and unequal tension in the individual conductors. Simpson & Flower (1977) extended Rawlins' transfer matrix method to include their own aerodynamic modelling, as described in the previous section, and sample stability results are given for one twin- and one quad-conductor bundle.

As an alternative to the transfer matrix approach, Tsui (1986) developed a finite-element analysis of a three-dimensional transmission line. He used a simplified version of the Hrudey et al. (1973) structural elements, neglecting the effects of torsion and conductor motion along the span; in addition, only one element was used for each sub-span. Tsui used two different aerodynamic models: one where the windward conductor was held fixed, and a second accounting for the windward conductor motion using the simple Tsui & Tsui (1980) model. Results are presented for a span with three sub-spans of equal length; however, rather than determining the critical wind velocity required to cause the bundle to become unstable, he indicates whether the bundle is stable or unstable at a wind velocity of 21 m/s for certain combinations of structural parameters (including spacer stiffness, spacer mass, conductor separation and inclination of the wind to conductor bundle – effectively simulating conductor blow-back).

---

* Wardlaw et al. (1975) do report some qualitative three-dimensional experiments on a small-scale experimental three-sub-span system in the Canadian National Research Council's 30-ft-wide wind tunnel. However, no quantitative results are given and, hence, it is not possible to perform any detailed comparison between experimental and theoretical stability boundaries.

Aerodynamic finite elements have also been developed by Lilien & Snegovski (2004) based on Simpson's (1971a, b) aerodynamic model; however, no stability results are presented.

A simple attempt at analyzing the stability of full-scale transmission lines was undertaken by the U.K. Central Electricity Generating Board (CEGB) and reported by Price, Allnutt & Tunstall (1979) and Allnutt *et al.* (1980). The transmission line frequencies, damping levels and mode shapes were first calculated via a finite-element analysis and then applied in a two-dimensional aeroelastic stability calculation. Although crude, this approach overestimates the effect of the aerodynamic forces and so gives a conservative estimate (in terms of design criteria) of the aeroelastic effects. It was estimated that for a standard CEGB twin bundle (span 366 m, six sub-spans of equal length with "rigid" spacers, conductor diameter 28.6 mm and separation between conductors of 10.7 diameters) the minimum wind speed required to cause instability was 7.9 m/s. In addition, using meteorological data, giving hourly mean wind speeds and directions, recorded in an area where transmission lines were known to be susceptible to wake-induced flutter, it was estimated that this standard bundle may execute up to 6×10^6 oscillations per annum. This was an upper bound, but it demonstrated the potential for wake-induced flutter to cause fatigue damage in transmission lines in a very short lifespan.

It was also suggested that the minimum wind speed required to cause instability could be increased substantially by using unequal sub-span lengths, with the spacer positions being chosen to optimise the frequency difference in all modal combinations which may potentially coalesce and become unstable. For example, considering the same standard CEGB twin bundle, it was estimated that using unequal sub-span lengths of 37.0, 83.1, 83.4, 68.1, 59.7 and 34.8 m, as opposed to six equal sub-spans, the minimum wind speed required to cause instability could be raised from 7.9 m/s to 11.9 m/s*. In addition, using a very crude nonlinear model (the aerodynamic force coefficients were assumed not to change as the amplitude of oscillation varied) it was shown that adding damping elements to the spacers, and so increasing the conductor modal damping, was not particularly effective in either raising the wind speed required to cause instability or reducing the amplitude of oscillation once instability had occurred – this, of course, is in agreement with results later obtained by Price and co-workers (Price & Piperni 1988, Price & Abdallah 1990, Price & Maciel 1990). This appeared to contradict results previously reported by Hearnshaw (1974) who presented experimental evidence suggesting that "spacer dampers" were beneficial. However, to increase the transmission line modal damping requires that the spacer dampers be positioned at unequal lengths, and thus, Price *et al.* (1979) and Allnutt *et al.* (1980) hypothesised that the beneficial effect of the spacer dampers came from frequency detuning and not from the increase in structural damping.

The most extensive set of field trials available in the literature related to wake-induced flutter comes from Hardy & Van Dyke (1995). They report experiments from a specially designed test-line built in a windy location on the shore of the Îles-de-la-Madeleine in the Gulf of St Lawrence, Canada. The test line consists of five spans, with a total length of approximately 1.5 km, of which the central span,

---

* Although it was reported that field trials had been initiated to test this hypothesis, the authors are not aware of any published results.

366 m long, is extensively instrumented; in addition, instantaneous values of wind speed and direction are monitored via anemometers located on the towers at both ends of this span. A number of significant conclusions were obtained by Hardy & Van Dyke and are summarised in the following.

- When experiencing wake-induced flutter, all sub-spans had frequency peaks at both their own fundamental frequency as well as the fundamental frequencies of the other sub-spans; this is indicative of the considerable coupling between sub-spans. In addition, full-span snaking occurred concurrently with wake-induced flutter.
- Equal velocity *magnitudes* normal to the conductors, but arising from different combinations of *total* velocity and yaw angle, did not yield the same instability conditions for wake-induced flutter (either minimum speed required to cause instability or amplitude of oscillation), indicating that the yaw angle is an important independent parameter (for Aeolian vibration this was not the case). As far as the authors are aware, yaw angle has never been investigated under laboratory conditions.
- Twin and quad bundles were more susceptible to wake-induced flutter than triple bundles (three conductors per bundle); this is consistent with the leeward conductors extracting energy from the wind and the windward conductors being dissipative.
- The effect of spacer positions along the span length, and hence sub-span lengths, was shown to have a significant effect on the instability. Mismatching the sub-span lengths increased the minimum wind speed necessary to cause wake-induced flutter *vis-à-vis* when the spacers equally divided the span. The unequal sub-span lengths also reduced the severity of the oscillation once it had occurred; no attempt was made to determine the "optimum" spacer positions.
- The vibration amplitude for spans fitted with either "spacer dampers" or just "spacers" were approximately equal when the sub-span lengths were kept the same, indicating that increased damping has little beneficial effect in alleviating wake-induced flutter.

4.3 Fluidelastic Instability of Offshore Risers

Oil and gas exploration in deep waters has necessitated the use of very long and flexible multitube risers, where it is common to have one large-diameter central tube surrounded by a number of smaller diameter peripheral or satellite tubes; see Figure 4.22. For example, Overvik, Moe & Hjort-Hansen (1983) show riser configurations where up to 12 satellite tubes, with diameters $\frac{1}{5}$ of that of the central tube, are positioned circumferentially on a radius of 1 or 1.5 times the central tube diameter.

In addition to vibrations resulting from wave actions, risers are sometimes subject to a steady current or cross-flow normal to their longitudinal axes. One consequence of this cross-flow is the possibility of vortex-induced vibrations. The basic physics of this is discussed in Chapter 3; however, it should be appreciated that the close spacing between the riser tubes, as well as the extremely large length-to-diameter ratio typically found in deep-water risers, does produce some unique problems: see, for example, Mittal & Kumar (2001), Trim *et al.* (2005) and Lucor, Mukundan & Triantafyllou (2006). The main topic of interest in this section, however, is the

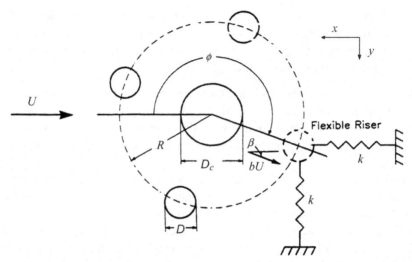

Figure 4.22. Schematic sketch of the cross-section of a five-tube riser; $D_c/D = 1.6$, $R/D = 2.95$.

possibility of multitube risers suffering from fluidelastic instability, with a mechanism similar to wake-induced flutter discussed in Section 4.2. The mechanism producing this fluidelastic instability has considerable commonality with other examples of clustered groups of cylinders in cross-flow: for example heat-exchanger tube banks, discussed in Chapter 5*, and overhead transmission lines, discussed in the previous section. Some experimental evidence for the occurrence of fluidelastic instability in riser bundles is reviewed in Section 4.3.1 and a number of different analytical approaches are considered in Section 4.3.2.

4.3.1 Experimental evidence for the existence of fluidelastic instability in riser bundles

The first study to examine the possibility of fluidelastic instability for multitube risers was performed by Moe & Overvik (1982) and Overvik *et al.* (1983) who investigated experimentally the bulk motion of a number of riser configurations subject to a steady current. Their experimental models were constructed such that no relative motion was allowed between the individual tubes in the riser bundle. In almost all their experiments, vortex-induced vibrations were obtained, but for a more limited set of configurations large-amplitude, single-degree-of-freedom instabilities occurred which could not be attributed to vortex shedding. Although no theoretical analysis was conducted, they concluded that this large-amplitude instability was due to "galloping" of the "bulk" riser bundle – as previously mentioned, the individual tubes in the riser bundle were not allowed to move relative to each other.

In a series of papers, Bokaian & Geoola (1984a, 1984b, 1987, 1989) performed experimental investigations on the fluidelastic instability of a pair of closely spaced circular cylinders in cross-flow. Only one of the two cylinders was free to oscillate and this was restricted to the direction transverse to the flow. Initially Bokaian & Geoola (1984a, b) considered cylinders of equal diameter, and the longitudinal and

---

* A discussion of the similarities and differences between fluidelastic instabilities in multitube risers and heat-exchanger arrays is given by Zdravkovich (1991).

transverse spacing between the cylinders was limited to the range $1.09 \le X \le 5.0$ and $0.0 \le Y \le 3.0$, respectively. Vortex-induced vibrations were obtained over the whole range of cylinder separation, irrespective of whether the flexibly mounted cylinder was in the upstream or downstream position. In addition, for a limited set of cylinder spacings they also obtained a fluidelastic instability, which they called galloping, for both upstream ($X \le 1.75$ and $Y \le 0.5$) and downstream ($X \le 3.0$ and $Y \le 1.0$) flexibly mounted cylinders. In most cases the galloping instability was independent of the vortex-induced vibrations; however, in some cases, the two vibration mechanisms were coupled. In two later papers Bokaian & Geoola considered the cases where the flexibly mounted of the two cylinders had a diameter of either half (Bokaian & Geoola 1987) or twice (Bokaian & Geoola 1989) that of the fixed cylinder; qualitatively the results were very similar to those obtained with equal-diameter cylinders.

Païdoussis, Price & Mark (1988b) and Price et al. (1989) accounted for the relative motion between tubes in a multitube riser and considered the stability of one of the peripheral tubes. Initially, experiments were conducted in airflow (Païdoussis et al. 1988b) and later in waterflow (Price et al. 1989). In both sets of experiments the central and three of the peripheral tubes were "rigid", while the fourth so-called flexible tube was flexibly mounted (see Figure 4.22). The riser bundle could be rotated and, hence, the flexible tube positioned at any desired orientation, ϕ, to the free-stream flow. Figure 4.23(a) shows a set of results for $\phi = 258°$, where vibrations due to vortex shedding are obtained at $U \approx 6$ m/s. The vibrations abate as U is increased past the vortex-shedding resonance, and the vibrational acceleration remains relatively low for velocities up to the maximum velocity of the wind tunnel. In fact, depending on the value of ϕ, the peripheral tube displayed two and sometimes three resonances, corresponding to three different Strouhal numbers, as the velocity was increased. (The origin of the three different Strouhal numbers is discussed in Païdoussis et al. (1988b) and Price & Serdula (1995).) In contrast, the results of Figure 4.23(b) for $\phi = 231°$, show that as U is increased beyond the point of recovery from vortex-shedding resonance, large-amplitude vibrations occur at $U = 18.5$ m/s. It was concluded that these large-amplitude vibrations, which did not subside as U was increased, were due to fluidelastic instability. Whether or not fluidelastic instability occurred strongly depended on the orientation of the riser to the flow and was limited to the incidence ranges of $\phi = 194$–$199°$ and $\phi = 225$–$241°$.

The experiments exemplified by Figure 4.23 were done in air- rather than waterflow, and the nondimensional mass, $m/\rho D^2 = 295$, was much higher than for real risers – where it is typically of order 1. To obtain more realistic values of $m/\rho D^2$, experiments were conducted in waterflow, with $m/\rho D^2 = 5.8$ (Price et al. 1989). Unfortunately, two troublesome problems were encountered during the waterflow experiments. First, the vibrational response from vortex shedding was so great that clashing occurred between the flexible tube and its supporting mechanism; thus, it was not possible to make measurements within the lock-on range (see Figure 4.24(a)). The second problem was that if fluidelastic instability did occur, then it did so within, or close to, the velocity range for amplified vortex-shedding response. Thus, it was not possible to separate the two mechanisms (see Figure 4.24(b)); consequently, it was very difficult to say definitively whether or not fluidelastic instability occurred and, if so, at what value of U. However, and most importantly, the fact that, at $\phi = 236°$, the vibration did not abate as U was increased well beyond the

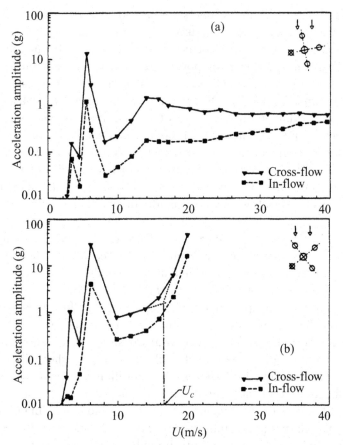

Figure 4.23. The acceleration amplitude of vibration for the flexible riser versus the upstream airflow velocity U (Païdoussis *et al.* 1988b); $\delta = 0.011$, $m/\rho D^2 = 295$, $f_n = 40$ Hz. (a) $\phi = 258°$, (b) $\phi = 231°$.

flow-velocity range of expected vortex-shedding lock-on, in contrast to the vibration response shown in Figure 4.24(a), and that this value of ϕ was within the range where fluidelastic instability occurred for airflow, suggests strongly that the riser suffered from fluidelastic instability in this case also.

Other examples of combined fluidelastic instability and vortex-induced vibrations for one cylinder in the wake of another (both cylinders having the same diameter) have also been observed by King & Johns (1976), Ruscheweyh (1983) and Brika & Laneville (1999); for the experiments of King & Johns and Ruscheweyh, both the upstream and downstream cylinders were free to vibrate, whereas in the experiments reported by Brika & Laneville only the downstream cylinder vibrated.

Huse (1996) also conducted experiments on a multitube riser consisting of 12 tubes as shown in Figure 4.25. The experiments were performed at the entrance to a Norwegian fjord on a 1/30th scale model of a prototype riser with a proposed length of 1333 m. The drilling pipe of the model riser, denoted by R5 in Figure 4.25, had a diameter of 0.03 m, whereas the other eleven pipes had diameters of 0.01 m; the centre-to-centre separation between pipes at the ends of the riser bundle was 0.3 m in

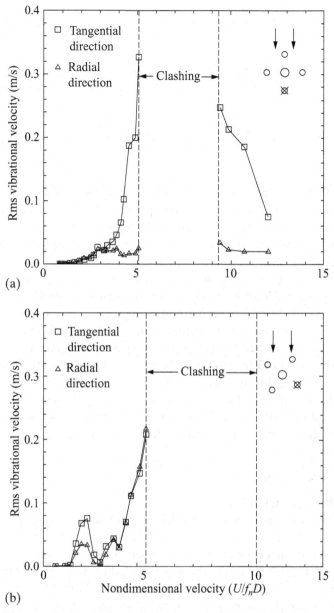

Figure 4.24. The amplitude of vibrational velocity for the flexible riser versus the non-dimensional upstream waterflow velocity U (Price *et al.* 1989); $\delta = 0.025$, $m/\rho D^2 = 5.8$. (a) $f_n = 38.6$ Hz, $\phi = 180°$, (b) $f_n = 28.3$ Hz, $\phi = 236°$.

both the longitudinal and transverse directions. It was observed that at low currents the risers exhibited a "high-frequency" vibrational response which was attributed to vortex-induced vibrations. However, as the magnitude of the current increased, "collisions" occurred between the individual risers; for example, R2 and R5 or R1 and R4 collided when the surface current was less than 0.75 m/s. As stated by Huse (1996), "in addition to the high-frequency VIV response of amplitudes up to one

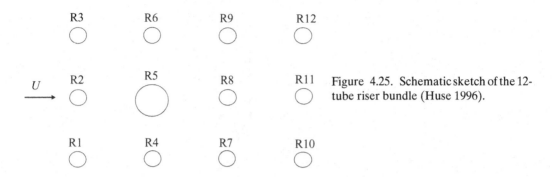

Figure 4.25. Schematic sketch of the 12-tube riser bundle (Huse 1996).

half diameter, there occurred low-frequency, in-line oscillations of an apparently very irregular or chaotic type. The peak-to-peak stroke of these oscillations could be 30 to 40 diameters or more". Almost certainly these low-frequency oscillations were due to fluidelastic instability – albeit, as will be discussed in Section 4.3.2(b), possibly a static instability.

Clashing between risers was also reported in some experiments described by Sagatun, Herfjord & Holmås (2002). In these experiments, conducted in an 85-m-long towing tank, five 10.0-m-long stainless steel pipes of outer diameter 0.01 m and wall thickness 1.0 mm were mounted in a straight line with a fixed centre-to-centre separation of 15 D at their ends. Clashing of the pipes was observed to occur at towing velocities of between 0.43 and 0.48 m/s when the incidence of the five-pipe riser with respect to the towing direction was varied between 0° and 10°. In agreement with Huse (1996), Sagatun et al. (2002) also concluded that the occurrence of clashing was determined by "wake-induced oscillations" although the accompanying vortex-induced oscillations accounted for most of the energy in the collisions.

4.3.2 Analytical models

4.3.2 (a) Dynamic instabilities
An analytical model was developed by Price, Païdoussis & Al-Jabir (1993) with the aim of predicting the current velocity at which fluidelastic instability would occur for the peripheral tube in a compact multitube riser; "bulk" motion of the riser bundle was not considered. The central tube was assumed to be considerably stiffer than the peripheral ones, and thus was taken as being rigid. Furthermore, because of the separation between peripheral tubes, it was assumed that they could be analysed independently. Hence, a single flexible peripheral tube in a cluster of otherwise rigid tubes was considered, as in Figure 4.22. The analytical model borrowed heavily from previous work on sub-span oscillation, as discussed in the Section 4.2 of this book, and is summarised in the following.

The "flexible" riser was, in fact, a rigid tube with the system flexibility being provided by an orthogonal spring system supporting it. The equations of motion may easily be written as

$$D\left([M]\ddot{z} + [C]\dot{z} + [K]z\right) = \boldsymbol{F}, \qquad (4.57)$$

where D is the diameter of the flexible riser, $[M] = ml\,[I_2]$ is the mass matrix, m being the modal mass per unit length; $[C] = c_s\,[I_2]$ is the structural damping matrix,

c_s being the structural damping constant (assumed to be equal in the x- and y-directions); $[K] = k_s [I_2]$ is the structural stiffness matrix, k_s being the structural stiffness (again, assumed to be equal in the x- and y-directions); $z = \{x, y\}^{\mathrm{T}}$ is the vector of nondimensional tube displacements; and F represents the fluid forces acting on the riser.

The fluid forces acting on the flexible tube were determined using a quasi-steady analysis in a manner very similar to that used for bundled conductors in overhead transmission lines (see Section 4.2) or for tubes in a heat-exchanger array (see Chapter 5). However, one complication which does arise for riser bundles is that neither the magnitude nor direction of the local approach flow velocity on the tube are known. (For bundled conductors, the intercylinder distances are too large for this uncertainty to arise. In the case of heat-exchanger tubes, it is the closeness of the tubes to one another, and especially the geometric repeatability (symmetry), which removes this uncertainty – at least for tubes in the central region of a tube array.) Thus, the magnitude of the local velocity was written as bU at angle β to the free-stream direction (see Figure 4.22) where b and β are functions of ϕ. (Clearly, b may be substantially different from 1 for $|\phi| > 30°$. approximately.) The resultant velocity vector is then obtained using a modified form of equation (4.13), giving (note that, to be consistent with the rest of this chapter, the sign convention for x and y employed here is different than that given by Price et al. (1993))

$$U_r = bU\{1 + (D\dot{x}/Ub)\cos\beta - (D\dot{y}/Ub)\sin\beta\}, \qquad (4.58a)$$

and

$$\alpha = \sin^{-1}\{(D\dot{y} - bU\sin\beta)/U_r\}. \qquad (4.58b)$$

The fluid forces on the flexible tube may now be written in exactly the same manner as for an overhead transmission line, and are given by equations (4.17). The lift and drag coefficients, $\overline{C}_L$ and $\overline{C}_D$, which are based on the local wake velocity $\overline{U} = bU$, can be transformed into C_L and C_D, based on the free-stream velocity, U, using equation (4.19).

Realizing that the ultimate objective is a linearised stability analysis, the riser motion can be restricted to small displacements; hence, C_L and C_D may be expressed in a linearised form, as given by equations (4.20). However, recognizing that at low values of U/f_nD, as typically obtained in offshore applications, there will be a phase lag between the tube motion and resulting fluid forces, it is necessary to modify equations (4.20). In a similar manner to that employed in the analysis of overhead transmission lines, the phase lag was accounted for via consideration of the flow retardation as it approaches a tube. Thus, a time delay of $\mu(D/bU)$ was introduced between tube motion and the fluid forces generated thereby; using the same arguments as presented in Section 4.2.2, μ was taken to be of order 1. Price & Païdoussis (1984a, 1986b) showed that this time delay may be accounted for by multiplying the x- and y-displacements by $\overline{g} = \exp(-\lambda\mu D\omega/bU)$, where λ is the nondimensional eigenvalue of the tube motion. Thus, equations (4.20) become

$$\begin{aligned} C_L &= C_{L0} + \overline{g}(xC_{Lx} + yC_{Ly}), \\ C_D &= C_{D0} + \overline{g}(xC_{Dx} + yC_{Dy}). \end{aligned} \qquad (4.59)$$

The principal effect of $\bar{g}$, and hence the phase lag, is to transform the fluid stiffness terms C_{Lx}, C_{Ly}, C_{Dx} and C_{Dy} into combined damping and stiffness terms.

Making use of equations (4.59) and neglecting all second-order terms in $\dot{x}$ and $\dot{y}$, the fluid forces acting on the flexible tube in the rise bundle may be written as

$$F = \{F_x, F_y\}^{\mathrm{T}} = -\tfrac{1}{2}\rho U^2 Dl(F_o + \bar{g}[E]z + D/(bU)[B]\dot{z}), \qquad (4.60)$$

where F_o, is the static fluid-force vector, $[E]$ is the fluid stiffness matrix, and $[B]$ is the fluid damping matrix, given in full as follows:

$$F_o = \{C_{Do}\cos\beta + C_{Lo}\sin\beta, \; C_{Lo}\cos\beta - C_{Do}\sin\beta\}^{\mathrm{T}},$$

$$[E] = \begin{bmatrix} C_{Dx}\cos\beta + C_{Lx}\sin\beta & C_{Dy}\cos\beta + C_{Ly}\sin\beta \\ C_{Lx}\cos\beta - C_{Dx}\sin\beta & C_{Ly}\cos\beta - C_{Dy}\sin\beta \end{bmatrix},$$

$$[B] = \begin{bmatrix} C_{Do}(1+\cos^2\beta) + C_{Lo}\cos\beta\sin\beta & -C_{L0}(1+\sin^2\beta) - C_{D0}\cos\beta\sin\beta \\ C_{L0}(1+\cos^2\beta) - C_{D0}\cos\beta\sin\beta & C_{D0}(1+\sin^2\beta) - C_{L0}\cos\beta\sin\beta \end{bmatrix}.$$

In addition to the terms given in equation (4.60), the so-called added mass terms must be included; thus, the final expression for the fluid forces is

$$F_o = \{F_x, F_z\}^{\mathrm{T}} = -\tfrac{1}{2}\rho U^2 Dl(F_o + \bar{g}[E]z + D/(bU)[B]\dot{z}_o + (\pi D^2/2U^2)[A]\ddot{z}_o), \qquad (4.61)$$

where $[A]$ is the added mass matrix. The added mass matrix could be evaluated using potential flow theory; however, because of the relatively large separation between risers, and realizing that the added mass is typically small compared with the true mass of a riser, the added mass matrix was taken to be that in unconfined fluid, and $[A] = [I_2]$.

Combining equations (4.57) and (4.61), the final equations of motion of the riser are obtained, which written in nondimensional form are

$$\{(1+\pi/4\overline{m})[I_2]\}z'' + \{(\delta/\pi)[I_2] + (V/2b\overline{m})[B]\}z' + \{[I_2] + (V^2/2\overline{m})\bar{g}[E]\}z = -F_o, \qquad (4.62)$$

where $\overline{m} = m/\rho D^2$ is the nondimensional mass, $\delta \simeq c_s\pi/ml\omega_n$ is the structural logarithmic decrement, $\omega_n = (k_s/ml)^{1/2}$ is the undamped natural frequency in vacuum, $(\;)' = \partial/\partial\tau$, $\tau = \omega_n t$ being the nondimensional time, and $V = U/\omega_n D$ is the nondimensional velocity. The stability of the system described by equation (4.62) may be assessed by ignoring the right-hand side, which merely changes the equilibrium position; a standard eigenvalue analysis is performed for increasing values of the flow velocity, until a stability boundary, given by an eigenvalue with a positive real part, is crossed. However, it should be noted that an iterative technique is required, because $\bar{g}$ cannot be determined until λ is known.

Before the analytical model can be employed, certain experimental inputs are required: the force coefficients, their derivatives with respect to the in- and cross-flow displacements and the direction and magnitude of the local approach velocity. This data was obtained from static experiments by Price *et al.* (1993) on a riser bundle in airflow, with geometry as shown in Figure 4.22.

To obtain the fluid-force coefficients, a two-dimensional model of the riser, very similar to that employed in the dynamic experiments was used. However, in these experiments the flexibly mounted tube was replaced by another, rigidly attached to a force balance, which enabled the in-flow (drag) and cross-flow (lift) forces to be measured simultaneously. In addition, the force balance was mounted on a mechanism which enabled the instrumented tube to be displaced from its equilibrium position in either the in-flow or cross-flow direction. Thus, the variation C_L and C_D with x- and y-displacement could be obtained giving the fluid stiffness terms C_{Lx}, C_{Ly}, C_{Dx} and C_{Dy}; numerical values of these stiffness terms are given in Price et al. (1993) for several values of ϕ.

The angle of the local approach velocity in the riser bundle with respect to the free-stream velocity, β, was assumed to be given approximately by the incidence of the stagnation point. The instrumented riser was equipped with a pressure tap, the orientation of which could be measured. The pressure was measured over a reasonable incidence range, and the orientation of maximum pressure, indicating stagnation, obtained. These results are also summarised in Price et al. (1993). An estimate of the magnitude of the approach velocity was obtained by assuming that the "true" drag coefficient of the peripheral tube remained constant as the orientation of the bundle was changed, and that the variation in C_D with ϕ is due to a variation in the local approach velocity. Thus, taking the true drag coefficient as $C_{D\infty} = 1.2$ one can write $(bU)^2 C_{D\infty} = U^2 C_D$; hence, using the measured C_D, b can be evaluated.

Using the analytical model and the experimental data, it was determined that for the riser positioned at an orientation of $\phi = 196°$, with $\overline{m} = 295$ and $\delta = 0.011$ (the same values as in some of the wind-tunnel experiments (Païdoussis et al. 1988b)), the stability boundary is given by $U_c/\omega_n D = 1.33$ compared with the experimental result of $U_c/\omega_n D = 2.6$; thus, although the comparison cannot be said to be good, it was encouraging that at this orientation the theoretical model predicted instability – in agreement with the experiments. The imaginary part of the eigenvalue showed very little variation with velocity, indicating that instability would occur at the riser natural frequency, in agreement with the experimental results (Païdoussis et al. 1988b).

For $\phi = 235°$ ($\overline{m} = 295$ and $\delta = 0.011$), $U_c/\omega_n D = 2.92$ was obtained; again, when compared with the experimental result of $U_c/\omega_n D = 0.8$, the agreement, although not good, was encouraging. Even more encouraging was that theoretical results for $\phi = 235, 270$ and $315°$ all indicated that instability would never occur, again in agreement with the experimental results.

The results presented in the preceding paragraph suggest that there is qualitative, but not quantitative, agreement between theory and experiment. Thus, although it cannot be concluded that this analysis can be used to give predictions of the actual flow velocity at which a riser will become unstable, it does appear that it is able to predict the orientations at which the riser bundle is likely to suffer fluidelastic instability. Furthermore, the reasonable success of this analysis demonstrates that the analytical model adequately represents the phenomena producing fluidelastic instability in riser bundles. In an attempt to determine which were the critical parameters in the analytical model, and so possibly improve the model, the effect of changing their magnitude on $U_c/\omega_n D$ was investigated; this was done for both $\phi = 196$ and $235°$. Because the two sets of results were very similar, only the results for $\phi = 196°$ will be discussed here.

First, the effect of b and β was investigated; it was found that for a reasonably wide variation in either of these parameters there was little change in $U_c/\omega_n D$ (see Price et al. (1993) for detailed results), indicating that the difference between theory

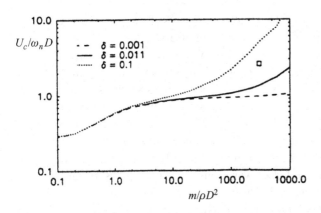

Figure 4.26. Variation of nondimensional critical flow velocity $U_c/\omega D$ for fluidelastic instability with nondimensional mass (Price et al. 1993); $\phi = 196°$; $\square$, experimental data point ($m/\rho D^2 = 295, \delta = 0.011$).

and experiment was likely not due to inaccuracies in their estimation. The effect of inaccuracies in the measurement of the various force coefficients was also examined; this was done by individually varying the force coefficients by $\pm 50\%$ of their nominal values (C_{Dx}, which was zero, was varied between -0.5 and $+0.5$). It was found that varying these force coefficients had little effect on the magnitude of $U_c/\omega_n D$, with the exception of C_{Ly}; even for C_{Ly}, however, a 50% increase or decrease caused only a 13% decrease or 37% increase, respectively, in $U_c/\omega_n D$.

The most sensitive parameter was the phase lag between the tube motion and the resulting fluid forces, accounted for by the term μ; this is in agreement with previous studies on heat-exchanger tube arrays (Price & Païdoussis 1986b), as discussed in Chapter 5. This shows that the unsteady nature of the fluid flow around the riser bundle has a dominant effect on the fluidelastic stability of the bundle. Price et al. (1993) showed that $\mu = 1$ gives almost the minimum value of $U_c/\omega_n D$ and, furthermore, a relatively small change in μ can produce a large increase in $U_c/\omega_n D$. Thus, it is suggested that the differences between the theoretical and experimental values of $U_c/\omega_n D$ are due to the uncertainty in the estimation of the phase lag, demonstrating its importance in the phenomena producing this instability. Unfortunately, it is *still* not possible to determine this phase lag exactly, indicating the continuing need to obtain a better understanding of the unsteady nature of the flow around an oscillating bluff body.

Although the agreement between theory and experiment can only be described as being qualitatively good, it is still of interest to study the theoretically predicted effect of varying either the structural mass or damping on $U_c/\omega_n D$. Detailed results are presented in Price et al. (1993); however, sample results showing the effect of $m/\rho D^2$ and δ are shown in Figure 4.26. Possibly the most significant observation from these results is that for low values of nondimensional mass, typical of those expected on a production riser, increasing the structural damping has little effect on $U_c/\omega_n D$.

Prior to the development of the analytical model of Price et al., Bokaian & Geoola (1984b) had considered the stability of two circular cylinders in close proximity to each other and subject to cross-flow, which is representative of a "bundle" of two riser tubes. The model developed by Bokaian & Geoola (1984b) was restricted to transverse motion of one of the risers, and in their original paper it was the dynamic instability of the upstream riser which was considered. The ultimate objective of Bokaian & Geoola was to predict not only the velocity required to initiate instability

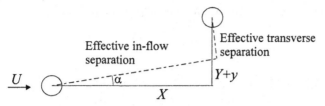

Figure 4.27. Schematic showing "effective" in-flow and transverse displacement used by Bokaian & Geoola (1984b).

but also the amplitude of oscillation of the ensuing instability; hence, they retained nonlinear terms in the derivation of the fluid forces acting on the riser and the equations of motion were solved using the first approximation of Krylov & Bogoliubov. A summary of the analytical model developed by Bokaian & Geoola (1984b) is given in the following.

The equation of motion for transverse vibration of the riser is given by equation (4.1), where the mass term, m, now includes the effect of the added mass. Using a standard quasi-steady approach, the fluid force acting on the oscillating riser is given by equation (4.2). However, realizing that only transverse vibration is allowed, the expressions for U_r and α become

$$U_r = \left[U^2 + (D\dot{y})^2 \right]^{1/2} \quad \text{and} \quad \alpha = \sin^{-1}(D\dot{y}/U_r),$$

where it is recalled that y is a nondimensional displacement. Bokaian & Geoola (1984b) restricted their analysis to small α and neglected the $\dot{y}^2$ terms; hence, $U_r = U$. Then, equations (4.1) and (4.2) can be combined to yield

$$\ddot{y} + 2\omega_n \zeta \dot{y} + \omega_n^2 y = -(nU/D)[C_L U/D + C_D \dot{y}], \tag{4.63}$$

where $n = \rho D^2/2m$, $\omega_n^2 = k/ml$ and $\zeta = c/(2ml\omega_n)$.

Bokaian & Geoola (1984b) linearised the C_L and C_D terms, but instead of linearizing by making use of the displacements of the oscillating riser, they did so using the *change in effective* wake position: furthermore, they linearised about the displaced position of the oscillating riser. Hence, the linearised terms for C_L and C_D can be expressed as

$$\begin{aligned} C_D &= C_{Do} + C_{Dx_0}\delta X + C_{Dy_0}\delta(y+Y), \\ C_L &= C_{Lo} + C_{Lx_0}\delta X + C_{Ly_0}\delta(y+Y), \end{aligned} \tag{4.64}$$

where $C_{Dx_0} = \partial C_D/\partial X$ (and similarly C_{Dy_0}, C_{Lx_0} and C_{Ly_0}), X and Y are the original nondimensional in-flow and transverse separations between the risers, y is the displacement of the upstream riser in the transverse direction and δX and $\delta(y+Y)$ are the changes in "effective" separation of the risers accounting for both the displacement of the upstream riser and the inclination of the wake. As previously mentioned, the linearisation is about the displaced position of the riser; hence, the term $\delta(y+Y)$ rather than δY. This approach is rather unusual and requires that the values of C_{Do}, C_{Lo}, C_{Dx_0}, C_{Dy_0}, C_{Lx_0} and C_{Ly_0} be evaluated about this displaced position.

As shown in Figure 4.27, neglecting $\dot{y}^2$ terms and assuming that the wake behind the oscillating upstream riser is shed at an inclination of $\alpha = \sin^{-1}(D\dot{y}/U)$ with respect to the free-stream velocity, then δX, which is the "effective in-flow separation" minus the original separation (X), may be expressed as

$$\delta X = \alpha (Y + y) = \frac{\dot{y}D}{U}(Y + y), \tag{4.65a}$$

and similarly

$$\delta (y + Y) = -\alpha X = -\frac{\dot{y}D}{U}X. \tag{4.65b}$$

Using equations (4.64) and (4.65) the equation of motion for transverse vibration of the riser (equation (4.63)) may now be written as

$$\ddot{y} + \omega_n^2 y = -nV\omega_n \left[V\omega_n C_{Lo} + (Y + y)\dot{y}C_{Lx_0} - X\dot{y}C_{Ly_0} + \dot{y}C_{Do} + \frac{2\varsigma\dot{y}}{Vn} \right], \tag{4.66}$$

where again $\dot{y}^2$ terms have been ignored, and $V = U/(\omega_n D)$. It should be noted that although $\dot{y}^2$ terms have been ignored, the nonlinear $y\dot{y}$ terms have been retained in equation (4.66) – this is somewhat surprising considering that for realistic values of ω the $\dot{y}^2$ terms will be larger than the $y\dot{y}$ terms.

From equation (4.66) it is apparent that the total damping (structural plus fluid) for the oscillating riser is equal to

$$nV\omega_n \left[(Y + y)C_{Lx_0} - XC_{Ly_0} + C_{Do} + \frac{2\varsigma}{Vn} \right] \dot{y}, \tag{4.67}$$

and instability will occur when this is negative.

The method of solution employed by Bokaian & Geoola (1984b) was to first express C_L and C_D in terms of power series of X and $Y + y$, which were of the form

$$C_L = \sum_{i=0}^{5} \sum_{j=0}^{4} A_{(2i+1,j)}X^j (y + Y)^{2i+1} \quad \text{and} \quad C_D = \sum_{i=0}^{6} \sum_{j=0}^{3} C_{(2i+1,j)}X^j (y + Y)^{2i}.$$

The coefficients A and C were obtained from their own experimental force-coefficient measurements. Then, after considerable algebra, equation (4.66) is expressed as

$$\ddot{z} + \omega_n^2 z = -n\omega_n Vf(z, \dot{z}), \tag{4.68}$$

where $z(t)$ is purely the time-varying component of y about its new static position, accounting for the static fluid forces acting on the riser, and the function $f(z, \dot{z})$ includes all terms associated with the fluid forces as well as the structural damping.

Equation (4.68) was solved using the first approximation of Krylov & Bogoliubov, assuming the solution to be of the form

$$z = a \cos(\omega_n t + \phi),$$

where

$$\frac{da}{dt} = K(a) = \frac{nV}{2\pi} \int_0^{2\pi} f(a \cos \psi, - a\omega_n \sin \psi) \sin \psi d\psi \tag{4.69}$$

and $\psi = \omega_n t + \phi$. A limit-cycle oscillation (LCO) occurs when $da/dt = 0$, and thus solving for $K(a) = 0$ gives the magnitude of a as a function of V; whether the LCO is stable or not can be determined from $dK(a)/da$ which is negative for a stable LCO

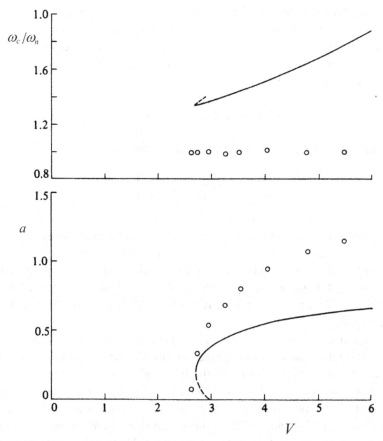

Figure 4.28. Variation of limit-cycle amplitude, a, and frequency of oscillation, ω_c/ω_n, with nondimensional velocity $V = U/\omega_n D$ for the upstream of two risers with $X = 1.25$, $Y = 0.0$ and $m\delta/(\rho D^2) = 3.723$ (Bokaian & Geoola 1984b): $\circ$, experimental data; —, stable LCO; --, unstable LCO.

and positive for an unstable one. Furthermore, the frequency of oscillation for the limit cycle is

$$\omega_c = \left(\omega_n^2 + \frac{nV\omega_n}{\pi a} \int_0^{2\pi} f(a\cos\psi, -a\omega_n\sin\psi)\sin\psi d\psi \right)^{1/2}. \qquad (4.70)$$

A typical set of results for $X = 1.25$, $Y = 0.0$ and $\zeta/n = 1.185$ $[m\delta/(\rho D^2) = 3.723]$ is presented in Figure 4.28, where a comparison is shown with their own experimental results. The riser loses stability at $V = 3.0$ via a subcritical Hopf bifurcation; the lower branch of the LCO curve is unstable for $2.75 \leq V \leq 3.0$ approximately, and a stable LCO exists for $V \geq 2.75$ approximately. It is evident that although there is reasonable agreement for the critical velocity required to initiate instability, the agreement in terms of magnitude of the LCO is not so good. Two potential sources of error in the magnitude of the LCO are: first, that all $\dot{y}^2$ terms were neglected, and second, that C_L and C_D have been linearised, via equation (4.64).

As stated earlier in this section the instability is damping-controlled, and the instability boundary can be obtained by setting the total damping, given by equation

(4.67), equal to zero. To gain further insight into the mechanism producing this instability, it is useful to consider a linearised form of this expression (which involves setting $y = 0$) and also considering the specific case where the downstream riser is initially on the wake centreline of the upstream one, or $Y = 0$. The total damping may then be written as

$$ nV\omega_n \left[-XC_{Ly_0} + C_{Do} + \frac{2\varsigma}{Vn} \right] \dot{y}. $$

Hence, the critical velocity required to initiate instability is given by

$$ \frac{U_c}{\omega_n D} = \left\{ \frac{2/\pi}{-C_{Do} + XC_{Ly_o}} \right\} \frac{m\delta}{\rho D^2}. \tag{4.71} $$

This expression is very similar to that given at the beginning of this chapter in equation (4.9). In both cases $U_c/\omega_n D$ is proportional to the mass-damping parameter, $m\delta/(\rho D^2)$; also, assuming C_{Do} to be positive, it is evident that instability can only occur when C_{Ly} is positive (when $Y = 0$, C_{Lyo} is identical to C_{Ly}). As noted earlier in this section, Bokaian & Geoola (1984b) obtained dynamic instabilities only for small values of X and Y, where C_{Lyo} is positive and large, which is consistent with this conclusion.

One significant difference between equations (4.9) and (4.71), however, relates to the C_{Ly} term. In equation (4.9) the reason why C_{Ly}, which is a fluid stiffness term, contributes to the damping is due to the assumed phase lag between the cylinder displacement and the resulting fluid forces, accounted for via the flow-retardation term μ. In Bokaian & Geoola's analysis flow retardation is not considered. However, it is assumed that the oscillating upstream riser sheds its wake at inclination $\alpha \approx \dot{y}D/U$ with respect to the upstream direction. Hence, the apparent transverse location of the downstream riser depends on $\dot{y}$ and X, and the contribution to the fluid damping from the C_{Ly} term is proportional to XC_{Ly_o}.

One weakness of the analysis presented by Bokaian & Geoola (1984b) is that it cannot explain the origin of dynamic instabilities experienced by the downstream of a pair of risers (Bokaian & Geoola 1984a), where there is no wake inclination to be accounted for. In a later paper, Bokaian (1989) considered this case and concluded that "self-excited galloping motion occurs only in the region where the static drag force on the downstream cylinder is negative". This is in agreement with equation (4.71), but as shown at the beginning of this chapter (equation (4.9)) and earlier in this section, retardation of the flow as it approaches the riser can cause negative fluid damping, even when the riser is restricted to purely transverse motion.

4.3.2 (b) Riser clashing

Apparently motivated by the clashing observed on multitube flexible risers in a steady current, Huse (1996), Sagatun et al. (2002), Wu, Huang & Barltrop (2002, 2003) and Blevins (2005) examined the stability of one flexible riser tube in the wake of another.

In the analyses presented by Wu et al. (2002, 2003) and Blevins (2005) the hydrodynamic forces acting on an oscillating downstream tube are written using Simpson's (1971a) quasi-steady analysis for overhead transmission lines; thus, the fluid forces are given by equation (4.24). In addition, it was assumed that there is no structural coupling either between the in-flow and cross-flow directions of the tubes or between individual tubes in a riser bundle; hence, the structural equations for each tube are given by equation (4.30b) with all off-diagonal terms set to zero. The equations of motion for the downstream tube can then easily be obtained from a modified form of equation (4.32), and are given by the following:

$$[I]\begin{Bmatrix} \ddot{x} \\ \ddot{y} \end{Bmatrix} + \frac{1}{ml}\begin{bmatrix} c_x & 0 \\ 0 & c_y \end{bmatrix}\begin{Bmatrix} \dot{x} \\ \dot{y} \end{Bmatrix} + \frac{\frac{1}{2}\rho DU}{mb}\begin{bmatrix} 2C_{D0} & -C_{L0} \\ 2C_{L0} & C_{D0} \end{bmatrix}\begin{Bmatrix} \dot{x} \\ \dot{y} \end{Bmatrix} + \frac{1}{ml}\begin{bmatrix} k_x & 0 \\ 0 & k_y \end{bmatrix}\begin{Bmatrix} x \\ y \end{Bmatrix}$$

$$+ \frac{\frac{1}{2}\rho U^2}{m}\begin{bmatrix} C_{Dx} & C_{Dy} \\ C_{Lx} & C_{Ly} \end{bmatrix}\begin{Bmatrix} x \\ y \end{Bmatrix} = -\frac{\frac{1}{2}\rho U^2}{m}\begin{Bmatrix} C_{D0} \\ C_{L0} \end{Bmatrix}. \qquad (4.72)$$

In fact, Wu *et al.* (2002, 2003) neglected the structural damping terms, whereas Blevins (2005) included them but neglected the fluid damping.

Because it is assumed that there is no structural coupling between the downstream and upstream tubes, the upstream tube will not oscillate; hence, the stability of the two tubes can be evaluated by considering purely motions of the downstream one using equation (4.72). However, the upstream tube *is* subject to a steady drag force, the effect of which must be accounted for when determining the equilibrium positions of the two tubes and the relative separation between them.

One significant problem which arises in the analysis of flexible risers *vis-à-vis* overhead transmission lines is that the equilibrium wake position of the tubes strongly depends on the current velocity. In particular, the downstream tube is subject to a reduced drag force compared with the upstream one, as well as a force in the transverse direction. Although this is also the case for overhead transmission lines, the extremely flexible nature of very long risers exacerbates this effect, resulting in a significantly reduced separation, and possible clashing, between individual risers. Consequentially, there is a need to evaluate C_L and C_D, as well as the fluid-stiffness terms C_{Lx}, C_{Ly}, C_{Dx} and C_{Dy}, as continuous functions of the downstream tube position over a wide range of tube separation. In addition, the upstream and downstream tubes can have different diameters.

In general, the C_L and C_D data available in the literature is given only at discrete values of wake position and is restricted to the upstream and downstream cylinders having the same diameter (Counihan 1963; Ko 1973; Cooper 1974; Price 1975a, b; Wardlaw *et al.* 1975). To overcome these difficulties, both Wu et al. (2002) and Blevins (2005) obtained approximate expressions for C_L and C_D as a function of wake position.

Blevins (2005) started from Schlichting's (1968) analysis of fully developed wakes for nonlifting sections and obtained the following expression for the mean velocity at any position in the wake of the upstream tube:

$$\overline{U}(X, Y) = U(1 - 1.2(C_{Du\infty}/X)^{1/2})\exp(-13Y^2/(C_{Du\infty}X)), \qquad (4.73)$$

where $C_{Du\infty}$ is the drag coefficient of the upstream tube and X and Y are the nondimensional in-flow and transverse wake positions (the nondimensionalisation being done using the upstream tube diameter, D_u). Then, assuming that the downstream tube has a constant drag coefficient of $C_{Dd\infty}$ based on the local velocity (this would be the value of drag coefficient measured for that tube in a uniform flow), and that the reduction in drag for the downstream tube is purely due to the reduction in wake velocity, an approximate expression for the drag coefficient of the downstream tube, C_{Dd}, can be obtained. In fact, Blevins replaced the constants 1.2 and 13 in equation (4.73) by values obtained by fitting his predictions of C_{Dd} to the experimental data of Price (1975a, b) and Price & Païdoussis (1984b), giving the following expression:

$$C_{Dd}(X, Y) = C_{Dd\infty}[1 - (C_{Du\infty}/X)^{1/2} \exp(-4.5Y^2/(C_{Du\infty}X))]^2, \qquad (4.74)$$

(Blevins used $C_{Du\infty} = C_{Dd\infty} = 1.15$). For the lift coefficient, Blevins started from Rawlins' (1974a) assumption that the lift is proportional to the transverse gradient of the drag, then making use of equation (4.74), and again adjusting the constants to best fit the data of Price (1975a, b) and Price & Païdoussis (1984b), he obtained the following for the lift coefficient on the downstream tube:

$$C_{Ld}(X, Y) = -10.6 \left(\frac{YC_{Dd\infty}D_d}{XC_{Du\infty}D_u} \right) \left(\frac{C_{Du\infty}}{X} \right)^{1/2}$$

$$\times \left[1 - \left(\frac{C_{Du\infty}}{X} \right)^{1/2} \exp \left(\frac{-4.5Y^2}{C_{Du\infty}X} \right) \right] \exp \left(\frac{-4.5Y^2}{C_{Du\infty}X} \right), \text{ for } Y \geq 0. (4.75)$$

Blevins (2005) presented comparisons between the values of C_L and C_D given by equations (4.74) and (4.75) and the experimental data of Price (1975a, 1976) and Price & Païdoussis (1984b). In general, the agreement is excellent, even when the upstream and downstream tubes have different diameters; however, equation (4.74) does not capture the negative values of C_D known to exist close to the wake centreline for $X < 3$ (see, for example, Price & Païdoussis (1984b) or Zdravkovich (1977)). In addition, as shown by Price & Païdoussis (1984b) the variation of C_L with Y can be very complex for $X \leq 3$, with double peaks in C_L occurring on each side of the wake; this is not captured by equation (4.75).

Wu et al. (2002) also used Schlichting's (1968) analysis of fully developed wakes to predict the magnitude of the velocity in the wake of the upstream tube; however, they then employed a free-streamline approach to model the wake boundary for the downstream tube, with this separating streamline being represented by a number of vortex elements. There are a number of simplifying assumptions required in the methodology employed by Wu et al. (2002) which are fully described in their paper; however, comparisons shown between the results obtained using their model for $X \geq 6$ and the experimental data of Price (1976) are very impressive.

Using their own individual models for the fluid-dynamic forces on the downstream of two tubes, both Wu et al. (2002, 2003) and Blevins (2005) investigate the stability of the downstream tube. Although both sets of authors show that dynamic instability is possible, they both conclude that the most serious problem is a static instability of the downstream tube, and that the clashing observed by Huse (1996) may be due to this static instability. Both Wu et al. and Blevins show that, as the current velocity increases, the equilibrium position of the downstream tube moves closer to the upstream one; this, of course, is because the downstream of the two tubes experiences a smaller static drag force than the upstream tube. What is more surprising is that at a critical value of current the static equilibrium position of the downstream tube bifurcates into two equilibrium positions, only one of which is stable. This is

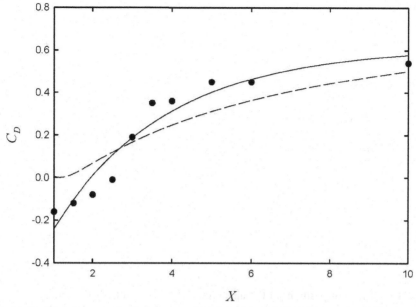

Figure 4.29. Variation of drag coefficient, C_D, for the downstream riser as a function of wake position, X: ●, experimental data from Zdravkovich & Pridden (1977) at Re $= 3.1 \times 10^4$; —, equation (4.77); - -, equation (4.76).

then followed, at a slightly higher magnitude of current, by the nonexistence of any equilibrium position for the downstream tube.

The reason for the behaviour of the equilibrium position of the downstream tube can be explained by considering the simple case where the two tubes have the same diameter and the downstream tube is positioned on the centreline of the upstream one (in fact, because of the extremely flexible nature of the risers, as the current increases the transverse fluid force acting on the downstream tube will tend to push it very close to the wake centreline). With the downstream tube on the wake centreline, its drag coefficient as predicted by equation (4.74) becomes

$$C_{Dd} = C_{Dd\infty}\left(1 - (C_{Du\infty}/X)^{1/2}\right)^2, \tag{4.76}$$

and the lift coefficient is zero. As mentioned previously this expression does not capture the negative values of the drag coefficient for small values of separation between the two tubes. As an alternative to using this expression, starting from the drag coefficient data given by Zdravkovich & Pridden (1977) for Re $= 3.1 \times 10^4$ and using a simple curve fit to this data, a crude approximation of the drag coefficient on the wake centreline is given by

$$C_{Dd} = -0.5941 + 1.2097\left[1 - \exp\left(-0.3461X\right)\right]. \tag{4.77}$$

A comparison between these two expressions for C_{Dd} and the experimental data of Zdravkovich & Pridden (1977) is given in Figure 4.29 showing that (4.77) does capture the negative values of C_{Dd} and gives a reasonable representation of its variation with X.

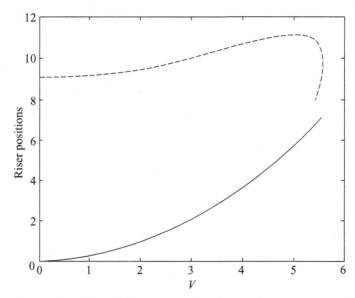

Figure 4.30. Locus of positions of the upstream $(-x_u)$, ——, and downstream $(X_o - x_d)$, - - -, risers, as measured from the no-current position of the upstream riser, as a function of V: $X_o = 9, a = 0.2$.

Given the drag coefficients for the two tubes, their static displacements may be obtained as a function of the current velocity and written as

$$x_u = -aV^2C_{Du} \quad \text{and} \quad x_d = -aV^2C_{Dd}, \tag{4.78}$$

where $V = U/\omega_n D$ and $a = \rho D^2/(2m)$.

Taking $C_{Du} = 1.15$ and C_{Dd} as a function of $X = X_o - x_d + x_u$ (as given by equation (4.77)), where X_o is the initial no-current separation between the two tubes, then for $a = 0.2$ and $X_o = 9$ the static positions of the upstream $(-x_u)$ and downstream tubes $(X_0 - x_d)$ – as measured from the no-current position of the upstream tube – are shown in Figure 4.30 as a function of V; the separation between the tubes, X, is simply the distance between the two curves. For $V < 5.4$, approximately, both tubes have one equilibrium position only – which in both cases is stable. However, when V is increased beyond 5.4 the variation of the structural restoring force and drag force with tube displacement for the downstream tube is as exemplified by the results of Figure 4.31 for $V = 5.5$ (the forces presented are normalised with respect to the structural stiffness, hence, the slope of the structural restoring force variation with riser displacement is unity). It is apparent that there are now two positions of the downstream tube (approximately 1.76D and $-0.35D$) where the drag force is just balanced by the structural restoring force; hence, there are two equilibrium positions. For the equilibrium position at 1.76D (i.e. 1.76D downstream of the no-current equilibrium position), a positive increment in the tube displacement results in the restoring force being greater than the drag force; hence, the tube returns to its equilibrium position which is stable. However, for the second equilibrium position $(-0.35D)$ corresponding to a forward movement of the downstream riser, a further movement upstream results in the magnitude of the negative drag force being greater than the structural restoring force, and hence this position is unstable. As shown in

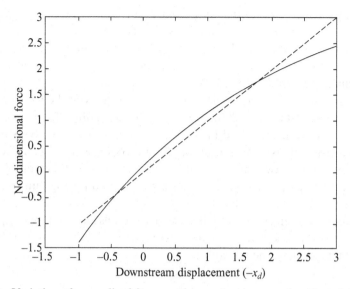

Figure 4.31. Variation of normalised forces acting on the riser as a function of downstream displacement, $-x_d$, from its no-flow equilibrium position for $V = 5.5$: $X_o = 9$, $a = 0.2$: - -, structural restoring force; ——, drag force.

Figure 4.31, two equilibrium positions can only exist when the slope of the drag curve changes from being less than 45° to greater than 45° over the range of displacement of the tube. The locus of the stable equilibrium position for the downstream tube is as given by the upper branch of the curve in Figure 4.30, whereas the locus of the unstable equilibrium is given by the lower branch – extending from $5.4 \leq V \leq 5.6$ approximately.

Once V is increased above approximately 5.6, the downstream tube is close enough to the upstream one to cause its drag coefficient to be sufficiently large and negative to result in it being "sucked" onto the upstream one – inducing "clashing" – and, hence, no equilibrium positions exist for the downstream tube.

As shown by both Wu et al. (2002, 2003) and Blevins (2005), if the downstream tube is positioned off the wake centreline then the resulting lift force acting on the downstream riser complicates matters and there is the possibility of more than two equilibrium positions existing.

The instability scenario presented in the preceding is based purely on a static analysis. In addition, there is the possibility of dynamic instability occurring. Wu et al. (2002, 2003) and Blevins (2005) consider this, and Wu et al. (2003) present dynamic stability boundaries as a function of wake position for zero structural damping. As previously mentioned, the occurrence of dynamic instability can be investigated using equation (4.63) – assuming that the tube equilibrium position has first been evaluated. However, as indicated in Section 4.2.1, "undamped" theory (ignoring both the structural and fluid damping) indicates that a necessary condition for instability to occur is given by equation (4.40); for no structural coupling between the two directions ($\varepsilon = 0$) this reduces to

$$C_{Lx}C_{Dy} \leq 0,$$

which is only satisfied on the edge of the wake. The undamped theory also indicates that, when $\omega_x = \omega_y$ or $k^2 = 1$, instability will occur at zero velocity, but as shown by the results of Figure 4.12 the presence of damping, either structural or fluid, will eliminate this instability – at least for values of a typically found in overhead transmission lines (of order 2×10^{-4}).

Sagatun *et al.* (2002) used an "in-house" CFD code (with the Reynolds number being held constant at Re = 200) to predict the static fluid forces acting on the two tubes. These were then used in two different quasi-static analyses of the riser motion (the effect of riser motion on the fluid forces was not accounted for); one analysis being based on two-dimensional strip theory and the other employed a three-dimensional finite-element approach. Results presented by Sagatun *et al.* indicate that clashing will occur and that the agreement, in terms of current required to cause clashing, between either of their analytical models and their own experimental results is remarkably good; unfortunately, no indication is given whether the clashing is due to a static or dynamic instability.

5 Fluidelastic Instabilities in Cylinder Arrays

5.1 Description, Background, Repercussions

A cylinder array is an agglomeration of cylinders parallel to one another in a geometrically repeated pattern. Typical examples are tube arrays in heat exchangers, involving hundreds or thousands of tubes; one fluid flows within the tubes, while another flows around them, partly normal to their long axis (cross-flow). Here, by heat exchangers we understand also steam generators and boilers, where the outer fluid boils and produces steam, and condensers, where the reverse process is involved. Figure 5.1 shows sectional views of two different types of steam generators. In both cases, the outer fluid flows axially in some regions and transversely as a cross-flow in others (Païdoussis 1981, 1983, 2004). It is well known that cross-flow promotes heat transfer, but at the expense of higher vibration levels and the possibility of cross-flow-induced instabilities, the subject matter of this chapter. Figure 5.2 shows the winding pattern of the outer flow in another kind of heat exchanger, involving cross-flow nearly everywhere.

What distinguishes heat-exchanger arrays compared with the groups of cylinders discussed in Chapter 4 is that (i) we have many more cylinders here and (ii) they are more closely spaced. The significance of (i) is that, other than the cylinders cross-sectionally on the periphery, the flow around any given cylinder is confined by the presence of adjacent cylinders, rather than being unconfined at least on one side. The repercussions of (ii) are that the fluid-dynamic forces can be quite large and the flow structure quite distinct. In heat exchangers, the centre-to-centre distance between tubes, or pitch, is typically 1.25 to 2.00 diameters, whereas for overhead transmission lines it is typically 12 to 20 conductor diameters.

Cylinder arrays are also found in nuclear reactor fuel channels; see for example Figure 5.3. In some designs, e.g. the CANDU design, the fuel rods (containing uranium oxide) are even more closely spaced, with a pitch of ~ 1.1 diameters. In this case, to avoid excessive vibration the flow is mainly axial, but some cross-flow near the inlet and outlet of the fuel channel is almost unavoidable.

Typical geometric patterns of heat-exchanger tube (cylinder) arrays are shown in Figure 5.4, namely (a) normal triangular, (b) rotated triangular, (c) in-line square and (d) rotated square arrays. If the flow in (a) were from the left, then this would become a rotated triangular array. There are surprisingly important differences in dynamical

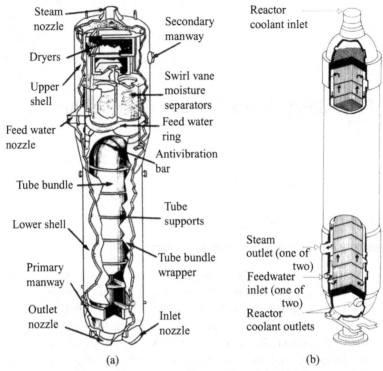

Figure 5.1. Sectioned views of (a) a U-tube type steam generator and (b) a 'once-through' type steam generator (Païdoussis 1980, 2004).

behaviour between these types of array, even between the normal triangular and the rotated square ones – which only differ by the layout angle being 30° or 45°. More obvious differences are expected between (i) the rotated triangular and in-line square arrays where there is a virtually unimpeded straightthrough flow path around the tubes, no matter how close they are, and (ii) the normal triangular and rotated-square patterns where the flow path must be sinuous, in particular for small pitch, P/D. However, other geometrical patterns are sometimes used, not

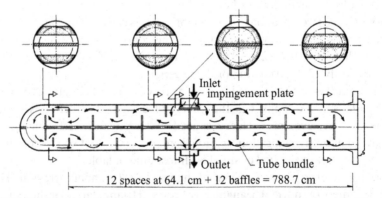

Figure 5.2. Cross-sectional view of a water-water heat exchanger of unusual design; the tubes in this shell-in-tube heat exchanger are not shown for clarity (Païdoussis 1980).

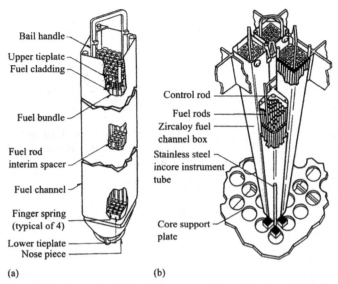

Figure 5.3. (a) A BWR (boiling-water reactor) fuel bundle in a fuel channel; (b) part of the core of the BWR, showing four fuel bundles in their fuel channel boxes (Païdoussis 1980, 2006).

involving equilateral triangles or squares, e.g. rectangular patterns in boilers where the longitudinal and lateral pitches are different.

The dynamical behaviour of individual cylinders in an array in cross-flow may be summarised, in idealised form, as in Figure 5.5 (Païdoussis 1981, 1983). At all flow velocities the cylinders are subject to turbulence-induced vibration or "buffeting". The amplitudes are generally small, leading to long-term problems only, e.g. when impacting with supports promotes fretting wear or because of fatigue. Vortex shedding occurs in arrays in a modified form of vortex shedding from a single cylinder. Hence, resonance is possible in this case also (Chapter 3), leading to vibration of considerable amplitude and possible short-term failures. At generally higher flow velocities, so-called fluidelastic instability may occur, with large-amplitude vibration

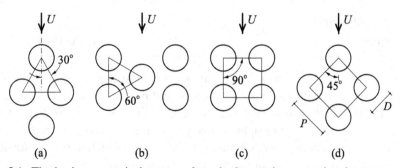

Figure 5.4. The basic geometrical patterns for tube layout (cross-sectional patterns of tube geometry) in heat exchangers: (a) normal-triangular, (b) rotated-triangular (or "parallel-triangular"), (c) normal or in-line square, and (d) rotated-square.

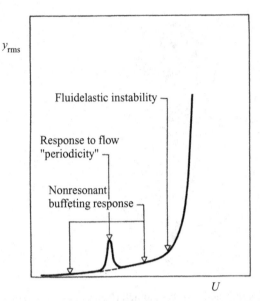

Figure 5.5. Idealised diagram of the response of a cylinder in an array subjected to cross-flow (Païdoussis 1981).

causing impact with supports and adjacent cylinders and short-term failure (in days rather than years); see Païdoussis (1980, 1983, 2006).*

Notation used in Chapter 5	
Cylinder diameter:	D
Free-stream flow velocity:	U
Relative flow velocity:	U_r, see equation (5.3)
Pitch velocity:	U_p, see equation (5.25)
Reduced flow velocity:	U/fD, f being the frequency in Hz $U/\omega D$, ω being the radian frequency
Dimensionless streamwise displacement:	x (positive downstream), see Figure 5.7
Dimensionless transverse displacement:	y (positive upwards), see Figure 5.7
Pitch between cylinders:	P, see Figure 5.4
Mass per unit length:	m
Mass ratio:	$m/\rho D^2 = \overline{m}$, ρ being the fluid density

The existence of fluidelastic instability in cylinder arrays was not discovered till the 1960s, although failures because of it occurred before but were erroneously attributed to "vortex shedding" (Païdoussis 1980, 2006). Moreover, insufficient knowledge, at a time when new designs of nuclear steam generators were being built, with even higher flow velocities, caused a proliferation of failures worldwide. In approximately a decade, the cumulative damages to the power-generating industry

---

* In Naudascher & Rockwell's (1994, 2005) classification, buffeting is an extraneously induced excitation (EIE), and vortex-shedding resonance is an instability-induced excitation (IIE); fluidelastic instability involves a movement-induced excitation (MIE), which is the very subject of this book.

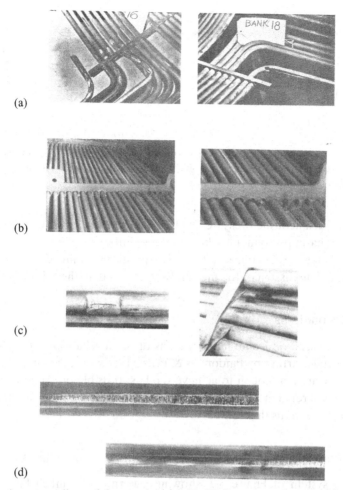

Figure 5.6. A compendium of damage in (a) a Na-H_2O steam generator, (b) a steam condenser, (c) a heat exchanger due to fretting at the location of supports, and (d) a heavy-to-light water heat exchanger. In (a), (b) and (d) the cause was likely or definitely due to fluidelastic instability (Shin & Wambsganss 1977; Pettigrew 1977; Pettigrew *et al.* 1978; Païdoussis 1980).

(including power replacement costs) are estimated at 1000 M\$. A compendium of practical cases of flow-induced vibration in nuclear reactors and heat exchangers may be found in Païdoussis (1980); in many of the cases analysed (14 of 33 involving cross-flow) the likely or definite cause of the problem was fluidelastic instability. The reader is also referred to Shin & Wambsganss (1977), Pettigrew, Sylvestre & Campagna (1978), Axisa (1993) and Au-Yang (2001).

The damages to heat exchangers caused by fluidelastic instability may sometimes involve a few tubes; but in other cases they can be devastating and even spectacular, as seen in Figure 5.6(a, b). Those in (a) involved a Na-H_2O steam generator; hence, when the tubes were sufficiently thinned out by fretting, the explosive Na-H_2O reaction virtually destroyed the system.

Another serious problem that may arise in tube arrays of heat exchangers with gaseous outer flow is that of acoustic resonance, involving very strong sound (160–176 dB in the heat exchanger, according to Blevins (1990), and only 20–40 dB

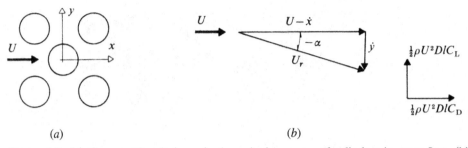

(a)   (b)

Figure 5.7. (a) Cross-sectional view of a kernel of an array of cylinders in cross-flow. (b) Velocity vector diagram and forces acting on the central cylinder, which is flexibly supported, while the others are immobile.

less outside). Such acoustic pressure levels are sufficiently high to damage the heat exchanger and harm personnel working in the vicinity.

In this chapter, we are concerned with fluidelastic instability exclusively, starting with a simplified discussion of the mechanisms involved, in the following section.

5.2 The Mechanisms

This section is based on a simplified analysis of the mechanisms of fluidelastic instability in cylinder arrays by Païdoussis & Price (1988); its main virtue is simplicity and therefore clarity, although to achieve the best possible predictions of the phenomenon, several refinements (different ones in different theories) are necessary, as described in the sections that follow.

5.2.1 The damping-controlled one-degree-of-freedom mechanism

Consider the system of Figure 5.7, showing a kernel of a much larger array of cylinders. For simplicity, consider all the cylinders to be immobile, except the central one which is also rigid, but flexibly mounted. Conventionally, in contrast to the work in Chapters 2 and 4, its displacements x and y are positive upwards and to the right, as shown.

Assuming no mechanical coupling between the x- and y-directions, the equation of motion of the cylinder in the y-direction is

$$(ml\ddot{y} + c\dot{y} + ky) D = F_y, \qquad (5.1)$$

where l is the length of the cylinder of mass m per unit length, D is the cylinder diameter, c and k are the effective mechanical damping and stiffness, F_y is the fluid-dynamic force and y the nondimensional displacement of the cylinder.[*] In terms of the lift and drag forces shown in Figure 5.7(b), where C_L and C_D are the lift and drag coefficients, using a quasi-steady approach we can write

$$F_y = \tfrac{1}{2}\rho\, U_r^2\, lD\{C_L \cos(-\alpha) - C_D \sin(-\alpha)\}, \qquad (5.2)$$

---

[*] Equation (5.1), involving both dimensional and dimensionless quantities (and hence D), may appear awkward, but it will prove to be very convenient in what follows.

where U_r and α are defined in Figure 5.7(b), such that

$$U_r = [(U - D\dot{x})^2 + (D\dot{y})^2]^{1/2}, \quad -\alpha = \sin^{-1}(D\dot{y}/U_r). \tag{5.3}$$

For small motions, we can write $C_L = C_{L_0} + (\partial C_L/\partial x)\, x + (\partial C_L/\partial y)\, y$, and similarly for C_D. Then equation (5.2) may be linearised to give

$$F_y = \frac{1}{2}\rho U^2 \, lD \left[C_{L_0} - 2C_{L_0} \left(\frac{\dot{x}D}{U} \right) + \left(\frac{\partial C_L}{\partial x} \right) x + \left(\frac{\partial C_L}{\partial y} \right) y - C_{D_0} \left(\frac{\dot{y}D}{U} \right) \right]. \tag{5.4}$$

For symmetric geometrical patterns of the cylinders, $C_{L_0} = 0$ and $\partial C_L/\partial x = 0$ and equation (5.4) simplifies to

$$F_y = \frac{1}{2}\rho U^2 lD \left[\left(\frac{\partial C_L}{\partial y} \right) y - C_{D_0} \left(\frac{\dot{y}D}{U} \right) \right]. \tag{5.5}$$

Now, as discussed in Section 4.1, a time lag exists between cylinder displacement and the fluid-dynamic forces generated thereby. This, like the discussion relating to the limits of quasi-steady fluid-dynamic theory, has been introduced via differing physical reasoning by different researchers. Simpson & Flower (1977) associate this time delay with the retardation experienced by the fluid as it nears the cylinder, particularly in the vicinity of the stagnation point, whereas intercylinder positions have changed meanwhile as a result of cylinder motions. This explanation was adapted for cylinder arrays by Price & Païdoussis (1984a, 1986a, b). On the other hand, Lever & Weaver (1982) associate this time delay with the time taken for the two fluid streams on either side of the cylinder to readjust to the changing flow-channel configuration as a cylinder oscillates. Païdoussis, Mavriplis & Price (1984) more generally conceived this time lag as a delay in the viscous wake adjusting continuously to the changing conditions due to cylinder motions. Last, Granger & Païdoussis (1996) related this time lag to the necessary reorganisation of the flow, driven by diffusion-convection of the vorticity layer generated on the surface of the body, following each and every change in its velocity. Whatever the explanation, in particular because a simplified first-order model will be used here, the time delay τ, may be expressed as (Price & Païdoussis 1984a)

$$\tau = \mu D/U, \tag{5.6}$$

where $\mu \sim \mathcal{O}(1)$, and D/U is the time taken for the fluid to travel a distance equal to one cylinder diameter.

Hence, with time delay taken into account, equation (5.5) may be written as

$$F_y = \frac{1}{2}\rho U^2 \, lD \left[e^{-i\omega\tau} \left(\frac{\partial C_L}{\partial y} \right) y - C_{D_0} \left(\frac{\dot{y}D}{U} \right) \right]. \tag{5.7}$$

Substituting in the equation of motion, we obtain

$$\ddot{y} + \left[\left(\frac{\delta}{\pi} \right) \omega_n + \frac{1}{2} \left(\frac{\rho U D}{m} \right) C_{D_0} \right] \dot{y} + \left[\omega_n^2 - \frac{1}{2} \left(\frac{\rho U^2}{m} \right) \left(\frac{\partial C_L}{\partial y} \right) e^{-i\omega\tau} \right] y = 0, \tag{5.8}$$

where ω_n is the natural frequency of the cylinder and δ is the logarithmic decrement, both *in vacuo*. For harmonic motions, $y = y_0 \exp(i\omega t)$, the total damping is proportional to

$$\left[\left(\frac{\delta}{\pi} \right) \omega \omega_n + \frac{1}{2} \left(\frac{\rho U D}{m} \right) \omega C_{D_0} + \frac{1}{2} \left(\frac{\rho U^2}{m} \right) \left(\frac{\partial C_L}{\partial y} \right) \sin\left(\mu \omega D/U\right) \right] \dot{y} \tag{5.9}$$

and instability is associated with $[\quad] = 0$; if $\mu\omega D/U$ is small, $\sin(\quad) \simeq (\quad)$, and

$$\frac{U_c}{f_n D} = \left\{ \frac{4}{-C_{D_0} - \mu(\partial C_L/\partial y)} \right\} \frac{m\delta}{\rho D^2}, \tag{5.10}$$

where $f_n = \omega_n/2\pi$. Hence, instability is possible only if

$$-C_{D_0} - \mu(\partial C_L/\partial y) > 0, \tag{5.11}$$

i.e. if $\partial C_L/\partial y < 0$ and large.

For staggered arrays of cylinders, it is normal to have $C_{D_0} > 0$; therefore, if there were no time delay ($\mu = 0$), it would be impossible to have fluidelastic instability via this mechanism, because the bracketed quantity should be positive. The same conclusion was reached via the analytical model of Païdoussis et al. (1984), where the fluid-dynamic forces are derived via potential flow theory; oscillatory instability was found to be possible only when a phase lag between cylinder motions and fluid-dynamic forces is heuristically introduced.

Three further items of interest are the following.

(i) For small values of $U/\omega D$, the approximation of the sine function by its argument is no longer valid; setting the transcendental expression (5.9) to zero admits an infinite set of solutions for neutral stability, as the sine oscillates between -1 and 1. Some of the solutions represent the threshold from stability to instability, and some the reverse. Thus, this leads to a spectrum of stable and unstable zones as $U/f_n D$ decreases, as obtained by Lever & Weaver (1982, 1986a, b), Chen (1983a, b) and Price & Païdoussis (1984a, 1986a, b). It is of interest that in such cases instabilities may arise not only for $\partial C_L/\partial y$ negative, but also positive and large. Typical solutions from the full form of expression (5.9) are presented in region 1 of Figure 5.8 (mechanism I).

(ii) The threshold of instability according to the approximations leading to (5.10) is insensitive to the frequency of oscillation in the fluid medium concerned, ω, depending only on the in vacuo frequency, ω_n (and hence f_n) – something that has perplexed researchers in the past (see discussion by Païdoussis (1980, 1983), for example).

(iii) As in Section 4.1, criterion (5.11) may be compared with the Glauert and Den Hartog inequality for galloping, $C_{D_0} + \partial C_L/\partial\alpha < 0$, which, accounting for the time lag, may be written as

$$C_{D_0} + \mathcal{R}e[e^{-i\omega\tau}(\partial C_L/\partial\alpha)] < 0. \tag{5.12}$$

If y is expressed as a function of α, namely $y = -(U/i\omega D)\alpha$, then $\partial C_L/\partial\alpha$ may be rewritten as $-(U/i\omega D)(\partial C_L/\partial y)$; further, assuming $\omega\tau$ to be small and τ to be given by (5.6), inequality (5.12) may be expressed as

$$C_{D_0} + \mu(\partial C_L/\partial y) < 0,$$

which is identical to inequality (5.11).

In the foregoing, motions in the y-direction were considered. Proceeding in a similar manner, for x-direction motions, it is found that for small $\omega D/U$,

$$\frac{U_c}{f_n D} = \left\{ \frac{4}{-2C_{D_0} - \mu(\partial C_D/\partial x)} \right\} \frac{m\delta}{\rho D^2}, \tag{5.13}$$

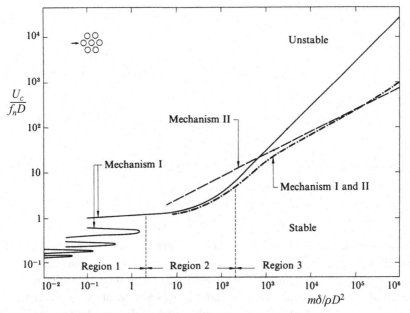

Figure 5.8. Stability diagram for a rotated triangular array with pitch-to-diameter ratio 1.375 (see elemental part of the array on the upper left-hand corner). Mechanism I refers to the damping-controlled one-degree-of-freedom mechanism (Section 5.2.1); mechanism II refers to the two-degree-of-freedom "wake-flutter" mechanism (Sections 5.2.3 and 5.2.5). Region 1 is where $U/\omega D$ is small enough for $\sin(\omega D/U) \neq \omega D/U$; region 2 is where $\sin(\omega D/U) \simeq \omega D/U$; region 3, for larger $U/\omega D$, is where both mechanisms I and II contribute to the instability.

so that for instability

$$-2C_{D_0} - \mu\,(\partial C_D/\partial x) > 0 \qquad\qquad (5.14)$$

must be satisfied. Although $\partial C_D/\partial x < 0$ often arises in several array configurations (Price & Païdoussis 1986b), it is generally smaller in magnitude than $|\partial C_L/\partial y|$, implying that the threshold of instability is more likely to be associated with y-direction motions, as observed experimentally. For *larger* $\omega D/U$, as for y-direction motions, multiple zones of instability may arise; furthermore, the instability may also be associated with $\partial C_D/\partial x$ positive and large.

According to relationships (5.10) and (5.13), $U_c/f_n D$ is proportional to $m\delta/\rho D^2$, or $\overline{m}\delta$. This appears to be correct in terms of experimental observations, in the middle range of $\overline{m}\delta$ (region 2 of Figure 5.8). For sufficiently high $\overline{m}\delta$, however, characteristic of gaseous flows, experimental evidence suggests that either $U_c/f_n D \propto (\overline{m}\delta)^{1/2}$ or $\overline{m}^a\delta^b$, where $a \neq b$ and $a, b \leq 1/2$, but certainly *not* $\propto \overline{m}\delta$. This is related to the increasing importance of the second mechanism associated with fluidelastic instability, to be discussed in Section 5.2.3.

5.2.2 Static divergence instability

Examining equation (5.8), it may be verified that a single flexible cylinder in the array may become unstable by divergence (nonoscillatory static instability) if

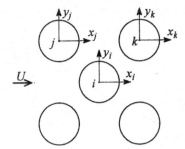

Figure 5.9. Cross-sectional view of the same kernel of an array of cylinders as in Figure 5.7(a) in cross-flow, for the analysis of wake flutter.

the stiffness term vanishes, which, for $\omega D/U$ sufficiently small, occurs at $U = U_{cd}$ given by

$$\frac{U_{cd}}{f_n D} = \left\{ \frac{8\pi^2}{(\partial C_L/\partial y)} \right\}^{1/2} \left\{ \frac{m}{\rho D^2} \right\}^{1/2} \tag{5.15}$$

in the y-direction;* similarly for the x-direction, with $\partial C_D/\partial x$ replacing $\partial C_L/\partial y$. This implies that for divergence, either $\partial C_L/\partial y$ or $\partial C_D/\partial x$ must be positive, which is opposite to the normal requirement for oscillatory instability. Hence, arrays subject to the usual, oscillatory fluidelastic instability will not be prone to divergence. Because most types of cylinder arrays are subject to oscillatory fluidelastic instability, it is perhaps not surprising that divergence has rarely been reported. Nevertheless, for an array resistant to oscillatory instability, divergence has in fact been observed (Païdoussis *et al.* 1989).

The mechanism of divergence is the same as that of buckling of a column subjected to axial load; if, when the column is flexed, the load-related lateral force exceeds the flexural restoring force, then the system behaves as if it had a negative net stiffness. For the problem at hand, writing $\omega_n^2 = k_{\text{eff}}/m$, the restoring force is clearly proportional to $k_{\text{eff}} y$, whereas the lift-related force (for $\cos \omega\tau \simeq 1$) is proportional to $\frac{1}{2}\rho U^2 l \,(\partial C_L/\partial y)\, y$; hence, if $\partial C_L/\partial y$ is positive, i.e. if the lift increases with lateral displacement, then, for sufficiently large U, the latter term will become larger than the restoring term, precipitating divergence.

5.2.3 The stiffness-controlled wake-flutter mechanism

There is another mechanism capable of causing (oscillatory) fluidelastic instability, namely the wake-flutter mechanism, introduced for pairs and small groups of cylinders in Section 4.2. This is a displacement-dependent, stiffness-controlled mechanism, a characteristic of which is that at least two degrees of freedom must be involved in the motion.

In general, both mechanisms will be at work for a given array, although one or the other may be dominant, depending on geometry and other system parameters. Here, however, we shall discuss wake flutter in isolation, as if it were the only mechanism at play.

Consider again the kernel of a large array, in Figure 5.9. We can consider two of the cylinders to be flexible, e.g. cylinder i and cylinder j, and in the interest of

---

* This expression is really independent of the mass of the cylinder, as may easily be verified and as it should be. The way it is written, however, in terms of the "standard" nondimensional groups, makes it appear otherwise.

keeping things simple only two degrees of freedom: motions x_j and y_i; alternatively, one could consider motions in the y-direction only, y_j and y_i. Thus, the coupled equations of motion may be written as

$$([M] + [M]_f)\{\ddot{\eta}\} + ([C] + [C]_f)\{\dot{\eta}\} + ([K] + [K]_f)\{\eta\} = \{0\}, \qquad (5.16)$$

where the unsubscripted matrices are associated with the mechanical system and those with suffix f are associated with the fluid-dynamic terms; $\{\eta\}$ is the displacement vector, so that for motion of two cylinders in the y-direction, for instance, $\{\eta\} = \{y_i, y_j\}^{\mathrm{T}}$.

Concerning the mechanical system itself, the following set of demonstrably reasonable and, for heat-exchanger arrays, realistic assumptions are made:* the modal masses, as well as the mechanical and damping terms, are equal in the two degrees of freedom concerned; also, there is no mechanical coupling between the two degrees of freedom. Thus, $M_{11} = M_{22} = ml$, $C_{11} = C_{22} = c$, $K_{11} = K_{22} = k$; and all off-diagonal terms in the mechanical matrices are zero.

Concerning the fluid-dynamic matrices, a set of at first sight more tenuous assumptions are made, which are nevertheless justifiable *a posteriori*, some by numerical computations, as will be discussed in the next paragraph: (i) the flow-retardation terms are entirely removed; (ii) the virtual (added) mass terms are neglected; (iii) the fluid-dynamic damping terms are neglected, compared with the mechanical damping terms. Thus, equation (5.16) is simplified to

$$\begin{bmatrix} ml & 0 \\ 0 & ml \end{bmatrix} \{\ddot{\eta}\} + \begin{bmatrix} c & 0 \\ 0 & c \end{bmatrix} \{\dot{\eta}\} + \begin{bmatrix} k & 0 \\ 0 & k \end{bmatrix} \{\eta\} + \frac{1}{2}\rho U^2 l \begin{bmatrix} \kappa_{11} & \kappa_{12} \\ \kappa_{21} & \kappa_{22} \end{bmatrix} \{\eta\} = \{0\}, \quad (5.17)$$

where the κ_{ij} are of the form $-\partial C_F/\partial \eta_j$, in which C_F stands for either C_D or C_L accordingly as $\eta = x$ or y.

The removal of the flow-retardation terms ($\mu = 0$) ensures that the generator of the negative-damping mechanism discussed in Section 5.2.1 is removed, so that instability, if it can arise at all, will have to be due to some other mechanism. Assumption (ii) above is justified partly for simplicity and partly because the mechanism under discussion is dominant for gaseous flows, where virtual mass effects are negligible. The ratio of the fluid-dynamic to the mechanical damping terms (with flow retardation removed) may be shown to be $(C_{D_0}/4\overline{m}\delta)(fD/U)$ and $(C_{D_0}/2\overline{m}\delta)(fD/U)$, respectively, for motions in the y- and x-direction (Price & Païdoussis 1984a, 1986a). Clearly, if instability occurs at sufficiently large values of U/fD, these ratios become small, typically 1/40 and 1/20, respectively, and the fluid-dynamic damping may be neglected; hence, assumption (iii) is also reasonable.

Let us now consider the dynamical implications of the form of the simplified equation of motion (5.17). Introducing the notation $\bar{t} = \omega_n t$, the equation of motion may be written in dimensionless form as follows:

$$\begin{Bmatrix} \eta_1'' \\ \eta_2'' \end{Bmatrix} + \frac{\delta}{\pi} \begin{bmatrix} 1 & 0 \\ 0 & 1 \end{bmatrix} \begin{Bmatrix} \eta_1' \\ \eta_2' \end{Bmatrix} + \begin{bmatrix} 1 & 0 \\ 0 & 1 \end{bmatrix} \begin{Bmatrix} \eta_1 \\ \eta_2 \end{Bmatrix} + \frac{\overline{U}^2}{2\overline{m}} \begin{bmatrix} \kappa_{11} & \kappa_{12} \\ \kappa_{21} & \kappa_{22} \end{bmatrix} \begin{Bmatrix} \eta_1 \\ \eta_2 \end{Bmatrix} = \{0\}, \quad (5.18)$$

where $\overline{U} = U/\omega_n D$, $\overline{m} = m/\rho D^2$ and $(\ )' = \mathrm{d}/\mathrm{d}\bar{t}$.

---

* Excluding the U-bend region in heat exchangers.

Assuming harmonic solutions of the form $\{\eta\} = \{\eta_0\} \exp(\lambda \bar{t})$, the characteristic equation, a quartic in λ, is obtained. The condition of zero total damping, i.e. the boundary of an oscillatory instability, may be obtained by application of the Routh criterion, or $p_1 p_2 p_3 - p_4 p_1^2 - p_0 p_3^2 = 0$, where the p_i are the coefficients of λ_i. This leads to the following quadratic expression for $\overline{U}^2/2\overline{m}$:

$$(\overline{U}^2/2\overline{m})^2[(\kappa_{11} - \kappa_{22})^2 + 4\kappa_{12}\kappa_{21}] + 2(\overline{U}^2/2\overline{m})[(\delta/\pi)^2(\kappa_{11} + \kappa_{22})] + 4(\delta/\pi)^2 = 0. \tag{5.19}$$

In general, the solution to this equation is cumbersome, but, if $(2\pi/\delta)[(\kappa_{11} - \kappa_{22})^2 + 4\kappa_{12}\kappa_{21}]^{1/2}/(\kappa_{11} + \kappa_{22})$ is sufficiently large compared with unity (which is the case for small enough δ), then an approximate solution of relatively simple form may be obtained, namely,

$$\overline{U}_c \simeq \left\{ \frac{-16/\pi^2}{(\kappa_{11} - \kappa_{22})^2 + 4\kappa_{12}\kappa_{21}} \right\}^{1/4} (\overline{m}\delta)^{1/2}, \tag{5.20a}$$

where it was assumed that a real $\overline{U}_c$ exists; in more conventional terms, this equation may be written as

$$\frac{U_c}{f_n D} = \left\{ \frac{-256\pi^2}{(\kappa_{11} - \kappa_{22})^2 + 4\kappa_{12}\kappa_{21}} \right\}^{1/4} \left(\frac{m\delta}{\rho D^2} \right)^{1/2}. \tag{5.20b}$$

Furthermore, for cylinders deep enough in the array (i.e. away from the first or last few rows) $\kappa_{11} = \kappa_{22}$, and equation (5.20b) may be simplified further to

$$\frac{U_c}{f_n D} = \left\{ \frac{-64\pi^2}{\kappa_{12}\kappa_{21}} \right\}^{1/4} \left(\frac{m\delta}{\rho D^2} \right)^{1/2}. \tag{5.20c}$$

Despite the simplifying assumptions leading to (5.20c), this expression does give results in excellent agreement with those obtained from the complete set of equations. Three important conclusions may be drawn directly from equations (5.20a, b, c):

 (i) for instability to occur, the stiffness matrix should not only be asymmetric ($\kappa_{12} \neq \kappa_{21}$), but the signs of the off-diagonal terms must be opposite;
 (ii) the dependence of $U_c/f_n D$ on the mass-damping parameter is that of a *square-root* relationship, rather than a linear one as was the case for the one-degree-of-freedom damping-controlled mechanism discussed in Section 5.2.1;
 (iii) if the diagonal terms κ_{11} and κ_{12} are unequal, this leads to an increase of $U_c/f_n D$; hence, (5.20c) is conservative *vis-à-vis* (5.20b).

The requirement in (i) above is often met in practice. Thus, for a rotated triangular array with pitch-to-diameter ratio $P/D = 1.375$ (see Figure 5.10), measurements have given $\kappa_{12} = -(\partial C_{L_1}/\partial y_2) = 16.7$ and $\kappa_{21} = -(\partial C_{L_2}/\partial y_1) = -26.6$, where cylinder 1 is in one row and cylinder 2 is diagonally adjacent to it in the row immediately upstream, as shown in Figure 5.10.

Physically, the nonequality of κ_{12} and κ_{21} is an attribute of the nonconservativeness of the system. Thus, if instead of fluid-dynamic coupling there were mechanical coupling involving springs, then the forces could be derived from a potential function and clearly $k_{12} = k_{21}$ would have been obtained: the force on cylinder 1 due to the motion of cylinder 2 would be equal to the force on cylinder 2 due to the motion

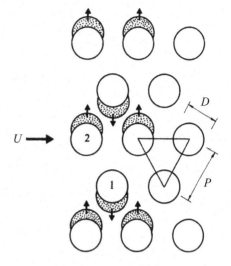

Figure 5.10. Part of a rotated triangular array, showing representative cylinders 1 and 2, as well as the pattern of motions during flutter; motions of neighbouring rows are 90° out of phase.

of cylinder 1, in the same directions. This is the well known *reciprocity principle* in solid mechanics. Significantly, if the flow field is modelled as a purely potential flow (Païdoussis *et al.* 1984), then the fluid-dynamic stiffness matrix obtained analytically is symmetric and, hence, this instability cannot materialize. In reality, of course, the flow field is rotational, with separated viscous flow regions (the existence of the wakes cannot be overlooked), and it is because of this that $\kappa_{12} \neq \kappa_{21}$ and, indeed, that $\kappa_{12}\kappa_{21} < 0$.

Another attribute of the viscous, rotational nature of the flow field and the non-conservativeness (asymmetry) of the stiffness matrix is that the energy derived from the flow is path-dependent; i.e. the phase between the motion of the two cylinders is of importance, as in the scenario proposed by Connors (1970), which is fundamentally the same as that described above. Indeed, comparing equations (5.20b, c) with Connors' expression, $U_c/f_n D = K(m\delta/\rho D^2)^{1/2}$, where K is an experimentally determined constant, the similarity becomes self-evident. Also, the expression obtained by Blevins (1977b), who generalised Connors' model, is in fact identical to the simplified expression derived here, equation (5.20c), although the notations differ. More will be said about these other theories in the following sections.

5.2.4 Dependence of the wake-flutter mechanism on mechanical damping

An apparent paradox in the results obtained for the wake-flutter mechanism is that the critical flow velocity is proportional to the square root of the logarithmic decrement of mechanical damping, δ. This contrasts to the results normally obtained for classical coupled-mode flutter instabilities, where the effect of mechanical damping is so insignificant that it is usually ignored; see for example Bisplinghoff, Ashley & Halfman (1955) for the flutter analysis of aircraft wings, and Simpson (1971), Price (1975a) and Price & Piperni (1988) for analysis of overhead transmission bundles.

To resolve this paradox it should first be realised that, in the aforementioned classical aeroelastic analysis, the two modes, associated with the two degrees of freedom involved, have *distinct* (non-equal) natural frequencies under "wind-off"

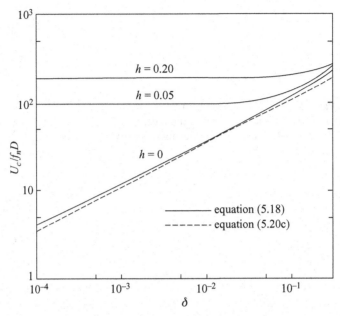

Figure 5.11. The effect of differences in the *in vacuo* modal stiffnesses, h, in the two-degree-of-freedom system of Sections 5.2.3 and 5.2.4 on the functional relationship between $U_c/f_n D$ and δ; for a rotated triangular array (pitch-to-diameter ratio $P/D = 1.375$, $\overline{m} = 10^5$).

(or *in vacuo*) conditions. Thus, the mechanical stiffness in equation (5.17) may in this case be written as

$$k \begin{bmatrix} 1 & 0 \\ 0 & 1+h \end{bmatrix}, \tag{5.21}$$

where h represents the difference in modal stiffness in the two degrees of freedom. The second point to realise is that even small values of h can have a profound effect on the relationship between $U_c/f_n D$ and δ, as will presently be shown.

Consider the same rotated triangular array, associated with Figures 5.8 and 5.10. Solutions of equation (5.17), but with the mechanical stiffness as given by equation (5.21), were obtained for various values of h and are shown in Figure 5.11. It is seen that for $h = 0$, i.e. when the two modes have equal frequency, $U_c/f_n D$ is sensibly proportional to $\delta^{1/2}$ – and the solution is very closely approximated by equation (5.20c). However, even for $h = 0.05$, which corresponds to a 2.5% difference in natural frequencies, the functional dependence of $U_c/f_n D$ on δ is entirely different: for $\delta \leq 0.02$, $U_c/f_n D$ is insensitive to δ. For $h = 0.2$, this effect extends to $\delta \leq 0.10$.

This resolves the apparent paradox referred to at the beginning of this section and, in this respect, reconciles the difference in dynamical behaviour of cylinder arrays and overhead transmission lines. However, this is also of direct interest to the dynamics of arrays in heat exchangers, for example in the so-called U-bend region where there exist substantial differences in natural frequency of adjacent cylinders. It is important to be aware that this "detuning" of adjacent cylinders leads to a considerable reduction in the efficacy of raising the threshold for fluidelastic instability by means of increased mechanical damping. There are also many examples in the literature illustrating how increased structural damping can be destabilising; see, for

Table 5.1. *Comparison between the critical flow velocities for fluidelastic instability obtained via (i) the full, constrained-mode solution and (ii) the simplified solution of (5.20a, b, c) and (5.22), for a rotated triangular array with $P/D = 1.375$, and $\delta = 0.01$ throughout*

$m\delta/\rho D^2$	10^5	10^4	10^3	10^2
$U_c/f_n D$, 'full solution'	236	70.4	17.3	2.91
$U_c/f_n D$, 'simplified solution'	244	77.9	24.7	7.79

example, the work of Broadbent & Williams (1956), Done (1963) and Benjamin (1963).

5.2.5 Wake-flutter stability boundaries for cylinder rows

Of course, the results of Section 5.2.3 may be applied directly, by considering a system of two flexible cylinders in an array of otherwise rigid ones. Numerical results were obtained in this manner and were found to be similar to those to be discussed below, which were obtained by a generalisation of the foregoing, whereby the results are more representative of the stability of a fully flexible array. This is done by considering an infinitely long double row of flexible cylinders, within a larger array of rigid cylinders.

It is realised at the outset that, realistically, all the cylinders in the array are fluid-dynamically coupled in a chain-like manner, so that the correct choice of a two-cylinder kernel representative of the array should take this fact into consideration, as well as the relative phase in the motions of cylinders in the same row; this leads to the so-called "constrained-mode solution" described by Price & Païdoussis (1986a), whereby the analysis of the two-cylinder kernel becomes representative of an infinite array of flexible cylinders. It is shown that this results in

$$\kappa_{12} = -(\partial C_{L_1}/\partial y_2)(1 + \delta_1) \quad \text{and} \quad \kappa_{21} = -(\partial C_{L_2}/\partial y_1)(1 + \delta_2), \tag{5.22}$$

where the δs can take on the values of 1, 0 or -1; similarly, if the $\bar{\kappa}$s involve C_D and x. If $\delta_1 = \delta_2 = 1$, this signifies that cylinders in the same row are presumed to move in phase; if they are equal to -1, then in antiphase. All possible δs are utilised, and the minimum U/fD obtained thereby is considered to be the critical one; significantly, the same set of δs is found to give the best agreement with the full, unconstrained analysis of multi-degree-of-freedom long rows of flexible cylinders (Price & Païdoussis 1984a). In the results presented in what follows, this is achieved with $\delta_1 = \delta_2 = 1$.

Solutions obtained by equations (5.18) and (5.22) are represented in region 3 of Figure 5.8, marked as due to mechanism II. Of course, in general, both mechanisms I and II are at work, i.e. both the single-degree-of-freedom negative-damping mechanism of Section 5.2.1 and the stiffness-controlled mechanism discussed here. The results obtained by the full solution of the equations of motion, without the introduction of the simplifications leading to equation (5.17), are also shown in region 3 of Figure 5.8, marked as due to both mechanisms I and II.

The results shown in Table 5.1 give numerical comparisons of the critical flow velocities obtained by the full constrained-mode solution of the equations of motion to the "simplified solution" according to equations (5.20a, b, c) and (5.22). It is clear that, for large values of $\overline{m}$ (and hence $\overline{m}\delta$), there is not much difference between the two; this signifies that the instability is predominantly due to the position-dependent "wake-flutter" mechanism discussed here. As $\overline{m}$ is diminished, however, the differences become more pronounced, indicating the increasing contribution of the damping-controlled one-degree-of-freedom mechanism (mechanism I) to the destabilisation of the system.

The instability in Table 5.1 involves y-direction motions (see Figure 5.10), where motions in adjacent rows are $90°$ out of phase. If, instead, an array of only two rows is considered, then the instability is found to involve predominantly x-direction motions. Indeed, in principle, there is no *a priori* reason why wake-flutter instabilities in some arrays should not involve only one flexible cylinder in the array, instead of two, the two requisite degrees of freedom being associated with x- and y-motions of the cylinder.

5.2.6 Concluding remarks

The foregoing represents an attempt to elucidate the mechanisms underlying fluidelastic instability of cylinder arrays in cross-flow, as well as providing links to the well known classical galloping and two-degree-of-freedom wake-flutter mechanisms.

It is shown that, provided the mass-damping parameter $\overline{m}\delta$ is sufficiently small, the instability is a modified form of the galloping mechanism proposed by Den Hartog for the limit-cycle motions of iced transmission lines. The essence of the difference between this "classical" galloping and what occurs in cylinder arrays is that, in the latter case, the mechanism is intimately connected to a time delay, which is associated with the time taken for the wake flow to adjust to cylinder motions.

This is the mechanism of instability predominating in regions 1 and 2 of Figure 5.8. Stability boundaries may be obtained by satisfying expression (5.9) set to zero for y-motions, or the equivalent one for x-motions. If $U/\omega D$ is sufficiently large, then stability boundaries may be obtained by the even simpler relation (5.10) – or the equivalent one for x-motions.

The main objective of this introductory section (Section 5.2) was to elucidate, rather than predict, the instability. In this respect, it should be stressed that setting expression (5.9) to zero is not sufficient for predicting instability, in that C_{D_0} and $\partial C_L/\partial y$ cannot at present be predicted analytically but must rather be obtained empirically.

In the higher range of $\overline{m}\delta$, region 3 of Figure 5.8, a second destabilising mechanism comes into play; for sufficiently high $\overline{m}\delta$ this mechanism becomes predominant, while the damping-controlled one contributes less. This is a position-dependent, fluid-dynamic stiffness-controlled mechanism, similar to that long known to be responsible for wake flutter of bundled electrical transmission lines. This mechanism requires at least two degrees of freedom, and has been demonstrated here in terms of y- or x-motions of two adjacent cylinders in the array. It is intimately related to the fluid-dynamic coupling between motions in these two degrees of freedom, which, because of the rotational and viscous nature of the flow field, does not exhibit reciprocity, as would be the case if coupling were of a mechanical nature. In matrix notation the coupling manifests itself in nonzero off-diagonal fluid-dynamic terms,

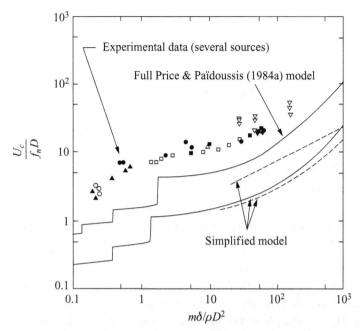

Figure 5.12. The stability chart according to the simplified theory of Païdoussis & Price (1988) presented in this section and the more elaborate model of Price & Païdoussis (1984a), compared to experimental data from several sources.

which are not equal because of the nonconservative nature of the flow field and hence of the system as a whole.

Based on a set of reasonable assumptions, the problem was simplified and it was possible to obtain simple relationships for the onset of instability according to this second mechanism, equations (5.20a, b, c). It was found, *a posteriori* that these relationships are similar to that obtained previously by Connors (1970) and, in their simplest form, identical to that obtained by Blevins (1977b). The dominance of this mechanism for high $\overline{m}\delta$ was recognised in Chen's (1983a, b) and later in Price and Païdoussis's (1984a, 1986a, b) work.

Significantly, the dependence of $U_c/f_n D$ on $\overline{m}\delta$ is found to be different in different ranges of the latter parameter, as shown in Figure 5.8, and in agreement with experimental observations (see Price & Païdoussis (1986a, b), for example). For low values of $\overline{m}\delta$ the principal instability boundary is insensitive to $\overline{m}\delta$ (region 1 of Figure 5.8), although a set of secondary instability zones exist below that boundary. In the middle range of $\overline{m}\delta$, $U_c/f_n D$ depends more or less linearly on that parameter (region 2). Finally, for sufficiently high $\overline{m}\delta$, $U_c/f_n D$ depends on the square root of $\overline{m}\delta$ (region 3) – an attribute of the wake-flutter mechanism. These are of course generalisations; quantitatively, the extent of these three regions depends on the geometry of the array and, for regions 1 and 2, not only on the product $\overline{m}\delta$, but also on the specific values of $\overline{m}$ and δ – two independent dimensionless parameters which are only combined by convention and sometimes for convenience.

A final question is this: just how good are the predictions of the simplified theory presented in this section? The answer is that they are very good qualitatively; but, as seen in Figure 5.12, not very good quantitatively. Not surprisingly, one of the more

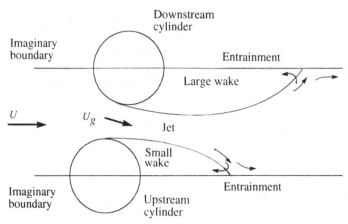

Figure 5.13. Idealised model of the jet-flow between two cylinders in a staggered row of cylinders (Roberts 1962).

elaborate and sophisticated theories, to be discussed in the sections that follow, achieves much better agreement. Nevertheless, it can be said that the foregoing elucidate fairly well the physics of fluidelastic instabilities in cylinder arrays.

5.3 Fluidelastic Instability Models

A review of all models of fluidelastic instability in cylinder arrays known to the authors is presented in this section; this represents an updated version of a summary previously given by Price (1995). Comparisons with experimental data and a review of the state of the art are given in Section 5.4. As discussed in Section 5.2, an essential component of any model attempting to predict the critical velocity at which fluidelastic instability will occur is that the mutual interaction between the structural displacement and resulting fluid forces must be accounted for. It is not sufficient to determine, either theoretically or experimentally, the fluid forces on an individual cylinder when it is statically mounted at its equilibrium position; instead, the effect of cylinder motion must be taken into account. This is one of the big challenges associated with modelling fluidelastic instability in cylinder arrays, the magnitude of this challenge becomes apparent when one considers the complexities and apparent paradoxes of the interstitial flow, as reviewed by Zdravkovich (1993).

5.3.1 Jet-switch model

The first semi-analytical model purporting to analyse the fluidelastic stability of cylinder arrays subject to cross-flow was by Roberts (1962, 1966), who considered both single and double rows of cylinders normal to the flow. Roberts' preliminary experiments indicated that instability was purely in the in-flow direction, with adjacent cylinders moving out of phase with each other; hence, his analysis was limited to in-flow motion. Furthermore, based on photographic evidence, it was assumed that the flow downstream of two adjacent cylinders could be represented by two wake regions, one large and the other small, and a jet flow between them (see Figure 5.13).

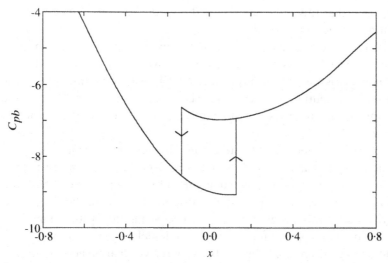

Figure 5.14. Variation of the theoretical base pressure coefficients for two adjacent cylinders in a row with $P/D = 1.5$ as a function of nondimensional in-flow displacement (Roberts 1962).

Further photographic evidence of this has since been supplied by Bradshaw (1965) and Ishigai & Nishikawa (1975).

Because only in-flow motion was allowed, Roberts suggested that a hypothetical channel flow involving two half cylinders and imaginary boundaries, as shown in Figure 5.13, was representative of the flow through a row of cylinders. Roberts considered this as being similar to a jet issuing close to a parallel flat plate. This jet flow had previously been analysed by both Sawyer (1960) and Bourque & Newman (1960) who had shown that fluid entrained by the jet from the wake region produces a pressure difference across the jet, causing it to curve and strike the plate; whereupon a portion of the jet flows upstream resupplying the wake region and so maintaining the jet entrainment. In Roberts' analysis there are two wake regions supplied from the two sides of the spreading jet.

The main assumptions employed by Roberts are as follows. First, flow separation from the cylinders occurs at the minimum gap between their centres. Second, the cylinder wakes are regions of constant pressure; hence, the pressure difference across the jet does not vary in the flow direction. Finally, the flow upstream of the separation points and in the jet region is inviscid. Employing basic fluid dynamics and geometrical considerations Roberts obtained the base pressure coefficient for both cylinders as a function of in-flow cylinder displacement, x, and a typical variation is presented in Figure 5.14, showing hysteresis around $x = 0$. Roberts also gave some experimental results showing excellent agreement with the theory.

The hysteresis shown in Figure 5.14 is of considerable importance and can be explained as follows. Consider a downstream cylinder moving upstream; as the two cylinders cross, insufficient fluid flows into the larger of the two wake regions to maintain the entrainment, causing the wake to shrink and the jet to switch directions. However, a finite pressure difference is required to initiate this jet switch and, if the cylinder moves in the opposite direction, the jet switch will not occur at the same position; thus, around $x = 0$ the flow has two stable configurations. A detailed analysis of these configurations was later given by Zdravkovich & Stonebanks (1990).

It is apparent that this phenomenon, and thus the resulting fluidelastic instability mechanism, is inherently nonlinear.

Roberts also determined the shape of the separating streamlines emanating from the cylinders. The flow upstream of the separation points and between the separating streamlines was assumed to be inviscid and the Laplace equation was solved, giving the pressure distribution around the surface of the cylinders upstream of the separation points; hence, the pressure distribution around the complete cylinder surface was known.

The drag force, obtained by integrating the pressure distribution, is dominated by the base pressure, and the variation of drag coefficient, C_D, with x is remarkably similar in shape to the curve shown in Figure 5.14. It is apparent that, for a cylinder oscillating with sufficient amplitude to cause the jet switch, the drag force aiding the downstream motion is greater than that opposing the upstream motion; hence, energy is extracted from the flow, resulting in fluidelastic instability.

Another significant contribution by Roberts was that he accounted for the unsteady nature of the jet-switch process. Realising that a finite time is required for the wake to shrink and cause the jet to switch directions, Roberts suggested that jet switching is possible only if this required time is less than half the period of cylinder motion. For a row of cylinders with $P/D = 1.5$ it was concluded that the requirement for jet reversal to occur is approximately $U/\omega D \geq 2$.

Accounting for all of the above phenomena, Roberts obtained the following equation of motion for a cylinder in a row:

$$\frac{d^2x}{d\tau^2} + 2\zeta\frac{dx}{d\tau} + x = \frac{\rho U^2}{2m\omega_n^2}\left\{0.717\left[1 - C_{pb}(x, \tau)\right] - 2\left(\frac{\omega_n D}{U}\right)(1 - C_{pb})_{\text{mean}}\frac{dx}{d\tau}\right\},$$

$$(5.23)$$

where τ is nondimensional time $(t\omega_n)$. The first expression on the right-hand side is effectively a fluid stiffness term (C_{pb} varies with x) due to the jet-switch mechanism – it is also unsteady (C_{pb} varies with τ). The second expression is an aerodynamic damping term due to the average drag coefficient.

Equation (5.23) was solved using the method of Krylov & Bogoliubov giving for any $m\delta/\rho D^2$ the velocity, $U/\omega_n D$, required to initiate a limit-cycle motion. In fact, three theoretical solutions were given, which are reproduced in Figure 5.15. The first is the exact solution of equation (5.23); in the second and third solutions the jet reversal is assumed to take place instantaneously; for the third solution, fluid damping is ignored. When both the unsteady terms and fluid damping are neglected, the solution reduces to

$$U_c/\omega_n \varepsilon D = K(m\delta/\rho D^2)^{1/2}, \qquad (5.24)$$

where ε is the ratio of fluidelastic frequency to structural frequency, which is approximately 1. Also presented in Figure 5.15 are Roberts own experimental data, showing good agreement with this theoretical model.

The above discussion indicates that this instability is stiffness-controlled with the energy input arising from the hysteretic drag variation caused by jet reversal. Furthermore, the importance of the unsteady fluid terms is demonstrated; for example, it is shown that instability cannot occur for $U_c/\omega_n \varepsilon D \leq 2$. Later Chen (1980)

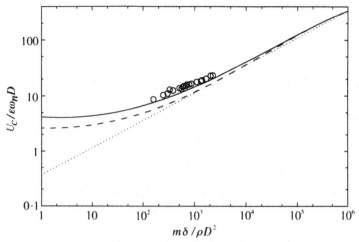

Figure 5.15. Theoretical stability boundary obtained by Roberts (1962) for a single flexible cylinder in a row of cylinders with $P/D = 1.5$: ——, solution including time for jet reversal and aerodynamic damping; - - -, solution assuming instantaneous jet reversal but still including aerodynamic damping; $\cdots\cdots$, solution assuming instantaneous jet reversal and neglecting aerodynamic damping. $\circ$, Roberts' experimental results.

also concluded that jet switching was the mechanism responsible for initiating this instability.

The theoretical variations of $U_c/f_n D$ obtained by Roberts are compared with all available experimental data for cylinder rows subject to cross-flow in Section 5.4, where it shown that, in general, the agreement is poor. Most probably this is because Roberts assumed that the cylinder motion was in the in-flow direction, whereas most experimental results indicate that it is predominantly normal to the flow. Also, Roberts' solution is for a single flexible cylinder, whereas all the experimental data except his own are for multiple flexible cylinders. In addition, there are a number of factors which limit the applicability of this analysis. First, it is suitable for analysing the stability of a downstream row only; second, it is only capable of predicting in-flow instabilities.

A somewhat similar approach was taken by Nakamura *et al.* (1992a) who considered pure cross-flow motion. However, instead of attempting to predict the fluid forces and unsteady effects, these are given in terms of a number of unknown constants which are then inferred from matching predicted instability velocities with experimental ones.

5.3.2 Quasi-static models

Possibly the most famous expression predicting fluidelastic instability for cylinder arrays subject to cross-flow is the one usually attributed to Connors (1970), who derived it using a quasi-static analysis. However, as shown in the preceding section this expression had previously been obtained by Roberts (1962). Similarly to Roberts, Connors also considered a single row of cylinders normal to the flow.

Connors did not attempt to determine the fluid forces analytically, but instead measured them. Having observed that many different unstable intercylinder modal

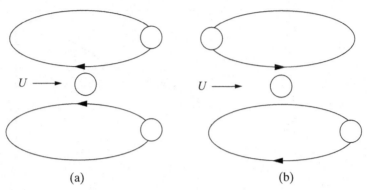

Figure 5.16. Cylinder motion of neighbouring cylinders employed by Connors (1970) during force measurements on the central cylinder: (a) symmetric motion, (b) antisymmetric motion.

patterns exist, Connors suggested that the most dominant was when the cylinders executed thin elliptical motions (whirling), with alternate cylinders moving predominantly in either the in- or cross-flow directions. This was simulated experimentally for a cylinder in a row with $P/D = 1.41$, and its lift and drag forces were measured while statically displacing its two neighbouring cylinders in either a symmetric or antisymmetric manner, as illustrated in Figure 5.16.

For the symmetric motion there was no change in the lift force on the central cylinder. The drag force, however, did vary, executing a hysteretic jump caused by a jet-switching mechanism similar to that observed by Roberts (1962). Realising that this jet switch cannot occur for low U/fD, Connors concluded that jet switch was not the dominant mechanism leading to instability. Thus, he subtracted the jet-switch-induced drag from the total measured drag. The work done by the airflow on the cylinder per cycle of motion (minus any contribution form the jet-switching process) is then equal to the area enclosed by this new drag curve; measurements made at different ratios of x/y enabled Connors to conclude that the work done depended only on the magnitude of y, giving an in-flow directional fluid-force coefficient which is linearly dependent on the transverse displacement, $C_x = 0.202y$.

During the antisymmetric motion, Connors measured the lift force only, apparently assuming the drag to be constant – this is a little surprising because this motion will almost certainly introduce a change in the drag. He again subtracted the jet-switch contribution from the original lift and obtained a fluid-force coefficient in the cross-flow direction which in linearly dependent on the in-flow displacement, $K_y = 0.330x$.

Using the measured fluid stiffnesses, Connors obtained energy balances in the in- and cross-flow directions, which must be satisfied simultaneously. Solving these equations gave $x/y = 0.78$, and more importantly

$$\frac{U_{pc}}{f_n D} = K \left(\frac{m\delta}{\rho D^2} \right)^{1/2}, \qquad (5.25)$$

where K is what later became known as the Connors constant. The magnitude of K is a function of C_x and K_y and, for this particular geometry, $K = 9.9^*$. In

---

* In general, the velocity used to compare theory and experiments in this chapter is the so-called pitch velocity U_p, where $U_p = UP/(P - D)$, P being the centre-to-centre intercylinder pitch.

general, Connors' equation is used very loosely, with different definitions being employed for how the mass, damping and frequency are measured (in vacuum, in stagnant fluid or in flowing fluid)*. However, strictly these should be purely the structural values, or values measured in vacuum. It is also clear that equation (5.25) is identical to that obtained by Roberts (equation (5.24)) when the unsteady terms and fluid damping are neglected and the fluidelastic frequency is assumed equal to the structural frequency. Despite this, possibly because Roberts' work is somewhat obscure, equation (5.25) is usually attributed to Connors.

Blevins (1974) rederived equation (5.25) by assuming that the fluid forces on a cylinder are purely due to the relative displacements between itself and its neighbouring cylinders[†]. The analysis for a row of N cylinders resulted in a set of $2N$ coupled equations; to decouple the cylinder motions, and so reduce the size of the resulting equations from $2N$ to 4, Blevins assumed a specific intercylinder modal pattern, with the cylinders adjacent to any other cylinder moving antiphase to each other. Based on Connors' measurements Blevins also assumed that relative motion in the in-flow direction produces no change in the drag force, and relative motion in the cross-flow direction similarly produces no change in the lift force. Using the Routh stability criteria the resulting equations yielded

$$\frac{U_{pc}}{f_n D} = \frac{2 (2\pi)^{0.5}}{(\overline{C}_x \overline{K}_y)^{0.25}} \left(\frac{m\delta}{\rho D^2} \right)^{1/2}, \tag{5.26}$$

where $\overline{C}_x = \partial C_x / \partial y$ and $\overline{K}_y = \partial K_y / \partial x$ are the fluid-stiffness terms. If the values of $\overline{C}_x$ and $\overline{K}_y$ given by Connors are employed, then equations (5.26) and (5.25) are identical. In a later paper, Blevins (1977a) attempted to predict the magnitude of the fluid-force coefficients, C_x and K_y, for both arrays and rows of cylinders.

Savkar (1977) criticised Blevins' analysis, claiming that it gave a static instability as opposed to a dynamic one; however, as pointed out by Gibson et al. (1979, 1980) this criticism is not correct; the instability predicted by Blevins is dynamic and the frequency of oscillation at instability is f_n.

Blevins (1979) modified his original analysis to account for flow-dependent fluid damping. Using a quasi-steady approach, the total damping factors in the in- and cross-flow directions are

$$\zeta_x = \zeta_o + \frac{\rho D^2}{4\pi m} \frac{U_p}{f_n D} C_D \quad \text{and} \quad \zeta_y = \zeta_o + \frac{\rho D^2}{8\pi m} \frac{U_p}{f_n D} C_D, \tag{5.27}$$

where ζ_o is the purely structural component. Using these damping factors leads to

$$\frac{U_{pc}}{f_n D} = K \left[\frac{m}{\rho D^2} 2\pi(\zeta_x \zeta_y)^{0.5} \right]^{0.5}. \tag{5.28}$$

Blevins' analysis was later generalised by Whiston & Thomas (1982) who used a more general form of intercylinder modal pattern, allowing any value of phase

---

* The authors recall at least one tongue-in-cheek, but very accurate, presentation where it was suggested that none of f_n, m, ρ and δ are unambiguously defined in (5.25); indeed the only one that is definitely known is the cylinder diameter D; refer to Section 5.4.2.

[†] As demonstrated by the fluid-force measurements of Price & Païdoussis (1986a), in general, this assumption is not correct and it is the motion of the individual cylinders, rather than the relative motion between the cylinders, which is important.

angle, ϕ, between the motion of adjacent cylinders (the specific intercylinder modal pattern assumed by Blevins (1974) is obtained with $\phi = \pi/2$). Following essentially the same procedure as Blevins, Whiston & Thomas obtained

$$\frac{U_{pc}}{f_n D} = K_c (\phi) \left(\frac{m\delta}{\rho D^2} \right)^{1/2}, \tag{5.29}$$

where $K_c (\phi)$ is a function of the phase angle.

Whiston & Thomas also extended Blevins' analysis to account for full arrays of cylinders. They considered a cylinder array to be made up of "effective rows" normal to the local velocity (these included diagonal rows not at 90° to the upstream flow). To decouple a small group of cylinders from the rest of the array, and thus make the analysis computationally practical, it was necessary to relate via a phase difference the motion between cylinders in different rows and columns. This enabled the stability of an array of flexible cylinders to be considered by use of a kernel of two cylinders.

Using the force coefficients given by Blevins (1979), the minimum value of Connors' constant, K, for a normal triangular array with $P/D = 1.41$ was shown to be 8.6. Whiston & Thomas concluded that this was too high, giving U_{pc} well above that observed in practice, and suggested that the analysis should be modified to take account of the wake regions behind the cylinders, so increasing the gap velocity in the diagonal rows. They then obtained $K = 3.4$ and 8.2 for sub- and supercritical Reynolds numbers, respectively, corresponding to large and small wake regions.

For in-line arrays the diagonal rows are less important, and the value obtained for K was approximately the same as for a row of cylinders, which is far too high. Whiston & Thomas suggested that for these arrays the instability is a combination of Connors' mechanism and a wake-induced effect from upstream cylinders. An analysis was presented accounting for this wake-induced effect in terms of an unknown fluid-stiffness term $\partial C_L/\partial y$. They chose appropriate values of $\partial C_L/\partial y$, to match $U_{pc}/f_n D$ with the results of Gibert et al. (1978). Unfortunately, when this was done, the comparison with other experimental results was not good.

The main criticism of the Whiston & Thomas analytical model relates to the manner in which the fluid-dynamic forces are accounted for, and is basically the same as that raised in connection with Blevins' model. However, this analysis made significant contributions in a number of ways. It was the first model to account for a general intercylinder modal pattern between cylinders, and it was also the first to attempt to deal with some of the intricacies of the interstitial flow through an array of cylinders.

Since Connors' original analysis, which was for a row of cylinders, the use of equation (5.25), or variations thereof, to predict the fluidelastic instability velocity in cylinder arrays has become ubiquitous. The implicit assumption is that this equation models the "physics" of fluidelastic instability, and all that is required to predict U_{pc} for cylinder arrays is to find the "correct" value of K or possibly to modify the exponent of $m\delta/\rho D^2$. Indeed, based on experimental results, Connors (1978) suggested that for cylinder arrays the constant K is given by $K = [0.37 + 1.76(T/D)]$ for $1.41 < T/D < 2.12$, where T is the separation between cylinders in a row. Since then, there have been a series of "practical" design guidelines based on then current data, with different values of K. For example, Païdoussis (1980), Au-Yang et al. (1991)

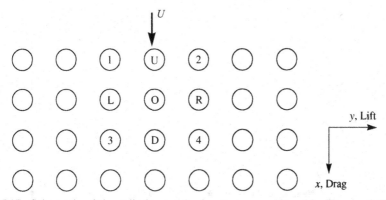

Figure 5.17. Schematic of the cylinder-numbering system employed by Tanaka & Takahara (1980, 1981).

and Pettigrew & Taylor* (1991) suggested $K = 0.8, 2.4$ and 3.0, respectively; whereas more complex correlations, with some account being taken of array geometry, were given by Chen (1984, 1987), Weaver & Fitzpatrick (1988) and Schröder & Gelbe (1999b). The appropriateness of using "Connors' equation" for design guides was discussed by Price (2001), and is summarised in Section 5.4 of this book.

5.3.3 Unsteady models

In the models presented in this section the unsteady forces on the oscillating cylinder are obtained directly from experiments. Tanaka & Takahara (1980, 1981) measured the unsteady forces on an in-line square array with $P/D = 1.33$, and Tanaka et al. (1982) made similar measurements for $P/D = 2.0$; results were also presented for a single row of cylinders with $P/D = 1.33$ by Tanaka (1980).[†] For the array measurements it was assumed that the fluid forces on cylinder O (see Figure 5.17) are affected by its own motion and that of cylinders U, R, L and D only, but not by cylinders 1–4. Hence, assuming that the fluid dynamics is linear and using arguments based on the assumed symmetry of the flow, the change in the lift and drag coefficient for cylinder O due to its own motion as well as that of the neighbouring cylinders was written as

$$\Delta C_L = C_{Ly_o} y_o + C_{Ly_L} (y_L + y_R) + C_{Lx_L} (x_L - x_R) + C_{Ly_U} y_U + C_{Ly_D} y_D,$$
$$\Delta C_D = C_{Dx_o} x_o + C_{Dy_L} (y_L - y_R) + C_{Dx_L} (x_L + x_R) + C_{Dx_U} x_U + C_{Dx_D} x_D, \quad (5.30)$$

where, for example, $C_{Ly_U} \equiv \partial C_L / \partial y_U$ is a coefficient relating the lift force on cylinder O to the y-directional displacement of cylinder U, and similarly for the other terms[‡]; to determine C_{Ly_U} the oscillatory force on cylinder O was measured while cylinder U was subject to a harmonic oscillation of frequency f and amplitude $0.1D$, with f being

---

* More recent papers by Pettigrew and co-workers still indicate that the appropriate value of Connors' constant is $K = 3.0$; see, for example, Pettigrew & Taylor (2003).
† The unsteady force coefficient data of Tanaka & Takahara for $P/D = 1.33$ was later presented in tabular form by Tanaka et al. (2002).
‡ Tanaka & Takahara use x to signify transverse displacement of the cylinders and y for in-flow displacements, which is the reverse of that used here. To be consistent with the rest of this book, Tanaka & Takahara's definitions of x and y have been changed.

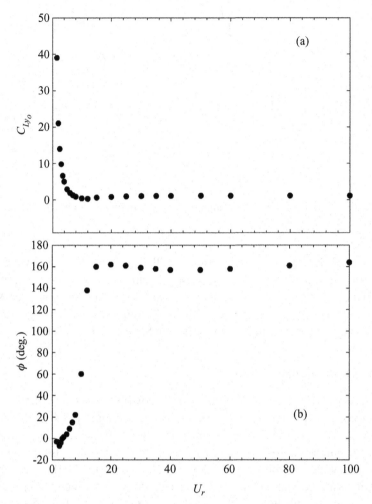

Figure 5.18. Variation of the fluid force coefficient C_{Ly_o} with nondimensional flow velocity $U_r = U_p/fD$ measured by Tanaka & Takahara (1980) for an in-line square array with $P/D = 1.33$ [reproduced using the tabulated data of Tanaka *et al.* (2002)]: (a) magnitude; (b) phase lead with respect to cylinder displacement.

varied over a wide range of frequencies. The force measurements were conducted in waterflow with cylinder U in the upstream row of the array.

Considering, for example, the coefficient C_{Ly_O} (the lift force coefficient on cylinder O due its own y-direction displacement) the force per unit length, F, was measured as U_p/fD was varied and it was expressed as

$$F = \tfrac{1}{2}\rho U_p^2 C_{Ly_o} y_o D \sin(2\pi f t + \varphi),$$ (5.31)

where y_o is the magnitude of the imposed oscillation, C_{Ly_o} is the magnitude of the nondimensional force coefficient, and ϕ is the phase lead of the fluid force with respect to the cylinder displacement. The variation of C_{Ly_o} and ϕ with $U_r = U_p/fD$ for an in-line square array with $P/D = 1.33$ is presented in Figure 5.18 (it should be remembered that the coefficient is based on the gap velocity $U_p = 4.03\,U$, and so the coefficients are approximately 16 times smaller than if they were based on U).

In common with the other coefficients, C_{Lyo} and ϕ are strongly dependent on U_r; although they do tend asymptotically to values which are independent of U_r as U_r is increased, this is not the case for many of the other coefficients.

As an alternative to presenting the unsteady data in terms of a magnitude, C_{Lyo}, and phase angle, ϕ, they can be presented in terms of added mass, damping and stiffness coefficients, c_m, c_d and c_k, respectively, giving

$$F = -\frac{\pi \rho D^3}{4} c_m \ddot{y} - \frac{1}{2} \rho U_p D^2 c_d \dot{y} - \frac{1}{2} \rho U_p^2 D c_k y; \qquad (5.32)$$

the negative signs in equation (5.32) imply that positive values of c_m, c_d and c_k cause an increase in the total (structural plus fluid) mass, damping and stiffness terms. At this point it is worth emphasising that the concept of *added mass* employed here, and generally in the fluidelastic-instability community, is much simpler than that discussed in Section 3.4.1 of this book with respect to vortex-induced vibrations. Here, the added mass coefficient, c_m, is a positive constant independent of U_r, whereas, as discussed in Chapter 3, in some models of vortex-induced vibrations the added mass coefficient is considered to be a function of U_r and, indeed, it may be negative*. Using the present understanding of c_m it may either be calculated or measured in quiescent fluid, and the value given by Tanaka *et al.* (2002) for an in-line square array with $P/D = 1.33$ is $c_m = 1.28$. Using equations (5.31) and (5.32) the following expressions are obtained for the damping and stiffness coefficients

$$c_d = -C_{Lyo} \frac{U_r}{2\pi} \sin\phi \quad \text{and} \quad c_k = -C_{Lyo} \cos\phi + c_m \frac{2\pi^3}{U_r^2}. \qquad (5.33)$$

The variation of c_d and c_k with U_r obtained using the data of Figure 5.18 is shown in Figure 5.19. Quasi-steady fluid mechanics suggests that c_d and c_k should be independent of U_r, or at least that they should tend to constant values as U_r becomes large. It is evident from Figure 5.19 that this is so for the stiffness term c_k which tends to a value of approximately 1.1; however, this is not true for c_d, even when U_r approaches 100. At first sight this implies that for this case the quasi-steady assumption is not applicable; however, as will be discussed later in this section, this conclusion is not necessarily correct.

Using their measured unsteady force coefficient data, Tanaka & Takahara (1981) investigated the stability of an array of flexible cylinders consisting of three rows and four columns. They employed a standard eigenvalue analysis to investigate stability, although this was somewhat complicated because the fluid coefficients are functions of f, which in turn is part of the solution of the final equations. However, standard iterative methods are available to solve this type of problem, and a flow chart showing a typical procedure is given by Ohta *et al.* (1982).

The variation in $U_{pc}/f_n D$ with $m/\rho D^2$ obtained by Tanaka & Takahara is presented in Figure 5.20, showing a discontinuity in the stability curves at $50 \leq m/\rho D^2 \leq 500$ depending on the value of δ. For $m/\rho D^2$ below the discontinuity, the intercylinder modal pattern at instability is almost pure cross-flow, whereas for $m/\rho D^2$ above the discontinuity there is also considerable in-flow motion. This suggests that the

---

* To some extent this is just a matter of convenience, the component of fluidelastic force in-phase with displacement can be represented as either a stiffness or inertia term, and thus, any influence of U_r on the total fluidelastic force can always be represented via the fluidelastic stiffness term c_k.

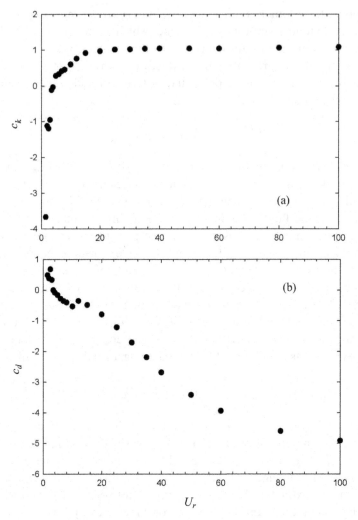

Figure 5.19. Variation of the fluid stiffness and damping coefficients with nondimensional flow velocity $U_r = U_p/fD$ for an in-line square array with $P/D = 1.33$ obtained using the data of Tanaka *et al.* (2002): (a) stiffness coefficient, c_k; (b) damping coefficient, c_d.

instabilities on either side of the discontinuity are due to different phenomena; this was pursued further by Chen (1983b), as is discussed later in this section.

Tanaka *et al.* (1982) also considered a single flexible cylinder in an array of rigid cylinders free to oscillate in the cross-flow direction only. Instability occurred at a slightly higher U_{pc}/f_nD than when there were 12 flexible cylinders; however, the difference was much less for $P/D = 2.0$ than for $P/D = 1.33$. The results obtained for $P/D = 1.3$ are presented in Figure 5.21 (these results were obtained by Price (1995) using the data of Tanaka & Takahara). For $m\delta/\rho D^2 \approx 2$ there is a discontinuity in the stability curve, similar to that observed in Figure 5.20; because these results are for a single flexible cylinder the instability is always damping-controlled (as shown in Figure 5.19(b), for approximately $U_r \geq 4$ the fluid-damping coefficient, c_d, is destabilising), and thus, the discontinuity is not necessarily due to a change from a damping- to a stiffness-controlled instability as suggested by Chen (1983b). A more

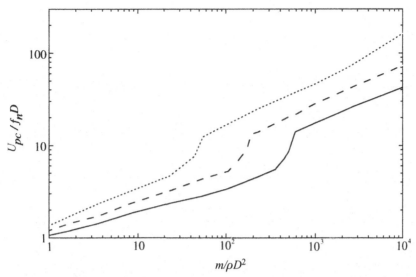

Figure 5.20. Theoretical stability boundary for an in-line square array, $P/D = 1.33$, obtained by Tanaka & Takahara (1980, 1981): ——, $\delta = 0.01$; - - -, $\delta = 0.03$, $\cdots\cdots$, $\delta = 0.1$.

likely explanation for the discontinuities in the stability curves is the rapid shift in phase angle between the cylinder motion and resulting fluid forces, as illustrated in Figure 5.18(b).

The fluid-force coefficients measured by Tanaka *et al.* were also employed by Chen (1983a, b) as empirical input in previously derived equations of motion for a cylinder array in cross-flow (Chen 1978).

Chen (1983a) first obtained analytical expressions, in terms of these coefficients, for the stability of a number of simple cases: these included a single flexible cylinder with freedom to move in one direction only and a row of cylinders oscillating in the mode shape prescribed by Connors (1970). For the single flexible cylinder it was demonstrated that the instability is damping-controlled, whereas for the row of cylinders the instability is stiffness-controlled and will occur even if the fluid damping is positive. Thus, Chen (1983a) demonstrated the existence of two distinct instability mechanisms in cylinder arrays.

In general, instability is due to a combination of these two mechanisms. However, as shown by Chen (1983a), for high U_p the cylinder vibrational velocity, $\dot{y}D$, is negligible compared with U_p; hence, the velocity-dependent damping will also be negligible. For low U_p, on the other hand, the cylinder vibrational velocity may be comparable with U_p, and thus, the velocity-dependent damping terms dominate. However, it is known from experimental investigations (Price *et al.* 1987; Price & Kuran 1991) that for some arrays the damping-controlled terms are never destabilising, and thus, if instability does occur, it is due to stiffness effects over the complete range of $U_p/f_n D$.

Chen (1983b) also conducted an unconstrained analysis (the mode shape at instability was not specified *a priori*) for the stability of a row and an in-line square array of cylinders. For a row of cylinders with $m\delta/\rho D^2 \leq 4$ (see Figure 5.22) there are three stability boundaries. The cylinders lose stability at the lower boundary, regain it at the second and finally remain unstable for all U_p above the third. This was the first

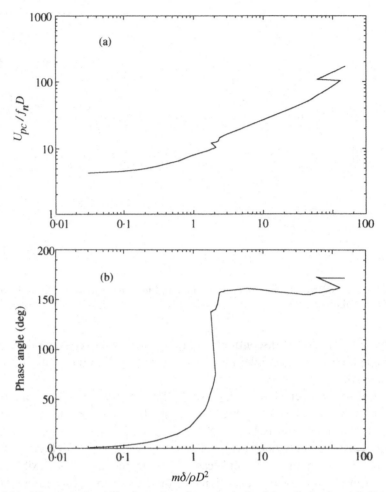

Figure 5.21. Theoretical stability boundary for a single flexible cylinder in an in-line square array with $P/D = 1.33$ obtained by Price (1995) using the unsteady force coefficient data of Tanaka & Takahara (1980, 1981): (a) critical flow velocity; (b) phase angle between cylinder motion and fluid force.

time that the existence of multiple stability boundaries had been suggested. Because of nonlinear effects it is unlikely that the restabilisation indicated in Figure 5.22 will occur in practice; thus, a practical stability boundary is as shown by the solid line in the figure. For the in-line array of cylinders, the solution is almost identical to that obtained by Tanaka & Takahara (1981) (see Figure 5.20); (this is to be expected, because the same unsteady data were used as input).

Chen suggested that discontinuities in the stability curve, similar to that shown in Figure 5.22, demarcate the boundary between damping- and stiffness-controlled instabilities for velocities less and greater than the discontinuity, respectively. However, as shown in Figure 5.21, a discontinuity, although not as significant, also exists for a single flexible cylinder, where the instability is necessarily damping-controlled; thus, it is questionable whether it is associated with a change in stability mechanism or just a rapid change in the phase angle between cylinder motion and resulting fluid forces.

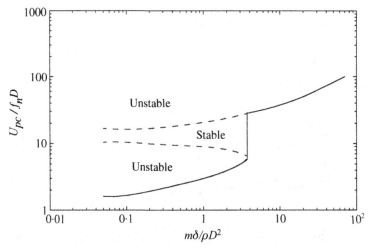

Figure 5.22. Theoretical stability boundary for fluidelastic instability predicted by Chen (1983b) for a row of cylinders with $P/D = 1.33$: - - -, theoretical solution showing multiple instability boundaries; ——, practical stability boundary.

Chen & Jendrzejczyk (1983) further pursued the analysis of Chen (1983b) and showed that at low values of U_p/f_nD, where the instability is dominated by damping-controlled effects, it is not permissible to combine the two nondimensional parameters δ and $m/\rho D^2$ into one parameter as is done traditionally. Chen & Jendrzejczyk (1983) also showed that, for those cases where instability is dominated by stiffness effects, detuning the natural frequencies of adjacent cylinders can have a significant stabilising effect on a row of cylinders; it is not surprising that frequency detuning had sensibly no effect when the instability was primarily damping-controlled.

A finite-element analysis of Chen's model was developed by Eisinger et al. (1991) for the U-bend region of a heat-exchanger tube array, showing good agreement with experimental results.

Eisinger & Rao (1998) used Tanaka & Takahara's unsteady force coefficient data to investigate the effect of tying together flexible cylinders in both cylinder rows and in-line arrays, in both cases for $P/D = 1.33$. For cylinder rows, tying either two or three flexible cylinders together caused a small increase in U_{pc}/f_nD for values of $m\delta/\rho D^2$ below the discontinuity in the stability curve shown in Figure 5.20. For values of $m\delta/\rho D^2$ above the discontinuity, however, they predicted that tying the cylinders together, in groups of either two or three, would prevent instability from occurring. For the cylinder arrays, the effect of tying cylinders together in either the streamwise or transverse directions was investigated. Tying the cylinders together in the transverse direction had a relatively modest stabilising effect on the fluidelastic instability of the array, but if the cylinders were tied together in the streamwise direction a significant stabilising effect of up to an order of magnitude was predicted. The effect of tying cylinders together in the streamwise direction was qualitatively verified experimentally for arrays of cylinders by Weaver et al. (2000).

As mentioned earlier, the unsteady data of Tanaka & Takahara seems to imply that quasi-steady fluid mechanics is not applicable in cylinder arrays. With the use of quasi-steady fluid mechanics the fluid forces on an oscillating body can be written as in equation (5.32); furthermore, both c_d and c_k should tend to constant values as U_r

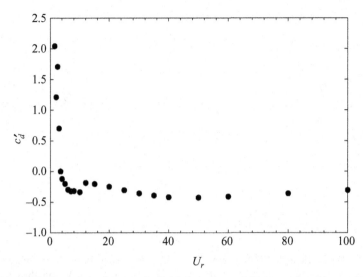

Figure 5.23. Variation of the hysteretic fluid damping coefficient, c'_d, with nondimensional flow velocity $U_r = U_p/fD$ for an in-line square array with $P/D = 1.33$ obtained using the data of Tanaka & Takahara (1980).

becomes large. The results shown in Figure 5.19(b) suggest that this is not the case for c_d. This was noted by Chen & Jendrzejczyk (1983), who observed that a more convenient way of writing the fluid force is of the form

$$F = -\frac{\pi \rho D^3}{4} c_m \ddot{y} - \frac{\rho U_p^2 D}{2\omega} c'_d \dot{y} - \frac{1}{2} \rho U_p^2 D c_k y, \qquad (5.34)$$

where c'_d is now a hysteretic damping coefficient, as opposed to a viscous damping coefficient traditionally used in quasi-steady analysis, and is given by

$$c'_d = -C_{Lyo} \sin \phi. \qquad (5.35)$$

If this is done, then, as shown in Figure 5.23, c'_d tends to a constant value for large values of U_r; however, Chen & Jendrzejczyk gave no explanation why the fluid damping should be hysteretic rather than viscous, and it is difficult to think of any physical reason for this.

At this point it should be realised that as U_r goes from small to large values (which may be interpreted as the frequency of the system going from large to small values for a constant velocity) the fluid system switches from being dominated by inertia to stiffness forces, with a corresponding change in phase between the cylinder motion and fluid force of approximately 180°, as shown in Figure 5.18(b); hence, as U_r becomes large, $\sin \phi$ will tend to a small value. However, as shown in equation (5.33) the estimate of c_d obtained is proportional to both $\sin \phi$ and U_r; and hence, for large values of U_r, estimates of c_d will be extremely sensitive to errors in the phase angle ϕ. More importantly, if there is a systematic constant error in the phase, ϕ_e, as opposed to a random error, then no matter how small this systematic error is, the estimate of c_d will vary linearly with U_r with a slope of approximately $\sin \phi_e$. It should be noted that estimates of the stiffness term c_k do not suffer from this magnification of errors in the phase measurement, and indeed, neither will estimates of c'_d. This does not imply that the fluid damping is hysteretic; it merely shows that, using the unsteady data measured in this way, it is not possible to check on the applicability of the quasi-steady assumption. Furthermore, it should be appreciated that estimates of U_{pc} *are relatively insensitive* to errors in the phase measurements. As long as the true measured value of the unsteady fluid force is used at any value of U_r, no matter whether it is expressed in terms of C_{Ly_O} and ϕ, c_d and c_k or c'_d and c_k, then the error in fluid damping, and corresponding error in the instability velocity, will be proportional to $\sin \phi_e$, and thus typically very small; the error in phase measurements only becomes overwhelming when attempting to interpret the unsteady data in terms of a quasi-steady approach.

Although the analysis presented in this section gives excellent agreement with experimental results (see Section 5.4), extensive experimental effort is required to obtain the unsteady force coefficients. Since the initial pioneering work of Tanaka & Takahara (1980, 1981), a number of researchers have contributed to the data-base of unsteady force coefficient data.

Goyder & Teh (1984) used a slightly different approach from that employed by Tanaka & Takahara to measure the fluid-force coefficients on one cylinder in a normal triangular array with $P/D = 1.375$. They measured the force required to oscillate a flexibly mounted cylinder with a specified amplitude as a function of U_p/f_nD; this was done in either the in-flow or transverse directions, but the cylinder was restricted to oscillate in one direction only. To evaluate the effect of Reynolds number, Re, Goyder & Teh employed two cylinders with natural frequencies of 10 and 30 Hz; thus, the same U_p/f_nD was obtained at different Re. Virtually no difference was observed between the two sets of results suggesting that the Reynolds number is not an important parameter for this array; however, this range of Re is much smaller than typically found in heat-exchanger tube arrays.

An attempt to extend this method to fully flexible arrays is described by Goyder (1990). Rather than measure the individual force coefficients directly, Goyder considered a global coefficient which can be obtained from traditional plots of U_{pc}/f_nD versus $m\delta/\rho D^2$. As it stands this is not particularly useful, because one needs the stability boundary before it can be predicted! However, the main usefulness of this method becomes apparent when considering nonuniform flow over heat-exchanger spans.

A similar approach to that of Goyder (1990) was taken by Granger (1991) and Granger *et al.* (1993), who attempted to construct a "global model" with a single degree of freedom for the analysis of an array of cylinders. Granger showed that each cylinder in the array responds predominantly in one mode of vibration, and that an equation of motion for this global mode may be written which approximates the array in terms of global fluid inertia, damping and stiffness coefficients. These global coefficients were estimated from the measured vibrational response of a flexibly mounted cylinder using multi-degree-of-freedom modal identification techniques (Granger & Campistron 1988; Granger 1988). Although this method reduces the equations to one degree of freedom, it does take account of the motion of neighbouring cylinders – the coefficients are determined from the response of one cylinder in a group of flexible cylinders. The vibrational response of a cylinder in an array was estimated using the global equation, and found to agree very well with the measured vibrational response. However, if this model is to be employed to predict the vibrational response of a different array, it is necessary to first do the vibrational experiment so that the force coefficient can be estimated! Thus, although this analysis represents an interesting development, and may possibly lead to a better understanding of fluidelastic instability, it cannot be regarded as a predictive tool.

Thothadri & Moon (1998) also attempted to measure the unsteady fluid forces for a row of flexible cylinders with $P/D = 1.35$. They measured the direct coupling terms (those relating either x- or y-motion between neighbouring cylinders) directly via a transfer-function approach, and then inferred heuristically the cross-coupling terms (those coupling the x and y motion between neighbouring cylinders) so that their analytical model gave the correct $U_{pc}/f_n D$.

Chen and co-workers (Chen *et al.* 1998; Chen & Srikantiah 2001) measured the unsteady force coefficients for a number of different configurations; including cylinder rows with $P/D = 1.35$ and 2.7, normal triangular arrays with $P/D = 1.35$, and square arrays with $P/D = 1.35$, 1.42 and 1.46. Measurements were made on the oscillating cylinder only (hence, the effect of cylinder motion on the fluid forces acting on neighbouring cylinders was not determined); however, the effects of position in the array, changing the amplitude of oscillation and Reynolds number were investigated. It was demonstrated that, for almost all of these configurations, a single flexible cylinder, with freedom to oscillate only in the transverse direction, could go unstable at low values of $U_p/f_n D$ (hence, the instability was damping-controlled). Because the effect of a neighbouring oscillating cylinder was not considered, it was not possible to investigate the possibility of stiffness-controlled instabilities. One important conclusion from these results was that the Reynolds number is an important parameter, with the region of U_r where the fluid exhibited negative damping in the transverse direction being extremely strongly dependent on Re.

In addition to the measurements presented in the preceding paragraphs, a number of research teams have attempted to measure the unsteady force coefficients in two-phase flows. Inada *et al.* (1996) measured the unsteady fluid forces in the transverse direction for one flexible cylinder in a square array, $P/D = 1.42$, in an air-water mixture. They showed that if the homogeneous model was employed there was considerable scatter in the results as the void fraction was changed, but that this scatter could be eliminated using the drift-flux model of velocity. In a later paper Inada *et al.* (1997) oscillated the central cylinder in both the transverse and in-flow directions, and also measured the forces on the surrounding cylinders as well as on the central cylinder itself. Restricting the cylinder motion to the transverse direction and considering only the forces acting on the oscillating cylinder, Inada *et al.* (2002) investigated the effect of amplitude of oscillation on the fluidelastic forces. They found that for amplitudes of oscillation less than $0.136D$ the fluidelastic forces were linear with cylinder amplitude, but that for amplitudes above this value nonlinear effects became apparent.

The unsteady fluid forces were also measured in a two-phase water-freon mixture by Delenne *et al.* (1997) for a square array with $P/D = 1.44$. Only one flexible cylinder was employed, and the unsteady fluid-force coefficients, obtained via an inverse approach from the cylinder vibrational response, showed considerable variation with void fraction. Furthermore, this variation depended on whether the flow velocity was defined using a homogeneous model, a slip model or a drift-flux model.

Mureithi *et al.* (2002) measured the unsteady fluid forces on a square array, with $P/D = 1.46$, in a two-phase steam-water mixture. They showed that the variation of the force coefficients with U_p/f_nD depended on the specific values of U_p and f_n; hence, U_{pc}/f_nD no longer appears to be the only relevant criterion of similarity, and void fraction and flow pattern seem to be important independent parameters. The results obtained did indicate, however, that the fluidelastic forces acting on a tube were relatively insensitive to the motion of neighbouring tubes; thus suggesting that, in this case, only damping-controlled instabilities are possible. This data was used by Hirota et al. (2002) in a stability analysis, although only forces induced by the cylinders' own motion were accounted for. The methodology employed to overcome the dual values of fluid-force coefficients as a function of U_p/f_nD was to select the force coefficients which gave the least stable configuration. A comparison between the theoretical instability boundary and experimental data was remarkably good.

As previously mentioned, although the "unsteady method" of Tanaka & Takahara and Chen gives excellent agreement with experimental results (see Section 5.4), extensive experimental effort is required to obtain the unsteady force coefficients. It is not surprising that a full set of force coefficients, accounting for the motion of adjacent cylinders, has been measured for two arrays only (Tanaka & Takahara 1980, 1981; Tanaka et al. 1982). Moreover, it should also be appreciated that, in some arrays, U_{pc}/f_nD varies dramatically from row to row (Price & Zahn 1991); thus, the unsteady fluid forces should be measured for all rows of the array. Hence, the experimental effort required for these methods is probably too great for them to be considered as practical design tools. There is no doubt, however, that these analyses have added greatly to the present understanding of fluidelastic instability in cylinder arrays.

5.3.4 Semi-analytical models

A more analytical approach to modelling fluidelastic instability is given in a series of papers by Lever & Weaver (1982, 1986a, b) and Yetisir & Weaver (1988), where a single flexible cylinder in an array of rigid cylinders is considered.* The justification for using a single flexible cylinder comes from experiments by Lever & Weaver (1982) who obtained virtually the same U_{pc} for a single flexible cylinder in an array of rigid cylinders as for an array of 19 flexible cylinders. In its original form (Lever & Weaver 1982) the analysis was limited to cross-flow motion only. It was later modified to account for in-flow motion (Lever & Weaver 1986a, b), although the in- and cross-flow motions were analysed independently of each other. This was justified via the results of Weaver & Koroyannakis (1983) who showed that U_{pc} in one direction is only slightly affected by changes in cylinder frequency in the other.

In fact, Lever & Weaver (1986a) present three different stability analyses, one each for dynamic and static instabilities in the cross-flow direction, and a third for both dynamic and static instability in the in-flow direction. In the following paragraphs the dynamic analysis for cross-flow motion is discussed.

Based on experimental observations, Lever & Weaver concluded that the stability of an array of cylinders could be analysed via a single flexible cylinder positioned

---

* As discussed later in this section, Yetisir & Weaver (1992, 1993a, b) extended this model to include multiple flexible cylinders.

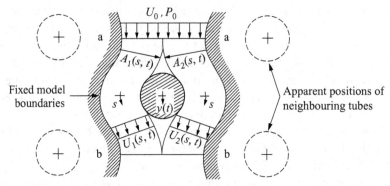

Figure 5.24. Schematic of the "unit cell" for the Lever & Weaver model [reproduced from Lever & Weaver (1982)].

in a so-called unit cell; see Figure 5.24.* The flow through the array is divided into "wake" and "channel" regions. The channel flow is assumed to be inviscid and one-dimensional, and its separation and reattachment positions are calculated approximately from the array geometry. The equilibrium cross-sectional areas of the two channel-flow streamtubes are taken as being constant and equal to the inlet areas. This should give constant pressure through the channels; however, a resistance term is included to account for frictional losses in the flow.

It is assumed that cylinder motion causes a redistribution of the streamtube area, proportional to its magnitude. For attached flow this redistribution is in-phase with the cylinder motion; however, because of fluid inertia, the flow upstream of the cylinder lags behind the cylinder motion, producing a phase lag between cylinder motion and the resulting pressure forces on the cylinder. As yet, no exact solution is available for the phase lag, and resort must be made to a simplified analysis. Lever & Weaver used the analogy of a flat plate in a channel aligned with the flow and oscillating normal to its plane, and obtained an approximate expression for the phase lag in terms of one empirical constant. The expression they arrived at had the physically reasonable property that the phase lag is zero at the attachment point and decreases as the distance from the oscillating cylinder increases. Then, making use of the unsteady continuity and momentum equations, a rather complex expression for pressure as a function of circumferential position on the cylinder was obtained. In fact, the fluid force in the cross-flow direction was approximated by integrating the pressure over the attached region only.

Assuming harmonic motion of the cylinder, the fluid force was decomposed into stiffness and damping terms, and the equation of motion for the cylinder is given by

$$lm\ddot{y} + [c + \tfrac{1}{2}C_D\rho DlU_g - F_o\sin\theta_0/(\omega y_o)]\dot{y} + [k - F_o\cos\theta_0/y_o]y = 0, \qquad (5.36)$$

where U_g is the "gap velocity"; m and c represent the mass per unit length and structural damping – those measured in stagnant fluid; the C_D term accounts for the flow-induced damping forces; and $F_o\sin\theta_0/(\omega y_o)$ and $F_o\cos\theta_0/y_o$ are, respectively, the fluid-damping and -stiffness terms.

---

* Although the analytical models presented in 1982 and 1986 are basically the same, there are some differences. The model discussed here is the 1986 version.

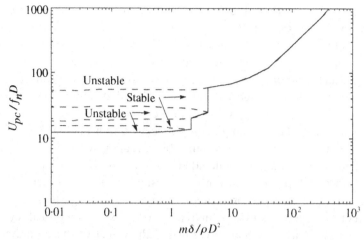

Figure 5.25. Theoretical stability boundary for fluidelastic instability obtained by Lever & Weaver (1986b) for a single flexible cylinder in a parallel triangular array with $P/D = 1.375$: - - -, theoretical solution showing multiple instability boundaries; ——, practical stability boundary.

The condition for dynamic stability is simply that the total damping is zero; this leads to the following criteria for instability:

$$\frac{m\delta}{\rho D^2} = C_1 V^2 \sin(1/V) + C_2 V \cos(1/V) + C_3 V^3[\cos(1/V) - 1] + C_4 V \quad (5.37a)$$

and

$$\frac{U_{pc}}{f_n D} = C_5 V, \quad (5.37b)$$

where C_1 to C_5 are constants which depend on the array geometry and on a number of other terms, including the following: the flow-resistance coefficient through the array; the drag coefficient, C_D; the positions of flow attachment and separation; and the unknown constant in the expression for the phase lag. Relating the pressure drop through the array to the flow-resistance coefficient, a relationship was given between C_D and this coefficient, and values of the coefficient were obtained from the data of Pierson (1937).

It should be noted that no account is taken of the fluid-stiffness terms; Lever & Weaver assume that instability occurs at the no-flow natural frequency, and thus, do not consider the effect of the fluid-dynamic stiffness. Interestingly, one of the conclusions obtained from this analytical model is that $m/\rho D^2$ and δ may be combined into one nondimensional parameter $m\delta/\rho D^2$. However, in a later paper, Rzentkowski & Lever (1992) showed that if the "true" fluidelastic frequency of oscillation (accounting for the fluid-stiffness terms) is used to predict the fluid forces, then $m/\rho D^2$ and δ must be considered as independent nondimensional parameters.

A typical stability boundary obtained by Lever & Weaver (for a parallel triangular array with $P/D = 1.375$) is reproduced in Figure 5.25. For low values of $m\delta/\rho D^2$, bands of instability exist, similar to those obtained by Chen (1983b); indeed, these do not cease to exist at the lowest band given by Lever & Weaver, but continue until much lower values of $U_{pc}/f_n D$. The reason for these bands can be understood from equation (5.37a), where the flow-induced damping is seen to be harmonically dependent on $1/V = C_5 (f_n D/U_{pc})$; thus, for low $U_{pc}/f_n D$ the damping oscillates between negative and positive values as $U_{pc}/f_n D$ is reduced. However, it should be appreciated that as $U_p/f_n D$ tends to smaller values the phase lag between cylinder motion and streamtube area becomes extremely large – in fact, it can exceed several periods of oscillation. Lever & Weaver concluded that minor flow perturbations, for example due to turbulence, will disrupt the pressure variations associated with these very long phase lags, making them physically unreasonable. Thus, in practice, probably only the upper and next two instability regions will exist. Furthermore, in a manner similar to Chen, they concluded that, once a cylinder has gone unstable, nonlinear effects will prevent it from restabilising as velocity is increased. Thus, a practical stability boundary as presented in Figure 5.25 is suggested.

Although this theory is mainly analytical, a number of empirical terms are required. Lever & Weaver (1986b) investigated the effect of these parameters and concluded that most of them have very little effect on the stability boundary. One exception to this was the empirical constant in the phase-lag expression; changing this by ±50% at low values of $m\delta/\rho D^2$ causes a change in $U_{pc}/f_n D$ of approximately ±50%. This indicates that it is the unsteady fluid terms, associated with the phase lag, which dominate the instability mechanism, and that to obtain accurate stability boundaries the phase lag must be known to a corresponding accuracy. This is in agreement with the conclusions obtained by Chen (1983b). Romberg & Popp (1998a) demonstrated experimentally that changing the separation points for the flow channel, accomplished by positioning trip wires on the cylinders, can have a very large effect on the stability of the array, suggesting, in contradiction to the conclusion of Lever & Weaver, that these parameters are also important.

In principle, it should be possible to analyse the static stability of a cylinder using equation (5.36); however, as the frequency of oscillation tends to zero, the phase lag between cylinder motion and fluid forces also tends to zero. To avoid this complication Lever & Weaver (1986a) developed a separate analytical model for cross-flow static instability. The need to have two analyses for static and dynamic instability represents one of the weaknesses of the model produced by Lever & Weaver; ideally, one model should predict both the fluid-damping and stiffness forces.

An attempt to refine the model of Lever & Weaver was made by Yetisir & Weaver (1988) who noted the inadequacies mentioned above, and also that at high $m\delta/\rho D^2$ the Lever & Weaver model predicts $U_{pc}/f_n D$ to be proportional to $m\delta/\rho D^2$ rather than $[m\delta/\rho D^2]^{0.5}$ as suggested by most experimental results. Although there are a number of minor differences between this later analysis and the original one, the main difference is the addition of a decay function in the area perturbation. In its original form, flow perturbations induced by cylinder motion were assumed to extend two cylinder rows upstream and downstream of the oscillating cylinder, and the magnitude of these perturbations was assumed constant over this distance. In the new analysis, the area perturbation gradually decreases as the distance from the oscillating cylinder increases. Unfortunately, the results presented suggest that by introducing this decay function agreement between theory and experiment deteriorates. Thus, at the present time, the original Lever & Weaver form of this model seems the most suitable.

Almost exactly the same approach as that taken by Yetisir & Weaver (1988) was also employed in a later paper by Marn & Catton (1992). As an extension of this approach Marn & Catton (1991) considered a column of flexible cylinders in an array of rigid cylinders. The flow through the array was still represented by a series of unit cells; however, the flow within the cells was considered to be two-dimensional. The variables, such as pressure and in- and cross-flow velocities, were split into mean and time-varying components. Furthermore, it was assumed that the time-varying terms arose solely because of the cylinder motion, and the mean terms were constant throughout the array. The fluid forces, obtained from integrating the pressure around the cylinder circumference, were then introduced into the in- and cross-flow vibrational equations of motion, giving five equations and five unknowns. To solve these equations it was assumed that "the mean direction in which information travels" is "parallel with the incoming flow" or x-direction. The variables were then written in terms of normal modes and a wavenumber associated with motion of adjacent cylinders in the column. Their representation implies that the pressure, velocities and cylinder displacements are all in-phase with one another. To obtain the stability boundary it was necessary to assume a specific modal pattern, or wavenumber, between cylinders; however, results presented by Marn & Catton suggest that the stability boundary is not sensitive to the particular intercylinder modal pattern chosen. A typical stability boundary was given by Marn & Catton for a rotated square array with $P/D = 1.5$. Although there was good agreement with experiments for $m\delta/\rho D^2 \geq 5$, the agreement was very poor for $m\delta/\rho D^2 < 5$; furthermore, the instability predicted by Marn & Catton is in the in-flow direction, as opposed to the cross-flow motion which is observed experimentally. Instability was predicted in the cross-flow direction, but at much higher $U_{pc}/f_n D$. Most likely, the reason why the theory overestimates $U_{pc}/f_n D$ for low $m\delta/\rho D^2$ is that the cylinder motion and resulting velocity and pressure terms are assumed to be in-phase with each other.

Marn & Catton (1993) also used the unit cell approach to analyse the stability of a full array of flexible cylinders, but in this case the flow within the cell was analysed using the vorticity transport equation, and the final equations were solved via a finite-difference scheme. It was found that there was a considerable difference in $U_{pc}/f_n D$ between when the first and last cylinders in the array went unstable. Marn & Catton suggest that, for design purposes, the stability boundary should be given by an average of these two extremes. Also, $U_{pc}/f_n D$ was approximately proportional to $m\delta/\rho D^2$, rather than to the power 0.5 as is more characteristic of experiment results.

Marn & Catton (1996) attempted to extend this vorticity-transport approach to include two-phase flow. To the authors' knowledge, this was the first theoretical attempt to obtain fluidelastic instability boundaries in two-phase flow; however, bearing in mind the extremely complex behaviour of cylinder arrays in two-phase flow, see for example Pettigrew & Taylor (1994, 2003), this analysis may be too simplistic to be of any practical use.

Yetisir & Weaver (1992, 1993a, b) extended their original analytical model (Yetisir & Weaver 1988) to account for the effect of multiple flexible cylinders in the array. The method is essentially the same as that described earlier, except that the effect of neighbouring cylinder motion on the area and pressure perturbations is accounted for. For example, considering an in-line array and using the Tanaka & Takahara cylinder numbering scheme, as shown in Figure 5.17, motion of cylinder

U will induce a change in the pressure distribution around cylinder O in exactly the same manner as it induces a change in the pressure distribution around itself. Thus, the effect of the combined motion of cylinders U, D, L and R on the fluid force acting on cylinder O can be obtained using the principle of superposition; similarly, the effect of the motion of cylinder O on the fluid force acting on cylinders U, D, L and R can also be obtained. Hence, for pure cross-flow motion, as considered by Yetisir & Weaver, a coupled 5×5 matrix equation is obtained for the stability of the five-flexible-cylinder group (O, U, D, L and R), which may easily be solved.

Yetisir & Weaver considered both in-line and staggered arrays, assuming a group of five flexible cylinders to be representative of a complete flexible array. For in-line arrays, the flexible cylinders are as shown in Figure 5.17 and are reasonably representative of a fully flexible array. However, for staggered arrays using only five flexible cylinders means that the motion of cylinders directly upstream and downstream of the central cylinder are not accounted for. Bearing in mind that the fluid forces acting on a cylinder in an array are typically strongly affected by the motion of an upstream cylinder, this would seem to represent a potential error in the analysis, but, as subsequently shown by Li & Weaver (1997), accounting for the flexibility of these two cylinders had a negligible effect on the array stability. Numerical results presented by Yetisir & Weaver for parallel triangular and in-line square arrays with $P/D = 1.375$ show that for $m\delta/\rho D^2 < 200$ there is very little difference between $U_{pc}/f_n D$ for an array with either one or multiple cylinders. However, when $m\delta/\rho D^2 \geq 200$, the variation of $U_{pc}/f_n D$ with $m\delta/\rho D^2$ for multiple flexible cylinders is such that the exponent on $m\delta/\rho D^2$ tends to 0.5 as $m\delta/\rho D^2$ becomes large, rather than tending to 1 as obtained with the original Lever & Weaver (1982, 1986a, b) model. This is in agreement with the models proposed by Chen (1983a, b) and Price & Païdoussis (1984a), (presented in the next section). A very similar analysis accounting for multiple flexible cylinders was proposed by Parrondo *et al.* (1993) who reached essentially the same conclusions as those outlined above.

5.3.5 Quasi-steady models

As discussed earlier in this book, the quasi-steady assumption states that, for an oscillating body, such as a cylinder, the effect of its motion on the resulting fluid forces is solely to modify the velocity vector relative to the body, with the resultant lift and drag forces being normal and parallel, respectively, to this relative velocity vector. Furthermore, it is assumed that C_L and C_D are unaltered by the oscillation and can be obtained from measurements, or calculations, made on a stationary body. This implies that the flow distribution around the body follows immediately, with no attenuation, the motion of the relative velocity vector. For motion of the cylinder which is slow compared with the free-stream velocity this is reasonable; however, when the cylinder velocity becomes comparable with the flow velocity, this assumption breaks down. Based on different physical arguments, Fung (1955) and Blevins (1977b) concluded that, for an isolated bluff body, quasi-steady fluid dynamics is valid provided $U_{pc}/f_n D \geq 10$; however, whether or not this limitation needs to be modified for cylinders in an array is open to question. Motivated by the success of quasi-steady aerodynamics in applications such as galloping and wake-induced flutter of overhead transmission lines, a number of researchers have attempted to use the quasi-steady assumption to analyse the stability of heat-exchanger tube arrays.

The first quasi-steady analysis of cylinder arrays subject to cross-flow was by Gross (1975), who concluded that instability in cylinder arrays is due to two distinct mechanisms, damping- and stiffness-controlled. Gross developed a quasi-steady analysis for the cross-flow motion of a single cylinder, considering only the damping-controlled mechanism. He assumed a linear variation of C_y with α; then, taking $\alpha \approx \dot{y}D/U_p$ he obtained an aerodynamic damping force proportional to $\partial C_L/\partial \alpha$. Instability occurs when the sum of the fluid and structural damping is zero, giving

$$\frac{U_{pc}}{f_n D} = \frac{m\delta}{\rho D^2(-\partial C_L/\partial \alpha)}. \tag{5.38}$$

This is similar to the classical Den Hartog (1932) expression, except that the fluid drag is incorporated in the coefficient $\partial C_L/\partial \alpha$. This analysis predicts U_{pc} to be linearly proportional to $m\delta/\rho D^2$, in contrast to experiments (Chen 1984; Weaver & Fitzpatrick 1988) where the exponent on $m\delta/\rho D^2$ is much less than 1. However, Gross suggested that, if fluidelastic stiffness effects were accounted for, then the exponent may be less than 1.0.

The next attempt at a quasi-steady analysis was by Price & Païdoussis (1982, 1983) who considered a double row of cylinders in cross-flow. The transverse fluid force on a cylinder was written as

$$F_y = \frac{1}{2}\rho Dl U_g^2 \left(C_L - \frac{\dot{x}D}{U_g}2C_L - \frac{\dot{y}D}{U_g}C_D \right), \tag{5.39}$$

where U_g is the gap velocity, and C_L and C_D are based on the local gap velocity, U_g, and not the free-stream velocity; a similar expression was also given for the in-flow direction; it is recalled that, as in the foregoing, x and y are dimensionless with respect to D. Price & Païdoussis assumed the force coefficients to vary linearly with displacement of the cylinder itself as well as that of its two neighbouring cylinders. Thus, the analysis of a row of flexible cylinders results in a fully coupled system, which was analysed using a constrained-mode approach. Price & Païdoussis also assumed the fluid-force coefficients to be functions of the incidence, α, of the resultant velocity vector *vis-à-vis* the free-stream direction. However, rather than measuring the stiffness terms $\partial C_L/\partial \alpha$ and $\partial C_D/\partial \alpha$, an attempt was made to relate them to the coefficients $\partial C_L/\partial y$ and $\partial C_D/\partial y$. The authors later realised that the manner in which this had been done was incorrect and attempted to refine this analysis in a series of papers (Price & Païdoussis 1984a, 1985, 1986a, b; Price *et al.* 1990). The main differences between the latter analysis and the earlier work are as follows.

Considering either in-line or staggered arrays, then for any cylinder, C (see Figure 5.26), it is assumed that the fluid forces acting on it are directly affected by its own motion and the motion of eight adjacent cylinders only. Then, the drag coefficient on cylinder C, C_{Dc}, may be expressed as a linear function of the displacements of cylinders C and 1 to 8 as follows:

$$C_{Dc} = C_{Dco} + \sum_{j=c,1}^{8} \left[\frac{\partial C_{Dc}}{\partial \xi_i}\xi_i + \frac{\partial C_{Dc}}{\partial \eta_i}\eta_i \right], \tag{5.40}$$

and similarly for C_{Lc}. In the above equation, ξ_i and η_i are the "apparent" cross- and in-flow displacements of the cylinders, taking into account the time delay between

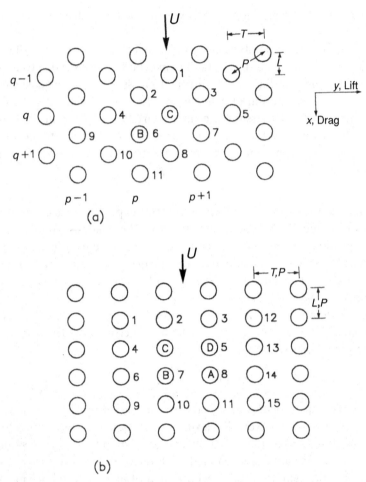

Figure 5.26. Schematic of the cylinder array numbering system for the Price *et al.* (1990) constrained-mode analysis; (a) a staggered array, (b) an in-line array.

motion of any cylinder and its effect being felt by cylinder C, and also the inclination of the wake shed from a cylinder due to its transverse motion. Expressions for the apparent displacements of the neighbouring cylinders are given by Price & Païdoussis (1984a).

One final effect accounted for by Price & Païdoussis, following the work of Simpson & Flower (1977), was that of retardation of the flow approaching a cylinder. Because flow slows down as it approaches a bluff body, it arrives at a later time than it would have done in a constant velocity flow. This effect is important in cylinder arrays where large changes in fluid force occur as a result of small displacements of a cylinder. Price & Païdoussis showed that flow retardation could be accounted for by multiplying the cylinder displacements by a factor $\exp(-\lambda\mu D/U_g)$, where μ was referred to as the flow-retardation parameter and λ is the eigenvalue. As yet, μ cannot be determined exactly, but, based on reasonable assumptions and consideration of the flow around a bluff body, Price & Païdoussis suggested μ of $O(1)$. The importance of such an unsteady effect had previously been suggested by Ruscheweyh (1983).

It should be appreciated that even though the effect of only the eight immediate neighbouring cylinders is accounted for in the expressions for the lift and drag coefficients acting on cylinder C, there is fluid coupling between all of the flexible cylinders in the array. For example, considering a staggered array as shown in Figure 5.26(a), motion of cylinder 9 will induce changes in the fluid forces on cylinders 4, 6 and 10, and thus, although indirectly, the fluid forces on cylinder C. Hence, significant computational effort is required to analyse the stability of a large array of flexible cylinders. To minimise this, Price *et al.* (1990) employed a constrained-mode analysis using essentially the method of Whiston & Thomas (1982). This was done by expressing the motion of any cylinder in row q and column p, see Figure 5.26(a), as

$$x_{p,q} = x_0 \exp(\lambda t + ip\phi_x + iq\theta_x),$$
$$y_{p,q} = y_0 \exp(\lambda t + ip\phi_y + iq\theta_y),$$
(5.41)

where $i = \sqrt{-1}$, ϕ_x and θ_x are the phase differences between the x-motion of adjacent cylinders in any row and column, respectively, and ϕ_y and θ_y are the corresponding y-motion phase differences; these phase differences are assumed constant throughout the array. Then, for example, the x-displacement of cylinder 1 of Figure 5.26(a) may be expressed as

$$x_1 = x_{p,q-1}^c$$
(5.42)

where superscript c and subscripts p, $q - 1$ denote that cylinder 1 is at position p, $q - 1$ relative to cylinder C. Hence,

$$x_1 = x_{p,q-1}^c = x_0^c \exp\left[\lambda t + ip\phi_x + i(q-1)\theta_x\right],$$
$$= x_0^c \exp\left[\lambda t + ip\phi_x + iq\theta_x\right] \exp(-i\theta_x),$$
(5.43)
$$= x_{p,q}^c \exp(-i\theta_x).$$

Similar procedures can be employed for cylinders 4, 5, 8 and 10; however, for cylinders 2, 3, 7, 9 and 11, the displacements must be written in terms of cylinder B (where the (p, q)-axis system is now centred on cylinder B and not cylinder C), giving

$$x_{11}^B = x_{p,q}^B \exp(i\theta_x).$$
(5.44)

Hence, the x- and y-displacements of cylinders 1 to 11 may be written in terms of the displacements of cylinders C and B and the two-cylinder kernel B – C is totally decoupled from the rest of the array. In a similar manner, for an in-line array, the displacements of cylinders 1 to 15 in Figure 5.26(b) may be written in terms of the displacements of cylinders A, B, C and D, decoupling the four-cylinder kernel from the rest of the array. It was later shown by Price *et al.* (1992a) that for an in-line array the kernel could be reduced further to one cylinder.

One problem with this analysis is that 38 fluid-force coefficients are required before the lift and drag coefficients for cylinder C can be evaluated. However, because of the geometric symmetry of the array, a number of equalities exist which reduce the number of required force coefficients from 38 to 21. Measured force coefficients are presented by Price *et al.* (1990) for three different arrays, and the stability of these arrays investigated.

For each array, it was first necessary to determine the specific intercylinder modal pattern in the constrained mode analysis which gave the minimum U_{pc}. This was characterised by the four phase angles (ϕ_x, θ_x, ϕ_y and θ_y) between the motion, either in-flow or cross-flow, of adjacent cylinders in a row or column. The two phase angles relating the in-flow motion (ϕ_x and θ_x), had no effect on $U_{pc}/f_n D$, the most important phase angle being θ_y which relates the cross-flow motion between adjacent cylinders. The value of this phase angle which produced the minimum $U_{pc}/f_n D$ was not constant but depended on $m\delta/\rho D^2$. It was suggested that this is because the instability mechanism changes from being predominantly stiffness-controlled at high $m\delta/\rho D^2$ to damping-controlled at low $m\delta/\rho D^2$.

A specific analysis for a single flexible cylinder in an array of otherwise rigid cylinders was also developed (Price & Païdoussis 1984a, 1986b) where only damping-controlled instabilities are possible. The analysis showed that instability is possible in the cross-flow direction, and that it is primarily due to the phase lag, caused by flow retardation, between the cylinder motion and resulting fluid force. Indeed, for sufficiently large values of $U_p/f_n D$, such that $\sin(-\omega\mu D/U_p)$ may be taken as $-\omega\mu D/U_p$, the following was obtained (where, C_L and C_D are based on the pitch velocity, U_p):

$$\frac{U_{pc}}{f_n D} = \frac{4m\delta}{\rho D^2(-C_D - \mu \partial C_L/\partial y)}. \tag{5.45}$$

This shows the importance of the flow-retardation parameter μ, and also that (assuming the drag coefficient C_D to be positive) $\partial C_L/\partial y$ must be negative for a damping-controlled instability to occur. It is also apparent that, as $m\delta/\rho D^2$ becomes large, $U_{pc}/f_n D$ is proportional to $m\delta/\rho D^2$; this, of course, is the same result as that obtained by the single-flexible-cylinder analyses of Chen, Lever & Weaver and Gross.

At low $m\delta/\rho D^2$, both the single-flexible-cylinder and multiple-flexible-cylinder analyses give multiple instability regions similar to those obtained by Lever & Weaver (see Figure 5.25). These are a consequence of the very large phase lags between cylinder motion and resulting fluid forces (possibly several periods long) which occur for low $U_p/f_n D$, causing the fluid damping to oscillate between negative and positive values. For exactly the same reasons proposed by Lever & Weaver it is very doubtful whether these long phase lags can exist in practice. Hence, Price & Païdoussis recommended that only two of the lower unstable regions be accounted for in a practical stability boundary. Also, again similarly to Lever & Weaver, they suggest that the cylinder will not restabilize for increasing flow velocity once it has gone unstable.

Of the three different phenomena which lead to the unsteady fluid terms in this analysis, by far the most important is the flow retardation. Its importance was investigated by varying the flow-retardation parameter, μ, for a single flexible cylinder; typical results presented in Figure 5.27 show it has a significant effect on $U_{pc}/f_n D$, particularly at low $m\delta/\rho D^2$. This clearly indicates the need to obtain a better understanding of the unsteady fluid dynamics in cylinder arrays so that more accurate predictions of $U_{pc}/f_n D$ can be made.

One of the disadvantages of this analysis, compared with that of Lever & Weaver (1986a, b) is the amount of experimental data required as input – although this is considerably less than that required in the analysis of Tanaka & Takahara (1981)

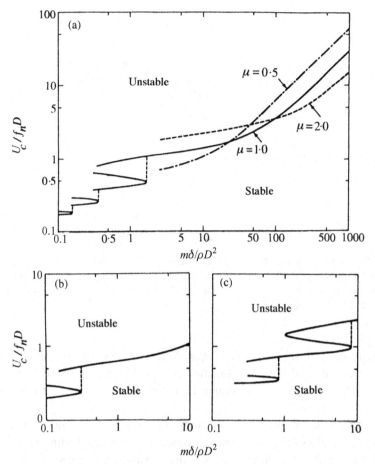

Figure 5.27. Effect of flow retardation parameter μ on the fluidelastic stability boundaries obtained via the analysis of Price & Païdoussis (1984a) for a single flexible cylinder in a parallel triangular array with $P/D = 1.375$: (a) $\mu = 0.5$, 1.0 and 2.0; (b) expanded view for $\mu = 0.5$; (c) expanded view for $\mu = 2.0$.

or Chen (1983a, b). From a sensitivity analysis, it was determined that the two coefficients which have the most effect on $U_{pc}/f_n D$ are $\partial C_L/\partial y$ terms, with the y-displacement being either that of the cylinder itself or the cylinder directly upstream. Indeed, it was found that fluid coupling between adjacent cylinders in a row could be ignored with little loss in accuracy, but that if the fluid coupling between cylinders in a column was ignored there was a large change in $U_{pc}/f_n D$.

Granger & Païdoussis (1996) proposed an improvement to the quasi-steady model, referred to as a *quasi-unsteady* analysis, which uses the continuity and Navier-Stokes equations to model the unsteady fluid-dynamic forces. These unsteady terms are due to memory effects, more specifically the diffusion of vorticity generated on the surface of the cylinder, as opposed to a time lag as suggested by Price & Païdoussis. The memory effect is modelled as a combination of decaying exponentials via the following function:

$$1 - \sum_{i=1}^{N} \alpha_i \exp(-\delta_i t), \tag{5.46}$$

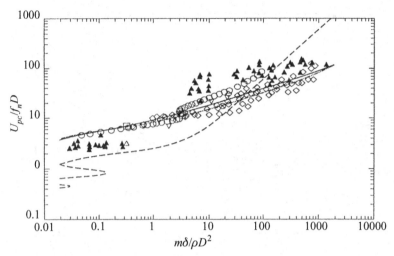

Figure 5.28. Stability boundaries for a single flexible cylinder in in-line square arrays [repro-duced from Granger & Païdoussis (1996)]: – –, Price & Païdoussis quasi-steady model; —, quasi-unsteady model ($N = 1$); – · –, quasi-unsteady model ($N = 2$). For experimental data see Granger & Païdoussis (1996).

with N being either 1 or 2 (for $N = 2$ this is of exactly the same form as the commonly used approximation of Wagner's function for oscillating airfoils; see Fung (1955)). The constants α_i and δ_i were obtained by matching the response to experimental data over the required range of $U_p/f_n D$. Fullana & Beaud (1999) pointed out that this procedure is ill-conditioned but proposed a numerical scheme to overcome it.

The advantages of the quasi-unsteady analysis, compared with the quasi-steady one, are that, provided accurate values of the constants are known, it gives superior predictions for the variation of fluidelastic damping and frequencies with U_p. Also, whereas for a single flexible cylinder the quasi-steady model predicts U_{pc} to be proportional to $m\delta/\rho D^2$ at high $m\delta/\rho D^2$, the quasi-unsteady analysis predicts U_{pc} to be proportional to $(m\delta/\rho D^2)^{1/2}$, which, as shown in Figure 5.28, agrees much better with experimental data.

There are, however, several disadvantages to the quasi-unsteady analysis. First, there are four new constants to be determined, which are array dependent – as pre-viously mentioned, to obtain these constants requires that the vibrational response of the array be known. Second, the analysis is restricted to a single flexible cylinder; hence, it cannot account for stiffness-controlled instabilities. It should be appreciated that, for some arrays, damping-controlled instabilities do not exist, whereas stiffness-controlled instabilities do; for example, Price & Kuran (1991) and Païdoussis et al. (1989) show that a single flexible cylinder in a rotated square array, with either $P/D = 1.5$ or 2.12, is always stable, whereas, for the same arrays, multiple flexible cylinders do become unstable. A final weakness of the quasi-unsteady model is that multiple instability regions are never obtained, whereas for low $m\delta/\rho D^2$ experimental evid-ence (see Section 5.4.2) strongly suggests that they do exist. Granger & Païdoussis (1996) suggest that the multiple instability regions may be due to vortex shedding, as opposed to fluidelastic instability, but again experimental evidence contradicts this.

Meskell (2005) used a simple wake model in an attempt to predict analytically the memory function (equation (5.46)) proposed by Granger & Païdoussis (1996).

It was assumed that the memory function could be represented by the normalised instantaneous bound circulation on a vibrating cylinder due to a sudden change in its transverse position; effectively the indicial response of the circulation. Hence, the instantaneous flow field in the array was modelled as a circular cylinder with a bound circulation, giving the static lift and drag forces, and a trailing vortex sheet which represented the transient nature of the fluidelastic forces. Pursuing an analysis very similar to that of an airfoil impulsively set into motion, estimates of the coefficients in equation (5.46), assuming $N = 1$, were obtained for a normal triangular array with $P/D = 1.375$, and a stability analysis performed. It should be appreciated that this analysis gives the memory effect only; thus, in common with the analysis of Granger & Païdoussis, the static force coefficients C_D and $\partial C_L/\partial y$ must still be supplied as empirical inputs. However, the stability boundary obtained by Meskell (2005) compares very favourably with both experimental data and the second-order model of Granger & Païdoussis (1996).

Hémon (1999b) also developed a quasi-steady model very similar to that of Price & Païdoussis, except that the time delay between cylinder motion and resulting fluid forces is written as L/V_c, instead of $\mu D/U_p$, where V_c is the array convection velocity. Unfortunately, V_c is unknown, and Hémon approximates V_c by matching static force coefficients measured in an array of cylinders with those for an isolated cylinder. This amounts to assuming that the pressure distribution around a cylinder in an array has the same shape as for an isolated cylinder; almost certainly this is not correct.

Recently, Mureithi and co-workers (Shahriary et al. 2007; Mureithi et al. 2008) used the quasi-steady analysis of Price & Païdoussis (1984a) to investigate the stability of an array of cylinders in two-phase flow. The quasi-static force coefficients were measured for a rotated triangular array with $P/D = 1.5$ in an air-water flow for a series of void fractions between 0% (single-phase waterflow) to 95%, and also for a number of different flow rates. The array consisted of 13 rows and 3 columns of cylinders surrounded by columns of $\frac{1}{2}$ cylinders on the walls of the tunnel. The central cylinder in the array could be statically displaced in either the in-flow or transverse directions, while the static lift and drag forces on itself and the surrounding cylinders were measured. One extremely surprising result from this work was the degree to which the force coefficients depended on void fraction; in particular, $\partial C_L/\partial y$ for the cylinder being displaced was extremely sensitive and varied from positive values for void fractions of $0, 21$ and 31% to being negative for void fractions of between 40 and 95%. In addition, it was shown that some of the coefficients, most notably those associated with the drag of the displaced cylinder, varied significantly with Re. Using the measured force coefficients in the quasi-steady analysis, reasonable agreement was obtained between theoretical predictions of U_{pc} and the previously measured experimental values of Violette et al. (2006) for the same array. However, it was shown that the theoretical predictions depended strongly on the flow-retardation parameter, μ, as well as the void fraction and Reynolds number.

5.3.6 Computational fluid-dynamic models

Although simulation of the interstitial fluid flow in oscillating cylinder arrays seems formidable, especially bearing in mind its unsteady and highly turbulent nature, there have been considerable recent advances in computational fluid dynamics (CFD), and several groups of researchers have used CFD in an attempt to simulate the vibrational response of arrays subject to cross-flow. A brief review of some of this work is given in the following. However, it should be appreciated that because the emphasis of this book is on fluidelastic instability, only those analyses which account for cylinder motion in arrays are discussed. There are a number of other analyses of cross-flow over statically fixed arrays of cylinders (for example, Liang & Papadakis (2007)); although these analysis are capable of predicting the steady flow field as well as the turbulent buffeting and vortex-induced forces, because no account is taken of cylinder motion, they are not directly applicable to the prediction of fluidelastic instability.

Despite the obviously viscous nature of the interstitial flow through arrays of cylinders, the compactness of some arrays suggests that the cylinder wake regions are small, especially for normal triangular arrays with small P/D – see, for example, the flow visualisation of Wallis (1939) and Mavriplis (1982). Based on this type of reasoning, a number of researchers have neglected the wake regions in the array and treated the flow as being inviscid. The methodology of all the inviscid analyses is similar. The velocity potential is approximated, typically in series form, and the unknown coefficients obtained from the known boundary conditions – the most important of which is impermeability at the cylinder surface. Once the velocity potential is known, Bernoulli's equation can be used to obtain the unsteady pressure distribution around the cylinders, leading to the unsteady fluid forces.

The first example of an inviscid analysis for a cylinder array in cross-flow was by Balsa (1977) who employed a matched-asymptotic-expansion technique. For each cylinder, the velocity potential was written as the sum of two terms – one in the vicinity of the cylinder and a second far away. The velocity potentials were written as series expansions of D/P including terms up to $(D/P)^3$, and the stability of an infinite array of cylinders in cross-flow considered. Unfortunately, the analysis predicts that only static instabilities will occur.

Chen (1975, 1978) assumed that the velocity potential associated with the motion of each cylinder could be expressed as a Fourier series. The calculated added mass coefficients compared well with experiments, but the comparison for the damping and stiffness terms was poor. Païdoussis *et al.* (1984) pursued a method very similar to this, except that care was taken to apply the impermeability condition on the moving cylinder surface. Extensive comparisons were presented between their own results and the previous ones of Chen and Balsa, and it was concluded that the numerical differences between the various methods were either due to the boundary conditions not being applied on the moving cylinder surface or, in the case of Balsa's analysis, ignoring higher order terms in the asymptotic expansion. In an attempt to account for the viscous nature of the flow, a phase lag between the cylinder motion and resulting fluid forces was incorporated heuristically. Dynamic instabilities were obtained only for nonzero phase lags, and U_{pc} was very sensitive to the value of the phase lag.

Following on from their previous work, Païdoussis *et al.* (1985) developed a semi-potential model where the velocity potential was determined from the linear superposition of two potentials. The first, from the steady viscous flow around a cylinder at its equilibrium position, was inferred from Bernoulli's equation and the measured pressure distribution around a cylinder in an array. The flow associated

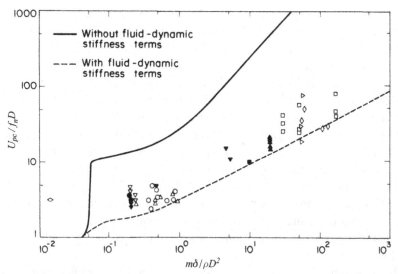

Figure 5.29. Comparison of the stability boundaries with and without fluid stiffness terms for a normal triangular array with $P/D = 1.375$ and $\delta = 0.01$ obtained using the semi-potential flow analysis of Païdoussis *et al.* (1985); experimental data identified therein.

with the cylinder motion, taken as being small, was considered to be inviscid, and hence, the second velocity potential was obtained in a similar manner to that of Païdoussis *et al.* (1977). However, because the effect of cylinder motion on the steady viscous flow was not accounted for, no fluid-stiffness terms were obtained. Realising their importance, measured fluid-stiffness terms were included in the analysis. Results for normal and parallel triangular arrays with $P/D = 1.375$, both with and without the measured fluid-stiffness terms, were presented and compared with experimental data; the results for normal triangular arrays are shown in Figure 5.29. At sufficiently high $m\delta/\rho D^2$ there is considerable difference between $U_{pc}/f_n D$ obtained with and without the fluid-stiffness terms, and much better agreement with experiments is obtained when the stiffness terms are included. This suggests that the instability is dominated by stiffness effects; however, it is important to realise that no account was taken of the phase lag between cylinder displacement and the resulting fluid forces.

Van der Hoogt & van Campen (1984) divided the fluid forces into two groups, one due to the flow through the undeformed array, and a second due to deformation caused by the cylinder displacements. A stability analysis was undertaken for two identical cylinders in cross-flow, but only static instability was obtained. However, if the symmetry of the cylinders was in any way destroyed, for example by having different diameters or different natural frequencies, then dynamic instability was predicted.

Delaigue & Planchard (1986) transformed the heterogeneous flow through an array of cylinders into the flow through a homogeneous medium. They assumed the array pitch to be periodic, the distance between cylinders to be small compared with the overall dimension of the array, and that neighbouring cylinders move in almost the same pattern. The velocity potential was then written as an asymptotic expansion in terms of P/D, the first term corresponding to the homogenised solution, whereas the remaining terms were correctors to this solution. Unfortunately, no results are presented; however, as stated by Delaigue & Planchard, this method is suitable for predicting tube response to transient motion only and not for predicting instabilities.

The results obtained from potential flow analysis are somewhat discouraging and, in general, do not give good predictions for dynamic instability. Also, it should be noted that more recent flow visualisation by Weaver & Abd-Rabbo (1985), Abd-Rabbo & Weaver (1986), Cheng & Moretti (1989) and Price and co-workers (Price et al. 1991, 1992b, 1995) suggests that, even though the wake regions are small, the interstitial flow is more complex than that accounted for in these analyses. Thus, it seems that, no matter what method is employed, inviscid flow theory is inadequate for a stability analysis of cylinder arrays in cross-flow. Hence, any realistic solution for fluidelastic instability in cylinder arrays must take into account the viscous nature of this flow. In its most general form this involves solving the Navier-Stokes equations subject to moving boundary conditions: some examples of this approach are reviewed in brief in the following paragraphs.

Direct numerical simulation of the Navier-Stokes equations for the flow through a row of cylinders was employed by Singh et al. (1989), who investigated the static stability of cylinders free to move in the transverse direction only. However, this simulation was limited to Re $\leq$ 100. Ichioka et al. (1994) presented a solution for the flow-induced vibration of a row of cylinders via a finite-difference analysis of the two-dimensional Navier-Stokes equations. The critical flow velocity for instability is presented for one value of $m\delta/\rho D^2$ for a row of cylinders with $P/D = 1.35$, showing excellent agreement with the experimental results of Chen (1984).

Kassera & Strohmeier (1997) used a finite-volume method to solve the two-dimensional Navier-Stokes equations with a k-ω turbulence model. The model was verified using a comparison of pressure distributions and wake velocities behind a stationary cylinder for Re $\leq 2 \times 10^5$. Simulations of the vibrational response for square and triangular arrays with $P/D = 2.4$, 1.6 and 1.2 are presented along with experimental data. Although the agreement between simulation and experiment for subcritical response is excellent, it is not clear that the numerical simulation correctly predicts the onset of fluidelastic instability.

Schröder & Gelbe (1999a) used a finite-element method to solve both the two- and three-dimensional Navier-Stokes equations with three different turbulence models. They also compared their results with experimental data from stationary circular cylinders, and concluded that the k-ω turbulence model gave the best agreement. However, they also concluded that the k-ω model could not be used when the cylinders were flexible, and hence, employed a two-layer k-ε model in their simulations for flexible cylinders – even though this model had been found unsatisfactory for a single stationary cylinder. Simulations were done for a row of five flexible cylinders, $P/D = 1.42$, and an excellent comparison, in terms of U_{pc}, was obtained with experimental results. Simulations are also presented, along with experimental results, for a flexible triangular array with $P/D = 1.25$; although the agreement is very good, it does not include any examples of fluidelastic instability.

Longatte et al. (2003) performed a numerical study for a single flexible cylinder in an in-line array of cylinders with the cylinder being free to oscillate in the cross-flow direction only. They accounted for the coupling between the structural motion and resulting fluid forces using an Arbitrary Lagrange Euler (ALE) formulation, which enables the computational mesh to remain regular even in the presence of large structural displacements. The computational mesh was limited to the interstitial flow between the oscillating cylinder and the eight surrounding static cylinders, with periodic boundary conditions being assumed for the inlet and outlet flows. A limited comparison is given with the experimental data of Granger *et al.* (1993) who considered an experimental array with 7 rows and 7 columns. The comparison is restricted to frequency measurements (no comparison of damping or instability velocities is given) for gap velocities of between 0 and 3.1 m/s. Both experimental and numerical results show a decreasing trend of frequency with increasing gap velocity, with reasonable agreement in the magnitude of the frequencies; however, the decrease in fluidelastic frequency over this small range of velocity is only 3%, so this does not provide a very severe test of the computational approach. Huvelin *et al.* (2008) extended this approach to include larger groups of cylinders, although only one of them was flexible, and also considered higher flow velocities such that instability was induced for the flexible cylinder; however, no comparison with experimental results is given.

Gillen & Meskell (2008) used a finite-element approach to solve the Reynolds averaged Navier-Stokes equations with a k-ε turbulence model for a normal triangular array with $P/D = 1.32$. They did not attempt to account for motion of cylinders in the array, but instead statically displaced transversely a cylinder in the third row of the array and then calculated the static fluid forces acting on it. Hence, it was possible to determine C_D and $\partial C_L/\partial y$ which can then be employed in a quasi-steady analysis. Very favourable comparisons are given with experimental data for the pressure distribution around the cylinder as well as the pressure drop through the array. The effect of changing Reynolds number in the range $13 \times 10^3 \le \text{Re} \le 52 \times 10^3$ was investigated, and both C_D and $\partial C_L/\partial y$ were shown to be Re-dependent, although the effect on C_D is minor. Using this data and a previously developed quasi-unsteady model (Meskell 2005) stability boundaries were calculated and compared with experimental data. Re is shown to be an important parameter; it is also demonstrated that $m/\rho D^2$ and δ need to be considered as independent parameters. This approach is interesting, because it avoids the complications in the CFD analysis associated with moving boundary conditions on the cylinder surface and it provides an alternative to experiments for determining the fluid-force and stiffness coefficients. However, once the coefficients are employed in a quasi-steady or unsteady analysis, all of the uncertainties associated with the phase lag between the cylinder motion and resulting fluid forces are automatically included in the analysis.

Although CFD analyses show considerable promise, it must be concluded that, as yet, they are not capable of giving accurate predictions of $U_{pc}/f_n D$ in cylinder arrays.

5.3.7 Nonlinear models

Numerous researchers have investigated the post-instability behaviour of cylinder arrays in cross-flow. The motivation for this research was twofold. First, because of manufacturing tolerances and thermal constraints there are likely to be small clearances between heat-exchanger tubes and tube supports, antivibration bars or baffle plates, resulting in "inactive", i.e. ineffectual supports. Hence, large lengths of unintentionally unsupported tubes may exist, having very low natural frequencies. These low-frequency modes may suffer from fluidelastic instability at relatively low U_{pc} (Weaver & Schneider 1983). Once the tubes go unstable, they will impact at the initially inactive supports, exciting higher modes of vibration, which are positively damped, and so limit the amplitude of vibration. An important practical consequence of this intermittent contact at the supports is the resulting impact/sliding which can produce a more severe wear mechanism than the pure fretting associated with permanent contact (Frick et al. 1984; Goyder & Teh 1989; Chen 1991a). A second, and more academic motivation for these nonlinear analyses has been to investigate the post-instability dynamics, and, in particular, the possibility of chaos.

The first nonlinear analytical model of fluidelastic instability was that of Roberts (1962, 1966), who employed the Krylov & Bogoliubov method of averaging to solve the nonlinear equations associated with the jump in C_D on an oscillating cylinder. However, from that time until approximately the mid 1980s all the analyses were essentially linear.

The more recent nonlinear investigations initially concentrated on impacting (e.g. with baffle plates). A typical example of this is given by Axisa et al. (1988), where the impacting is represented via an equivalent bilinear spring stiffness. For tube motion less than the clearance the spring stiffness is zero, but for amplitudes greater than the clearance a nonzero value of spring stiffness is taken. Appropriate values of spring stiffness are given by Axisa et al. based on the tube deformation, and experimental values of impact stiffness between tubes and flat antivibration bars are reported by Yetisir & Weaver (1985). To represent fluidelastic instability Axisa et al. extended Connors' quasi-static analysis, assuming the destabilising fluid damping to be proportional to U_p^2. Hence, the modal damping at any velocity, U_p, was replaced by

$$\zeta_n^* = \zeta_n[1 - (U_p/U_{pc})^2], \tag{5.47}$$

where ζ_n is the damping in stagnant liquid and U_{pc} is the critical velocity at instability. Thus, for $U_p \le U_{pc}$ the damping is positive, but for $U_p \ge U_{pc}$ the damping is negative, supplying energy to the system. The equations of motion were written in terms of assumed modes and then solved numerically using a time-marching numerical integration. Estimates of the wear rate, associated with the impacting, were made using a modified form of Archard's wear rate equation (Archard & Hirst 1956).

A specific example of this analysis was given for a pinned-pinned tube with one inactive support at midspan. For a range of U_p, time histories were computed and FFTs of the vibrational response obtained. For $U_p/U_{pc} \le 2$ the tube motion was essentially harmonic, with frequency close to the tube fundamental natural frequency. However, for $U_p/U_{pc} \ge 2$ the motion became what appeared to be chaotic, characterised by a broadening of the frequency-response spectra and a nonrepeatable phase-plane locus of the vibrational response. Of most significance from a practical point of view was a 350% increase in the calculated wear rate associated with this chaotic behaviour. Antunes et al. (1992b) showed that very similar results can be obtained when only a few modes are considered.

An experimental verification of the numerical methods discussed above was undertaken by Antunes et al. (1990, 1992a). However, rather than subject a

heat-exchanger tube to "true" fluid-dynamic excitation, they employed an elec-
tromechanical shaker with a feedback mechanism which simulated the negative
fluid damping suggested by equation (5.47). This excitation was applied at the free
end of a cantilever beam which passed through a clearance hole midway along its
length; furthermore, only planar motion was considered. The main motivation of
these experiments was to investigate the impact dynamics rather than the fluid-
dynamic excitation. The results obtained agreed reasonably well with the theoretical
model. This experimental verification was later extended to include motion in two
directions (Vento *et al.* 1992). Similar experimental investigations have also been
undertaken by France & Connors (1991), Langford & Connors (1991) and Connors
& Kramer (1991).

An analysis similar to that of Axisa *et al.* was developed by Fricker (1991, 1992),
who obtained the following expression for the total fluid damping:

$$\zeta_n^* = \zeta_n \left[1 - \frac{\omega_n}{\omega} \left(\frac{U_p}{U_{pc}} \right)^2 \right]. \tag{5.48}$$

This is similar to the expression given by Axisa *et al.*, except that it allows for a
variation in the frequency of vibration. The validity of this model was evaluated
by comparing theoretical and experimental results for a cantilever beam in cross-
flow. The agreement was excellent, showing very similar increases in vibrational
frequency as U_p was increased past U_{pc}. To the authors' knowledge, this was the
first time that a comparison between theoretical and "true" experimental post-critical
results has been given. Having verified the theoretical model, Fricker analysed the
U-bend region of a steam generator with a number of loose antivibration bars. As
U_p increased past U_{pc} a number of jumps in the frequency of vibration occurred
which were coincident with large increases in the impact force.

Cai *et al.* (1992) and Cai & Chen (1993a, b) attempted a theoretical investigation of a two-span tube using the unsteady fluid-dynamic analysis of Chen (1983a).
The tubes were pinned at two supports and had a clearance hole at the third. For
velocities less than that required to cause impacting at the clearance hole the tube
was considered as being pinned-pinned. However, as U_p increased, the tube-support-
inactive modes became unstable and impacting occurred; the tube was then treated as
being pinned-pinned-spring mounted and the natural modes of vibration and modal
damping altered accordingly. For one particular value of clearance, at $U_p/U_{pc} = 1.2$
chaotic motion occurred, as verified via the use of bifurcation diagrams, fractal dimensions and Lyapunov exponents. The predicted behaviour agrees very well with
both previous and subsequent experiments on the same system (Chen *et al.* 1984,
1995b). However, once the tube-support-active modes become unstable there is no
other dissipation mechanism and the theory predicts that the amplitude increases
indefinitely.

One of the problems with the above methods is that it is necessary to account for
several modes to represent the real system adequately; this precludes an analytical
solution. To enable analytical methods to be employed, de Langre *et al.* (1990)
considered a simplified model of a flexibly mounted rigid cylinder free to oscillate in
the cross-flow direction only. The impact dynamics and fluid-dynamic excitation were
modelled in a manner similar to that of Axisa *et al.* (1988). As expected, because

this system had two degrees of freedom only, the calculated response was purely periodic. However, adding a forced harmonic excitation, representative of vortex shedding or narrow-band turbulent buffeting, produced a chaotic response as the magnitude of the forcing function was increased.

Chen & Chen (1993) considered the chaotic response of three flexibly mounted rigid tubes in an otherwise rigid tube row, each impacting against rigid support plates. The fluid forces were modelled using Chen's (1983a) unsteady model. Chaotic motion, implied from bifurcation diagrams and Lyapunov exponents, was predicted for velocities somewhat higher than that required for fluidelastic instability.

In the analysis given above the fluid dynamics was linear; thus, if the impact forces were removed, the vibrational amplitude would become infinitely large. The first analysis after Roberts (1962, 1966) which attempted to account for nonlinearities in the fluid dynamics was by Price & Valerio (1990). Price & Valerio considered a single flexibly mounted cylinder free to oscillate in the transverse direction only; thus, only damping-controlled instabilities were considered. The analysis employed was an extension of the Price & Païdoussis (1986b) model, accounting for two types of nonlinearity. Second-order terms in the resultant velocity were considered, and more importantly, measured nonlinear variations in C_L and C_D were included which were approximated using fifth-order polynomials. The resulting equations were solved using the Krylov & Bogoliubov method of averaging.

Austermann et al. (1992) attempted to combine the nonlinear elements of the Price & Valerio model with the linear unsteady terms of Lever & Weaver's analysis (1986a). However, given the very different physical assumptions on which these two analyses are based, there seems to be little justification in doing this.

Païdoussis & Li (1992) attempted a three-dimensional analysis of a clamped-clamped beam, with an inactive support at its midspan and free to oscillate in the transverse direction only, and employed the linear fluidelastic instability model of Price & Païdoussis (1986b). The effect of impacting was modelled by use of either the bilinear stiffness proposed by Axisa et al. (1988) or a cubic spring stiffness. The cubic spring is less representative of the true behaviour but has the advantage of being analytic, so enabling analytical solutions to be obtained for the Lyapunov exponents – which give a quantitative measure of the onset of chaos. For the cubic spring, Païdoussis & Li presented results in terms of a bifurcation diagram as shown in Figure 5.30. As U is increased, instability is lost at a Hopf bifurcation, leading to a limit-cycle oscillation. This is followed by a post-Hopf bifurcation period-doubling sequence and eventually chaos.

A two-dimensional model accounting for nonlinearities in both the fluid and structural dynamics was presented by Païdoussis et al. (1991a, 1993) for a single flexible cylinder in the third row of a rotated triangular array with $P/D = 1.375$. The fluid dynamics was accounted for via a nonlinear quasi-steady analysis similar to that employed by Price & Valerio, with the exception that cylinder motion was allowed in both the in- and cross-flow directions. The cylinder impacting was modelled assuming that the cylinder vibrated in a circular clearance hole, and values of the coefficient of restitution and Coulomb friction were employed to relate the vibrational velocities before and after the impact. Impacting occurred for U_p slightly greater than U_{pc}, and as U_p was further increased a series of bifurcations were obtained, eventually leading to chaos.

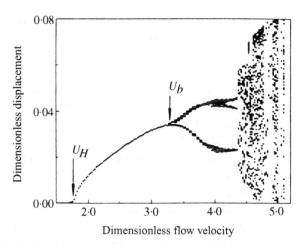

Figure 5.30. Bifurcation diagram, based on dimensionless flow velocity, for a single flexible cylinder in an in-line square array of cylinders (Païdoussis & Li 1992); $P/D = 1.5, m/\rho D^2 = 3.0, \delta = 0.06$.

A detailed analysis of a cantilevered flexible cylinder and comparison with experiments was presented by the same authors (Mureithi *et al.* 1994a, b; Païdoussis *et al.* 1992), showing that not only did the experiments and theory yield the same series of bifurcations but that they occurred at virtually the same flow velocities. The transition mechanisms leading to chaos were also investigated by Mureithi *et al.* (1995).

Rzentkowski & Lever (1992) maintained the nonlinear terms in the continuity equation of the original Lever & Weaver (1982) model. The resulting equations were solved using both the Krylov & Bogoliubov method of averaging and numerical integration – both methods yielded essentially the same results. Unfortunately, for the two arrays examined (parallel triangular with $P/D = 1.375$ and in-line square with $P/D = 1.433$) only unstable limit cycles were obtained, in contrast to their own experimental results (Lever & Rzentkowski 1989). In an attempt to change this behaviour they considered the time lag to be linearly dependent on the cylinder displacement. This resulted in stable limit-cycle oscillations for high $m\delta/\rho D^2$ (50 and 100 for the parallel triangular and in-line arrays, respectively), but for lower values of $m\delta/\rho D^2$ they still obtained unstable limit cycles. These results demonstrate that changes in the unsteady fluid dynamics in cylinder arrays may have a profound effect on the post-instability response.

Rzentkowski & Lever (1998) extended their earlier nonlinear analysis (Rzentkowski & Lever 1992) to account for the effect of upstream turbulence. They show that for those cases where there is a hysteretic tube response, accounting for free-stream turbulence can cause a reduction in U_{pc}/fD.

An interesting study by de Langre *et al.* (1992) considered the effect of using three different fluidelastic instability models: the negative-fluid-damping model given by equation (5.48), the Blevins (1974) coupled mode analysis and the Price & Païdoussis (1984a) quasi-steady mode – all three of which consider linear fluid-dynamic effects only. The most interesting conclusion from this study is that all three analyses yield the same types of bifurcation as the flow velocity is increased past its critical value; however, the bifurcations occur at different velocities for the three different models. This suggests that the sequence of bifurcations obtained in the nonlinear post-instability response is governed by the impacting of the tube with the loose supports, rather than by the fluid dynamics. However, it also shows that to predict accurately the velocity at which the bifurcations occur or the work rate on the tubes, it is necessary to have a good estimate of the energy input to the system – requiring an accurate model of the nonlinear fluid dynamics. Furthermore, it illustrates that when comparing experiment and theory it is not sufficient to obtain the same series of bifurcations, but the velocity at which these bifurcations occur is also of importance.

Meskell & Fitzpatrick (2003) performed an experimental study on normal triangular tube arrays with $P/D = 1.32$ and 1.58 with a single flexible cylinder free to vibrate in the transverse direction only. Based on detailed measurements of the free-vibrational response as a function of flow velocity, an empirical model for the nonlinear fluidelastic damping and stiffness forces was proposed. It was assumed that the fluidelastic force, $f(y, \dot{y}, U)$, could be decomposed into three functions: one of which depends on tube displacement; a second on tube vibrational velocity; and a third which depends on both the tube displacement and velocity (all three functions depend on flow velocity); thus,

$$f(y, \dot{y}, U) = -N_1(y, U) - N_2(y, \dot{y}, U) - N_3(\dot{y}, U). \tag{5.49}$$

Using measured free-response data for the tube at different flow velocities along with the force-state mapping technique (Masri & Caughey 1979; Meskell *et al.* 2001) it was proposed that functions N_1 and N_3 could be represented by cubic polynomials, $N_1 = \eta y^3 + k_f y$ and $N_3 = \beta \dot{y}^3 + c_f \dot{y}$, while $N_2 \approx 0$; k_f and c_f being, respectively, the linear fluid-stiffness and damping terms. Considering the variation with flow velocity it was shown that k_f, c_f and η were linear functions of U, whereas the cubic fluid-damping term β was represented by a third-order polynomial in U. The fact that the fluid-stiffness term, k_f, is a function of U, as opposed to U^2, is a little unexpected, and unfortunately, Meskell & Fitzpatrick do not comment on this. Using the inferred functions N_1 to N_3, Meskell & Fitzpatrick then predict the vibrational amplitude of the flexible cylinder as a function of flow velocity. Although reasonably good estimates of the critical velocity were obtained, the predicted limit-cycle amplitudes for the post-instability behaviour were of an order 100% greater than those observed experimentally. Given that the functions N_1 to N_3 were evaluated from this same post-instability data, this overestimate in the limit-cycle amplitudes is somewhat disappointing.

5.3.8 Nonuniform flow

Application of the two-dimensional theoretical models, presented in the preceding sections, to three-dimensional heat-exchanger tubes requires a little care. If the flow over the tubes is uniform, then the models can be used with the equivalent modal terms. However, there are many applications where the flow velocity is not uniform, and it is necessary to modify the stability analysis to account for this. Also, if an attempt is made to use experimental correlations of $U_{pc}/f_n D$ versus $m\delta/\rho D^2$, such as those given by Chen (1984), Weaver & Fitzpatrick (1988), Pettigrew & Taylor (1991) or Schröder & Gelbe (1999b), a similar problem is encountered. The experiments on which these correlations are based were, in general, conducted in uniform flow, and it is necessary to know how to modify these results to account for nonuniform flows. The method suggested in several design guides is to employ the concept of a so-called effective velocity (Au-Yang 1984; Blevins 1984; Wambsganss *et al.* 1984; Au-Yang *et al.* 1991; Pettigrew & Taylor 1991), which enables a two-dimensional stability analysis or experimental correlation to be employed to predict the stability boundary in three-dimensional nonuniform flow. The "effective velocity" is defined as follows:

$$U_{\text{eff}} = \left(\frac{\int_0^l U^2(z)\phi^2 dz}{\int_0^l \phi^2 dz} \right)^{0.5}, \tag{5.50}$$

where $U(z)$ is the velocity distribution along the span and ϕ the vibrational mode shape. A number of authors have evaluated experimentally the use of this effective velocity, and these results are discussed later in this section. However, first, this concept is given a more firm theoretical basis via the application of the two-dimensional quasi-steady model for the analysis of a three-dimensional heat exchanger. A similar methodology can also be employed for the other theoretical models – see, for example, Ohta *et al.* (1982) and Chen & Chandra (1991) who developed three-dimensional versions of the unsteady analysis of, respectively, Tanaka & Takahara (1981) and Chen (1983a).

Assuming all tubes in the array to be of the same length and have the same boundary conditions, and modelling each tube as an Euler-Bernoulli beam gives the following equations of motion in the in- and cross-flow directions for tube j:

$$EI_j \frac{\partial^4 \xi_j}{\partial z^4} + \bar{c}_j \frac{\partial \xi_j}{\partial t} + m_j \frac{\partial^2 \xi_j}{\partial t^2} = H_j,$$

$$EI_j \frac{\partial^4 \eta_j}{\partial z^4} + \bar{c}_j \frac{\partial \eta_j}{\partial t} + m_j \frac{\partial^2 \eta_j}{\partial t^2} = G_j, \tag{5.51}$$

where EI_j, $\bar{c}_j$ and m_j are the flexural rigidity, damping coefficient and mass per unit length, respectively; $\xi_j(z, t)$ and $\eta_j(z, t)$ are the displacements in the in- and cross-flow directions; and $H_i(z, t)$ and $G_i(z, t)$ are the fluid forces per unit length acting on the tubes in the in- and cross-flow directions. In general, $H_i(z, t)$ and $G_i(z, t)$ depend on the fluid velocity, density and viscosity, and on the motion of both the cylinder itself and its neighbours. Using a linearised quasi-steady analysis, the fluid forces per unit length may be written in terms of the added mass, damping and stiffness coefficients (ignoring the time-averaged fluid forces). Thus, for example

$$H_j(z, t) = -\tfrac{1}{2}\rho[D^2\{A_j\}\ddot{\chi} + DU\{B_j\}\dot{\chi} + U^2\{E_j\}\chi], \tag{5.52}$$

where χ is the vector of both the in- and cross-flow dimensional displacements of the cylinder itself and the surrounding cylinders; similarly for $G_i(z, t)$.

To illustrate the procedure necessary to reduce equations (5.51) and (5.52) to two dimensions, a simplified problem of a single flexible cylinder, with freedom to move in the cross-flow direction only, is considered. Hence, in equation (5.52), χ may be replaced by η, and the matrices $\{A_j\}$, $\{B_j\}$ and $\{E_j\}$ by scalar quantities a, b and e, respectively; also, the subscript j may be dropped.

The equations may be solved via the modal-analysis method (Bishop & Johnson 1960) where $\eta(z, t)$ is expressed as

$$\eta(z, t) = D \sum_{q=1}^{N} \phi_q(z) y_q(t), \tag{5.53}$$

in which y_q is the dimensionless generalised displacement, and ϕ_q is a set of assumed mode shapes which must satisfy the geometric boundary conditions and be orthogonal, such that

$$\int_0^l \phi_f \phi_g \, dz = l \quad \text{if} \quad f = g$$

$$= 0 \quad \text{if} \quad f \neq g, \tag{5.54}$$

where it has been assumed that m does not vary along the tube length. We consider a tube subject to a nonuniform velocity $U(z) = U_p \psi(z)$, and assume constant $\bar{c}$ and EI along the tube span. Then, equations (5.51) and (5.52) may be simplified, via the use of assumed modes (equation (5.53)) and the orthogonality condition (equation (5.54)), after multiplying all terms by ϕ_i and integrating over the length of the span, to give for the ith mode of vibration

$$[m l \ddot{y}_i + c \dot{y}_i + k y_i] D = -\frac{1}{2} \rho D \left[D^2 l a \ddot{y}_i + D b U_p \int_0^l \psi \sum_{q=1}^{N} (\phi_i \phi_q \dot{y}_q) dz \right.$$

$$\left. + e U_p^2 \int_0^l \psi^2 \sum_{q=1}^{N} (\phi_i \phi_q y_q) dz \right], \tag{5.55}$$

where m, c and k are the modal mass per unit length, damping and stiffness, respectively; similarly for the x-direction. It is apparent that the damping and stiffness terms on the right-hand side of (5.55) introduce coupling between the various modes of the heat-exchanger span; thus, it is not possible to consider the ith mode in isolation, and the equation cannot be solved analytically. Only if the velocity is constant over the span can the right-hand side be simplified to yield

$$-\tfrac{1}{2} \rho D l \left[D^2 a \ddot{y}_i + D b U_p \dot{y}_i + e U_p^2 y_i \right], \tag{5.56}$$

which is of a more convenient form. It should be remembered that this simplified analysis considered a single flexible cylinder with one degree of freedom only; in the more general case, the scalar terms in (5.55) will be matrices. Thus, it is not surprising that many authors have taken a simpler approach and resorted to the use of an "effective velocity" as outlined below.

The concept of an effective velocity was originally developed by Franklin & Soper (1977), Connors (1978) and Pettigrew et al. (1978), who reconsidered Connors' original analysis (1970). Goyder (1990) also arrived at a similar idea from the point of view of a global fluid coefficient. Various degrees of complexity were considered, but if the tube mass is constant over the span length, then

$$\frac{U_{pc}}{f_n D} = K \left(\frac{\delta m l}{D^2 \int_0^l \rho \psi^2 \phi^2 dz} \right)^{1/2}, \tag{5.57}$$

where ψ is the velocity shape function and ϕ is the structural mode shape. The denominator of (5.57) is similar to the fluid-stiffness term in equation (5.55). Hence, it should be noted that an effective velocity is strictly applicable to Connors' instability model only – or, at least, instability models where the excitation mechanism is due to fluid-stiffness effects and is proportional to both the dynamic head of the flow and the cylinder displacement.

Numerous authors have evaluated experimentally the effectiveness of using equation (5.57). For example, Connors (1978) tested a U-bend model in airflow with half the span subject to flow; he obtained a deviation in U_{pc} from experimental results of less than 15%. Franklin & Soper (1977) did experiments on a normal triangular array with $1.25 \leq P/D \leq 1.78$ in airflow. They found that equation (5.57) could either underestimate or overestimate U_{pc} by as much as 21% (in both of these cases the factor K had been adjusted to be correct for uniform flow). Eisinger et al. (1989) used the effective velocity in a finite-element analysis to predict U_{pc} for a multispan tube array and found reasonable agreement with experimental data in the literature.

A series of experimental verifications of equation (5.57) has also been conducted by Weaver and co-workers. Waring and Weaver (1988) did experiments in airflow on a parallel triangular array with $P/D = 1.47$. Experiments were conducted with uniform flow over one of the spans of one-, two- and three-span tubes, as well as uniform flow over a portion of a single-span tube. They found that the theory could be substantially nonconservative in some cases; moreover, in some circumstances the theory predicted incorrectly which was the unstable mode.

Weaver & Goyder (1990) modified the theory to account for different modes of vibration for a multispan tube. For the ith mode of an N-span array with velocity distribution $U_p \psi_n$, the critical velocity, U_{pci} was given by

$$U_{pci} \sum_{n=1}^{N} \left(\frac{\psi_n}{fD_i}\right) S_{in}^{0.5} = K \left(\frac{m\delta_i}{\rho D^2}\right)^{0.5}, \tag{5.58}$$

where S_{in} is the so-called energy fraction given by

$$S_{in} = \int_{l_1}^{l_2} \phi^2 dx \Big/ \int_0^l \phi^2 dx, \tag{5.59}$$

and $l_2 - l_1$ represents the length of the nth span. Experiments were done on a normal triangular array, $P/D = 1.4$, with airflow over either one, two or three spans of a three-span tube. Provided K was first corrected for uniform flow, equation (5.58) gave a maximum error of 11% in U_{pc} compared with experimental results. The reason why the effective velocity worked well in the present case, but not for the experiments of Waring & Weaver on parallel triangular arrays, was that the empirical relationship relating $U_{pc}/f_n D$ with $m\delta/\rho D^2$ for normal triangular arrays has an exponent closer to 0.5 than for parallel triangular arrays.

The energy fraction idea was taken further by Weaver & Parrondo (1991), who suggested that the expression for the critical velocity be modified to

$$U_{pci} \sum_{n=1}^{N} \left(\frac{\psi_n}{fD_i}\right) = K \left(\frac{m}{\rho D^2 S_{in}}\right)^{\alpha} \delta_i^{\beta}, \tag{5.60}$$

with the exponents α and β to be determined from experiments.

Weaver & Parrondo did seven different sets of experiments with airflow over one span of a three-span parallel triangular array with $P/D = 1.47$. They compared experimental results with equation (5.60) for $K = 3.3$ and $\alpha = \beta = 0.5$ (suggested by Pettigrew et al. (1978)), $K = 4.8$ and $\alpha = \beta = 0.3$ (suggested by the experimental correlation of Weaver & Fitzpatrick (1988)), and $K = 4.6$, $\alpha = 0.29$ and $\beta = 0.21$ (suggested by the experimental correlation of Weaver & El-Kashlan (1981)). For all seven cases, using either the Weaver & Fitzpatrick or Weaver & El-Kashlan correlations the theory was conservative and it predicted correctly which mode went unstable. The error in U_{pc} was less than 37% for the Weaver & El-Kashlan correlation, but it was as much as 77% for the Weaver & Fitzpatrick correlation. Using the Pettigrew correlation, the error was less than 57%, but the theory could either underestimate or overestimate U_{pc}; more importantly, the mode that went unstable was sometimes not predicted correctly. Based on these results, Weaver & Parrondo suggested that equation (5.60) should be used with $\alpha = \beta = 0.30$, 0.40, 0.48 and 0.48, and $K = 4.8$, 3.2, 2.5 and 4.0 for parallel triangular, normal triangular, in-line square and rotated square arrays, respectively.

A different approach was taken by Chen & Chandra (1991) who conducted a numerical investigation on the use of an effective velocity based on Chen's (1983a) unsteady analysis. For high values of $U_{pc}/f_n D$ (where fluid-inertia effects can be ignored and the fluidelastic frequency taken as the no-flow natural frequency), Chen & Chandra showed that the instability boundary is well approximated by equation (5.57), justifying the use of the effective velocity. Considering the complete range of $U_{pc}/f_n D$, however, things are a little more complex. Chen & Chandra suggest that the effective velocity may be employed provided the in-vacuum frequency and damping are used. However, even if this is done, the results presented by Chen & Chandra suggest that the value of $m\delta/\rho D^2$ at which the "jump" in stability boundary occurs is not accurately predicted. Furthermore, for low $m\delta/\rho D^2$, δ may be dominated by fluid effects, and it is not convenient to use in-vacuum values of damping.

The results presented in this section suggest that although the effective velocity concept may be a quick way of accounting for a nonuniform flow, it should be used with some caution, and preferably a more exact modal analysis should be employed.

5.4 Comparison of the Models

Although equation (5.25), "Connors' equation", was derived for a single row of cylinders with $P/D = 1.41$, it is often used as the basis for correlations of experimental U_{pc} as a function of $m\delta/\rho D^2$ for complete arrays of cylinders, the assumption being that it is merely necessary to obtain the correct value of "K". For example, using experimental correlations with various categories of data being excluded, the following values of K have been suggested: for in-line square arrays, Connors (1978) obtained $K = (0.37 + 1.67T/D)$ provided $1.41 \leq T/D \leq 2.12$; for all array geometries, $K = 3.3$, 3.3, 0.8, 2.4 and 3.0 (independent of P/D) have been suggested by Gorman (1978), Pettigrew et al. (1978), Païdoussis (1980), Au-Yang et al. (1991) and Pettigrew & Taylor (1991), respectively; in addition, more complex correlations for individual array geometries have been suggested by Chen (1984) and Weaver & Fitzpatrick (1988), where both K and the exponent on $m\delta/\rho D^2$ are functions of $m\delta/\rho D^2$.

A discussion of the experimental support for and against the use of Connors' equation is given in Section 5.4.1; this is followed by an evaluation of other models of fluidelastic instability in Section 5.4.2. Finally, some concluding comments on the present "state of the art" for prediction of fluidelastic instability are given in Section 5.4.3.

5.4.1 Experimental support for and against Connors' equation

At this point it is worthwhile mentioning that there are at least two different prevalent motivations for researchers investigating fluidelastic instability in cylinder arrays. On the one hand stands the practitioner with the pragmatic need to produce as simple a design guide as possible, while on the other hand are those more concerned with the physics causing this instability. Although it may appear that these motivations are mutually exclusive, they need not be so. If the physics of the instability mechanism is not sufficiently understood, then it is very unlikely that a safe, but not overly conservative, design guide can be achieved. Hence, a complete understanding of the physics is an essential prerequisite for a good design guide.

When assessing the validity of Connors-type equations, it should be appreciated that these expressions imply a number of basic physical phenomena for fluidelastic instability. First, Connors' equation states that $U_{pc}/f_n D$ is proportional to $(m\delta/\rho D^2)^{1/2}$; implicit in this statement is that the nondimensional mass, $m/\rho D^2$, and damping, δ, may be combined into one parameter and are totally interchangeable. Finally, Connors' equation gives one stability boundary only; hence, there is no possibility of multiple stability boundaries. As discussed in the previous section, most of the other models of fluidelastic instability contradict these statements. The majority of the other theoretical models predict multiple instability boundaries, and most of them suggest that $m/\rho D^2$ and δ should be treated as independent parameters, and that, in general, $U_{pc}/f_n D$ is not proportional to $(m\delta/\rho D^2)^{1/2}$ over the complete range of $m\delta/\rho D^2$.

With respect to the existence of multiple instability regions there is conflicting experimental evidence. Chen & Jendrzejczyk (1983) and Popp and co-workers (Andjelić *et al.* 1990; Austermann & Popp 1995; Andjelić & Popp 1989; Romberg & Popp 1998a, b; Rottmann & Popp 2003) show very clearly that multiple instability regions do exist for a single flexible cylinder in a variety of different arrays. Furthermore, Andjelić *et al.* and Rottmann & Popp show that multiple instability boundaries can also exist when there are multiple flexible cylinders in the array.

Despite the clear evidence that multiple instability regions can be detected in carefully controlled experiments it is apparent that they have not been observed in operating heat exchangers. Possible reasons for this apparent contradiction are as follows. To obtain experimentally the stable regions for velocities greater than the lowest critical velocity usually requires some care. First, nonlinear effects may prevent the cylinders from restabilising, and thus, to obtain the interleaved stable regions it is necessary to restrain the cylinders while increasing the velocity through an unstable region. In addition, based on extensive fluid-force measurements, Païdoussis *et al.* (1996) showed, in agreement with the work of Austermann & Popp, that whether or not multiple instability regions will occur is extremely sensitive to small misalignments of cylinders in the array. For example, for a parallel triangular array with

$P/D = 1.375$ a misalignment of as little as $0.02D$ is sufficient to quench the multiple instability regions, but leave the main instability boundary intact. Also, Romberg & Popp (1998b), Rzentkowski & Lever (1998) and Rottmann & Popp (2003) show that the multiple instability regions can be stabilised by the presence of elevated turbulence levels upstream of the array, as is likely to be found in real heat-exchanger arrays. A final possible explanation why the multiple instability regions are rarely observed in practice comes from the work of Nakamura et al. (1997), who showed that their existence requires near-perfect correlation of the fluidelastic forces along the cylinder span; this analysis was specific to two-phase flow, but there is no reason why it should not be correct for single-phase flow.

With respect to the independence of $m/\rho D^2$ and δ there are two types of experimental evidence. A number of authors have conducted carefully controlled laboratory experiments where the effect of varying either $m/\rho D^2$ or δ on $U_{pc}/f_n D$ has been investigated. For example, using their own data as well as that of Weaver & Grover (1978) and Weaver & El-Kashlan (1981), Weaver & Koroyannakis (1982) show that for a single flexible cylinder in a rotated triangular array with $P/D = 1.375$, $U_{pc}/f_n D$ is proportional to $(m/\rho D^2)^{0.29} \delta^{0.21}$. For a rotated square array, with four flexible cylinders and $P/D = 2.12$, Price & Kuran (1991) showed $U_{pc}/f_n D$ to be approximately proportional to $\delta^{0.06}$ for $m/\rho D^2 = 280$ and 490. For one flexible cylinder in a square array, $P/D = 1.5$, Price & Païdoussis (1989) suggest $U_{pc}/f_n D$ is proportional to $\delta^{0.05}$ for $m/\rho D^2 = 3.79$, and $\delta^{0.24}$ for $280 < m/\rho D^2 < 2380$.

The second type of experimental evidence comes from a number of different authors who have correlated all existing data – including that from operating heat exchangers as well as laboratory experiments – in the form of plots of $U_{pc}/f_n D$ as a function of $m/\rho D^2$ and δ. It is not surprising that different conclusions have been obtained from these correlations when it is realised that different definitions of mass (with or without the added mass), frequency (values determined in vacuum (practically this implies experiments done in air), or in stagnant fluid)), and damping (in vacuum, in stagnant fluid or in flowing fluid) have been employed.

The first of these correlations was developed by Païdoussis (1980) who, using all existing data, presented correlations of $U_{pc}/f_n D$ as functions of $(m/\rho D^2)^{0.5}$, $(m/\rho D^2)^{0.5}\delta^{0.25}$ and $(m\delta/\rho D^2)^{0.5}$. Païdoussis showed that the least scatter was obtained for the second of these correlations ($\delta^{0.25}$), but that the first (δ^0) and third ($\delta^{0.5}$) contained approximately the same amount of scatter.

Chen (1984) correlated the critical flow velocity as a function of $m\delta/\rho D^2$ for the four different array geometries (in vacuum structural values were used if available; if not, stagnant-fluid values). Furthermore, for each array geometry the correlation was based on a velocity, which was pitch dependent, the functional dependence being such as to minimise the scatter of the experimental data. For example, for the triangular array the appropriate velocity was $(U/f_n D)(2.105(P/D) - 0.9)$. Although Chen chose to combine $m/\rho D^2$ and δ into one parameter, he showed very clearly that there was a discontinuity in the slope of the stability boundary, and that the velocity was only proportional to $(m\delta/\rho D^2)^{1/2}$ for velocities above this discontinuity. For example, again considering the triangular array, U_{pc} was proportional to $(m\delta/\rho D^2)^{0.1}$ for $0.1 < m\delta/\rho D^2 < 2$, and $(m\delta/\rho D^2)^{0.5}$ for $2 < m\delta/\rho D^2 < 300$.

Using additional data to that employed by Chen (1984), Weaver & Fitzpatrick (1988) correlated the pitch velocity, U_{pc}, as a function of mass-damping for the four different array geometries. No account was taken of different pitch ratios, and they employed structural damping and frequencies measured in air. For all four array geometries a discontinuity in the slope of the stability boundary was obtained. For example, considering the triangular array, they found U_{pc} to be independent of $(m\delta/\rho D^2)$ for $m\delta/\rho D^2 < 0.3$, and to be proportional to $(m\delta/\rho D^2)^{1/2}$ for $m\delta/\rho D^2 > 0.3$.

Pettigrew & Taylor (1991) presented a global correlation for all four array geometries. They restricted their database to exclude experimental results with single flexible cylinders. In addition, they employed damping values measured in the appropriate fluid; if these values were not available, the in-fluid damping was estimated using the methodology of Pettigrew *et al.* (1986). They found that a reasonable lower limit for the stability boundary of this particular data set was given by Connors' equation with an exponent of 0.5. However, it is interesting to note that using the same data set they also presented a second correlation between $U_{pc}/f_n D$ and $(m/\rho D^2)\delta^{0.5}$, and they concluded that for this second correlation "the agreement with the data is no better than for the simpler model"; however, as shown in Figure 5.31, neither is it any worse!

The most recent attempt to correlate experimental stability data comes from Schröder & Gelbe (1999b). They used the same data sets as Chen (1984), Weaver & Fitzpatrick (1988) and Pettigrew & Taylor (1991), but their correlations account for the effect of pitch ratio. Although they found the minimum error between correlation and actual data to be when $m/\rho D^2$ and δ were combined into one parameter, they also predicted discontinuities in the slope of critical velocity as a function of $m\delta/\rho D^2$.

Although there is no general consensus from the presently available experimental data and the correlations obtained from this data, it is clear that, in general, experimental evidence does not support the use of Connors' equation. In particular, it appears that $m/\rho D^2$ and δ should be treated as independent parameters. There is also considerable experimental evidence implying that the dependence of $U_{pc}/f_n D$ on δ is less than $\delta^{1/2}$, and that the manner in which $U_{pc}/f_n D$ varies with $m/\rho D^2$ and δ strongly depends on array geometry and the individual values of $m/\rho D^2$ and δ.

5.4.2 Comparison of theoretical models with experimental data

In this section a number of the theoretical models presented in Section 5.3 are compared with each other and with available experimental data. The comparisons are not only in terms of U_{pc}, but also in terms of the pre-instability fluidelastic damping as a function of U_p.

Comparisons of the stability boundaries obtained from the theoretical analysis with available experimental data are presented in Figure 5.32 for rows of cylinders and in Figures 5.33 to 5.36 for four different array geometries. As discussed by Pettigrew & Taylor (1991, 2003), there are many problems in making this type of comparison. There is still no universally accepted definition of what values of m, δ and f_n should be employed; different experimenters use purely structural values (practically speaking, this means values measured in still air) or those values measured in either stagnant or flowing fluid. This lack of uniformity is particularly important for liquid or two-phase flows where there may be considerable differences in m, δ and f_n, depending on the conditions under which these structural parameters are measured.

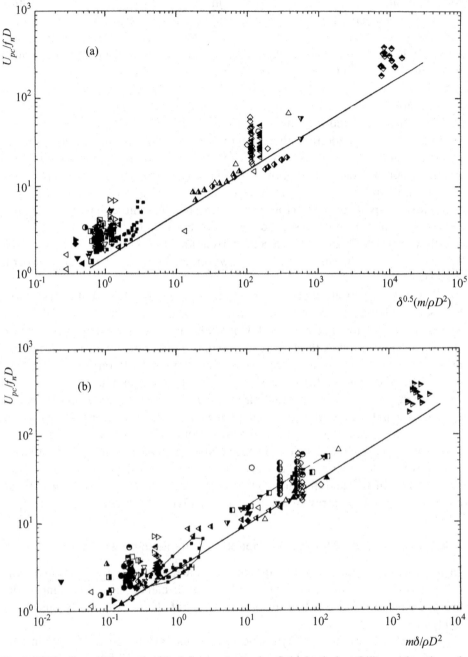

Figure 5.31. Correlations of critical flow velocity for fluidelastic instability as function of mass and damping, reproduced from Pettigrew & Taylor (1991), with the experimental data identified therein: (a) $U_{pc}/f_n D$ as a function of $(m/\rho D^2)\delta^{0.5}$; (b) $U_{pc}/f_n D$ as a function of $m\delta/\rho D^2$.

This problem is not only confined to the experimental results; the theoretical models also have different definitions of what values of m, δ and f_n should be used.

Pettigrew & Taylor (1991) give some very convincing arguments for using values measured in stagnant fluid for those experiments conducted in liquid or two-phase

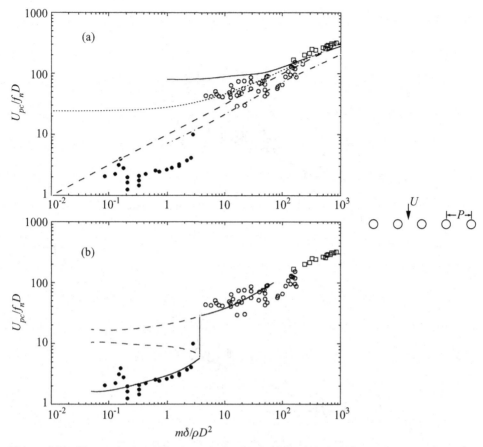

Figure 5.32. Theoretical stability boundaries for fluidelastic instability and comparison with experiments for *a row of cylinders*: ●, multiple flexible cylinders in liquid flow; ○, multiple flexible cylinders in gaseous flow; □, single flexible cylinder in gaseous flow [experimental data from Blevins *et al.* (1981), Chen & Jendrzejczyk (1981), Connors (1970, 1978). Gross (1975), Halle & Lawrence (1977), Hartlen (1974), Heilker & Vincent (1981), Roberts (1962), Southworth & Zdravkovich (1975) and Tanaka (1980)]. (a) —, Roberts' (1962) solution accounting for time taken by jet reversal and fluid damping, $P/D = 1.5$; – · – ·, Roberts' (1962) solution ignoring time taken for jet reversal and fluid damping, $P/D = 1.5$; - - -, Connors' (1970) solution, $P/D = 1.41$; · · · · · ·, Blevins' (1979) solution including fluid damping; $P/D = 1.41$. (b) - - -, Chen's (1983b) and Tanaka & Takahara's (1980, 1981) theoretical solution showing multiple instability boundaries; —, practical stability boundary.

flow. However, in the comparisons given here this has not been done. If the theoretical model correctly predicts all the fluid-dynamic effects, then the change in frequency and damping caused by the fluid will be accounted for automatically, and modificaton of the experimental data would result in a double accounting of these effects.

Other difficulties encountered in making these comparisons are that the experimental data (i) are for a wide range of P/D, (ii) come from experiments with different numbers of flexible cylinders in the array and (iii) come from a mixture of single- and two-phase flows.* Thus, it should not be expected that perfect agreement

---

* The figures do indicate which data points come from single-flexible- or multiple-flexible-cylinder experiments, and whether they come from liquid, gaseous or two-phase flows.

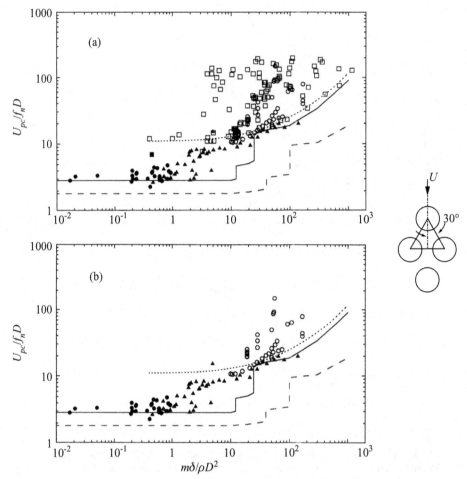

Figure 5.33. Comparison of theoretical stability boundaries for fluidelastic instability with experimental data for *normal triangular arrays*: ●, multiple flexible cylinders in liquid flow; ○, multiple flexible cylinders in gaseous flow; □, single flexible cylinder in gaseous flow; ■, single flexible cylinder in liquid flow; ▲, multiple flexible cylinders in two-phase flow [experimental data from Andjelić & Popp (1989), Chen & Jendrzejczyk (1981), Connors (1978, 1980), Gay *et al.* (1988), Gibert *et al.* (1981), Gorman (1978), Granger *et al.* (1991), Gross (1975), Halle & Lawrence (1977), Halle *et al.* (1988), Hartlen (1974). Heilker & Vincent (1981), Minakami & Ohtomi (1987), Pettigrew *et al.* (1978, 1985, 1989), Price & Zahn (1991), Soper (1980), Teh & Goyder (1988), Yeung & Weaver (1983), Weaver & Yeung (1984) and Žukauskas & Katinas (1980)]; - - -, theoretical solution from Lever & Weaver (1986b) for $P/D = 1.2$; —, theoretical solution from Lever & Weaver (1986b) for $P/D = 2.0$; · · · · · ·, theoretical solution from Teh & Goyder (1988) for $P/D = 1.35$. (a) Includes all experimental data, (b) contains data from multiple-flexible-cylinder experiments only.

will be obtained between theory and experiment. However, despite this, it is felt that the comparisons do give some indication of the validity of the theoretical models.

The theoretical variations of $U_{pc}/f_n D$ obtained from two of Roberts' solutions (1962, 1966), assuming the frequency of oscillation at instability to be equal to the natural frequency ($\varepsilon = 1$), are compared with the available experimental data for rows of cylinders in Figure 5.32(a). Except for $m\delta/\rho D^2 \geq 200$ (which includes

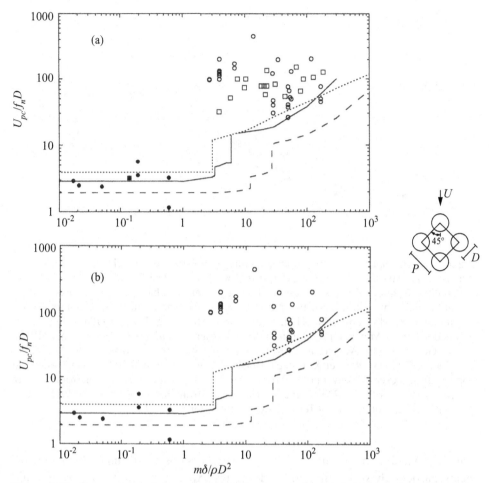

Figure 5.34. Comparison of theoretical stability boundaries for fluidelastic instability with experimental data for *rotated square arrays*: ●, multiple flexible cylinders in liquid flow; ○, multiple flexible cylinders in gaseous flow; □, single flexible cylinder in gaseous flow; ■, single flexible cylinder in liquid flow [experimental data from Abd-Rabbo & Weaver (1986), Gorman (1978), Hartlen (1974), Halle *et al.* (1988), Heilker & Vincent (1981), Païdoussis *et al.* (1989), Price & Kuran (1991), Soper (1980) and Weaver & Yeung (1984)]; - - -, theoretical solution from Lever & Weaver (1986b) for $P/D = 1.2$; ——, theoretical solution from Lever & Weaver (1986b) for $P/D = 2.0$; ······, theoretical solution from Price *et al.* (1990) for $P/D = 2.12$. (a) Includes all experimental data, (b) contains data from multiple-flexible-cylinder experiments only.

Roberts' own experimental data), Roberts' full solution, including jet reversal and fluid damping, overestimates $U_{pc}/f_n D$, especially for $m\delta/\rho D^2 \leq 2$. This is probably because Roberts assumed that the cylinder motion is in the in-flow direction, whereas most experimental results indicate that the motion is predominantly normal to the flow.

Also presented in Figure 5.32(a) is the traditional Connors (1970) equation, and the modified Blevins (1979) expression accounting for fluid damping (a value of $C_D = 0.7$ was assumed to obtain this curve). For $m\delta/\rho D^2 \geq 5$, the agreement between either the Connors solution or the modified Blevins solution and the experimental

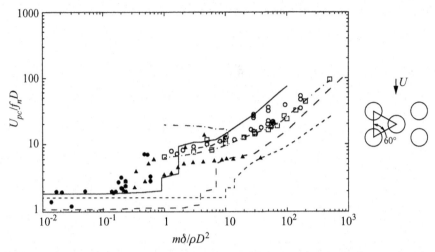

Figure 5.35. Comparison of theoretical stability boundaries for fluidelastic instability with experimental data for *rotated triangular arrays*: ●, multiple flexible cylinders in liquid flow; ○, multiple flexible cylinders in gaseous flow; □, single flexible cylinder in gaseous flow; ▲, multiple flexible cylinders in two-phase flow [experimental data from Connors (1980), Gorman (1978), Hartlen (1974), Halle *et al.* (1988), Heilker & Vincent (1981), Johnson & Schneider (1984), Minakami & Ohtomi (1987), Pettigrew *et al.* (1978, 1989), Soper (1980), Teh & Goyder (1988), Weaver & El-Kashlan (1981), Weaver & Grover (1978), Weaver & Koroyannakis (1983), Weaver & Yeung (1984) and Yeung & Weaver (1983)]; – –, theoretical solution from Lever & Weaver (1986b) for $P/D = 1.2$; —, theoretical solution from Lever & Weaver (1986b) for $P/D = 2.0$; - - -, theoretical solution from Price *et al.* (1990) for $P/D = 2.12$; - · - · , theoretical solution from Teh & Goyder (1988) for $P/D = 1.52$.

data is good; however, for $m\delta/\rho D^2 \leq 5$ both expressions overestimate $U_{pc}/f_n D$, this being particularly so for the modified Blevins expression accounting for the fluid damping.

It is evident that none of the models represented in Figure 5.32(a) can adequately predict the discontinuity in $U_{pc}/f_n D$ between low and high values of $m\delta/\rho D^2$ which is apparent in the experimental results. This is not the case for the unsteady analysis of Tanaka & Takahara (1980, 1981) and Chen (1983a)*, which are compared in Figure 5.32(b) with the same data-set as that presented in Figure 5.32(a). Clearly, the unsteady models give excellent agreement with experimental data, and they are able to predict the discontinuity at $m\delta/\rho D^2 \approx 4$. This indicates the importance of the unsteady fluid forces, and shows that to model fluidelastic instability in rows of cylinders properly, and by extension in cylinder arrays, it is necessary to be able to predict these unsteady fluid forces accurately.

For cylinder arrays the situation is less clear. Indeed, it should be noted that for normal triangular (Figure 5.33) and rotated square (Figure 5.34) arrays the experimental results themselves exhibit considerable scatter in $U_{pc}/f_n D$. If the single-flexible-cylinder experiments are removed, there is less scatter for the normal triangular array (Figure 5.33b).

---

* It is recalled that the solutions given by Chen (1983a) and Tanaka & Takahara (1980, 1981) are identical.

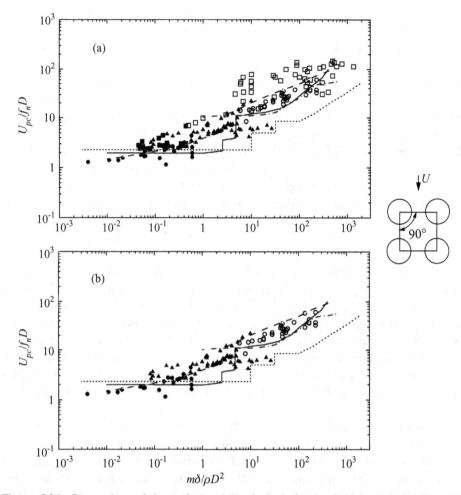

Figure 5.36. Comparison of theoretical stability boundaries for fluidelastic instability with experimental data for *in-line square arrays*: ●, multiple flexible cylinders in liquid flow; ○, multiple flexible cylinders in gaseous flow; □, single flexible cylinder in gaseous flow; ■, single flexible cylinder in liquid flow; ▲, multiple flexible cylinders in two-phase flow [experimental data from Axisa *et al.* (1984), Blevins *et al.* (1981), Connors (1978), Gorman (1978), Granger *et al.* (1991), Gross (1975), Hartlen (1974), Halle *et al.* (1988), Heilker & Vincent (1981), Lubin *et al.* (1986), Nakamura *et al.* (1991), Pettigrew *et al.* (1978, 1985, 1989), Price & Païdoussis (1989), Soper (1980), Tanaka & Takahara (1981), Teh & Goyder (1988), Weaver & Abd-Rabbo (1985), Weaver & Yeung (1984) and Žukauskas & Katinas (1980)]; ——, theoretical solution from Lever & Weaver (1986b) for $P/D = 1.2$; - - -, theoretical solution from Chen (1983b) for $P/D = 1.33$; ······, theoretical solution from Price *et al.* (1990) for $P/D = 1.5$; - · - · -, theoretical solution from Teh & Goyder (1988) for $P/D = 1.26$. (a) Includes all experimental data, (b) contains data from multiple-flexible-cylinder experiments only.

The theoretical models of Lever & Weaver (1986a, b)*, Price *et al.* (1990), Chen (1983a, b) and Tanaka & Takahara (1980, 1981) all exhibit jumps in the stability

---

* Strictly speaking, it would be better if the theoretical predictions of Yetisir & Weaver (1992), which account for multiple flexible cylinders, were used as opposed to those of Lever & Weaver (1986a, b) which only account for a single flexible cylinder. However, as pointed out by Yetisir & Weaver, the difference in $U_{pc}/f_n D$ is very small for $m\delta/\rho D^2 \leq 200$, and so the comparisons are given in terms of the Lever & Weaver results.

boundary. These jumps are due to the multiple instability regions occurring at low $m\delta/\rho D^2$. Following the reasoning given in Section 5.3, only the envelopes of these boundaries are given. The models of Lever & Weaver and Price & Païdoussis predict many more instability regions than shown in the figures, but as discussed in Section 5.3, only the upper three of these instability regions are included.

For the normal triangular array, see Figure 5.33, the theoretical model of Lever & Weaver with $P/D = 1.2$ tends to underestimate the experimental results. The theoretical curve for $P/D = 2.0$ does provide a reasonable lower limit to the experimental data, but this value of P/D is much larger than the experimental pitch values. The stability boundary given by Teh & Goyder (1988) provides a reasonable lower boundary to the experimental data for high $m\delta/\rho D^2$; however, this model does not predict the jump in $U_{pc}/f_n D$ which the experimental data exhibits. Comparing Figures 5.33(a) and (b) it is clear that many of the experimental data points with a single flexible cylinder tend to overestimate $U_{pc}/f_n D$ *vis-à-vis* the multiple-flexible-cylinder data points.

A comparison similar to that discussed above is also obtained for the rotated square array (see Figure 5.34). The theoretical stability curve from Lever & Weaver's model for $P/D = 2.0$ is very similar to the curve obtained by Price & Païdoussis (with multiple flexible cylinders) for $P/D = 2.12$; this, despite the fact that Lever and Weaver's results are for a single flexible cylinder, whereas the Price & Païdoussis model suggests that a single flexible cylinder will not go unstable in this array with $P/D = 2.12$. As can be seen by comparing Figures 5.34(a) and (b), very little of the experimental scatter is eliminated by removing the single-flexible-cylinder experiments.

In addition to their experimental dynamic instability results, shown in Figure 5.34, Païdoussis *et al.* (1989) also obtained a divergence instability for cylinders in a rotated square array with $P/D = 1.5$. The mechanism responsible for this type of instability is described in Section 5.2.2. The divergence was predominantly in the in-flow direction; but for cylinders in the second row of the array, cross-flow divergence was also observed. Although previously discussed theoretically this was the first time that the phenomenon was actually observed. In addition, to being of considerable academic interest, this instability is also of concern from a practical point of view. Once one, or more, of the cylinders diverge, they lodge themselves close to others that have not, or have done so in other directions; the cylinders then chatter and rattle against one another in this post-critical state, thus being subject to fretting and wear. Indeed, it should be noted that unless particular care is taken, as described in Païdoussis *et al.* (1989), it is very easy to confuse this static instability with a dynamic one, and it is possible that some of the "dynamic" instabilities indicated in Figure 5.34 may, in fact, be static ones.

For the rotated triangular array, see Figure 5.35, the comparison is similar to those discussed in the foregoing. For this array the results obtained from single- and multiple-flexible-cylinder experiments are very similar to each other, and thus, only one data set including all the experimental points is presented. The stability boundaries predicted by either Price & Païdoussis or Lever & Weaver with $P/D = 1.2$ tend to underestimate the experimental stability boundary, especially at high $m\delta/\rho D^2$, although reasonable agreement is obtained at low $m\delta/\rho D^2$. Interestingly, for this array, Teh and Goyder predict a lower and upper stability boundary for

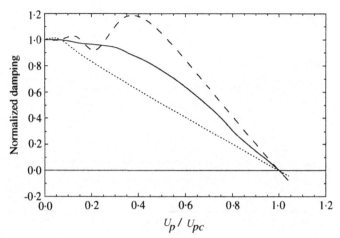

Figure 5.37. Variation of normalised fluidelastic damping for a single flexible cylinder in an in-line square array with flow velocity for $m/\rho D^2 = 6000$ and $\delta = 0.01$: —, solution using Tanaka & Takahara's (1980) data, $P/D = 1.33$; ······, solution using Price & Païdoussis' (1984a) model, $P/D = 1.5$; - - -, solution using Lever & Weaver's model (1986a), $P/D = 1.33$.

$P/D = 1.52$; this is similar to the multiple stability boundaries predicted by the other theoretical models.

For the in-line square array, excellent agreement is obtained between the experimental results and the theoretical predictions of Chen, Lever & Weaver and Price & Païdoussis (see Figure 5.36), especially when only the experimental data from multiple-flexible-cylinder experiments are considered (see Figure 5.36(b)). The stability boundary predicted by Teh & Goyder also agrees well with experiments, but does not extend to sufficiently low $m\delta/\rho D^2$ for a full comparison to be made.

The comparison of U_{pc} given in the previous paragraphs tests the "end result" of the theoretical stability models. However, it is possible to compare the models in other ways; for example, by comparing fluidelastic damping in the pre-instability range. This is done in the following for the theoretical models of Tanaka & Takahara (1980, 1981), Chen (1983a),[*] Lever & Weaver (1986a) and Price & Païdoussis (1984a). The specific array chosen is an in-line square array with $P/D = 1.33$ (for the Price & Païdoussis model $P/D = 1.5$ is used), and for the sake of simplicity one flexible cylinder only is considered.

Typical results for the variation of fluidelastic damping with flow velocity for high $m\delta/\rho D^2$ are shown in Figure 5.37. To make the comparison less dependent on the actual value of U_{pc}, the velocities are normalised with respect to U_{pc}. The fluidelastic damping predicted by the unsteady model varies in an approximately parabolic manner with velocity. The variations predicted by the Lever & Weaver and Price & Païdoussis models, however, are more complex. The Lever & Weaver model shows several oscillations in fluidelastic damping before it eventually decreases in a reasonably linear manner to the point of instability; the results predicted by Price & Païdoussis show only one small hump. Bearing in mind that the results obtained from the unsteady model come from the measured unsteady force coefficient data, it is to be expected that this variation will be the most accurate, and the results shown

---

[*] The models of Tanaka & Takahara and Chen are essentially the same, and for the sake of brevity will be referred to jointly as the "unsteady model".

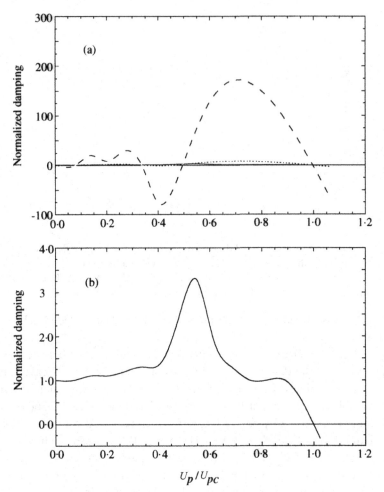

Figure 5.38. Variation of normalised fluidelastic damping for a single flexible cylinder in an in-line square array with flow velocity for $m/\rho D^2 = 10$ and $\delta = 0.01$. (a) —, Solution using Tanaka & Takahara's model (1980), $P/D = 1.33$; ······, solution using Price & Païdoussis' (1984a) model, $P/D = 1.5$; - - -, solution using Lever & Weaver's model (1986a), $P/D = 1.33$; (b) expanded view of Tanaka & Takahara's results.

in Figure 5.37 suggest that the variations predicted by either the Lever & Weaver or Price & Païdoussis models follow this behaviour in a reasonable manner.

Unfortunately, the comparison of fluidelastic damping with velocity for low $m\delta/\rho D^2$ is not nearly as good, as seen in Figure 5.38. The fluidelastic damping predicted by both the Lever & Weaver and Price & Païdoussis models oscillates much more than that given by the unsteady model (Figure 5.38(a)); this is particularly so for the Lever & Weaver model. An expanded view of the damping variation given by the unsteady model is presented in Figure 5.38(b), showing that, although the damping does oscillate with increasing velocity, the magnitude of these oscillations is not nearly as large as suggested by either the Lever & Weaver or Price & Païdoussis models.

Measurements of modal damping as a function of U_p for in-line arrays with $m\delta/\rho D^2 \approx 0.4$ have been reported by Chen & Jendrzejczyk (1981) and Granger *et al.* (1993). Both sets of experiments show oscillations of the modal damping in the transverse direction before instability is obtained; however, qualitative comparison with Figure 5.38 suggests that both the Lever & Weaver and Price & Païdoussis models severely overestimate the magnitude of these oscillations. The reason for the large oscillations in damping predicted by Lever & Weaver and Price & Païdoussis lies in the estimation of the phase lag between cylinder motion and resulting fluid forces. These results suggest that both of these theoretical models considerably overestimate the phase lag at low values of velocity. This does not have such a significant effect on U_{pc}, but it does produce this unrealistic behaviour for the pre-instability response of the cylinders. Thus, it is apparent that to better predict not only U_p, but also the pre-instability behaviour of cylinder arrays, a better understanding of the unsteady fluid forces is required. Granger *et al.* (1993) compared the variation of pre-instability damping with U_p obtained from the Lever & Weaver model with their own experimental results and showed that the agreement could be improved significantly if some of the parameters in the Lever & Weaver model (such as the phase lag, drag coefficient C_d or the positions of the separation and reattachment positions in the unit cell) were adjusted; however, no physical justification was given for these new values.

5.4.3 State of the art

In the previous section all known theoretical models of fluidelastic instability have been compared with one another and with available experimental data. Bearing in mind the complexity of the problem, the comparison between theory and experiment is encouraging. However, there is considerable room for improvement, and there are a number of issues which none of the models address.

Probably the most accurate of the theoretical models are those of Tanaka & Takahara (1981) and Chen (1983a), which employ measured unsteady fluid-force coefficient data. Unfortunately, this data is extremely difficult to obtain, and is presently available for one type of array and two values of P/D only. Thus, there is a need for more approximate analytical models, requiring less experimental data, such as those of Lever & Weaver (1982, 1986a, b), Price & Païdoussis (1984a, 1986a, b) and Yetisir & Weaver (1992, 1993a, b).

The theoretical models differ considerably, not only in the amount of experimental data required as input, but also in the assumptions employed. However, despite these differences, a number of similarities do exist, and it is possible to draw some broad conclusions concerning the models. It is apparent that, for most types of array, there is a distinct difference between the mechanism producing instability at high $m\delta/\rho D^2$ compared with that at low $m\delta/\rho D^2$; the boundary between high and low $m\delta/\rho D^2$ is somewhat ill defined, but is approximately 30. Chen (1983a) and Païdoussis & Price (1988) explain this difference in terms of a stiffness-controlled instability at high $m\delta/\rho D^2$ and a damping-controlled instability at low $m\delta/\rho D^2$. The most important difference from the practical viewpoint resulting from the distinction of these two mechanisms is the dependence of U_{pc} on damping. For low $m\delta/\rho D^2$, as typically found in liquid shell-side flows, U_{pc} is relatively insensitive to δ, a result which is commensurate with the experimental correlations of Chen (1984) and Weaver & Fitzpatrick (1988); whereas, for high $m\delta/\rho D^2$, typical of gaseous flows, increasing δ has a significant effect on U_{pc}. It should be noted that this conclusion is somewhat contradictory to the suggested nomenclature for the mechanisms. For example, at low $m\delta/\rho D^2$ the instability is referred to as being damping-controlled, which suggests that an increase in δ should have a significant effect on $U_{pc}/f_n D$; however, as stated above, this is not the case. Conversely, for high $m\delta/\rho D^2$, where the instability is referred to as being stiffness-controlled, the name suggests that an increase in δ should have much less effect on $U_{pc}/f_n D$; again, this is the reverse of what happens.

A second common theme of the theoretical models is the importance of the phase lag between cylinder motion and the resulting fluid forces on the cylinder itself and on its neighbouring cylinders. It is this phase lag which enables a single flexible cylinder in an array of rigid cylinders to become unstable, and for low $m\delta/\rho D^2$ it is the dominant destabilising mechanism in fully flexible cylinder arrays. The phase lag is also responsible for producing multiple regions of instability at low $m\delta/\rho D^2$. In retrospect, it should not be surprising that the phase difference between structural motion and resulting fluid forces is so important – this has been recognised in the aeronautical area since the time of Theodorsen (1935). Undoubtedly, the key to obtaining good simulations of the vibrational behaviour of cylinder arrays lies in the ability to predict these unsteady fluid forces. Almost certainly the reason for the greater success of the unsteady models given by Tanaka & Takahara or Chen is that this phase lag is measured rather than approximated analytically. Unfortunately, because of the viscous nature of the flow and the mutual interaction between cylinders in an array, obtaining reliable values of the unsteady fluid forces on an oscillating cylinder is extremely difficult. Hence, the predictions developed by Lever & Weaver and Price & Païdoussis should be regarded as approximations only. Hopefully, some of the CFD techniques discussed in Section 5.3.6 will eventually lead to more accurate predictions.

The possibility of using two-dimensional models for the analysis of three-dimensional heat-exchanger tubes subject to nonuniform flow has also been reviewed. It was shown that although approximate methods, such as the so-called "effective velocity" or "energy fraction", may yield reasonable results in most cases, they should be used with some care.

Finally, theoretical models of the post-instability response of cylinder arrays in cross-flow have been considered. It is shown that extremely complex dynamics, possibly leading to chaotic motion, can occur. Apart from the obvious academic challenges associated with the nonlinear dynamics, this has considerable practical importance associated with wear-rate predictions. Although there are many differences in the fluid-dynamic models used in these post-instability investigations, there is remarkable agreement in the type of behaviour obtained. This suggests that it is the nonlinear impacting forces which are dominant in producing the complex dynamical behaviour rather than the fluid dynamics, which essentially just provides a source of energy to drive the system. However, it is also apparent that, to be able to predict the flow velocity at which a particular type of behaviour occurs, a good estimate of the energy input is required, necessitating a good nonlinear model of the fluid dynamics. It should be noted that, if the stiffness-controlled mechanism plays a significant role in the instability, then obtaining a nonlinear model of the fluid dynamics will be extremely difficult. All of the existing instability models which account for the stiffness-controlled mechanism employ the principle of superposition, which is not valid if the fluid dynamics is nonlinear. Furthermore, it is not obvious how this superposition principle can be replaced in these models.

An aspect of predicting the onset of fluidelastic instability not discussed in this book relates to its interaction with vortex shedding. Fluidelastic instability and vortex shedding are typically assumed to be distinct. However, it is important to realise that this may not always be so, and resonance with a flow periodicity and fluidelastic instability may sometimes occur almost simultaneously; in which case, instead of there being two distinct vibration phenomena, there may be a complex interaction of the two mechanisms. An example of an combined fluidelastic instability and vortex-shedding instability for a parallel triangular array, $P/D = 1.375$ with low $m\delta/\rho D^2$, was suggested by the flow-visualisation results of Price $et\ al.$ (1992b).

To analyse combined instabilities of the type discussed above is much more difficult than analysing either mechanism individually – the interaction between them will almost certainly be nonlinear. The only such analysis known to the authors is that of Corless & Parkinson (1988, 1992) who attempted to analyse a combined vortex shedding and galloping instability. A combined analysis of fluidelastic instability and vortex shedding in cylinder arrays was undertaken by Sandstrom (1987) who developed a finite-element analysis of a group of cylinders in cross-flow; however, both the fluidelastic instability and vortex-shedding models were linear.

Another effect not addressed significantly by any of the theoretical models is that of the Reynolds number. There is now considerable evidence from the unsteady force-coefficient measurements of Chen $et\ al.$ (1998) and Chen & Srikantiah (2001), from the quasi-static force measurements in two-phase flow by Shahriary $et\ al.$ (2007) and Mureithi $et\ al.$ (2008), as well as from the CFD calculations of Gillen & Meskell (2008), that the Reynolds number has a significant effect. In addition, direct evidence of the effect of Re is given by Mewes & Stockmeier (1991), who made measurements of U_{pc} in liquid flows of different viscosities in the range $0.7 \leq \nu/\nu_{\text{water}} \leq 87.3$; this was achieved by using mixtures of water and organic substances (results are presented for the range 0.7 to 2.52 only). For a normal triangular array with $P/D = 1.28$ significant differences were obtained as ν was varied. In particular, fluidelastic instability was obtained for $\nu/\nu_{\text{water}} = 2.52$ but not for 0.7. All of this suggests that the Reynolds number is an important parameter which needs to be accounted for properly in the theoretical models.

Most of the theoretical models presently available, with the exception of Marn's (1996), implicitly assume a single-phase flow; however, in a large number of practical situations the flow is two-phase. The effect of two-phase cross-flow on the fluidelastic instability of cylinder arrays has been reviewed by Pettigrew $et\ al.$ (1989), Chen (1991b), Nakamura $et\ al.$ (1992a, 1995), Pettigrew & Taylor (1994, 2003) and Feenstra $et\ al.$ (2000). From these reviews it is clear that attempting to account for two-phase flows will considerably complicate the theoretical analysis. One approach, as discussed by Pettigrew & Taylor (2003, 2004), is to obtain the appropriate value of damping to use in a Connors-type expression. However, it should be appreciated that this damping value will "mask" the complex fluid dynamics associated with the two-phase flow, as the damping value employed is inherently due to the fluid dynamics – it is not possible to maintain a stagnant two-phase flow and so the damping must be measured at some flow rate. An alternative to this approach is suggested by Baj & de Langre (2003) who define a normalised fluid damping based on the density of the two-phase mixture, as opposed to the liquid density normally employed. They show that for "bubbly flows" with void fractions varying from 20% to 80% the normalised damping tends to the value measured in liquid as U_p/fD tends to zero. They also show that the damping strongly depends on U_p/fD, with increasing U_p/fD initially causing an increase in damping for low values of U_p/fD before reaching a maximum beyond which the damping decreases with increasing U_p/fD. The variation in damping with U_p/fD being so significant that the maximum damping can be twice as large as the zero-U_p/fD value – indicating the danger of using damping values measured in cross-flow as being representative of those found at low values of U_p/fD.

Attempts have been made to measure the unsteady force coefficients in two-phase flows by Inada et al. (1996, 1997, 2002), Delenne et al. (1997) and Mureithi et al. (2002), but the experiments proved to be very difficult and, in general, there was an unacceptable degree of scatter in the results. An alternative to this is the approach taken by Shahriary et al. (2007) and Mureithi et al. (2008) who have measured the quasi-static forces in two-phase flows. However, the remaining question is then whether or not the quasi-steady analysis of Price & Païdoussis (1984a, 1986a) is applicable to two-phase flows.

6 Ovalling Instabilities of Shells in Cross-Flow

6.1 A Historical Perspective

Ovalling oscillation, or simply "ovalling", of chimney stacks refers to the wind-induced shell-mode oscillation of thin metal stacks, involving deformation of the cross-section – strictly in the second circumferential mode (hence, the name) but, by common usage, in higher circumferential modes also. In a cross-section of the chimney the radial component of shell vibration varies proportionately to $\cos n\theta$, where n is the circumferential wavenumber (see Figure 6.1). Thus, for ovalling oscillations, $n \geq 2$; whereas $n = 1$ for conventional beam-like lateral, or "swaying", vibrations of the stack which are discussed only parenthetically here.

Ovalling as a technological problem first arose with the construction of thin-walled, tall and metallic chimney stacks; thin enough to easily deform as shells, tall enough to be unprotected by the earth's boundary layer near the top and with low internal damping in the metal construction. Dickey & Woodruff (1956) and Dockstader et al. (1956) describe some full-scale ovalling experiences: e.g. at Moss Landing Harbor, California, where a tall chimney ($L = 68$ m; $D = 3.44$ m, $h = 7.9$ mm at the top) developed ovalling in a $U = 40$ km/h wind with a frequency of about 1.47 Hz. Johns & Allwood (1968) describe a case of large-amplitude ovalling, which eventually led to a collapse of the chimney during a typhoon.

A number of experimental studies followed, notably by Heki & Hawara (1965) and Langhaar & Miller (1967). In the latter of these two studies it was found that ovalling oscillation precedes buckling collapse of the shell in a strong cross-wind. Wind-induced buckling will not be discussed here; the reader is referred to Kundurpi et al. (1975) and Johns (1983). But what causes the ovalling oscillation? That is the question that has exercised the Civil Engineering community since ovalling was first reported.

Based on the observations of the Moss Landing case, which is one of the few reasonably well documented ones, Dockstader et al. (1956) noted that (i) ovalling occurred at, or near, the natural frequency of the stack, involving the $n = 2$ mode, and (ii) the ovalling frequency corresponded to approximately twice the vortex-shedding frequency. Because the vortices are shed alternately from the two sides of the cylinder, and one cycle of vortex shedding involves the shedding of two vortices, it was proposed that, unlike the case of vortex-induced swaying, the chimney would go

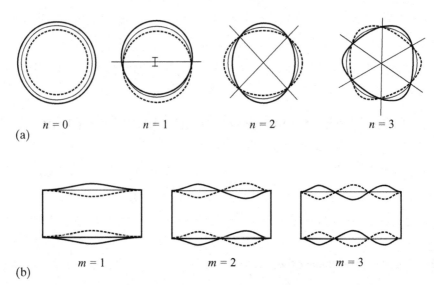

Figure 6.1. Modes of vibration of circular cylindrical shells, involving n circumferential waves and m axial half-waves in the case of a clamped-clamped shell. For a cantilevered shell, only the axial beam-like mode shape is different.

through one cycle of oscillation for each vortex shed. This explanation was sufficiently plausible for the matter not to be looked at again till fourteen years later, when Sharma and Johns (1970) conducted the first extensive experimental study of this problem. Sharma & Johns' results suggested modification to the Dockstader *et al.* explanation, but the "vortex-shedding hypothesis" was retained.

In what follows, the development of an alternative explanation for ovalling will be documented, leading to the demise of the vortex-shedding hypothesis. From a historical perspective, it is interesting that vortex shedding was suspected to be the cause of ovalling, just as it was in the case of fluidelastic instability of cylinder arrays in cross-flow (see Chapter 5 and refer to Païdoussis (1980, 2006)). Because vortex shedding was virtually the only fluid-elastic mechanism known to excite large-amplitude vibration of slender structures in cross-flow (effectively until 1970 for arrays of cylinders, and until 1982 for chimney stacks), it was the obvious candidate – even if considerable massaging was necessary to make the vortex-shedding hypothesis fit the observations. In this regard, a singularly Procrustean attitude to *forcing* data to fit a favourite hypothesis, not totally unknown in Science, was displayed.*

Before closing these introductory comments, it should be mentioned that some later work on ovalling was inspired by the anticipation of its occurrence, not in chimneys, but in liquid-storage tanks and silos, as construction of the latter becomes ever lighter, with the use of high-strength steel plate. Refer, e.g., to Katsura (1985), Uematsu & Uchiyama (1985) and Uchiyama *et al.* (1986).

---

* Prokrustes (Προκρούστης), an infamous brigand on the route to Megara in Attica, arrested travellers and tortured them to death by making them fit the size of his bed: the tall ones by cutting off their feet, and the short ones by "stretching them" to fit. He was finally slain by the classical hero Theseus, by Procrustes' own methods; this latter treatment is also not unknown to Science.

Notation used in Chapter 6	
Shell radius:	a
Shell diameter:	D
Shell-wall thickness:	h
Free-stream flow velocity:	U
Dimensionless flow velocity:	u, defined in equation (6.21)
Ovalling frequency:	$f_{n,m}$; n and m defined in Figure 6.1
Vortex-shedding frequency:	f_{vs}

6.2 The Vortex-Shedding Hypothesis

As stated already, Dockstader *et al.* (1956) observed that in the Moss Landing incident the shell-ovalling frequency, f_o, was close to its natural frequency $f_{n,m}$, in the $n = 2$, $m = 1$ mode (second circumferential wavenumber and first axial wavenumber; see Figure 6.1), and that this corresponded to approximately twice the estimated vortex-shedding frequency, f_{vs}; so that $r \equiv f_o / f_{vs} \simeq 2$. Because vortices are shed alternately from the two sides of the shell, it was supposed that the chimney executed one complete cycle of oscillation for each vortex shed (i.e. one oscillation cycle per half-cycle of vortex shedding), rather than the more usual case where $r = 1$. The explanation was sufficiently plausible for the matter not to be looked at again, till twelve years later, when Johns, Allwood and Sharma undertook an extensive experimental study of the problem (Johns & Allwood 1968; Sharma & Johns 1970; Johns & Sharma 1974).

Johns and his co-workers experimented with cantilevered metallic cylindrical shells, the upper portion of which was subjected to cross-wind in an open-working-section wind tunnel. They measured $f_{n,m}$ and the threshold flow velocity for ovalling, U_{thr}, and observed the modal form (n, m) of the oscillating cylinder; they did not measure f_{vs}, but inferred it by assuming a constant Strouhal number: either $S = 0.2$ or 0.166.

Here, it is recalled that $S = 0.2$ applies to long cylinders in two-dimensional cross-flow in the *subcritical* Reynolds number range (approximately $40 < \text{Re} < 2 \times 10^5$). In the *transitional* Reynolds number range ($1.5 \times 10^5 < \text{Re} < 3.5 \times 10^6$ approximately), vortex shedding becomes less regular and the Strouhal number varies, depending on the surface roughness for instance (see Blevins (1990)). For a chimney, the situation is further complicated by nonnegligible three-dimensional effects about the free end of the chimney, and so deviations from $S = 0.2$ are inevitable. Nevertheless, some data for model chimney stacks, obtained by Wootton (1968) for $10^5 < \text{Re} < 3 \times 10^6$, suggest an "effective" $S = 0.16$ in this Re range. Johns and his co-workers presumed that the real value of S for the chimneys in their experiments was somewhere between 0.16 and 0.20. Also for chimneys, values of $S = 0.165$ to 0.173 are reported by Païdoussis & Helleur (1979) for lower Re: $8 \times 10^4 < \text{Re} < 1.6 \times 10^5$.

The main findings of this research (Sharma & Johns 1970; Johns & Sharma 1974) were that

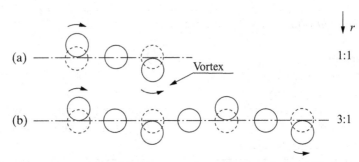

Figure 6.2. Alternative relationships between bending (swaying) oscillations and vortex shedding [after Johns & Sharma (1974)]; $r = f_{n,m}/f_{vs}$.

(i) cylinders vibrated at or near one of their natural frequencies, $f_{n,m}$,
(ii) at ovalling an integral relationship existed between $f_{n,m}$ and f_{vs}, namely $r \equiv f_{n,m}/f_{vs} = 1$ to 6, depending on the experiment.

The experimental results will be discussed further shortly. However, presently, more needs to be said about finding (ii) above, because it implies a subharmonic sustenance of the oscillations. How this could work according to Johns & Sharma (1974) for swaying ($n = 1$) oscillations is illustrated in Figure 6.2. In (a), a complete cycle of vortex shedding corresponds to a full cycle of oscillation ($r = 1$), whereas the situation for $r = 3$ is illustrated in (b). The same is ingeniously extended to ovalling oscillations with $n = 2$ in Figure 6.3, illustrating possible scenarios for sustaining $r = 1, 2, 3$ and 4 ovalling.

Obviously, if both $f_{n,m}$ and f_{vs} are measured in any given experiment, then $r = f_{n,m}/f_{vs}$ could be determined and, according to finding (ii) above, it should be an integer. If, however, as in the work by Sharma and Johns, f_{vs} was not measured, then r could be determined via (a) calculating the threshold flow velocity with $r = 1$, thus $U_{thr}^{calc} = f_{n,m} D/S$, presuming $S = 0.16$ (or 1/6) or 0.20, and then (b) comparing with

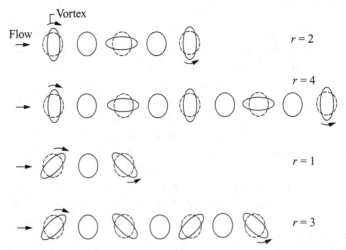

Figure 6.3. Alternative relationships between ovalling and vortex-shedding frequencies for inducing/sustaining ovalling according to Johns et al. [after Johns & Sharma (1974)]; $r = f_{n,m}/f_{vs}$.

Table 6.1. *Measured and predicted data of the natural frequencies $f_{n,m}$ (with $m = 1$ in all cases), and critical (threshold) flow velocities for ovalling, U_{thr}; from table 2 in Johns & Sharma (1974). The Reynolds number range in these results is $5 \times 10^4 < \text{Re} < 2.5 \times 10^5$*

Shell number	n	Predicted $f_{n,m}$ (Hz)	Ovalling $f_{n,m}$ (Hz)	$U_{\text{thr}}^{\text{calc}}$ (m/s) with $r = 1$		$U_{\text{thr}}^{\text{meas}}$ (m/s)	Deduced r
				For $S = \frac{1}{5}$	For $S = \frac{1}{6}$		
2	1	136.0	–	82.9	99.5	–	–
	2	62.0	60	37.8	45.4	12.2–14.9	3
	3	126.3	–	77.0	92.4	–	–
3	1	49.5	50	30.2	36.2	18.0	2
	2	46.8	45	28.5	34.2	11.9	3
	3	124.7	119	76.0	91.2	19.5	4 or 5
	4	238.6	–	145.5	174.5	–	–
6	1	137.3	–	139.5	167.4	–	–
	2	47.2	46	47.9	57.5	18.0 (19.8)	3
	3	49.6	48	50.4	60.5	11.6 (13.4)	5
	4	86.7	–	88.2	105.7	–	–
7	1	58.5	–	59.5	71.3	–	–
	2	24.5	23	24.9	29.9	9.1	3
	3	45.7	43	46.4	55.7	15.8	3
	4	86.1	–	87.4	104.8	–	–
8	1	89.0	–	135.7	162.8	–	–
	2	29.6	29	45.1	54.1	10.7	4
	3	24.2	22 (19)	36.8	44.3	7.9 (13.7)	5(3)
	4	39.0	36	59.5	71.3	11.6	6
	5	62.0	–	94.5	113.4	–	–

the measured value $U_{\text{thr}}^{\text{meas}}$, which is supposed to satisfy $(f_{n,m}/r) D/S$, presuming a subharmonic relationship; then,

$$r = \text{int} \left[U_{\text{thr}}^{\text{calc}}/U_{\text{thr}}^{\text{meas}} \right], \tag{6.1}$$

where int [   ] denotes the closest integer value.

In Sharma & Johns' (1970) experiments, one steel and eight aluminium shells were tested, varying in length L from 20 to 71 inches (50.8–180 cm) and radius a from 2.4 to 6.0 inches (6.10–15.2 cm); the thickness was $h = 0.01$ inch, in all cases; so that $8.3 \leq L/a \leq 29.6$ and $240 \leq a/h \leq 600$. Typical results, excerpted from Johns & Sharma (1974, table 2) are given in Table 6.1.

The interpretation of the information in Table 6.1 is made clear by discussing some of it in detail following Johns & Sharma (1974).

We start with shell 3, which was designed such that the frequencies in the $n = 1$ and $n = 2$ modes be nearly equal. Ovalling commenced in the $n = 2$ mode at $U_{\text{thr}}^{\text{meas}} = 11.9$ m/s. The calculated values $U_{\text{thr}}^{\text{calc}}$ based on the calculated $f_{n,m} = 46.8$ Hz and $S = 1/5$ and $1/6$ ($\simeq 0.16$) are 28.5 and 34.2 m/s, respectively. Clearly, therefore, $r \neq 1$, and the value of r can be estimated via equation (6.1): $U_{\text{thr}}^{\text{calc}}/U_{\text{thr}}^{\text{meas}}$ is 2.39 and 2.87 for the two Strouhal numbers. So, by (6.1), the closest integer is 3, and hence $r = 3$. At higher wind speed, the $n = 1$ mode was excited, at $U_{\text{thr}}^{\text{meas}} = 18.0$ m/s, but with some evidence of ovalling; in this case, $r = 2$ is estimated. Finally, at $U_{\text{thr}}^{\text{meas}} = 19.5$ m/s, the $n = 3$ mode was excited, alone, and in this case $r = 4$ or 5.

Surveying some of the other data in Table 6.1, we note that r as high as 5 or 6 can be found. The quantities in parentheses indicate results with some change in the experimental set-up. For example, for shell 6 the seams in the plate-formed shell were initially along the windward and leeward generators of the shell, and subsequently (the figures in parentheses) the seams were at $90°$ to the wind direction; there was sensibly no change in the measured frequencies, but the critical speeds increased slightly. In both sets of tests the $n = 3$ mode was excited well before the $n = 2$ mode even though the latter had the lower predicted critical speed, but this cannot be explained.

In retrospect, the main weakness in the vortex-shedding hypothesis (as described thus far) is that *it was based on an analysis of experimental data in which f_{vs} was not measured but inferred.* Moreover, the open wind tunnel introduced a number of uncertainties concerning the flow field and vortex shedding. Nevertheless, this model for the mechanism of ovalling remained unchallenged until Païdoussis & Helleur's (1979) work.

6.3 Ovalling with No Periodic Vortex Shedding

6.3.1 Païdoussis and Helleur's 1979 experiments

Johns & Sharma's physical explanation essentially claims that the ovalling oscillation of the shell was sustained by a helpful "push" by the shedding of a vortex; the shell would then undergo one, two or three full cycles of oscillation (for $r = 2$, 4 and 6, respectively) or a number of half-cycles (for r odd), before receiving another push from the next vortex shed. This is conceptually easy to accept for $r = 1$ and 2, but it becomes a little contrived for $r = 5$ and 6. Païdoussis & Helleur's (1979) study was motivated in part by lingering doubts about this explanation. The main motivation, however, had been to investigate the effect on ovalling of high-speed internal flow, exiting as a jet from the free end – as would be the case for tall metal chimneys. The first step was to validate the vortex-shedding hypothesis, without internal flow. The results obtained were so interesting that the second step (with internal flow) was never taken!

6.3.1 (a) First-phase experiments

The experiments were conducted in a low-speed, open-return, suck-through wind tunnel with a 0.9 m × 0.61 m closed working section. The cylindrical shell was mounted as in Figure 6.4. The shell was clamped at its base to a rigid cylindrical pipe support which was normally sealed; the upper end of the shell was free. In most experiments a 0.95-cm-thick plexiglas edge plate was placed over the free end of the shell in an attempt to reduce entrainment from above into the wake, for reasons to be explained shortly. The plate was mounted such that a gap of 0.64 cm (1/4 inch) was left between it and the top of the cylindrical shell. The supporting posts (1.27 cm in diameter) were sufficiently distant from the test shell so as to cause minimal disturbance to the flow about it. It is clear, therefore, that the aim of these

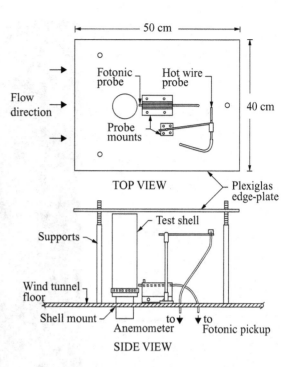

Figure 6.4. Top and side view of the cylindrical shell mounted in the wind tunnel test-section, showing also some of the measuring apparatus (after Païdoussis & Helleur (1979)).

experiments was to study the *mechanism* of ovalling, rather than to model ovalling in real chimney stacks.

Some preliminary tests were done with brass shells machined from stock (typically $L = 160$ mm, radius $a = 25$ mm and $h = 0.8$ mm), but their wall thickness was unacceptably nonuniform. Hence, subsequent experiments were conducted with low-viscosity epoxy cylindrical shells which were spin cast and possessed excellent uniformity (3% or less variations in wall thickness); what follows henceforth relates to the epoxy shells (typically, 65 mm in diameter, 230 mm long and 0.50 mm in wall thickness). Due to manufacturing difficulties the length-to-diameter ratio did not exceed 4. There is a disadvantage in this, because flow over the top of the cylindrical shell was found to be entrained into the wake, and to result in distortion or dispersal of the vortex street in the wake, rendering the measurement of periodicity therein difficult; indeed, this was the main reason for installing the edge plate over the free end of the shell. A photograph of one of the shells ovalling in the $n = 3, m = 1$ mode is shown in Figure 6.5.

The frequency of oscillation of the shell was measured by fibre optics. A "Fotonic" probe was mounted on the tunnel floor, behind and close to the base of the cylindrical shell. The modal shape when the shell was subject to ovalling oscillation could be determined visually and photographically by viewing from above, where there was a transparent observation port, with the aid of a synchronizable stroboscopic system. To further help identify the modes excited, the natural frequencies were calculated by a finite-element method based on SAPIV (Bathe *et al.* 1973). The frequencies of the visually identified modes in the wind tunnel were found to be in good agreement with the calculated values. The vortex-shedding frequency was determined by a hot-wire anemometer placed typically 200 mm behind the shell,

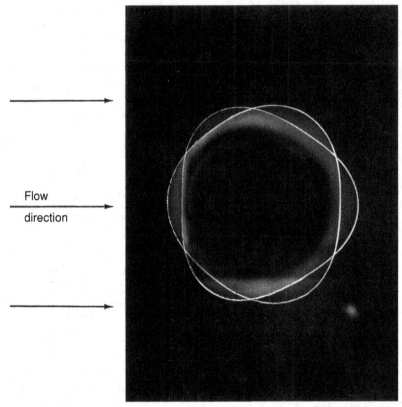

Flow
direction

Figure 6.5. Photograph, taken from above of Shell 7 ($L = 246$ mm, $D = 76.2$ mm, $h = 0.38$ mm) undergoing ovalling in the $(n, m) = (3, 1)$ mode at $U = 21$ m/s. Aluminium paint was applied to the free end of the epoxy shell to make it visible; the photograph was taken with stroboscopic light on, to show two positions of the shell (the thin white lines) with $180°$ phase difference (after Païdoussis & Helleur (1979)).

170 mm above the base of the shell. The signals of both the fibre-optic probe and the hot-wire anemometer were digitised via a high-speed A/D converter; then the frequency spectra were obtained via a FFT program.

Typical results for (a) the shell vibration and (b) the wake-velocity spectrum at the onset of ovalling are shown in Figure 6.6.

The vortex-shedding frequency is recognised as the dominant peak in the wake signal (Figure 6.6(b)), so that $f_{vs} = 60.5$ Hz; vortices on the other side of the wake are also sensed to some extent, giving rise to the peak at $2f_{vs} = 120$ Hz. The dominant component of shell response is at 242 Hz, which corresponds to the $n = 2, m = 1$ mode. The other important components of the response occur at 60.5 and 121 Hz and are clearly associated with direct excitation of the shell by the vortex-shedding process[*]. The visually observable ovalling was clearly in the third circumferential mode. Hence, taking $f_0 = f_{2,1} = 242$ Hz one obtains $r = f_0/f_{vs} = 4$ in this case. So far, the results agree with others reported previously.

---

[*] It is of interest to notice the small peak at 242 Hz in the wake spectrum, corresponding to $4f_{vs}$. It would be reasonable to suppose that this is the aeroelastic influence of the ovalling oscillation *on the flow*.

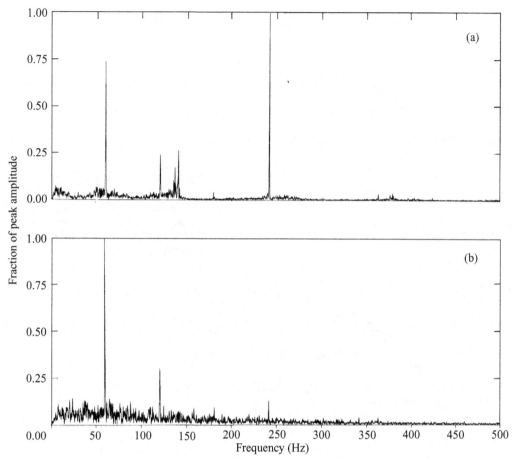

Figure 6.6. Frequency spectra for Shell 4 ($L = 230$ mm, $D = 65$ mm, $h = 0.51$ mm) at the onset of ovalling, at a wind speed of 22.6 m/s. (a) The shell vibration spectrum; (b) the wake velocity spectrum, 20 cm behind the shell (Païdoussis & Helleur 1979).

Upon increasing the flow velocity, even by substantial amount (10 to 20%) it was found that ovalling oscillation continued in the same mode and that its amplitude increased. Because one would expect that f_{vs} would normally be linearly proportional to the flow velocity, two plausible explanations could be proposed, consistent with the requirements of existing theory: (i) vortex shedding could have locked on the ovalling oscillation, such that f_{vs} and f_o remained constant and, of course, their ratio remained at $r = 4$; (ii) the vortex-shedding frequency could have increased with flow in the normal way, while the cylinder continued to be harmonically excited, not at one of its natural frequencies, but at a frequency always equal to f_{vs} (a less likely possibility).

In fact, neither of these occurred: the shell continued to oscillate at 242 Hz predominantly, whereas the vortex shedding frequency continued to increase with flow. As summarised in Figure 6.7,* the ovalling reached a maximum amplitude at $U \simeq 30$ m/s and then subsided at $U \simeq 34.5$ m/s, always at $f_o = f_{3,1} = 242$ Hz, while

* The nondimensional amplitude in this figure is equal to 10^4 times the amplitude measured by the Fotonic sensor (at the base of the shell) divided by the radius.

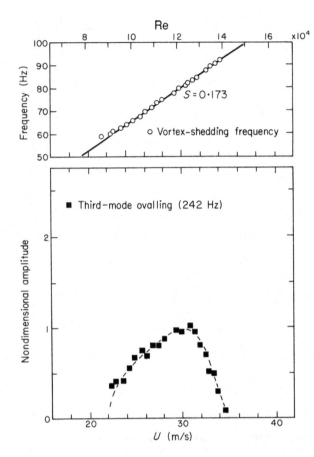

Figure 6.7. The vortex-shedding frequency (on top) and the amplitude of third-mode ($n = 3$) ovalling (below) as a function of the flow velocity U for Shell 4 – see Figure 6.6 (after Païdoussis & Helleur (1979)).

f_{vs} varied from about 60 to 90 Hz. Clearly, an integral relationship between f_o and f_{vs} did not appear to be necessary for the sustenance of ovalling!

Similar results were obtained for other shells (Païdoussis & Helleur 1979); some are shown here in Figure 6.8.* In this figure the nondimensional flow velocity u is utilised defined by $u = U/\{E/[\rho_s(1 - v^2)]\}^{\frac{1}{2}}$, where E is Young's modulus for the shell, v is the Poisson ratio and ρ_s is its density. In both cases shown, ovalling begins in the $n = 3$ mode, such that $r = 4$; in the case of Figure 6.6(a), second-mode ($n = 2$) ovalling succeeds third-mode ovalling, in the high u range. In all cases, however, with increasing flow the shell continues to oscillate in one (or more) of its natural frequencies, whereas the vortex-shedding frequency increases essentially linearly with flow velocity.

It was at this point that it was decided that vortex shedding might not be the underlying mechanism of ovalling oscillation after all. Although at the onset of ovalling there is a simple relationship between f_o and f_{vs}, the phenomenon persists to higher flows, where such a relationship no longer holds.

---

* It should be remarked that the Strouhal numbers obtained in the experiments ($S = 0.165$ to 0.173) may at first sight appear to be oddly low compared with the "conventional" figure of $S = 0.20$ for this Re range. This matter was considered carefully, but it was found that the values obtained are indeed correct, to within 5%, and, furthermore, that they are not unlike values obtained by others in the same range of Reynolds numbers (Païdoussis & Helleur 1979, appendix 1).

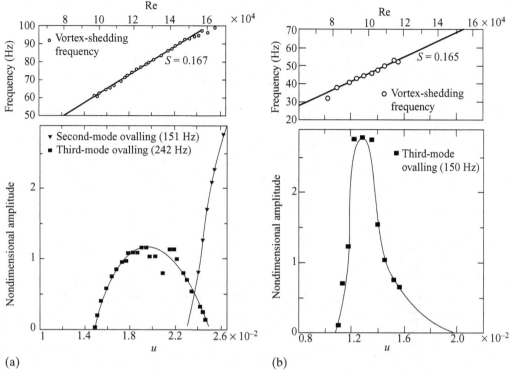

Figure 6.8. The vortex-shedding frequency (on top) and the amplitude of ovalling (below) as a function of the dimensionless flow velocity u: (a) for Shell 5 ($L = 210$ mm, $D = 65$ mm, $h = 0.51$ mm); (b) for Shell 7 ($L = 246$ mm, $D = 76.2$ mm, $h = 0.38$ mm); (after Païdoussis & Helleur (1979)).

6.3.1 (b) Second-phase experiments

To determine whether vortex shedding has any effect at all on the onset and sustenance of ovalling oscillation, the apparatus was modified by mounting a splitter plate behind the cylinder, parallel to the flow, as shown in Figure 6.9(a). Regular, periodic vortex shedding was indeed suppressed, as may be seen in Figure 6.9(c), with the probe mounted in its usual position ("position 1" of Figure 6.9(a)); yet ovalling was not suppressed as may be seen in the corresponding vibration spectrum of Figure 6.9(b).

The full results for this and for another shell are shown in Figure 6.10. It is seen that ovalling develops close to the flow velocity where it began with the splitter plate absent. With increasing flow, the amplitude *increases more precipitously**. In both cases the test was discontinued for fear of damaging the shells; it is, therefore, not known if at yet higher flows the oscillation subsides.

At this point it was concluded that this constituted the demise of periodic vortex shedding as a credible underlying mechanism for ovalling, and the search was on for the true cause – prematurely, according to some; see Section 6.5.

---

* Compare with similar observations with regard to galloping by Nakamura & Tomonari (1977).

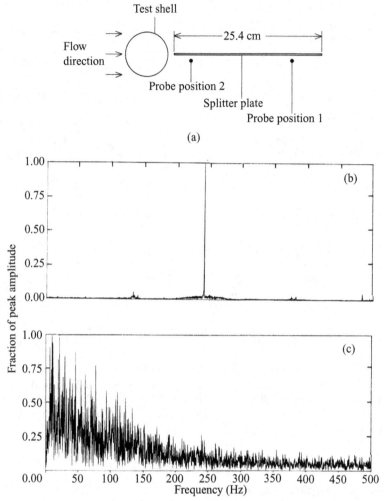

Figure 6.9. (a) Top view of the shell with the splitter plate behind it; (b) the shell vibration spectrum for Shell 4 at a flow velocity of 32 m/s; (c) the corresponding wake velocity spectrum at position 1 (Païdoussis & Helleur 1979).

6.3.2 In search of a new cause

One mechanism that was considered was that motion of the point of separation, caused by small flow-induced motions of the cylindrical shell, could result in energy being fed into the cylinder from the flow, in phase with the oscillation, eventually leading to the observed limit cycle. To test this hypothesis it was decided to force separation at a point well ahead of the natural point of separation by gluing flexible plastic tubes, 0.32 cm in diameter, on either side of the shell.

Initially the tubes were placed at 75° from the front stagnation point on one of the shells. These tests were, of course, conducted with the splitter plate removed. Upon following the same test procedure it was found that ovalling occurred as before, but with increased amplitude. With the tubes attached even further ahead, at 65° from the stagnation point, the amplitude of ovalling was greater still.

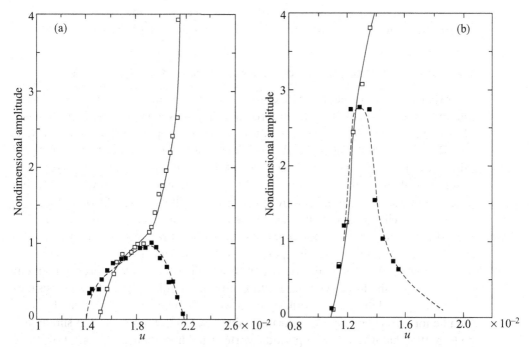

Figure 6.10. (a) Dominant-mode ovalling (third-mode response) amplitude versus flow velocity for Shell 4: □, with splitter plate; ■, without the splitter plate; (b) the same for Shell 7 (after Païdoussis & Helleur (1979)).

At this point, because all results were contrary to expectations, it was decided to rotate the cylinder 180° so that the tubes would now be at the back of the cylinder in the wake, where their effect was expected to be minimal. In this arrangement, with the tubes at either 105° or 115° from the stagnation point, no ovalling could be induced over the flow-velocity range available! In case attachment of the tubes had altered the structural characteristics of the system significantly enough to cause this peculiar phenomenon, the tubes were next attached on the interior surface of the cylinder. It was verified that in this arrangement the behaviour of the cylinder was essentially the same as when the tubes were entirely absent.

At this point the importance of the overall character of flow in the wake became evident, and it was decided to take a closer look at the effect of the flow between the edge plate and the shell. It is recognised that at its normal position the edge plate is insufficiently close to the shell to prevent a streaming flow over the top of the shell. A cylinder insert was utilised, which could be moved up and down; the radial clearance between the top of the insert and the cylindrical shell was 6.4 mm (1/4 in).

When the top of the insert was 50 mm below the free end of the shell, its effect on ovalling was insignificant; when it was flush with the top of the shell, the only effect was that the onset of ovalling occurred at slightly higher flow velocity. With the insert extending above the shell, halfway to the edge plate, this effect was more pronounced. Finally, when the insert was extended further, so that it pressed against the edge plate, ovalling could not be observed at all, over the range of flows where it had appeared heretofore. Clearly, therefore, the flow over the top of the shell plays an important role in the mechanism involved in ovalling.

Obviously, the above, although offering useful clues on the nature of the mechanism underlying ovalling oscillation, have not clarified the mechanism itself.

6.4 Further Evidence Contradicting Vortex-Shedding Hypothesis

The experiments by Helleur and Païdoussis described in Section 6.3 have established that there appears to exist an integral relationship between $f_{n,m}$ and f_{vs} at the onset of ovalling ($r = f_{n,m}/f_{vs}$ = integer). This is clearly in support of the hypothesis that periodic vortex shedding is the driving mechanism for ovalling. However, some other, disquieting observations were made at the same time, namely: (i) beyond the onset of ovalling r = integer ceases to hold, $f_{n,m}$ remaining constant with increasing speed, whereas f_{vs} increases in conformity with a sensibly constant Strouhal number; (ii) ovalling oscillations could be excited even when a long splitter plate was mounted behind the cylindrical shell, albeit at a slightly higher flow velocity than without the splitter plate, despite the fact that periodic vortex shedding had been suppressed.

Obviously, (i) above signifies that lock-in does not occur, which is in disagreement with the vortex-shedding hypothesis. More importantly, observation (ii) suggests that the ovalling mechanism has nothing to do with periodic vortex shedding, because the phenomenon can equally occur in its absence. On the other hand, it should not be forgotten that when there *is* periodic vortex shedding, then, at the onset ovalling, there exists an integral relationship between $f_{n,m}$ and f_{vs}. Thus, on the one hand, periodic vortex shedding appears to have nothing to do with the phenomenon and, on the other, it appears to have some organic relationship with it. Furthermore, it was demonstrated that the flow in the wake and over the top of the shell is important in determining the onset of ovalling and, indeed, whether it occurs or not.

To further explore ovalling, some new experiments were undertaken at McGill University in the 1980s (Païdoussis & Suen 1982a), which will be discussed next. However, it should be mentioned that, at the same time, a new theoretical model for ovalling was proposed by Païdoussis & Wong (1982) and refined by Païdoussis, Price & Suen (1982b); these will be discussed in Sections 6.6, 6.7 *et seq.*

The experiments were conducted in the same wind tunnel described in Section 6.3. However, the fibre-optic sensor was mounted inside the shell (Figure 6.11(a)), so as to be even less obtrusive, also allowing the measurement of the modal shapes. Moreover, to eliminate uncertainties associated with the flow field over the top of the shell (and between it and the edge or top plate, experiments were also done with the shell spanning the height of the wind tunnel, with both ends clamped (Figure 6.11(b)) – thus approximating two-dimensional flow conditions over the shell.

The shells used were again spin-cast epoxy shells, but were generally larger ($D = 76.2$ and 147.3 mm). Again, for some of the tests, a 500 mm × 400 mm "top plate" was mounted over the free (top) end of cantilevered shells, 0.05 to 0.35 diameters above their upper lip.

6.4.1 Further experiments with cantilevered shells

Some preliminary experiments, to verify those of Section 6.3, were done for four different cantilevered shells, without and with a splitter plate. The results are summarised in Table 6.2.

Table 6.2. *Critical threshold flow velocities for ovalling of four shells of different lengths (D = 76.2 mm), without and with a splitter plate; $f_{n,m}$ is the measured ovalling frequency; in these experiments, no top plate was used*

Shell	Length (mm)	Critical mode n, m	Without splitter plate			With splitter plate	
			U_{cr} (m/s)	$f_{n,m}$ (Hz)	$f_{n,m}/f_{vs}$ (–)	U_{cr} (m/s)	$f_{n,m}$ (Hz)
E	290	3, 1	27.2	213	4.0	27.6	213
F	290	3, 1	25.7	207	4.3	24.0	207
C	479	2, 1	20.8	76	1.7–1.8	21.5	76
D	479	2, 1	20.8	76	1.8	21.4	76

It is seen that ovalling occurs in the third circumferential mode for the shorter shells, but in the second circumferential mode for the longer ones. It was confirmed that ovalling oscillation does occur, even with a splitter plate behind the shell, when periodicity in the wake was found to have been suppressed. However, the most perplexing finding in the results of Table 6.2 is that, for the first time, in some cases $f_{n,m}/f_{vs}$ appreciably deviates from an integral value.

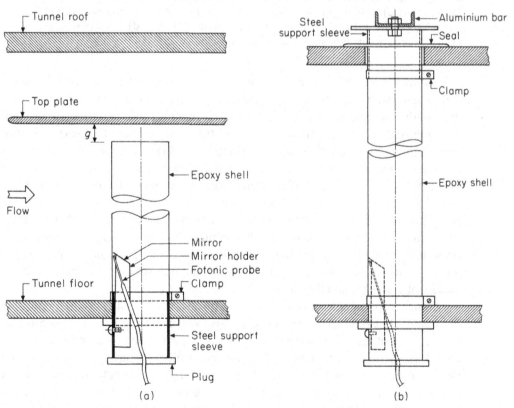

Figure 6.11. Schematic of the cylindrical shell mounted in the wind tunnel. (a) Clamped-free (cantilevered) shell showing the internally mounted Fotonic sensor and the sometimes used "top" plate; (b) clamped-clamped shell (Païdoussis, Price & Suen 1982a).

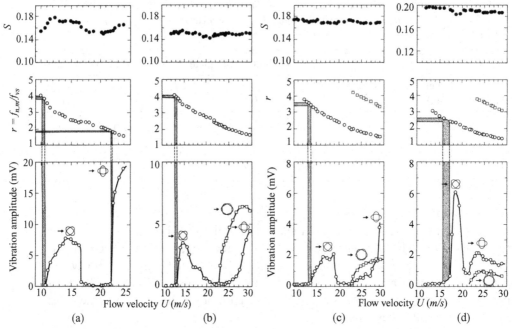

Figure 6.12. Measured vibration and wake characteristics of a 76.2 mm diameter cantilevered shell, showing the Strouhal number, the ratio r and the vibration amplitude of the shell (arbitrary scale), as functions of the wind velocity. (a) $L = 479$ mm, no top plate; (b) $L = 311$ mm, no top plate; (c) $L = 479$ mm, gap to top plate 24.4 mm; (d) $L = 479$ mm, gap to top plate 3.2 mm [after Païdoussis, Price, Fekete & Newman (1983)].

To investigate this observation further, it was decided to conduct additional experiments with one of the shells, varying systematically one parameter, namely the length, and determining the ratio $f_{n,m}/f_{vs}$ in each case. The results presented and discussed in detail in Païdoussis *et al.* (1982a), but not here for brevity, showed that: (i) the modal characteristics of ovalling oscillation change with length, as does the nodal orientation of the modes concerned *vis-à-vis* the free-stream vector; (ii) the Strouhal number varies with flow in each case, in general being higher at higher vibration amplitudes, which, in view of the attendant widening of the wake, may be said to be consistent with the concept of a universal Strouhal number (Roshko 1955); (iii) at the onset of second-mode ovalling with a node facing the free stream (as well as for third-mode ovalling), one obtains r close to 4, while at the onset of second-mode ovalling with an antinode facing the free stream one obtains r in the vicinity of, but substantially below, 2.

Similar experiments were conducted with the same shell ($D = 76.2$ mm, $L = 311.2$ mm), but now using the top plate and varying the gap, g, between it and the free end of the shell (see Figure 6.11(a)). Typical results may be seen in Figure 6.12(c, d), for $g = 25.4$ mm and 3.2 mm; the results in Figure 6.12(a), with no top plate, may be considered to correspond to the "top plate" actually being the roof of the wind tunnel. Once again, one may observe considerable differences in vibrational behaviour for different values of g; this indicates a strong three-dimensional flow effect on ovalling. Significantly, in both Figures 6.12(c) and (d), r at $U = U_{cr}$ is not close to an integer.

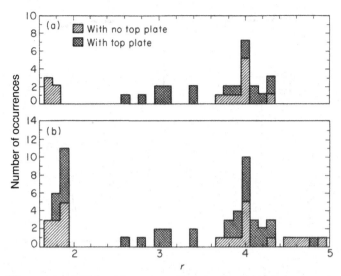

Figure 6.13. Histogram of all $f_{n,m}/f_{vs}$ measured at the threshold for onset of ovalling in all tests with clamped-free shells: (a) for the mode first encountered, at $U = U_{cr}$; (b) for all the modes excited, at various $U = U_{thr}$ (Païdoussis, Price & Suen 1982a).

Collecting all available results for cantilevered shells, similar to those shown in Figure 6.12, it is first noted that more than one circumferential mode may be excited concurrently and that the orientation of some modes *vis-à-vis* the wind can change with wind speed. However, the crucial point is that, whereas in many cases r is close to an integer, observations have been made for the first time where r at onset is quite far from an integer, e.g. Figure 6.12(c) and (d).

Figure 6.13 shows a histogram of frequency of occurrence of ovalling versus r, for all the tests conducted, where it is seen that, whereas in most cases r is close to an integral value, there are cases where it clearly is not. (It should be mentioned that in this figure the distinction is made between a threshold flow velocity, U_{thr}, for the onset of ovalling at any one mode, and the *critical* flow velocity, U_{cr}, which is the particular U_{thr} for the mode first encountered with increasing flow.)

6.4.2 Experiments with clamped-clamped shells

Ovalling, very similar to that for cantilevered shells, was observed for shells with clamped ends ("clamped-clamped" for short), some experimental observations for 76.2-mm-diameter shells are shown in Figure 6.14.

One important observation made in all such tests is that the second (circumferential) mode always occurred with a node facing the free stream, whereas the third mode was with an antinode facing the free stream.

Another important observation is that at the onset of ovalling oscillation the ratio r was quite distinctly a noninteger. Thus, in Figure 6.14(a) U_{cr} is associated with $r \simeq 3.3$ to 3.4, whereas in Figure 6.14(b) $r \simeq 3.3$ to 3.5, the uncertainty in r arising from the corresponding difficulty in pin-pointing U_{cr}.

The behaviour of the larger diameter shells (147.3 mm) was considerably more complex, as may be seen in Figure 6.15, where one may observe that several modes

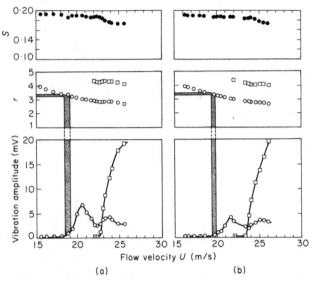

Figure 6.14. Measured vibration and wake characteristics for two nominally identical 76.2 mm diameter clamped-clamped shells for (a) Shell B and (b) Shell A: ○, second mode, $f_{2,1}$; □, third mode, $f_{3,1}$ (Païdoussis, Price & Suen 1982a).

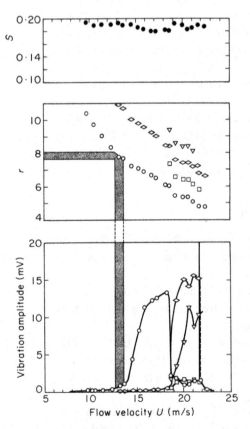

Figure 6.15. Measured vibration and wake characteristics of the 147.3 mm clamped-clamped shell G. ○, $f_{4,1}$; ◇, $f_{5,1}$; ▽, $f_{5,2}$; □, most likely $f_{3,1}$ (Païdoussis, Price & Suen 1982a).

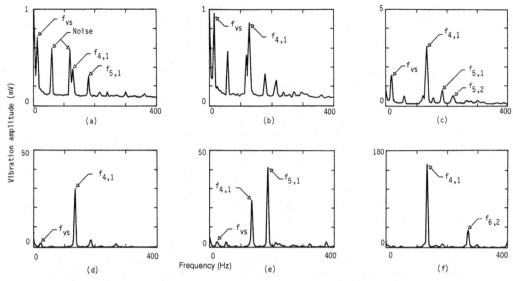

Figure 6.16. Frequency spectra of shell vibration of the clamped-clamped shell of Figure 6.15 at different flow velocities. (a) Sub-critical vibration, $U = 9.7$ m/s; (b)–(f) post-critical vibration of flow velocities: (b) $U = 13.3$ m/s, (c) $U = 15.8$ m/s, (d) $U = 17.9$ m/s, (e) $U = 19.1$ m/s, (f) $U = 21.3$ m/s (Païdoussis, Price & Suen 1982a).

are excited. Typically, the system first develops ovalling in the fourth circumferential mode, the frequency of which slightly increases with flow and amplitude of vibration ($f_{4,1} = 131$–139 Hz). At higher flow, ovalling in this mode subsides, while usually the fifth circumferential mode becomes predominant ($f_{5,1} = 188$–191 Hz), but several other modes could also be picked up, with frequencies, e.g. of 216–217 Hz ($f_{5,2}$)[*] and 166 Hz (most likely, $f_{3,1}$). Some of these lesser amplitude components of the vibration were not always excited, presumably because of the differences from test to test of the precise manner of clamping and mounting the shell, as well as variations in temperature and humidity.

The intricacy of shell behaviour with increasing flow for the results shown in Figure 6.15 may best be appreciated by examining the power-spectral-density plots, at some selected flow velocities, shown in Figure 6.16.[†] Unlike those in previous figures, these results were taken at the same setting of the Fotonic sensor, so that comparison in relative amplitude between Figures 6.16(a) to (f) is possible. Such complex response of shells in cross-wind has previously been noted by Nataraja and Johns (1977).

A number of clamped-clamped shell experiments without and with a splitter plate were conducted, similar to those with cantilevered shells. A summary of the results is given in Table 6.3. Where more than one entry appears for the same

---

[*] It may be of interest to note that, for the clamped-clamped shells, the frequencies increased with amplitude and flow – hence the frequency ranges for the different modes, as given above – indicating what appears to be a nonlinear hardening; on the other hand, the oscillation frequencies of canti-levered shells generally displayed softening nonlinear characteristics (as measured by K. M. Aaron at McGill University).

[†] The results displayed in Figures 6.15 and 6.16 were taken at different times. Therefore, the amplitudes, in mV, do not correspond. There were also some differences, albeit small, in the U_{thr} between the two experimental runs.

Table 6.3. *Threshold flow velocities for ovalling of three nominally identical 76.2-mm-diameter clamped-clamped shells without and with a splitter plate*

	Without splitter plate				With splitter plate			
	Mode $(n, m) = (2, 1)$		Mode $(n, m) = (3, 1)$		Mode $(n, m) = (2, 1)$		Mode $(n, m) = (3, 1)$	
Shell	U_{cr} (m/s)	$f_{n,m}$ (Hz)	U_{thr} (m/s)	$f_{n,m}$ (Hz)	U_{cr} (m/s)	$f_{n,m}$ (Hz)	U_{thr} (m/s)	$f_{n,m}$ (Hz)
O	17.9	155	21.4	225	20.3[a]	154	–	– [b]
O	17.3	154	21.4	227[c]	20.5	154	–	– [d]
O	17.9	154	20.8	221	20.0	155	24.5	219
B	18.4	156	20.5	235	–	–	25.7	235
B	16.8	155	21.1	235	19.3	156	–	– [e]
A	18.1	158	21.0	229	19.4	158	24.4	229

[a] Amplitude of ovalling with splitter plate is much higher than without.
[b] The $f_{3,1}$ signal is totally absent, but a frequency twice as high could be observed.
[c] A small component at 309 Hz could also be observed, but the mode involved could not be identified.
[d] The (3,1)-mode is absent but the 309-Hz frequency component is present.
[e] The 309-Hz signal could be observed throughout.

shell, this indicates a different test, in which the shell, having been dismounted, was remounted in the wind tunnel, usually on a different date. The repeatability of the results is reasonable, with the exception of those involving shell B; differences in clamping, possibly involving inadvertent axial compression of the shell while mounting it, are thought to be responsible.*

The most noteworthy feature of the results in Table 6.3 is that, as in the case of cantilevered shells, ovalling oscillation does occur even when periodic vortex shedding has been suppressed in the wake, but at slightly higher threshold flow velocities. Other important observations are that, with a splitter plate, the behaviour of the shell becomes more changeable from one test run to the other. Thus, in the first two runs with shell O, second-mode ovalling only is excited, whereas in the third run both second- and third-mode ovalling arise; similarly, for shell B, in the first run only third-mode ovalling occurs, whereas in the second only second-mode ovalling is observed. There are other, lesser differences noted in the footnotes to the table.

Finally, the possibility of synchronous, symmetric vortex shedding from the two sides of the shell was looked into, as yet another excitation mechanism via which ovalling oscillations could be induced by vortex shedding.† This type of vortex shedding has been found to occur in conjunction with in-line oscillation of cylinders in cross-flow. A special set of experiments was conducted in which two hot-wire anemometers were placed close to the separation points on either side of the shell. The phase difference in cross-spectral-density analysis of the two signals was found to be close to 180°, both before and after the onset of ovalling, indicating that vortex shedding is of the normal, antisymmetric type. When a splitter plate was used, no periodicity whatever could be found anywhere in the wake, except with the hot-wire

---

* It should also be noted that, because the wind tunnel used was a suction tunnel, as the flow increased there was a tendency for the steel-reinforced plywood wall of the tunnel to be "sucked in" slightly; although the mounting arrangement was designed to minimise this effect, this problem may not have been dealt with entirely successfully.

† This possibility was suggested by an anonymous referee to the paper by Païdoussis & Wong (1982).

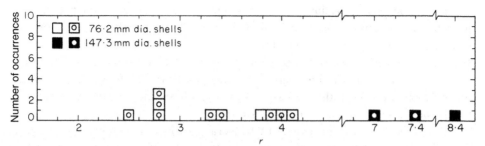

Figure 6.17. Histogram of $f_{n,m}/f_{vs}$ at the threshold of various types of ovalling oscillation of clamped-clamped shells. The symbols with an "o" in the middle are associated with the critical flow velocities for onset of ovalling (Païdoussis, Price & Suen 1982a).

sensor very close to the shell, when the perturbation to the flow generated by shell vibration could be picked up, at frequency $f_{n,m}$.

Additional tests were conducted in which damping was added to the shell by painting its inner surface with a damping compound called "Aquaplas", sometimes used to damp vibrations and sound emission from thin-walled ducts conveying air-flow. The procedure was to apply a certain maximum coating of the material, allow it to dry, and then conduct the usual tests in the wind tunnel. Then a certain amount of the material was washed off – the compound being a water-soluble suspension – and the tests repeated with progressively less and less of the material. In general, with more Aquaplas the critical flow velocity was higher, thus supporting the nascent belief that ovalling may be a flutter phenomenon. For that reason, these results will be discussed in Section 6.6.

Collecting all available ovalling data, as was done for cantilevered shells (Figure 6.13), a histogram of $r = f_{n,m}/f_{vs}$ was constructed for clamped-clamped shells, in Figure 6.17. it is seen that, even more so than for cantilevered shells, r is not always an integer; indeed, for all cases except two, where $r = 4$ and 7, respectively, r is not an integer.

6.5 Counterattack by the Vortex-Shedding Proponents and Rebuttal

Panesar & Johns (1985) countered the findings described in Section 6.4 with a two-pronged attack, as described in what follows.

6.5.1 The "peak of resonance" argument

Panesar and Johns modified the original vortex-shedding hypothesis, by asserting that the phenomenon is one of subharmonic *resonance* (Panesar & Johns 1985); hence, r = integer should apply not at the onset (as had hitherto been contended) but at the *peak* of ovalling oscillation – which could not be reached with the maximum available U in their experiments, but sometimes was in those of Païdoussis *et al.* (1982a). Of course, whether this "peak" corresponds to resonance or simply to a limit cycle is another question.

As may be seen in Figure 6.14, for instance, this would alter r from ∼3.3 to 3, which goes along with Panesar and Johns' argument. However, in cases where

$r \simeq$ integer at onset, the integral relationship is destroyed at resonance, i.e. $r \neq$ integer "at resonance"; e.g. in the case of Figures 6.12(a) and (b). Furthermore, for some cases, as in Figures 6.12(c) and (d), $r \neq$ integer, either at resonance or at onset. Indeed, applying this argument collectively to the results of Figures 6.13 and 6.17 amounts to shifting, in varying degrees, the data points to the left, which, if anything, makes matters worse for the vortex-shedding hypothesis: more ovalling occurrences would then not satisfy the $r =$ integer criterion.

Further evidence was provided by Katsura (1985) from experiments with polyester film shells, where ovalling occurred in high circumferential modes ($n = 5$ to 8). Although f_{vs} was unfortunately not measured, Katsura assumed $0.16 < S < 0.20$ and found that at the ovalling peak, for subharmonic "resonance" to exist, $r \simeq 8$ to 13 would have to be imagined! Moreover, from measurements of the fluctuating pressure on the shell, it was found that the frequency component corresponding to vortex shedding is "negligibly small". Hence, Katsura concluded that "it is hard to think that this small component of vortex shedding induces a resonant vibration at a frequency ratio of more than seven"; quite so!

6.5.2 Have splitter plates been ineffectual?

One of the strongest pieces of evidence against the vortex-shedding hypothesis is that ovalling can occur even when periodic vortex shedding is absent. The question of whether vortex shedding is really suppressed by a splitter plate had already been raised by Johns (1979). Now, Panesar and Johns (1985) provided surprising evidence that, even with a long splitter plate ($L_{sp}/D = 5.36$, $L_{sp}/h_{sp} = 240$, $g_{sp}/D \geq 0.027$), periodic vortex shedding actually occurred in their experiments! It was implicitly suggested that this failure of the plate to suppress vortex shedding was related to the small gap, g_{sp}, between the shell and the splitter plate – necessary to allow oscillation.

These results contradict those obtained previously (Païdoussis & Helleur 1979; Païdoussis, Price & Suen 1982; Ang 1981; Païdoussis, Price & Ang 1988a). In these authors' experiments the splitter plate was slightly shorter, but sturdier ($L_{sp}/D = 4.67$, $L_{sp}/h_{sp} = 55.6$, $g_{sp}/D = 0.032$).

Careful experiments were done (Païdoussis, Price & Ang 1988a) to test this question, with the hot wire in all the positions shown in Figure 6.18(a), as well as around the shell circumference, both with and without the splitter plate. However, no periodicity was found anywhere when the splitter plate was in place – for $g_{sp}/D = 0.032$, and also 0.083 and 0.167. Typical results are shown in Figures 6.18(b, c) without and with the splitter plate, respectively; the only visible peak in Figure 6.18(c) corresponds to 60-Hz electrical noise, its height exaggerated because of the changed ordinate scale, while the f_{vs} peak of Figure 6.18(b) is totally inconspicuous.

The authors also conducted experiments with a thinner splitter plate ($L_{sp}/h_{sp} = 224$) and otherwise the same dimensions. Interestingly, in some cases it was found that a weak periodic vortex-shedding signal could be picked up. However, it was discovered that this was due to sympathetic vibration of the splitter plate! Once the plate was stiffened, periodic vortex shedding disappeared. Evidently, a sufficiently flexible splitter plate, in conjunction with a small gap, cannot totally isolate the two halves of the wake, allowing periodic vortex shedding to be established.

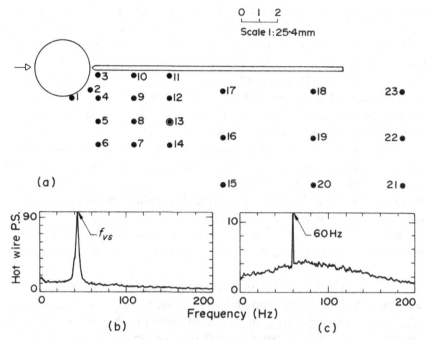

Figure 6.18. (a) Hot-wire locations to detect periodicity in the wake; (b) power spectrum of a hot-wire measurement of velocity fluctuations at location 13 without splitter plate; (c) same location and U, with splitter plate (Païdoussis, Price & Ang 1988a).

Now, Panesar and Johns' splitter plate was even more compliant (longer and slightly thinner). Hence, the foregoing offers the most likely explanation to their paradoxial observations. Additional contributing factors could be that the experiments were conducted in an open-section wind tunnel, where a part of the shell is "unprotected" by the splitter place, but still subject to some flow, albeit smaller than U, as well as being subject to flow over the open top, making the flow very three-dimensional. In contrast, in the Païdoussis, Price & Ang experiments with clamped-clamped shells, both the shells and the splitter plate spanned the wind-tunnel working section; for cantilevered shells, an edge plate near the free top was used in most cases to reduce three-dimensional effects.

In conclusion, a properly designed, stiff splitter plate, even when placed g_{sp} downstream of the shell, is capable of wholly suppressing periodic vortex shedding – but, as stated previously, ovalling occurs just the same.

The authors also investigated the possibility of acoustic effects destroying or altering any $r =$ integer relationship; but this was also negative.

6.5.3 Dénouement

Summarizing all of the foregoing, it may be stated that (i) the initiation of ovalling can be very abrupt, i.e. a small increment in U results in a steep rise in amplitude (Païdoussis et al. 1982a; Katsura 1985); (ii) if the amplitude in a particular mode is eventually decreased, giving rise to a resonance-like hump in the amplitude versus U curve, this hump can be uncharacteristically broad for this to be a true resonance in structures of such low internal damping – unless there were lock-in; (iii) there is no lock-in. Even if one still imagined this to be a resonance phenomenon, it would have to be a positive act of faith to accept that it is a subharmonic resonance with r anywhere between 2 and 13! In any case, r is *not necessarily* an integer either at onset or at "peak" amplitude of ovalling. Finally, the phenomenon can also occur in the total absence of the purported driving mechanism.

At this point, circa 1988, it may be said that the evidence against vortex shedding being the underlying case of ovalling was overwhelming.

The only question that remains to be answered is why at the onset of ovalling $r \simeq$ integer so frequently. Two plausible explanations may be offered, and both mechanisms involved are probably operative; they are described below.

The first proceeds as follows. Suppose that the ovalling threshold for a particular system is at $U = U_1$. Suppose also that at $U_1' = U_1 - \epsilon$, where ϵ is small, there should exist a relationship $r f_{vs} = f_{n,m}$ in the mode concerned. Then, it is possible that ovalling is triggered at U_1' by subharmonic reinforcement of the mechanism causing ovalling.

The second hypothesis suggests that ovalling occurs at its "natural" U_{thr} by some mechanism, but its onset may entrain f_{vs}, if sufficiently close to $f_{n,m}/r$, to yield $f_{n,m}/f_{vs} =$ integer by organizing the wake subharmonically; some evidence, albeit not conclusive, exists for this explanation (Païdoussis, Price & Suen 1982a).

In any case, an alternative explanation, based on another and more plausible mechanism for ovalling, has been available as of 1982, as described in Section 6.6.

6.6 Simple Aeroelastic-Flutter Model

As an alternative to the vortex-shedding hypothesis, an aeroelastic-flutter model was proposed by Païdoussis & Wong (1982). More refined versions of the model are presented in Sections 6.7 and 6.9. Here, the simplest form of this model is developed, so that the reader can easily grasp the essentials.

In this analysis, the shell is initially idealised to be infinitely long, so that the problem may be treated as two-dimensional. The cross-flow is further simplified by regarding the entire field as quasi-irrotational. The wake is separated from the outer flow by a dividing streamline. Within the wake, the von Kármán vortex street is totally ignored, giving a zero flow velocity in this region. Hence, the flow field is assumed to be steady, except for the disturbances arising from the vibration of the shell. Moreover, it is assumed that the time-averaged positions of separation of the mean flow from the body surface are unaffected by the small vibrations of the shell. The effect of turbulence is ignored throughout.

The shell is considered to be purely elastic, homogeneous and isotropic. To study its stability, a small-amplitude vibration in a given circumferential mode is imposed on the shell. As this study is principally concerned with the *onset* of ovalling, the vibration amplitude is taken to be sufficiently small to allow linear shell theory and linearised fluid mechanics to be utilised. In the same spirit, the boundary conditions on the shell surface will be applied at the equilibrium position.

In its essence, the analysis determines the perturbation flow field associated with shell motions, and then the resultant pressure fluctuations on the shell surface, the effect of which would indicate whether the initial vibration is attenuated or amplified.

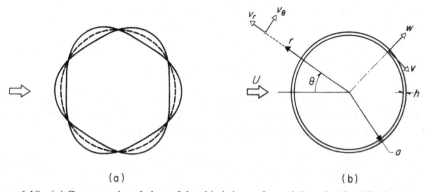

Figure 6.19. (a) Cross-sectional view of the third circumferential mode of oscillation, showing the two extreme positions of the middle surface of the shell in the course of a cycle of oscillation; (b) the system under consideration, defining various quantities used in the analytical model.

6.6.1 Equations of motion and boundary conditions

Consider a uniform, thin cylindrical shell of infinite length, mean radius a and thickness h, subjected to a uniform cross-flow of incident velocity U, as shown in Figure 6.19(b); the shell is filled with stationary fluid of the same density as the outer flow. Because the problem is two-dimensional, only planar displacements of the middle surface of the shell, in the plane of Figure 6.19, are considered: $v(\theta, t)$ and $w(\theta, t)$, in the circumferential and radial directions, respectively, measured from the equilibrium (circular) configuration.

In this case, the equations of motion of a shell of density ρ_s, Young's modulus E, and Poisson ratio v, according to Flügge (1957) reduce to

$$\frac{\partial^2 v}{\partial \theta^2} + \frac{\partial w}{\partial \theta} = \gamma \frac{\partial^2 v}{\partial t^2}, \tag{6.2}$$

$$\frac{\partial v}{\partial \theta} + w + \kappa \left\{ \frac{\partial^4 w}{\partial \theta^4} + 2\frac{\partial^2 w}{\partial \theta^2} + w \right\} = -\gamma \left\{ \frac{\partial^2 w}{\partial t^2} - \frac{q_r}{\rho_s h} \right\}, \tag{6.3}$$

where $\kappa = h^2/12a^2$, $\gamma = \rho_s a^2 (1 - v^2)/E$ and $q_r = p_i - p_e$, with p_i and p_e being, respectively, the internal and external pressure on the shell surface.

These equations being linear, the pressure difference q_r may be separated into two components: $q_{r_0}(\theta)$, which is due to the static loading of the mean flow, and $q_r^*(\theta, t)$, which is due to perturbations associated with shell deformations characterised by $v^*(\theta, t)$ and $w^*(\theta, t)$. Considering deformations associated with $q_{r_0}(\theta)$ to be negligible,[*] (6.2) and (6.3) may be viewed as relationships among q_r^*, v^* and w^*, from which the static components have been filtered out.

Since the flow field is assumed to be irrotational outside the wake, a velocity potential $\Phi(r, \theta, t)$ may be defined by

$$v_r = \frac{\partial \Phi}{\partial r}, \quad v_\theta = \frac{1}{r}\frac{\partial \Phi}{\partial \theta},$$

---

[*] For the shells used in the experiments, this may be considered to be supported by experimental evidence: the natural frequencies of oscillation of the shell with and without flow are essentially the same.

where $v_r(r, \theta, t)$ and $v_\theta(r, \theta, t)$ are, respectively, the radial and tangential velocities at position (r, θ). Because Φ must satisfy the Laplace equation, which is linear, one may further let

$$\Phi(r, \theta, t) = \phi_0(r, \theta) + \phi^*(r, \theta, t),$$

where $\phi_0(r, \theta)$ is associated with the mean flow, and $\phi^*(r, \theta, t)$ with the perturbation flow field due to oscillations of the shell. Once $\Phi(r, \theta, t)$ is known, the pressure may be determined from Bernoulli's equation for unsteady flow,

$$\frac{\partial \Phi}{\partial t} + \frac{1}{2} V^2 + \frac{p}{\rho} = 0, \qquad (6.4)$$

where p is the static pressure measured relative to the stagnation pressure of the free stream, ρ is the fluid density, and $V^2 = v_r^2 + v_\theta^2$; the $F(t)$ term normally appearing on the right-hand side of (6.4) may be suppressed for incompressible flow (Lamb 1957) – or alternatively may be considered to have been absorbed in $\partial \Phi / \partial t$.

It is noted that ϕ_0 is associated entirely with the external flow, and may be defined in terms of the radial and tangential velocities on the surface of a stationary cylinder, i.e.

$$\left. \frac{\partial \phi_0}{\partial r} \right|_{r=a} = 0, \quad \left. \frac{1}{a} \frac{\partial \phi_0}{\partial \theta} \right|_{r=a} = U f(\theta), \qquad (6.5)$$

where $f(\theta)$ may be determined empirically, or by semi-empirical methods such as that developed by Parkinson & Jandali (1969). Because all ovalling experiences have occurred in the Reynolds-number range $\text{Re} = 10^4$ to 10^6, the boundary-layer thickness on the shell prior to separation is sufficiently thin for the pressure change across it to be negligible. The local steady-state velocity just outside the boundary layer may then be related to the surface static pressure p by Bernoulli's equation for the steady flow, so that

$$f(\theta) \equiv \left\{ v_\theta \big|_{r=a} / U \right\} = (1 - C_p)^{\frac{1}{2}} \quad \text{for} \quad |\theta| \le \beta_s, \qquad (6.6)$$

where C_p is the pressure coefficient defined by

$$C_p = (p - p_\infty) / \tfrac{1}{2} \rho U^2,$$

in terms of the free-stream static pressure p_∞; β_s is the angular position of the point of separation. In the wake, the surface flow may be considered to be negligible, so that

$$f(\theta) = 0 \quad \text{for} \quad \pi \ge |\theta| > \beta_s. \qquad (6.7)$$

Here, $f(\theta)$ was determined empirically in terms of the measured C_p for a rigid cylinder in cross-flow. For example, using Roshko's (1954) data for $\text{Re} = 1.45 \times 10^4$ and employing a least-squares-fit technique, $f(\theta)$ may be approximated by the fourth-order polynomials

$$f(\theta) = 1.6073|\theta| + 0.5700|\theta|^2 - 0.9394|\theta|^3 + 0.1714|\theta|^4$$

$$\text{for} \quad |\theta| \le 1.484 \text{ rad } (85^o), \qquad (6.8)$$

$$f(\theta) = 1.5137|\theta| + 0.5128|\theta|^2 - 0.9418|\theta|^3 + 0.1733|\theta|^4$$

$$\text{for} \quad |\theta| \le 1.396 \text{ rad } (80^o), \qquad (6.9)$$

for a cylinder without and with a splitter plate, respectively. The flow is symmetric relative to the stagnation point ($\theta = 0$).

Turning next to the perturbation velocity potential ϕ^*, it is noted that it has components associated with both internal and external fluids, ϕ_i^* and ϕ_e^* respectively. In view of (6.4), they give rise to pressure fluctuations

$$p_i^* = -\rho \frac{\partial \phi_i^*}{\partial t}\bigg|_{r=a} \quad \text{and} \quad p_e^* = -\rho \left\{ \frac{\partial \phi_e^*}{\partial t} + \frac{Uf(\theta)}{a} \frac{\partial \phi_e^*}{\partial \theta} \right\}\bigg|_{r=a}, \quad (6.10)$$

correct to first order of small quantities, having introduced the approximation $V^2 \simeq U^2 f^2(\theta) + 2Uf(\theta)a^{-1}(\partial \phi_e^*/\partial \theta)|_{r=a}$.*

Finally, ϕ_i^* and ϕ_e^* may be related to the shell deformation. For a fluid particle on the outer surface of the shell, the radial velocity is given by

$$\frac{\partial \phi_e^*}{\partial r}\bigg|_{r=a} = \frac{Dw^*}{Dt} \simeq \frac{\partial w^*}{\partial t} + \frac{1}{a} v_\theta \bigg|_{r=a} \frac{\partial w^*}{\partial \theta} ;$$

assuming that $Uf(\theta) \gg (\partial \phi_e^*/a\partial \theta)|_{r=a}$, this may be simplified further to

$$\frac{\partial \phi_e^*}{\partial r}\bigg|_{r=a} = \frac{\partial w^*}{\partial t} + \frac{Uf(\theta)}{a} \frac{\partial w^*}{\partial \theta}. \quad (6.11)$$

Similarly, for the internal fluid

$$\frac{\partial \phi_i^*}{\partial r}\bigg|_{r=a} = \frac{\partial w^*}{\partial t}. \quad (6.12)$$

6.6.2 Solution of the equations

For harmonic shell motions, it may be assumed that

$$\phi^* = R(r) \, T(\theta) e^{i\omega t},$$

which substituted into the Laplace equation, in cylindrical coordinates, gives

$$\frac{r^2}{R} \frac{d^2 R}{dr^2} + \frac{r}{R} \frac{dR}{dr} = -\frac{1}{T} \frac{d^2 T}{\partial \theta^2} = \lambda^2,$$

where λ^2 has to be a positive integer in order to satisfy the condition $\phi^*(\theta) = \phi^*(\theta + 2\pi)$. Hence

$$T(\theta) = A_1 \cos \lambda\theta + A_2 \sin \lambda\theta,$$

for some constants A_1 and A_2, to be determined.

So far, all observations on the ovalling process have revealed that, independently of the circumferential mode excited, either a node or an antinode faces the free stream. Therefore the analysis will be restricted to conform to either of these two

---

* Of course, it is realised that the mean-flow component of (6.4), after (6.6) and (6.7) have been substituted in it, is incorrect, because it corresponds to a mean pressure distribution such that the pressure in the wake is effectively equal to the stagnation pressure. This relation, however, is not used (nor is it useful) in this analysis, the assumption having been made that the steady pressure field induces no appreciable deformation of, or stresses in, the shell.

observed conditions. For the type of *oscillation with an antinode facing the free stream*, the symmetry of the flow field would render

$$v_\theta \bigg|_{\theta=0} = U f(0) + \frac{1}{a} \frac{\partial \phi_e^*}{\partial \theta} \bigg|_{r=a, \theta=0} = 0,$$

and, in view of the symmetry of the steady cross-flow, $f(0) = 0$, leading to

$$(\partial \phi_e^* / \partial \theta) \big|_{r=a, \theta=0} = 0 .$$

This clearly implies that $T(\theta) = A_1 \cos \lambda \theta$ for such configurations. The cases with a node facing the free stream may be tackled in very similar fashion (Païdoussis & Wong 1982).

Now, it may easily be seen that

$$R(r) = D r^{-\lambda} + E r^\lambda,$$

for some constants D and E. Physically, one would expect $\phi_e^* \to 0$ as $r \to \infty$ and ϕ_i^* to be finite at $r = 0$. Hence, the perturbation potential, external and internal to the shell, must take the form

$$\phi_e^* = e^{i\omega t} \sum_{\lambda=0}^{\infty} D(\lambda) r^{-\lambda} \cos \lambda \theta, \quad \phi_i^* = e^{i\omega t} \sum_{\lambda=0}^{\infty} E(\lambda) r^{-\lambda} \cos \lambda \theta. \tag{6.13}$$

To proceed with determining $D(\lambda)$ and $E(\lambda)$, w^* is expanded in series form

$$w^* = e^{i\omega t} \sum_{l=0}^{\infty} B_l \cos(2l + 1) n\theta, \tag{6.14}$$

so chosen that the number of nodal points for the nth mode be $2n$, as required. Consequently, v^* must have the similar form*

$$v^* = e^{i\omega t} \sum_{l=0}^{\infty} A_l \sin(2l + 1) n\theta. \tag{6.15}$$

Now, substitution of (6.14) and (6.13) into (6.11) yields

$$\sum_{l=1}^{\infty} \lambda D(\lambda) a^{-\lambda-1} \cos \lambda \theta = \frac{Un}{a} f(\theta) \sum_{l=0}^{\infty} B_l (2l + 1) \sin(2l + 1) n\theta - i\omega \sum_{l=0}^{\infty} B_l \cos(2l + 1) n\theta.$$

For a particular $\lambda = \lambda_j$, one may solve for $D(\lambda_j)$ by multiplying both sides by $\cos \lambda_j \theta$ and integrating over θ from 0 to π, whereby it may be shown that

$$D(\lambda) = \frac{-2}{\pi} \frac{a^{\lambda+1}}{\lambda} \left\{ \frac{1}{2} i\omega\pi \sum_{l=0}^{\infty} B_l \, \delta_{\lambda b} - \frac{U}{a} \sum_{l=0}^{\infty} B_l b \int_0^\pi \sin b\theta \cos \lambda \theta f(\theta) d\theta \right\}, \tag{6.16}$$

for $\lambda = 1, 2, 3, \ldots$, where $b = (2l + 1)n$ and $\delta_{\lambda b}$ is the Kronecker delta. $D(0)$ is arbitrary, because only the derivative of ϕ_e^* is specified at the boundary.

By similar means one may obtain

$$E(\lambda) = \frac{i\omega}{\lambda} a^{1-\lambda} B_l \, \delta_{\lambda b}, \tag{6.17}$$

for $\lambda = 1, 2, 3, \ldots$.

---

* For a node facing the free stream ϕ_i^* and ϕ_e^* in (6.13) involve $\sin \lambda \theta$ instead of $\cos \lambda \theta$, and $\cos(2l + 1)n\theta$ is replaced by $\sin(2l + 1)n\theta$ in (6.14) and *vice versa* in (6.15).

From (6.13) and (6.16) it is obvious that p_e^* will be specified in terms of a double series, and some simplification is desirable before proceeding further with the analysis. Fortunately, the modal shapes are adequately described by the leading terms in the series of (6.14) and (6.15), suggesting

$$w^* = \mathrm{e}^{\mathrm{i}\omega t} B_0 \cos n\theta, \quad v^* = \mathrm{e}^{\mathrm{i}\omega t} A_0 \sin n\theta; \tag{6.18}$$

the validity of this approximation was tested *a posteriori* by repeating the analysis with a three-term approximation; the difference in the results was found to be negligible. Then, letting $p^* = P^* \mathrm{e}^{\mathrm{i}\omega t}$, one obtains

$$P_e^* = \rho B_0 \left\{ \frac{2}{\pi} \frac{U^2 n}{a} f(\theta) \sum_{\lambda=1}^{\infty} F_1(\lambda) \sin \lambda\theta - U \mathrm{i}\omega f(\theta) \sin n\theta \right.$$

$$\left. - \frac{2}{\pi} \mathrm{i}\omega \, Un \sum_{\lambda=1}^{\infty} \frac{F_1(\lambda)}{\lambda} \cos \lambda\theta - \frac{\omega^2 a}{n} \cos n\theta \right\} - \rho \mathrm{i}\omega \, D(0), \tag{6.19}$$

$$P_i^* = \rho B_0 \omega^2 \frac{a}{n} \cos n\theta - \rho \mathrm{i}\omega \, E(0), \tag{6.20}$$

where $F_1(\lambda) = \int_0^\pi f(\xi) \sin n\xi \cos \lambda\xi \, \mathrm{d}\xi$.

Finally, substituting (6.18) to (6.20) into (6.2) and (6.3) yields two equations of the form

$$a_{11} A_0 + a_{12} B_0 = 0, \quad a_{21} A_0 + a_{22} B_0 = 0,$$

where

$$a_{11} = n^2 - \gamma\omega^2, \quad a_{12} = n, \quad a_{21} = \tfrac{1}{2}n\pi,$$

$$a_{22} = \tfrac{1}{2}\pi\{1 + \kappa(n^2 - 1)^2 - \gamma\omega^2\}$$

$$+ \frac{\gamma}{\rho_s h} \left\{ -\rho \frac{a\pi}{n} \omega^2 - 2\rho U F_1(n) \mathrm{i}\omega + \rho U^2 \frac{2n}{\pi a} \sum_{\lambda=1}^{\infty} F_1(\lambda) F_2(\lambda) \right\},$$

and $F_2(\lambda) = \int_0^\pi f(\xi) \sin \lambda\xi \cos n\xi \, \mathrm{d}\xi$. For nontrivial solutions of A_0 and B_0, the associated determinant must vanish, giving the characteristic equation for the frequency ω – which will generally be complex.

The foregoing analysis applies to shell deformations such that an antinode faces the free stream. Using a similar approach, the case of a node facing the stream has also been investigated; the reader is referred to Païdoussis & Wong (1982).

6.6.3 Theoretical results and comparison with experiment

Calculations were conducted for an infinitely long shell of radius, thickness and material properties identical with one of the shells tested in the experiments to be discussed next, namely, $a = 38.1$ mm, $h = 0.51$ mm, $E = 0.28 \times 10^{10}$ N/m^2, $\rho_s = 1.29 \times 10^3$ kg/m^3 and $\nu = 0.4$. The complex frequencies $f (\equiv \omega/2\pi)$ (in Hz) of the second and third circumferential modes are shown in Figure 6.20 in the form of an

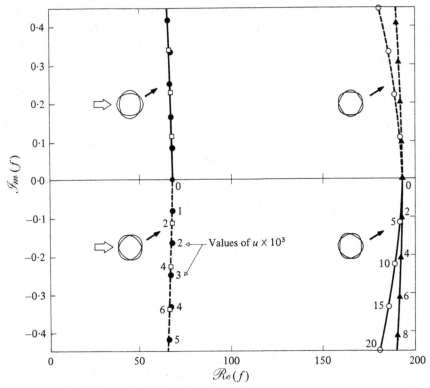

Figure 6.20. Argand diagram of the complex frequencies f of the second and third modes of an infinitely long shell in cross-flow, as functions of the dimensionless flow velocity u. ●, 2nd mode, no splitter plate; □, 2nd mode, with splitter plate; ▲, 3rd mode, no splitter plate; ○, 3rd mode, with splitter plate (Païdoussis & Wong 1982).

Argand diagram, for increasing dimensionless flow velocity u, defined by

$$u = U \left\{ \rho_s (1 - \nu^2)/E \right\}^{\frac{1}{2}}. \tag{6.21}$$

As internal dissipation (material damping) has not been included in the theoretical model, the frequencies in Figure 6.20 at $u = 0$ are wholly real. For $u > 0$, however, it is seen that second-mode oscillation ($n = 2$) with a node facing the stream is associated with a *negative* aerodynamic damping, i.e. $\mathcal{I}m(f) < 0$; if, on the other hand, an antinode faces the stream, the motion is aerodynamically positively damped. The opposite is true for third-mode ($n = 3$) oscillations. It is also seen that similar results are obtained in the presense of a splitter plate, except that in this case $|\mathcal{I}m(f)|$ is generally smaller.

These results may be interpreted as follows. With increasing flow velocity the vibration frequencies remain almost constant, as the values of $\mathcal{R}e(f)$ for $u > 0$ are only slightly lower than the corresponding ones for $u = 0$. The aerodynamic damping, however, conveniently characterised by the logarithmic decrement $\delta_{ad} = 2\pi\mathcal{I}m(f)/\mathcal{R}e(f)$ decreases almost linearly with u, as shown in Figure 6.21. For sufficiently high flow velocity, the negative aerodynamic damping associated with each of the two modes will eventually exceed the corresponding dissipative-modal-damping logarithmic decrement, δ_{md}; at that point the net energy transfer from the

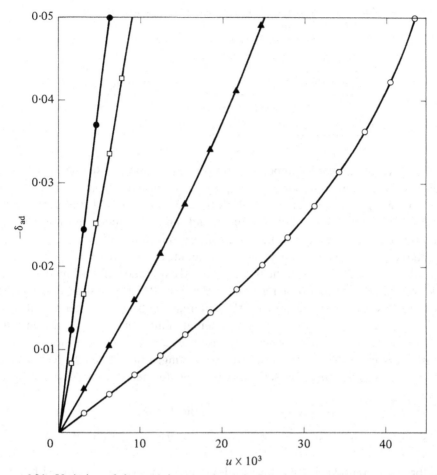

Figure 6.21. Variation of the negative aerodynamic damping $-\delta_{ad}$ with dimensionless flow velocity u for the $n = 2$ and $n = 3$ modes of an infinitely long shell in cross-flow. Symbols are as in Figure 6.20 (Païdoussis & Wong 1982).

fluid to the shell exceeds the energy lost through dissipation, and oscillations will be amplified, i.e. this is the threshold of instability in the mode concerned.

Thus, the mechanism of instability is recognised as being that of single-degree-of-freedom flutter, rather than the so-called classical two-degree, or coupled-mode flutter.

As an illustration of how Figure 6.21 may be utilised, consider the case of a shell without a splitter plate and with modal damping $\delta_{md} = 0.02$ for both $n = 2$ and $n = 3$. Then, as the threshold of flutter occurs when $\delta_{md} = -\delta_{ad}$, with the aid of Figure 6.21 one obtains the critical flow velocities associated with second- and third-mode flutter, namely, $u_c^{(2)} = 2.50 \times 10^{-3}$ and $u_c^{(3)} = 11.75 \times 10^{-3}$.

These theoretical results were compared with some experimental ones for clamped-clamped shells from Suen (1981), similar to those described in Section 6.4.2. It is obvious that measurements of δ_{md} are necessary to be able to compare with $-\delta_{ad}$. The manner in which these measurements were made is described in Appendix B.

Table 6.4. *The natural frequencies and modal damping of the shell tested. The natural frequencies given here are those associated with the first axial mode of the shell*

Circumferential mode number n	Natural frequency $\mathcal{R}e(f)$ [Hz]	Damping logarithmic decrement δ_{md}
2	155	0.031
3	225	0.034

For a typical clamped-clamped shell and ovalling behaviour similar to that shown in Figure 6.14, the measured modal damping and natural frequency in the $n = 2$ and $n = 3$ nodes are given in Table 6.4. With the aid of Figure 6.21, the experimental critical flow velocities $u_c^{(2)}$ and $u_c^{(3)}$ for second- and third-mode ovalling could be determined. Theory and experiment are compared in Table 6.5.

Agreement between theory and experiment, at least for $n = 2$, superficially appears to be good. However, the theoretical and experimental third-mode frequencies, $\mathcal{R}e(f)$, are in gross disagreement: 193 versus 225 Hz, casting doubt on the above results. This of course reflects the difference between the theoretical (infinite length, two-dimensional) and experimental (finite length, three-dimensional) systems. Moreover, a discrepancy was discovered between dimensional and dimensionless quantities at the time of this writing; thus, although the results in Table 6.5 are believed to be self-consistent, no further discussion will be given here.

In summary, the results obtained so far indicate that:

(i) shell flutter is possible in the absence of any periodicity in the wake of the shell;
(ii) for the two modes involved, second-mode flutter should develop with a node facing upstream, whereas third-mode flutter should develop with an antinode facing upstream;
(iii) the critical flow velocities, $u_c^{(2)}$ and $u_c^{(3)}$, are raised somewhat by the presence of a splitter plate in the wake.

6.7 A Three-Dimensional Flutter Model

An outline of a three-dimensional aeroelastic flutter model is given here, mainly differing from that in Section 6.6 in that, here, a finite-length shell is considered, with clamped ends. Details may be found in Païdoussis, Price & Suen (1982b).

Table 6.5. *Comparison of the negative aerodynamic damping $-\delta_{md}$ with modal damping δ_{md} at the measured critical-flow velocities*

	n	u_c	δ_{md} (expt)	$-\delta_{ad}$ (theory)
Without splitter plate	2	5.5×10^{-3}	0.031	0.044
	3	6.6×10^{-3}	0.034	0.011
With splitter plate	2	6.3×10^{-3}	0.031	0.033
	3	7.6×10^{-3}	0.034	0.005

6.7.1 The model and methods of solution

Flügge's (1957) equations of motion for a thin circular cylindrical shell are used,

$$a^2 \frac{\partial^2 u}{\partial x^2} + \frac{1-\nu}{2} \frac{\partial^2 u}{\partial \theta^2} + a \frac{1+\nu}{2} \frac{\partial^2 v}{\partial x \partial \theta} + a \nu \frac{\partial w}{\partial x}$$

$$+ \kappa \left\{ \frac{1-\nu}{2} \frac{\partial^2 u}{\partial \theta^2} - a^3 \frac{\partial^3 w}{\partial x^3} + \frac{1-\nu}{2} a \frac{\partial^3 w}{\partial x \partial \theta^2} \right\} = \gamma \frac{\partial^2 u}{\partial t^2},$$

$$a \frac{1+\nu}{2} \frac{\partial^2 u}{\partial x \partial \theta} + \frac{\partial^2 v}{\partial \theta^2} + a^2 \frac{1-\nu}{2} \frac{\partial^2 v}{\partial x^2} + \frac{\partial w}{\partial \theta}$$

$$+ \kappa \left\{ \frac{3}{2} a^2 (1-\nu) \frac{\partial^2 v}{\partial x^2} - \frac{3-\nu}{2} a^2 \frac{\partial^3 w}{\partial x^2 \partial \theta} \right\} = \gamma \frac{\partial^2 v}{\partial t^2},$$

$$a\nu \frac{\partial u}{\partial x} + \frac{\partial v}{\partial \theta} + w + \kappa \left\{ a \frac{1-\nu}{2} \frac{\partial^3 u}{\partial x \partial \theta^2} - a^3 \frac{\partial^3 u}{\partial x^3} - \frac{3-\nu}{2} a^2 \frac{\partial^3 v}{\partial x^2 \partial \theta} \right. \tag{6.22}$$

$$\left. + a^4 \frac{\partial^4 w}{\partial x^4} + 2a^2 \frac{\partial^4 w}{\partial x^2 \partial \theta^2} + \frac{\partial^4 w}{\partial \theta^4} + 2 \frac{\partial^2 w}{\partial \theta^2} + w \right\} = -\gamma \left\{ \frac{\partial^2 w}{\partial t^2} - \frac{q_r}{\rho_s h} \right\},$$

where

$$q_r = p_i - p_e; \tag{6.23}$$

u, v and w are the displacements of the middle surface of the shell, respectively, in the axial, circumferential and radial directions.

It is assumed that shell deformation in the course of vibration is small, and variations along x are sufficiently gradual, for the induced flow in the x-direction to be negligible. This allows the aerodynamic forces at each location x to be determined by strip theory. Hence, because it has been assumed that the flow field is irrotational outside the wake, a two-dimensional velocity potential may be defined, such that

$$v_r = \partial \Phi / \partial r, \quad v_\theta = (1/r)(\partial \Phi / \partial \theta), \quad v_x \simeq 0, \tag{6.24}$$

where $v_r(x, r, \theta, t)$, $v_\theta(x, r, \theta, t)$ and $v_x(x, r, \theta, t)$ are, respectively, the radial-tangential- and axial-flow velocities at each position (x, r, θ). Thus, the flow field is effectively modelled as being two-dimensional; hence, that part of the analysis given in Section 6.6 may be used here also. Specifically, equations (6.4) to (6.12) apply here too. Therefore, proceeding as in Section 6.6, the velocity potential, which must satisfy the Laplace equation, has internal and external components of the form

$$\phi_e^* = e^{i\omega t} \sum_{\lambda=0}^{\infty} D(\lambda) r^{-\lambda} \cos \lambda \theta, \quad \phi_i^* = e^{i\omega t} \sum_{\lambda=0}^{\infty} E(\lambda) r^{\lambda} \cos \lambda \theta, \tag{6.25}$$

where the oscillations are of radian frequency ω and symmetrically disposed about the $(\theta = 0, \theta = \pi)$-plane; i.e. for shell vibrations with an antinode facing the free stream.

The shell displacements may be expressed as follows:

$$u^* = e^{i\omega t} \sum_{n=0}^{\infty} \sum_{m=1}^{\infty} A_{nm} \cos n\theta \, \phi'_m, \quad v^* = e^{i\omega t} \sum_{n=0}^{\infty} \sum_{m=1}^{\infty} B_{nm} \sin n\theta \, \phi_m,$$

$$w^* = e^{i\omega t} \sum_{n=0}^{\infty} \sum_{m=1}^{\infty} C_{nm} \cos n\theta \, \phi_m, \tag{6.26}$$

where $\phi_m(x)$ are comparison functions for the axial mode shape of the shell; the prime denotes differentiation with respect to x. Applying now equation (6.11), in conjunction with (6.26), gives

$$-\sum_{\lambda=0}^{\infty} \lambda D(\lambda) a^{-\lambda-1} \cos \lambda\theta = \sum_{n=0}^{\infty} \sum_{m=1}^{\infty} C_{nm} \{i\omega \cos n\theta - (Un/a)f(\theta)\sin n\theta\}\phi_m.$$

Multiplying both sides by $\cos j\theta$ and integrating from 0 to π yields

$$D(j) = \left(\frac{a^{j+1}}{j}\right) \sum_{m=1}^{\infty} \left\{ -i\omega \, C_{jm} + \frac{2}{\pi}\frac{U}{a} \sum_{n=0}^{\infty} nF(n,j)C_{nm} \right\} \phi_m, \tag{6.27}$$

for $j = 1, 2, 3, \ldots$, where

$$F(n,j) = \int_0^{\pi} f(\theta)\sin n\theta \cos j\theta \, d\theta. \tag{6.28}$$

In a similar manner, we obtain

$$E(j) = \left(\frac{a^{1-j}}{j}\right) \sum_{m=1}^{\infty} i\omega \, C_{jm} \, \phi_m. \tag{6.29}$$

As mentioned above, these expressions apply to modal configurations with an antinode facing the free stream. (It is recalled that it was found experimentally that ovalling occurs with either an antinode or a node facing the free stream.) Similar expressions for the latter case may easily be obtained; indeed, the expressions for ϕ_e^* and ϕ_i^* are identical to those of equations (6.25), except that $\sin \lambda\theta$ replaces $\cos \lambda\theta$.

With $D(j)$ and $E(j)$ determined, ϕ_e^* and ϕ_i^* are now expressed in terms of shell displacements. Substitution of equations (6.27) and (6.29) into (6.10) yields

$$p_e^* = -\rho e^{i\omega t} \sum_{j=1}^{\infty} \left\{ \frac{a\cos j\theta}{j} \sum_{m=1}^{\infty} \left\{ \omega^2 C_{jm} + \frac{2}{\pi}\frac{U\omega i}{a} \sum_{n=0}^{\infty} nF(n,j)C_{nm} \right\} \phi_m \right.$$

$$\left. - Uf(\theta)\sin j\theta \sum_{m=1}^{\infty} \left\{ -i\omega C_{jm} + \frac{2}{\pi}\frac{U}{a} \sum_{n=0}^{\infty} nF(n,j)C_{nm} \right\} \phi_m \right\} - \rho e^{i\omega t} i\omega \, D(0), \tag{6.30}$$

$$p_i^* = -\rho e^{i\omega t} \sum_{j=1}^{\infty} \left\{ \frac{a\cos j\theta}{j} \sum_{m=1}^{\infty} \{-\omega^2 C_{jm}\} \phi_m \right\} - \rho e^{i\omega t} i\omega \, E(0). \tag{6.31}$$

Hence, $q_r^* \equiv p_i^* - p_e^*$ has now been determined.

It is recalled that ϕ_e^* and ϕ_i^* have been determined by two-dimensional analysis – in the (r, θ)-plane; yet, $q_r^* = q_r^*(x, \theta, t)$ in the expression above, as $\phi_m \equiv \phi_m(x)$. Thus, strip theory has implicitly been used to obtain an approximate three-dimensional expression for q_r^*; clearly, it remains valid only so long as changes in shell deformation with x are sufficiently gradual for the induced flow in the x-direction to be negligible.

Substituting now equations (6.30) and (6.31) into (6.23), and the resulting equation together with (6.26) into (6.22) yields the following set of equations:

$$\sum_{n=0}^{\infty}\sum_{m=1}^{\infty}\left\{\left\{a^2\,\phi_m''' - \frac{1}{2}n^2(1-v)(1+\kappa)\phi_m'\right\}\cos n\theta\,A_{nm} + \left\{\frac{1}{2}an(1+v)\phi_m'\right\}\cos n\theta\,B_{nm}\right.$$

$$+ \left\{av\,\phi_m' - a^3\kappa\phi_m''' - \frac{1}{2}an^2\kappa(1-v)\phi_m'\right\}\cos n\theta\,C_{nm} + \gamma\omega^2\,\phi_m'\cos n\theta\,A_{nm}\right\} = 0,$$

$$\sum_{n=0}^{\infty}\sum_{m=1}^{\infty}\left\{\left\{-\frac{1}{2}an(1+v)\phi_m''\right\}\sin n\theta\,A_{nm} + \left\{-n^2\phi_m + \frac{1}{2}a^2(1+3\kappa)(1-v)\phi_m''\right\}\sin n\theta\,B_{nm}\right.$$

$$+ \left\{-n\phi_m + \frac{1}{2}a^2n\kappa(3-v)\phi_m''\right\}\sin n\theta\,C_{nm} + \gamma\omega^2\,\phi_m\sin n\theta\,B_{nm}\right\} = 0,$$

$$\sum_{n=0}^{\infty}\sum_{m=1}^{\infty}\left\{\left\{-av\phi_m'' + \frac{1}{2}an^2\kappa(1-v)\phi_m'' + a^3\,\kappa\,\phi_m^{\mathrm{iv}}\right\}\cos n\theta\,A_{nm}\right. \tag{6.32}$$

$$+ \left\{-n\phi_m + \frac{1}{2}a^2\,n\,\kappa(3-v)\phi_m''\right\}\cos n\theta\,B_{nm}$$

$$+ \left\{-\phi_m - a^4\,\kappa\phi_m^{\mathrm{iv}} + 2a^2n^2\,\kappa\phi_m'' - (n^2-1)^2\kappa\phi_m\right\}\cos n\theta\,C_{nm} + \gamma\,\omega^2\,\phi_m\cos n\theta\,C_{nm}\right\}$$

$$+ \frac{\rho\gamma}{\rho_s h}\sum_{j=1}^{\infty}\left\{\cos j\theta\sum_{m=1}^{\infty}\left\{\frac{2\omega^2 a}{j}C_{jm} + \frac{2U\omega\mathrm{i}}{j\pi}\sum_{n=0}^{\infty}nF(n,j)C_{nm}\right\}\phi_m\right\}$$

$$-Uf(\theta)\sin j\theta\sum_{m=1}^{\infty}\left\{-\mathrm{i}\omega\,C_{jm} + \frac{2U}{\pi a}\sum_{n=0}^{\infty}nF(n,j)C_{nm}\right\}\phi_m\right\}$$

$$+ \rho\,\omega\,\mathrm{i}\{D(0) - E(0)\} = 0.$$

The prime denotes differentiation with respect to x, and "iv" is the fourth derivative.

To solve these equations, a variant of Galerkin's method is used. The first and third equations are multiplied by $\cos p\theta$ and the second by $\sin p\theta$ and integrated from 0 to π. Then the first equation is multiplied by ϕ_k' and the other two by ϕ_k and integrated from $x = 0$ to $x = L$. The eigenfunctions of a clamped-clamped beam are used as a suitable set of comparison functions ϕ_k. Eventually one obtains three sets of algebraic equations, which may be written in matrix form as

$$(-\omega^2[M] + \mathrm{i}\omega[C] + [K])\left\{\begin{array}{c}\mathbf{A}\\\mathbf{B}\\\mathbf{C}\end{array}\right\} = \{0\}, \tag{6.33}$$

where $\mathbf{A} = \{A_{11}, A_{12}, \ldots, A_{1M}; \ldots; A_{N1}, A_{N2}, \ldots, A_{NM}\}$, with N being the limit of n and M the limit of m in the summations of equations (6.32), and similarly for

B and **C**. It should be noted that the aerodynamic forces contribute terms to all $[M]$, $[C]$ and $[K]$, and will give rise to, respectively, the virtual mass effect, and aerodynamic damping and stiffness effects. Indeed, as in the equations of motion internal dissipation in the shell material is not taken into account, $[C]$ contains only aerodynamic terms: i.e. it is the aerodynamic damping matrix.

Now, equation (6.33) may easily be transformed into a standard eigenvalue problem, from which eigenfrequencies of the system may be obtained by the usual methods.

Three types of solutions were tried, as follows: (i) fully coupled solutions to the general set of equations, as described in this section; (ii) partially simplified solutions, where circumferential coupling between modes is considered to be inconsequential; (iii) simplified solutions where both axial and circumferential mode coupling is neglected.

In method of solution (ii), u^*, v^* and w^* of equations (6.26) reduce to

$$u^* = \mathrm{e}^{\mathrm{i}\omega t} \sum_{m=1}^{M} \overline{A}_m \cos n\theta \, \phi'_m, \quad v^* = \mathrm{e}^{\mathrm{i}\omega t} \sum_{m=1}^{M} \overline{B}_m \sin n\theta \, \phi_m,$$

$$w^* = \mathrm{e}^{\mathrm{i}\omega t} \sum_{m=1}^{M} \overline{C}_m \cos n\theta \, \phi_m, \tag{6.34}$$

and all subsequent equations are simplified accordingly. Then, one may choose n for each calculation and obtain the eigenfrequencies associated with that value of n only; i.e. $f_{n,1}, f_{n,2}, \ldots, f_{n,M}$. In this case the theory is much simplified.

Even greater simplification is achieved by method of solution (iii), where the series of equations (6.34) are truncated at $M = 1$. The elements of $[M]$, $[C]$ and $[K]$, which now become 3×3 matrices, are given by

$$M_{11} = -\gamma f_{11}, \quad M_{22} = -\gamma a_{11}, \quad M_{33} = -\left\{\gamma + [2(\rho\gamma)/(\rho_s h)](a/n)\right\} a_{11},$$

$$M_{ij} = 0 \quad \text{for} \quad i \neq j\,;$$

$$C_{33} = [(\rho\gamma)/(\rho_s h)] \left\{(4/\pi) U F(n, n)\right\} a_{11}, \quad \text{all other} \quad C_{ij} = 0\,;$$

$$K_{11} = a^2 g_{11} - \tfrac{1}{2} n^2 (1 - v)(1 + \kappa) f_{11}, \quad K_{12} = K_{21} = \tfrac{1}{2} an(1 + v) f_{11},$$

$$K_{13} = K_{31} = av f_{11} - a^3 \kappa g_{11} - \tfrac{1}{2} an^2 \kappa (1 - v) f_{11},$$

$$K_{22} = -n^2 a_{11} + \tfrac{1}{2} a^2 (1 + 3\kappa)(1 - v) c_{11}, \quad K_{23} = K_{32} = -na_{11} + \tfrac{1}{2} a^2 n\kappa(3 - v) c_{11},$$

$$
\begin{aligned}
K_{33} = &-a_{11} - a^4 \kappa e_{11} + 2a^2 n^2 \kappa c_{11} - (n^2 - 1)^2 \kappa a_{11} \\
&- \left\{(\rho\gamma)/(\rho_s h)\right\} \left\{(4/\pi^2)\left(U^2 n/a\right) \sum_{\lambda=1}^{\infty} F(\lambda, n) F(n, \lambda) a_{11}\right\},
\end{aligned} \tag{6.35}
$$

where

$$a_{jm} = \int_0^L \phi_j \, \phi_m \, \mathrm{d}x = L\delta_{jm}\,, \quad c_{jm} = \int_0^L \phi_j \, \phi''_m \, \mathrm{d}x\,,$$

$$e_{jm} = \int_0^L \phi_j \, \phi_m^{\mathrm{iv}} \, \mathrm{d}x \quad f_{jm} = \int_0^L \phi'_j \, \phi'_m \, \mathrm{d}x \quad g_{jm} = \int_0^L \phi'_j \, \phi'''_m \, \mathrm{d}x.$$

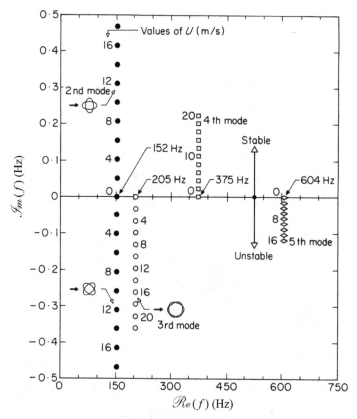

Figure 6.22. Argand diagram of the complex frequencies $f_{2,1}$, $f_{3,1}$, $f_{4,1}$ and $f_{5,1}$ of a clamped-clamped cylindrical shell, as functions of the flow velocity U. Shell parameters: $L = 565$ mm, $h = 0.508$ mm, $E = 0.28 \times 10^{10}$ N/m^2 $\rho_s = 1.29 \times 10^3$ kg/m^3, $\nu = 0.4$; fluid density: $\rho = 1.204$ kg/m^3. Both orientations of the (2,1)-mode are shown; in all other cases only one is arbitrarily given, where the missing ones are the mirror images of those shown (Païdoussis, Price & Suen 1982b).

As was stated at the outset, the specific expressions obtained here pertain to the case of oscillations with an antinode facing the free stream. The equivalent expressions may easily be found for the case of a node facing the free stream. One significant observation that may be made is that, in the latter case, [C] changes sign, but is otherwise identical to the one given here, while [K] and [M] remain exactly the same.

6.7.2 Theoretical results

Calculations were made for parameters close to those corresponding to the Païdoussis, Price & Suen (1982a) experiments, so that theory and experiments could be compared.

Typical results for one such case, with $f(\theta)$ as given by equation (6.8), and using the simplest method of solution (iii) just described, are shown in Figure 6.22. The results are similar to those in Figure 6.20. More specifically, we stress the following.

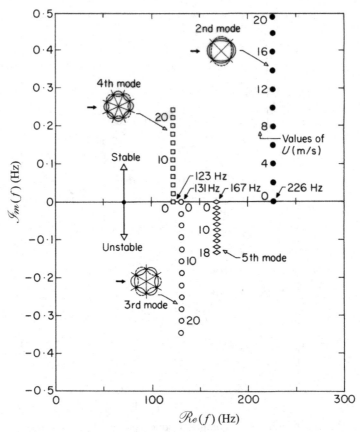

Figure 6.23. Argand diagram of the complex frequencies $f_{2,1}$, $f_{3,1}$, $f_{4,1}$ and $f_{5,1}$ of a clamped-clamped cylindrical shell, as functions of the flow velocity U. Shell parameters: as in Figure 6.22, except $D = 147.3$ mm. The modes shown are oriented with an antinode facing the free stream (Païdoussis *et al.* 1982b).

(i) For $U > 0$ (recall no dissipation is taken into account in the model), even-numbered modes are associated with negative (aerodynamic) damping – i.e. $\mathcal{I}m(f) < 0$ – if a node faces the free stream.* In the opposite case, where an antinode faces the free stream, free-shell motions are aerodynamically damped positively ($\mathcal{I}m(f) > 0$); i.e. the aerodynamic forces act in the same way as dissipative forces. This is consistent with the observations on the sign of $[C]$ made in the last paragraph of Section 6.7.1.

(ii) The opposite is true for odd-numbered modes, which are negatively damped – i.e. they could give rise to instabilities only if the orientation of the ovalling oscillation is such that an antinode faces the free stream.

(iii) The aerodynamic damping, positive or negative, increases almost linearly with U, while the oscillation frequencies, $\mathcal{R}e(f)$, are almost constant.

Similar results to those of Figures 6.22 are shown in Figure 6.23 for a shell of larger diameter, but the same wall thickness, so that h/a is considerably smaller.

---

* In all cases, except for the $(2, 1)$-mode locus, only results for one orientation of the mode are shown, to avoid cluttering the figure. The locus of the mode orientation omitted is, in each case, the mirror image, about the $\mathcal{R}e(f)$-axis, of the one shown.

Table 6.6. *Typical results obtained by the three methods of analysis (fully coupled, partly coupled and uncoupled) showing the effect of various approximations, for the system of Figure 6.22. n = N is the number of circumferential modes considered, and m = M is the number of axial modes*

Type of solution	N	M	$f_{2,1}(U = 0)$ (Hz)	$f_{2,1}(U = 18 \text{ m/s})$ (Hz)	$f_{3,1}(U = 18 \text{ m/s})$ (Hz)
Fully coupled	3	3	149.6	148.8-0.461i	203.6-0.284i
Partly coupled	4	1	152.5	151.4-0.462i	204.0-0.290i
Partly coupled	2	1	152.2	151.4-0.468i	–
Simplified (uncoupled)	1	1	152.2	151.4-0.468i	204.0-0.295i

In this case the lowest frequency at zero flow velocity is that of the fourth mode, followed in ascending order by that of the third, the fifth, and the second modes. As before, negative aerodynamic damping is associated with (i) an antinode facing the free stream for odd-numbered modes, and (ii) a node facing the free stream for even-numbered modes. Also, as in Figure 6.22, only some of the results are shown; e.g. only the locus of the (4, 1)-mode with an antinode facing the free stream is shown, whereas that with a node facing the stream, which is the mirror image about the $\mathcal{R}e(f)$-axis of that shown, is omitted. Without further elaboration, it may be stated that the dynamical behaviour of the systems of Figures 6.22 and 6.23 is qualitatively similar.

The results shown in Figures 6.22 and 6.23 were obtained with the so-called simplified method of solution (iii) of the previous section, where both axial and circumferential mode coupling are not taken into account: i.e. the calculations were done with equations (6.35). The effect of this rather drastic simplification was found, *a posteriori*, to be rather insignificant, as may be seen in Table 6.6. Comparing typical results obtained by the most general method of solution – method (i) – with $N = 3$, $M = 3$ and those of Figure 6.22 (method (iii): $N = 1, M = 1$), differences were found to be of the order of less than 2%.

The results of Table 6.6 effectively indicate that both circumferential- and axial-mode coupling are rather weak and quite negligible. It should be remarked, nevertheless, that this conclusion cannot be considered to be general; it has been reached on the basis of a limited set of calculations, with parameters expressly close to those pertaining in the experimental systems tested previously (Païdoussis & Helleur 1979; Païdoussis *et al.* 1982a), involving low-viscosity epoxy shells in airflow. Thus, this conclusion may not be valid, for instance, for shells in liquid cross-flow.

6.7.3 Comparison with experiment

As described in Païdoussis *et al.* (1982a), experiments were conducted with two sizes of shells: 76.2 and 147.3 mm in diameter.

The 76.2-mm-diameter shells developed ovalling in their second circumferential mode, with a node facing the free stream. At higher flow velocity, third-mode ovalling developed, with an antinode facing the free stream (cf. Figure 6.14). In terms of orientation of the modes, these observations are in qualitative agreement with the results of Figure 6.22.

Table 6.7. *Theory and experiment compared in terms of predicted and measured ovalling frequencies and the corresponding threshold flow velocities; the figures in parentheses, for* U_{thr}, *correspond to the case where a splitter plate has been used; the theoretical* U_{thr} *were computed on the basis of the minimum* $\delta_{n,m}$ *measured (i.e.* $\delta_{2,1} = 0.031$, $\delta_{3,1} = 0.034$; $\delta_{4,1} = 0.035$, $\delta_{5,1} = 0.024$)

		$f_{n,m}$ (Hz)		U_{thr} (m/s)	
Type of shell	Mode number	Theory	Experiments	Theory	Experiment
76.2-mm-diameter	(2,1)	151	155	28.5	16.8–18.4
shells				(36.0)	(19.3–20.5)
	(3,1)	204	225	67.5	21.0–22.5
				(87.2)	(24.4–25.7)
147.3-mm-diameter	(4,1)	122	131	62	13.2–14.2
shells	(5,1)	165	190	87	17.2–20.2

The 147.3-mm-diameter shells developed ovalling first in their fourth circumferential mode, followed by fifth-mode ovalling (cf. Figure 6.15), but other modes were also excited, although the fourth and fifth were the dominant ones. The orientation of the modes to the free stream in this case also was as predicted by theory (cf. Figure 6.23).

Comparison between theoretical and experimental threshold flow velocities, U_{thr}, for the onset of ovalling in the various modes observed in the experiments is made in Table 6.7 – the theoretical results having been obtained as described in the previous section. The corresponding ovalling frequencies, $f_{n,m}$, are also compared.

As may been seen in Table 6.7, agreement between theory and experiment, in terms of ovalling frequencies, is within 3 and 13%, which is reasonably good. The principal reasons for discrepancy are thought to be (i) inaccuracy in the value of E used in the calculations (which was the value supplied by the shell manufacturers), and (ii) lack of perfect uniformity, e.g. in the wall thickness of the shells.

Discrepancies in the values of U_{thr} are far greater. Thus, for the first instability encountered with increasing flow, the theoretical U_{thr} differ from the experimental values by 55% *at best*, for the 76.2-mm shells, and by a factor of 4 for the 147.3-mm shells. The higher ovalling modes, in each case, differ by correspondingly higher margins. Here, it should be noted that the experimental ranges of U_{thr} correspond to different experimental runs with nominally identical shells. The theoretical U_{thr} were determined with the lowest $\delta_{n,m}$ which was measured, because experimental errors generally result in overestimating the damping values. (In this connection, it should be noted that, although great care was exercised in determining $\delta_{n,m}$, the values of $\delta_{n,m}$ obtained could not be said to be more reliable than $\pm 20\%$.)

Clearly, agreement between theoretical and experimental U_{thr} is not good. Nevertheless, it should be noted that (i) the sequence of modes in which the system is predicted to flutter, with increasing flow, is the same as in the experiments; (ii) when a splitter plate is used, both theory and experiments indicate higher threshold flow velocities for flutter, and (iii) as indicated previously, theory and experiment agree that ovalling in even- and odd-numbered circumferential modes occurs with, respectively, a node and an antinode facing the free stream.

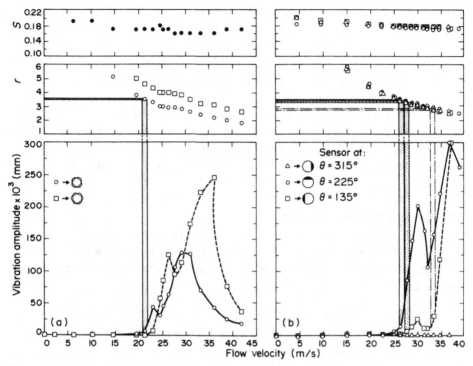

Figure 6.24. Experimental results for a clamped-clamped shell (a) without and (b) with rigid insert glued in (Païdoussis, Price & Ang 1988a).

6.7.4 Improvements to the theory

Some attempts at improvement of the aeroelastic model were made, with the aim of bringing the theoretical predictions closer to the measured ones. These attempts were largely successful, as seen in what follows, but equally interesting are the physical insights gained.

Before describing these improvements, the perplexing findings described in Section 6.3.2 are recalled, which suggest that "disrupting the flow" in the wake can have a major effect on ovalling – indeed, to the extent of suppressing it. The inevitable conclusion is that the near-flow in the wake plays an important role on ovalling.

Further experiments in this direction were performed (Païdoussis, Price & Ang 1988a, 1991), involving a shell clamped at both ends ($D = 76.2$ mm, $h = 0.495$ mm, $L = 533$ mm, $E = 2.8 \times 10^9$ N/m^2, $\rho_s = 1.29 \times 10^3$ kg/m^3, $\nu = 0.4$). A rigid (aluminium) insert was glued onto the whole length of a segment of the inner surface of the shell, as shown in the inset of Figure 6.24(b), immobilizing that segment of the shell (the edges of the insert subtended an angle $\theta = 74°$). By rotating the shell with respect to the wind, the rigid portion of the shell could be (i) in the front, where it was expected to have maximum effect, (ii) in the wake, where according to theory it should have no effect, or (iii) at an intermediate position.

The results without and with the insert are given in Figure 6.24. For this shell, without the insert, the thresholds of ovalling in the $n = 2$ and $n = 3$ modes are very close (the corresponding measured δ_{md} are: $\delta_{21} = 0.049$, $\delta_{31} = 0.052$). The results with the insert (Figure 6.24(b)) are, as nearly always, somewhat unexpected. When the rigid portion is in the forward part of the cylinder, the system is slightly stabilised: $U_{\text{thr}} = 27.8$ to 28.7 m/s, as opposed to 21.2 m/s with no insert. This effect is less pronounced when the centre of the rigid portion is at $\theta = 90°$. (Here, it should be mentioned that, with the insert glued on, the δ_{md} were increased: $\delta_{21} = 0.053$, $\delta_{31} = 0.11$). Finally, when the insert was totally in the wake, the effect was strongest and *no ovalling* took place at all – at least up to $U = 45$ m/s.

Hence, it would appear yet again that not only is the wake important, but near-wake effects may have a profound influence on ovalling. This has resulted in a reassessment of wake effects, as in Section 6.7.4(b).

We now present the improvements introduced by Païdoussis, Price & Ang (1988a, 1991) to the model described in Section 6.7.1.

6.7.4 (a) Moving boundary effects

The pressure on the shell may be obtained via the Bernoulli equation for unsteady flow more systematically, also applying the boundary condition on the moving boundary. Thus, the velocity components on the moving boundary may be expressed approximately as

$$V_r\big|_{r=a+w^*} \simeq V_r\big|_{r=a} + \frac{\partial V_r}{\partial r}\bigg|_{r=a} w^* \simeq \frac{\partial \phi^*}{\partial r}\bigg|_{r=a} + \frac{\partial^2 \phi_0}{\partial r^2}\bigg|_{r=a} w^*, \tag{6.36}$$

$$V_\theta\big|_{r=a+w^*} \simeq V_\theta\big|_{r=a} + \frac{\partial V_\theta}{\partial r}\bigg|_{r=a} w^* \simeq Uf(\theta) + \frac{1}{a}\frac{\partial \phi^*}{\partial \theta}\bigg|_{r=a} + \left[\frac{1}{a}\frac{\partial^2 \phi_0}{\partial r \partial \theta} - \frac{1}{a^2}\frac{\partial \phi_0}{\partial \theta}\right]\bigg|_{r=a} w^*. \tag{6.37}$$

The Bernoulli equation for unsteady flow is

$$\frac{p}{\rho} + \frac{1}{2}V^2 + \frac{\partial \Phi}{\partial t} - (\mathbf{v} + \boldsymbol{\omega} \times \mathbf{r}) \cdot \nabla \Phi = \frac{p_0}{\rho}, \tag{6.38}$$

in which $\boldsymbol{\omega} = \mathbf{0}$ because of the assumption of irrotational flow and $\mathbf{v} \simeq (\partial w^*/\partial t)\hat{\mathbf{e}}_r + (\partial v^*/\partial t)\hat{\mathbf{e}}_\theta$, $\hat{\mathbf{e}}_r$ and $\hat{\mathbf{e}}_\theta$ being the unit vectors in the radial and tangential directions, leading to

$$p \simeq p_0 - \rho\left\{\frac{\partial \phi^*}{\partial t} + \frac{1}{2}U^2 f^2(\theta) + \frac{Uf(\theta)}{a}\frac{\partial \phi^*}{\partial \theta}\right.$$

$$\left. + \frac{Uf(\theta)}{a}\frac{\partial^2 \phi_0}{\partial r \partial \theta} w^* - \frac{Uf(\theta)}{a^2}\frac{\partial \phi_0}{\partial \theta} w^* - Uf(\theta)\frac{\partial v^*}{\partial t}\right\}\bigg|_{r=a}. \tag{6.39}$$

The perturbation potential may be related to shell displacement by the impermeability boundary relationship, $V_r\big|_{r=a+w^*} = Dw^*/Dt$, where D/Dt denotes the substantial derivative. By utilizing equation (6.36) this leads to

$$\frac{\partial \phi^*}{\partial r}\bigg|_{r=a} = \frac{\partial w^*}{\partial t} + \frac{Uf(\theta)}{a}\frac{\partial w^*}{\partial \theta} - \frac{\partial^2 \phi_0}{\partial r^2}\bigg|_{r=a} w^*. \tag{6.40}$$

6.7.4 (b) Wake pressure and pressure fluctuations

The wake has hitherto been assumed to be a stagnant region, where $v_\theta(\theta) = 0$. As a result, in the analysis the wake pressure becomes equal to the stagnation pressure.

To correct for this, it is assumed that

$$f(\theta) = -f(\theta_s) \quad \text{for} \quad |\theta| > \theta_s, \tag{6.41}$$

which effectively suggests that there is a quasi-coherent reverse flow on the surface of the shell in the wake – without supporting physical evidence.

The pressure fluctuations p_e^* on the shell ahead of the separation point may be determined from equation (6.39) after the static components are filtered out:

$$
p_e^* = -\rho \left[\frac{\partial \phi_e^*}{\partial t} + \frac{Uf(\theta)}{a} \frac{\partial \phi_e^*}{\partial \theta} + \frac{Uf(\theta)}{a} \frac{\partial^2 \phi_0}{\partial r \partial \theta} w^* \right.
$$
$$
\left. - \frac{Uf(\theta)}{a^2} \frac{\partial \phi_0}{\partial \theta} w^* - Uf(\theta) \frac{\partial v^*}{\partial t} \right] \Bigg|_{r=a}, \quad \text{for} \quad 0 \le |\theta| \le \theta_s. \tag{6.42}
$$

Up to now, the wake was considered to be a "dead zone"; but, as shown in Section 6.3.2, it is not. It is presumed that shell motions would cause variations of flow pattern around the body and consequently changes in base pressure. Hence, equation (6.38) may be written as

$$
\frac{p}{\rho} + \frac{1}{2} V^2 + \frac{\partial \Phi}{\partial t} - (\mathbf{v} + \boldsymbol{\omega} \times \mathbf{r}) \cdot \nabla \Phi = \frac{1}{\rho} \left[p_b + \frac{\partial p_b}{\partial (w^*/a)} \left(\frac{w^*}{a} \right) e^{-i\psi} \right], \quad \text{for} \; |\theta| > \theta_s, \tag{6.43}
$$

where the term $[\partial p_b / \partial (w^*/a)](w^*/a)e^{-i\psi}$ has been introduced to account for changes in the base pressure associated with shell motions in a given mode. Because analytical expressions for $\partial p_b / \partial (w^*/a)$ are not available, this quantity was obtained quasi-statically by measurements on permanently deformed shells for different w^*/a, as described in Païdoussis *et al.* (1991b). Moreover, because it was found that these pressure changes lagged substantially behind deformation in the course of shell oscillations, the phase angle between the two, ψ, was measured as well; these measurements are also described in Païdoussis *et al.* (1991b).

To simplify the analysis, the shape-related base-pressure variations and ψ are assumed to be independent of θ within the wake, and in the calculations they are taken equal to the values measured at $\theta = 180°$. The measured base-pressure coefficients may be approximated by

$$
C_{pb} = \frac{p_b - p_\infty}{\frac{1}{2} \rho U^2} = H_0 + H \frac{w^*}{a},
$$

where H_0 and H are constants; it was found that, approximately,

$$
\begin{aligned}
C_{pb} &= -1.37 + 3.79(w/a) \quad \text{for} \; n = 2, \\
C_{pb} &= -1.47 - 2.38(w/a) \quad \text{for} \; n = 3,
\end{aligned} \tag{6.44}
$$

for a node and an antinode facing the free stream, respectively. Thus,

$$
\frac{\partial p_b}{\partial (w^*/a)} = \tfrac{1}{2} H \rho U^2, \tag{6.45}
$$

needed in (6.43) can be determined, with H as given in (6.44).

Table 6.8. *Theoretical (improved theory) and experimental (weighted average of all tests) values of* U_{thr} *(m/s)*

Mode (n, m)	Experiment	Theory
(2,1)	21.5	24
(3,1)	23.5	19

The experimentally determined phase ψ was found to be quite substantial:

$$\psi = 40^\circ \quad \text{for} \quad n = 2; \quad \psi = 270^\circ \quad \text{for} \quad n = 3, \tag{6.46}$$

for a node and an antinode facing the free stream.

6.7.4 (c) Comparison with experiments

New experiments were conducted with clamped-clamped epoxy shells ($D = 76.2$ mm, $h = 0.490$ and 0.495 mm). The results (see Figure 6.24) were quite similar to those reported heretofore and will not be elaborated upon; the main difference is that second- and third-mode ovalling thresholds were closer together.

A comparison between the improved theory and the experimental results is given in Table 6.8. It is obvious that theory now performs much better than heretofore.

Also agreement between theoretical and experimental frequencies, at zero flow and with flow, including the ovalling frequencies, is very good – being better than 6% for $n = 2$ ovalling and 11% for $n = 3$ ovalling.

It is of interest to identify which one of the improvements was mostly responsible for enhanced agreement between theory and experiment. Without a doubt, this was found to be the taking into account of base-pressure variations (equation (6.43) *et seq.*). Taking this effect into account alone would change the $n = 2$ and $n = 3$ entries in Table 6.8 to 26 and 28 m/s, respectively, not too far from the values with *all* improvements taken into account.

6.8 An Energy-Transfer Analysis

As a result of the improved agreement between theory and experiment, as in Section 6.7.4, it is now more evident than heretofore that ovalling is a flutter phenomenon, which may be adequately predicted by considering it to involve but one degree of freedom. According to that model, if the energy gained from the fluid by the aerodynamic force exceeds the energy lost by dissipation in the one mode considered, then flutter will ensue. This suggests the possibility of a more direct approach to the determination of the threshold of instability (Païdoussis, Price & Ang 1991b).

Thus, suppose that the dynamic pressure around the shell vibrating in the (n, m)-mode, and hence the work done by these forces, may be determined over a cycle of oscillation as a function of the flow velocity; this work would represent the energy input that would go towards sustaining oscillation. If, for a given flow velocity U, this work were to be just greater than the energy lost by dissipation, this U would clearly correspond to the threshold of flutter.

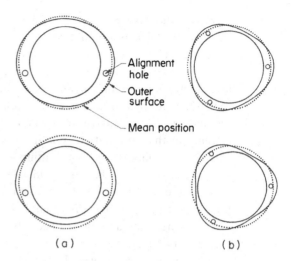

Figure 6.25. The "statically deformed" shapes (a) for $n = 2$ and (b) for $n = 3$ ovalling, tested in conjuction with the energy transfer analysis (Païdoussis, Price & Ang 1988a, 1991b).

An attempt was first made to obtain the fluctuating pressure on a vibrating shell via a surface-mounted miniature pressure transducer; however, although this transducer is very small, weighing only 0.3 g, it was found that the mode shape of the ovalling shell was substantially distorted by its presence. As a result, a more circuitous route had to be adopted for finding the same information.

Suppose that the static-pressure difference across the surface of the shell is known as a function of stationary deformation: $p_i(\theta, w/a) - p_e(\theta, w/a)$. Then, the force on an element of the shell may be expressed as

$$F = [p_i(\theta, w/a) - p_e(\theta, w/a)] (a + w)\delta\theta\, e^{-i\psi} \qquad (6.47)$$

per unit length. For a given amplitude of oscillation, because the oscillation is harmonic, this is equivalent to the force as a function to time:

$$F = [p_i(\theta, t) - p_e(\theta, t)] (a + w)\delta\theta\, e^{-i\psi}. \qquad (6.48)$$

Hence, the energy input per cycle of oscillation may be expressed as

$$W_i = \int_0^{2\pi/\omega_0} \int_0^{2\pi} Re\left[(p_i - p_e)(a + w)e^{-i\psi}\right] Re\,(\partial w/\partial t)\, d\theta\, dt. \qquad (6.49)$$

The energy dissipated, where damping is modelled via an equivalent viscoelastic damping coefficient, is given by

$$W_d = \int_0^{2\pi/\omega_0} \int_0^{2\pi} Re\left[\rho_s a h \frac{\delta_{md}}{\pi}\omega_0 \frac{\partial w}{\partial t}\right] Re\,(\partial w/\partial t)\, d\theta\, dt. \qquad (6.50)$$

It is noted that quasi-static aerodynamics cannot be applied here, because the dimensionless ovalling frequencies (reduced frequencies), $\omega/2aU$, are large, typically $\mathcal{O}(800)$; this makes the introduction of the phase angle ψ imperative.

The quantity $[p_i(\theta, w/a) - p_e(\theta, w/a)]$ was obtained by measuring the pressure distribution around stationary deformed two-dimensional shapes of the shell in the $n = 2$ and $n = 3$ modes, the cross-sections of which are shown in Figure 6.25. They were made of aluminium, approximately 300 mm long and equivalent radius of 76 mm. They were produced with a NC-milling machine. For the $n = 2$ mode, "frozen" amplitudes of 1.80 and 3.10 mm were obtained; for the $n = 3$ mode, the frozen amplitudes were of 2.54 and 3.30 mm. A circular cylinder of the correct radius was also available as a reference shape.

Each of the cylinders had 19 static pressure taps, spirally located around its circumference, and was mounted in the wind tunnel with appropriate boundary layer guards and dummy cylinders of the same blockage so as to span the wind-tunnel height. Measurements were taken in the case of $n = 2$ with the long axes of the ellipses (Figure 6.25(a)) either at 45° or at −45° to the free stream, to simulate ovalling in that orientation. For the $n = 3$ modes the cylinders were orientated with a protruding or receding antinode facing the free stream. The experiments were done for $\mathcal{R}e = 3.80 \times 10^4$ and 1.18×10^5. Including the circular cylinder, for each θ there were therefore five measurements of C_p as a function of w/a available for $-0.10 < w/a < 0.10$ approximately. For each θ, linear relationships of $C_p(\theta, w/a)$ were obtained, similar to those of equation (6.44), which could be used in conjunction with equation (6.49).

The phase angle ψ was determined as follows. In the wake, ψ was measured directly as the phase difference between the displacement of the ovalling cylinder (measured by an internally mounted fibre-optic probe) and the pressure close to the shell surface (measured by a miniature pressure transducer 9.5 mm away from the shell in the wake); it was found possible to do these measurements only for $120° < |\theta| < 180°$. Ahead of the separation point, $0 < |\theta| < 85°$, ψ was determined analytically, essentially by the method of Section 6.7.4 (see also Ang (1983) and Païdoussis *et al.* (1991b)).

By approximating $p_i - p_e$ in (6.49) by

$$p_i(\theta, t) - p_e(\theta, t) \simeq \tfrac{1}{2} \rho U^2 \left[1 - C_p(\theta, w^*/a) \right], \tag{6.51}$$

and approximating C_p as a first-order polynomial, $C_p \simeq H_0 + H(w^*/a)$, and invoking the quasi-static assumption, one obtains

$$p_i(\theta, t) - p_e(\theta, t) \simeq \tfrac{1}{2} \rho U^2 \left[1 - H_0(\theta) - H(\theta, w^*/a) \right]; \tag{6.52}$$

then, considering shell oscillations with an antinode facing the free stream, $w^* \simeq B_0 \cos n\theta \exp(i\omega_0 t)$, the energy gained over a small segment of the shell subtending an angle $\theta_2 - \theta_1$ may be obtained via equation (6.49),

$$W_i(\theta_2 - \theta_1) = -\tfrac{1}{4} B_0^2 (1 - H - H_0) \rho U^2 \pi \sin \psi \left[(\theta_2 - \theta_1) + \frac{1}{2n} (\sin 2n\theta_2 - \sin 2n\theta_1) \right], \tag{6.53}$$

with all parameters evaluated at $\theta = \tfrac{1}{2}(\theta_1 + \theta_2)$. A table listing $H(\theta)$, $H_0(\theta)$ and $\psi(\theta)$ as a function of U may be found in Païdoussis *et al.* (1991b). By summing over the circumference, the total energy gained, W_i, per unit length of the shell may be obtained.

Proceeding similarly, the energy dissipated per unit length is found to be

$$W_d = B_0^2 \rho_s a h \delta_{\mathrm{md}} \pi \omega_0^2. \tag{6.54}$$

This is clearly a single-mode analysis. However, previous work has shown that the effect of coupling – both circumferential and axial – is quite negligible (Païdoussis *et al.* 1982b), at least in airflow.

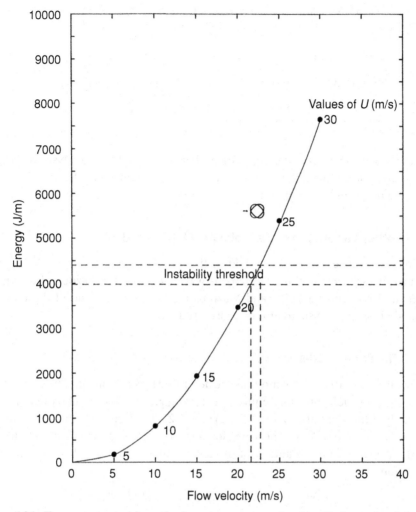

Figure 6.26. Energy extracted from the free stream, per cycle of oscillation and unit length, for the $n = 2$ mode with a node facing the free stream. At the instability threshold the energy lost by mechanical dissipation in the experimental shell is equal to the energy gained from the flow. The experimental critical flow velocity is 21.5 m/s, whereas the theoretical one is 22 m/s (Païdoussis *et al.* 1991b).

The results for the critical mode ($n = 2$) of the shell used in the experiments are shown in Figure 6.26. It is seen that energy balance, in terms of energy gained and lost, is achieved at approximately $U = 22$ m/s. This compares favourably with the measured value of 21.5 m/s, in fact, better than expected; i.e., although good agreement had been anticipated, the degree of agreement (being better than 1%) may well be fortuitous. The discrepancy for the $n = 3$ mode is considerably larger, with a predicted value of $U_{thr} = 35$ m/s compared with the experimental one of 23.5 m/s. Once again, however, it should be recalled that the shell is already oscillating in the $n = 2$ mode when the $n = 3$ mode begins to appear; hence, the predictions here, which are based on the assumption of pure single-mode ovalling and oscillations about the *undisturbed equilibrium* could not be expected to work well in the case of $n = 3$.

Table 6.9. *Critical flow velocity, U_{cr} (m/s), for onset of ovalling: comparison of the experimental threshold with the theory developed in this paper*

Critical mode	Improved flutter analysis	Experiment	Energy-transfer analysis
$n = 2$, $m = 1$	24	21.5	22

Table 6.9 summarizes the comparison between the flutter analysis via the improved analytical model, the present energy-transfer analysis (of Section 6.7.4) and experiment; it is judged to be satisfactorily good.

6.9 Another Variant of the Aeroelastic-Flutter Model

This model, in its essence, is a variant of the models of Païdoussis & Wong (1982) and Païdoussis, Price & Suen (1982b). It was developed by Mazouzi, Laneville & Vittecoq (1991) and Laneville and Mazouzi (1996); in the second of these papers, the mode in which the shell flutters is also predicted.

6.9.1 The flutter model

In Mazouzi *et al.* (1991), shell motions are described by the Donnell (1933) equations, which for thin shells are fully equivalent to Flügge's (Leissa 1993; Amabili 2008). Hence, the equivalent of equations (6.22) and (6.23) hold true. Also, the fundamental relations in (6.4) to (6.6) are the same, leading to the same ϕ_e^* and ϕ_i^* as in equations (6.13). Further, taking advantage of the comments leading to (6.18), the simplified expressions

$$\phi_e^* = e^{i\omega t} D_n r^{-n} \cos n\theta \quad \text{and} \quad \phi_i^* = e^{i\omega t} E_n r^n \cos n\theta \tag{6.55}$$

are adopted directly.

As in Sections 6.6 and 6.7, the flow field is presented to be two-dimensional; thus

$$v(r, \theta, t) = v^o(r, \theta) + \nabla \phi^*(r, \theta, t). \tag{6.56}$$

On the surface of the cylinder,

$$v_r^o(a, \theta) = 0, \quad v_\theta^o(a, \theta) = \frac{1}{a} \frac{\partial \phi^o}{\partial \theta}, \quad v_x^o(a, \theta) = 0, \tag{6.57}$$

where ϕ^o is the steady flow-velocity potential. Furthermore, utilizing (6.6) we can write

$$v_\theta^o(a, \theta) = U [1 - C_p(\theta)]^{1/2} \quad \text{for} \quad |\theta| < \theta_s, \tag{6.58}$$

θ_s being the angle, from the front, where separation occurs; it is also assumed that

$$v_\theta^o(a, \theta) = 0 \quad \text{for} \quad |\theta| > \theta_s. \tag{6.59}$$

The first divergence from the previous theoretical model is that, instead of using Roshko's measurements to determine $v_\theta^o(a, \theta)$ – i.e. $f(\theta)$ as in (6.8) – it is obtained

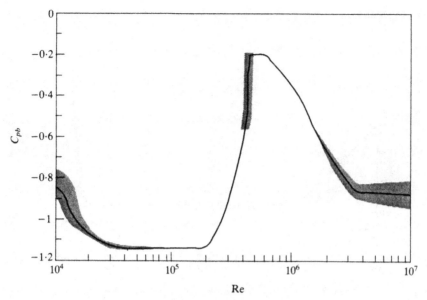

Figure 6.27. The base pressure coefficient as a function of the Reynolds number according to Roshko (1954, 1961), Bearman (1969) and Chen (1972); from Mazouzi, Laneville & Vittecoq (1991).

directly from Parkinson & Jandali (1969), namely

$$\frac{v_\theta^o}{U} = \frac{\sin\beta(1 - 2\cos\alpha\cos\beta + \cos^2\alpha)}{\cos\delta - \cos\beta}, \quad 0 \le |\theta| \le \theta_s, \tag{6.60}$$

where

$$\alpha = \frac{1}{2}(\pi - \theta_s) \quad \text{and} \quad \cos\delta = \cos\alpha + \frac{\sin^3\alpha}{K(\text{Re})},$$

in which

$$K(\text{Re}) = [1 - C_{pb}(\text{Re})]^{1/2}. \tag{6.61}$$

The azimuthal angle, θ, corresponding to v_θ^o is obtained from

$$\sin\theta = \cos\alpha \left[\frac{\sec\alpha - \cos\beta}{\frac{1}{2}(\sec\alpha + \cos\alpha) - \cos\beta} \right] \sin\beta. \tag{6.62}$$

Mazouzi *et al.* (1991) proceed as follows. By conformal transformation and by symmetry, the domain $0 \le \theta \le \theta_s$ is mapped in the domain $0 \le \beta \le \pi$. As expected, this potential flow model requires two experimental inputs: the mean base-pressure coefficient, C_{pb}, and the separation angle, θ_s, both functions of the Reynolds number, as observed by Roshko (1954, 1961), Bearman (1969) and Chen (1972). Figures 6.27 and 6.28 show a polynomial curve fitting their data. As recommended by Bearman (1969), Chen (1972) and Parkinson & Jandali (1969), θ_s should have a value between 80° and 83° for the Reynolds number range $10^3 \le \text{Re} \le 1.6 \times 10^5$. The great advantage in this model of knowing v_θ^o (or $f(\theta)$) as a function of Re must be stressed, because very high Re may obtain in real ovalling experiences.

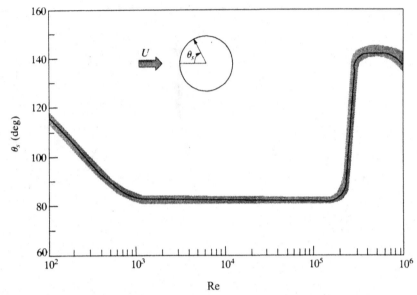

Figure 6.28. The separation angle, θ_s, as a function of the Reynolds number according to Parkinson & Jandali (1969), Bearman (1969) and Chen (1972); from Mazouzi *et al.* (1991).

Next, $p_i^* - p_e^*$ is determined, as in Section 6.6.1, leading to equations (6.10) to (6.12), and then to (6.19) and (6.20). Hence, so far, the main difference lies in the determination of $v_\theta^o(a, \theta)$, i.e. $f(\theta)$, and the simplification $\sum_{\lambda=1}^{\infty} F_1(\lambda) \simeq F_1(n)$.

Another difference is that here a dimensional definition of $F_1(n)$ is used; therefore, in view of the above simplification, we have

$$F_1(n) = \int_0^\pi v_\theta^o \cos n\theta \, \sin n\theta \, d\theta. \tag{6.63}$$

Invoking the experimental finding (Païdoussis *et al.* 1988a) that over half of the total work done by the pressure forces on the shell occurs in the wake, and noting that v_θ^o in (6.63) would be zero therein, the work done by the pressure forces is empirically taken to be symmetrical with respect to $\theta = \frac{1}{2}\pi$, and thus

$$F_1(n) = 2F_1^*(n), \quad F_1^*(n) = \int_0^{\pi/2} v_\theta^o \cos n\theta \, \sin n\theta \, d\theta. \tag{6.64}$$

The analysis then proceeds as in Section 6.6.

6.9.2 Typical results

Typical results for a cantilevered shell are shown in Figure 6.29. They are quite similar in nature to those obtained by Païdoussis *et al.* (1982a, 1988a).

However, agreement with Mazouzi's (1989) own experiments is much better than anything achieved heretofore, as can be seen in Tables 6.10 and 6.11 (excerpted from Mazouzi *et al.* (1991), where more results are given).

The question immediately springs to mind as to the source(s) for the improvement in this variant of the model. This was discussed by Mazouzi *et al.* (1991). The use of the Parkinson & Jandali model is discounted because, in the Re range concerned,

Table 6.10. *Comparison of experimental results with theory for clamped-clamped shells from Mazouzi* et al. *(1991)*

		$f_{n,m}$ (Hz)		U_{thr} (m/s)	
Test no.	Mode (n,m)	Theoretical	Experimental	Theoretical	Experimental
S1	(2,1)	152	155	12	16.8–18.4
S2	(3,1)	205	225	21.5	21.0–22.5
S3	(4,1)	122	131	15.6	13.2–14.2
S4	(5,1)	167	190	16.8	17.2–20.2
S5	(5,1)	167	190	20.4	19.3
S6	(2,1)	155	158	18.1	17.9
S7	(3,1)	202	223	21.4	22.5
S8	(2,1)	151	156	15.0	18.4
S9	(3,1)	196	235	19.6	20.5
S10	(3,1)	196	235	22.4	21.1
S11	(3,1)	196	223	21.9	22.5
S12	(2,1)	166	166	20.5	21.5
S13	(3,1)	203	223	27.9	23.5

Reynolds number effects are weak. The question of the "single-mode" approximation, referred to in the paragraph above the one involving equation (6.63), is credited with "matching of the solid boundary condition with the flow boundary condition". Here, we note that, by removing other modal components, which are damped, the negative damping associated with any given mode is enhanced. Finally, the assumption leading to (6.64) is considered to be important. Indeed it is; we should say here

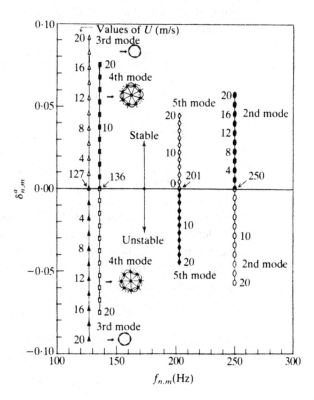

Figure 6.29. Predicted threshold flow velocity, frequency and aerodynamic damping for Shell L4 ($L = 410$ mm, $a = 60$ mm, $h = 0.127$ mm, $\rho_s = 2.64 \times 10^3$ kg/m³, $E = 68.96$ GPa); from Mazouzi et al. (1991).

Table 6.11. *Comparison with experimental results for the clamped-free cylindrical shells (tests with false ceiling), from Mazouzi et al. (1991)*

Shell no.	Test no.	Mode (n, m)	$f_{n,m}$ (Hz)		U_{thr} (m/s)	
			Theoretical	Experimental	Theoretical	Experimental
L1	1	(2,1)	50	47	7	7.8
	2	(3,1)	65.5	64.5	4.5	5
L2	3	(2,1)	62	–	7.5	–
	4	(3,1)	67.7	66.5	4.6	5.1
	5	(4,1)	121	118	9.1	12.6
L3	6	(2,1)	107	–	12	–
	7	(3,1)	78.7	76.5	6.5	6.6
	8	(4,1)	123	120.5	6.3	9.5
L4	9	(3,1)	127	122	17.5	18
	10	(4,1)	136	132	10.6	10.75
L5	11	(5,1)	203	195	–	18

that this empirical and not-too-closely metered modification is mostly responsible for the improvement; it is capable of almost doubling the negative aerodynamic damping, thus greatly reducing the theoretical U_{thr} – e.g. in the results of Table 6.7.

6.9.3 An empirical relationship for U_{thr}

A very useful semi-empirical collapse of much of the available data on ovalling was proposed by Laneville & Mazouzi (1996). Starting with a simplified form of the radial equation of motion while neglecting terms involving κ (see the third of equations (6.22)), one has

$$av\frac{\partial u}{\partial x} + \frac{\partial v}{\partial \theta} + w = a^2\frac{1-v^2}{E}\left(-\rho_s\frac{\partial^2 w}{\partial t^2} + \frac{p_i - p_e}{h}\right). \tag{6.65}$$

Then, proceeding as in the foregoing, one obtains

$$\left[-\omega^2\left(\rho_s\pi\frac{D}{2}Lh + \rho\frac{\pi}{2}D^2\frac{L}{n}\right) + (-1)^\zeta i\omega(4\rho\,DUL_e\,F_1(n))\right.$$

$$\left. + \left(2\pi\frac{hL}{D}\frac{E}{1-v^2} + 16\frac{\rho n}{\pi}U^2F_1^2(n)\,L_e\right)\right]C_{n,m} = 0, \tag{6.66}$$

where $\zeta = 1$ for an antinode facing the flow and $\zeta = 2$ for a node facing the flow, and L_e is the length of the shell exposed to flow; $C_{n,m}$ is the maximum shell deformation in the radial direction in the (n, m) mode. Then $F_1(n)$ is calculated, and it is found that it can be approximated by $(-1)^{n-1}/2n$ for $10^4 < \mathrm{Re} < 10^5$. Thus, the aerodynamic damping is determined as

$$(C_A)_{n,1} = (-1)^{\zeta+n-1}4\rho\,L_e\,DU|F_1(n)|, \tag{6.67}$$

with $F_1(n)$ as above. The next step is to express the structural damping coefficient as an equivalent viscoelastic-damping coefficient,

$$(C_{eq})_{n,m} = DhL\,\rho_s\,\delta_{n,m}\,\omega_{n,m}. \tag{6.68}$$

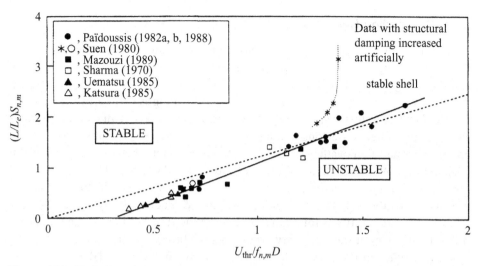

Figure 6.30. Variation of the Scruton number, $S_{n,m}$, as a function of the reduced onset flow velocity $U_{thr}/f_{n,m}D$. *, Experimental data for clamped-clamped cylindrical shells made stable by artificially increasing their damping (Suen 1980); in these cases, the abcissa is a reduced flow velocity determined by using the flow velocity of the tests (Laneville & Mazouzi 1996).

Then, the rather tenuous approximation is introduced that $F_1(n) \simeq F_1(1) \simeq 0.5$ for all n, since the shell is circular prior to the onset of ovalling. Collecting all results, one finds $(L/L_e)S_{n,m} = 1.22\,(U_{thr}/f_{n,m}D)$, where

$$S_{n,m} = 2\frac{\mu}{\rho D^2}\delta_{n,m} = 2\pi\frac{\rho_s}{\rho}\frac{h}{D}\delta_{n,m} \tag{6.69}$$

is the Scruton number. This corresponds to the dotted line in Figure 6.30. This is modified to $(L/L_e)S_{n,m} = 1.55\,(U_{thr}/f_{n,m}D) - 0.48$, giving a better fit. Thus, the following semi-empirical relationship is obtained for the threshold of ovalling:

$$U_{thr} = f_{n,m}D\,[0.65(L/L_e)S_{n,m} + 0.3]. \tag{6.70}$$

The next significant contribution by Laneville & Mazouzi (1996) is the prediction of the critical mode $(n, m)_{crit}$ for ovalling, and modal sequence thereafter.

Starting with equation (6.50) from Païdoussis *et al.* (1991b) and evaluating the integral for harmonic oscillations in a given mode, one obtains

$$W_d = \alpha\,\pi^2\,D\rho_s\,h^3\,\delta_{n,m}f_{n,m}^2, \tag{6.71}$$

αh being a "critical amplitude". For a given shell, the work due to structural damping is proportional to

$$E_{n,m} = \delta_{n,m}f_{n,m}^2. \tag{6.72}$$

Now, it is recalled that the work done by the aerodynamic forces is proportional to U. Because that work and the work dissipated by the structural damping have to balance at the onset of ovalling, it follows that the order of occurrence of the different ovalling modes will range from the lowest $E_{n,m}$ towards the highest as the flow velocity is increased. This allows, in a very simple manner, the determination of the critical mode and the sequence of ovalling modes that follow. This was tested against experimental data and excellent agreement was obtained.

6.10 Concluding Remarks

The reader will have realised that, by following a chronological accounting of the work on ovalling oscillation, the purpose has partly been tutorial, and partly to tell an interesting tale interestingly.

It was shown that the early hypothesis that ovalling is caused by regular vortex shedding is untenable, although the "battle" towards convincing the vortex-shedding proponents was long and hard. In the process, an alternative explanation emerged, namely that ovalling is an aeroelastic-flutter phenomenon. In its initial form (Païdoussis & Wong 1982), the aeroelastic flutter model was found to be in excellent qualitative, but unsatisfactory quantitative agreement with experiment. The model was subsequently improved by Païdoussis, Price & Suen (1982b) and Païdoussis, Price & Ang (1991). It was found that, to predict the threshold of ovalling satisfactorily, empirical input was necessary,* regarding the base-pressure variations and the phase difference in the rear of the shell relative to shell motion (Païdoussis *et al.* 1988a, 1991b). Excellent predictions were thereby possible. The key in this appears to be the phase difference just described, measured, as it cannot be predicted analytically. The same holds true in the good agreement achieved by an energy transfer analysis (Section 6.8); again, the empirical input of phase-difference information between the pressure acting on the shell and shell motion is essential.

A different approach was adopted by Mazouzi, Laneville & Vittecoq (1991) and Laneville & Mazouzi (1996), introducing what amount to helpful but sometimes arbitrary "simplifications" which aid in achieving a better agreement with experiment; this transforms the model to a semi-empirical one. The achievement, however, was in obtaining a model that (i) is in excellent agreement with experiments; (ii) leads to a very simple algebraic expression for predicting the ovalling threshold flow velocity U_{thr}; and (iii) identifies the critical mode $(n, m)_{crit}$ for ovalling, as well as the sequence of ovalling modes that follow.

In the process of invalidating the vortex-shedding hypothesis, it was found that a splitter plate in the wake of the shell, far from suppressing ovalling, it energizes it. This is similar to what was found for another aeroelastic phenomenon: the galloping of prisms (Section 2.5).

---

* Here, the surprising experimental findings described in Sections 6.3.2 and 6.7.4 are of interest for their own sake.

7 Rain-and-Wind-Induced Vibrations

Of all cross-flow induced instabilities, rain-and-wind-induced vibrations, referred to as RWIV, are a special case. They have been identified quite recently and are certainly a challenge to understand. For the purposes of this book, they somehow gather together several of the issues that have been presented: hence, they deserve some treatment, even if the relentless evolution of knowledge on the subject makes any review soon obsolete.

7.1 Experimental Evidence

7.1.1 Field cases

This surprising case of cross-flow-induced motion was identified in the 1970s on cables of cable-stayed bridges. First reported by Wianecki (1979), it was described in detail by Hikami & Shiraishi (1988) as follows. Large-amplitude motion of some cables occurred in the presence of wind with rain but disappeared when the rain stopped. This is illustrated in Figure 7.1, where the existence of motion of the cables, as recorded *in situ*, is clearly correlated to the occurrence of rain. Similar occurrences have been reported in various bridges over the years; see for instance Ruscheweyh & Verwiebe (1995), Matsumoto *et al.* (1989), Zuo *et al.* (2008), Ni *et al.* (2007) and Main & Jones (1999). Although the cables undergoing such vibrations differed from one bridge to another, some common features were identified. First, the motion was specific to the concurrent existence of significant rain and wind, but it stopped if the wind velocity exceeded some value. Second, only inclined cables (neither horizontal nor vertical) moved, and only when descending in the direction of wind, and preferably with the wind slightly skewed. Third, the amplitude of motion, in the vertical plane usually, was as large as 10 diameters. Finally, in some cases, an oscillating rivulet of rainwater was observed under the cable.

The practical consequences of these so-called rain-wind-induced vibrations, or RWIV, are obvious in terms of fatigue of cables and cable anchorage. Avoiding them is now a design concern in most large projects of cable-stayed bridges, in particular in regions subject to simultaneous rain and wind risks, which is the case in Southeast Asia. The use of damping systems is efficient but costly, and understanding the mechanism causing such motion is certainly a requirement to finding a proper cure.

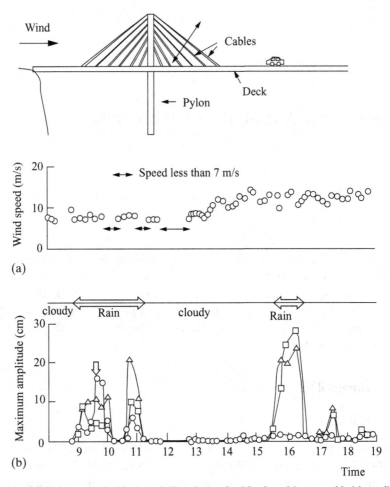

Figure 7.1. (a) Rain-and-wind-induced vibrations of cables in cable-stayed bridges. (b) Evidence that motion is correlated with the presence of rain, from field experiments by Hikami & Shiraishi (1988). For a given wind velocity, motion of the cable occurs only when rain is present.

Classical mechanisms such as VIV, wake galloping, galloping, or others have readily been discarded: the reduced velocity is generally above 20 (no VIV), the distance between cables is large (no wake galloping) and the cross-section of a cable with running water is almost identical to that of a dry cable (no section galloping).

7.1.2 Wind-tunnel experiments

A large set of wind-tunnel experiments has been conducted over the years to try to reproduce the phenomenon observed *in situ*; see for instance Hikami & Shiraishi (1988), Matsumoto *et al.* (1995), Bosdogianni & Olivari (1996), Verwiebe & Ruscheweyh (1998), Gu & Du (2005), Flamand (1995), Wang *et al.* (2005), Alam & Zhou (2007), Cosentino *et al.* (2003a), and Matsumoto *et al.* (2003b). These experiments differed principally from one another by the way the rain or its assumed effect was reproduced: this was done either by introducing water droplets in the airflow (using

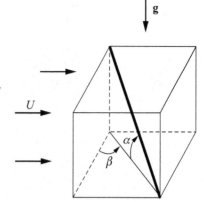

Figure 7.2. An inclined cable, with respect to gravity
and to wind.

an upstream shower in the tunnel), by directly creating a running-water rivulet on
the cable by pouring water at its upper end, or by adding an artificial fixed or moving
rivulet-like system on the cable surface. In some tests the rainwater was replaced by
oil to facilitate rivulet formation (Flamand 1995; Bosdogianni & Olivari 1996). They
also differed by the cable model utilised: a straight rigid cylinder has usually been
considered, either fixed or free to move through a spring support. Its position with
respect to the airflow direction is then defined by the yaw angle, β, and with respect
to gravity by the inclination angle α (see Figure 7.2). The cable surface was carefully
controlled, being either PE (Polyethylene) or PP (Polypropylene), which are typical cable casings. Even the effect of atmospheric pollution has been explored using
soot on the cylinder surface. To avoid scaling effects, most experiments have been
done with cable diameters identical or close to those *in situ*, typically $D = 150$ mm.
This led to most experiments being undertaken in large wind tunnels, capable of
withstanding a rain environment.

With all these experiments, a general agreement has been reached over the years
on the following points, showing that there is definitely a wind-and-rain-induced
vibration issue. First, motions of large amplitude, that do not exist without 'rain', are
in fact found in a limited range of wind velocity U and direction (defined by angles α
and β). These values are similar to those observed on bridges in natural conditions,
typically $U = 10$ m/s, $\beta = 35°$, $\alpha = 45°$. Second, the magnitude of reduced velocities
involved is more that 20. Third, it was observed that under the effect of rain in
the wind, water drops impact the cable and then flow downwards on its surface.
Depending on the flow velocity and angular disposition of the cable, the flow may
take the form of one or two rivulets, as seen in Figure 7.3. These rivulets are flat,
typically 1 mm high and 20 mm wide on a $D = 150$ mm cable. When two rivulets
exist, the one on the windward side is higher than that on the leeward side. If the
rivulets are artificially created by pouring water on top of the cable, their geometry
is not significantly different from the rivulets resulting from impacting drops. Finally,
RWIV occur only when an upper rivulet exists, and when it is able to move: an
artificial rigid rivulet generally does not cause motion; and if it does, this is over a
more limited range of parameters. These conclusions lead to questions of two kinds,
some related to the rivulets, which seem to play an important role, and some related
to the possible interactions between the flow, the rivulets and the cable: How do

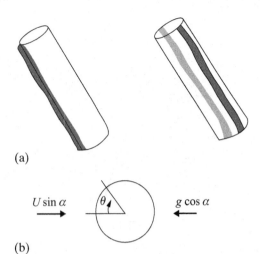

(a)

Figure 7.3. (a) Rivulets of rainwater on a cable. Left: a single gravitational rivulet, at low wind speed. Right: two wind-induced rivulets positioned near the separation points at higher wind speed. (b) A two-dimensional section of the cable with flow, gravity and the liquid film (Lemaitre *et al.* 2006, 2007).

$U \sin \alpha$ $\longrightarrow$ θ $g \cos \alpha$ $\longleftarrow$

(b)

rivulets form, and what is their position, shape and dynamics, depending on the parameters of the system? How does the triple system consisting of the cable, the rivulet and the flow couple in such a way as to produce motions of large amplitude?

7.2 Modelling Rainwater Rivulets

7.2.1 Development of rivulets

As rivulets have been found to be essential in the mechanism of RWIV, the first issue is to understand how they develop. The problem at hand consists of a film of water flowing on the surface of a cylinder which is inclined with respect to gravity and with respect to the airflow. Experimental evidence shows the existence of one rivulet or two, or none, at various positions around the cylinder. Moreover, the angular position of the rivulet may fluctuate along the cylinder, in a sort of meandering path (Wang *et al.* 2005). Rivulets have been found to be either simple bumps on a flat liquid film covering most of the cylinder (Cosentino *et al.* 2003a) or well defined with contact angles (Wang *et al.* 2005).

Three types of forces are at work on the liquid film: gravity, capillary forces, and pressure and friction forces due to the external airflow. The dimensionless numbers that govern their relative contributions are the Froude number $\mathrm{Fr} = U/\sqrt{gR}$ and the Weber number $\mathrm{We} = \rho U^2 R^2/\gamma$, where ρ is the air density, U is its velocity, R is the diameter of the cable, and γ is the surface tension. The Bond number may alternatively be used as a combination of these two. In the range of relevant values of these quantities, $U = 10$ m/s, $R = 0.1$ m, $\gamma = 0.1$ N/m, we have $\mathrm{We} = 100$ and $\mathrm{Fr} = 10$, so that the airflow is expected to play a significant role on the equilibrium of the film. To derive criteria for the existence and positions of rivulets, a simple two-dimensional configuration has been analysed by Lemaitre *et al.* (2006, 2007), then extended to more complex situations by Lemaitre *et al.* (2010). A similar approach may be found in Peil *et al.* (2003), and some further refinements in Robertson *et al.* (2008, 2010). Let us consider the case of a cylinder descending in the direction of flow only, $\beta = \pi/2$, with a uniform liquid film of thickness $h_0 = \varepsilon R$. Only a cross-section is analysed; see Figure 7.3(b). The film thickness $h(\theta, t)$ is expected to evolve in time under the effect of gravity and of airflow loading. By using standard lubrication theory applied to the case of a film on a cylinder (Reisfeld & Bankoff 1992) the linearised evolution of the film thickness is given by (Lemaitre *et al.* 2007)

$$H_{,\tau} = \cos\theta + \frac{1}{2}\text{M}\,\text{Fr}^2\frac{\sin^2\alpha}{\cos\alpha}\left(C_{P,\theta\theta} - \frac{3}{2\varepsilon}C_{F,\theta}\right), \tag{7.1}$$

where $H = h/h_0$ and $\tau = t(gh_0^2\sin\alpha/3v^2R)$ are the dimensionless film thickness and time, respectively, t being the time, $\text{M} = \rho_{\text{air}}/\rho_{\text{water}}$ is a mass ratio and C_P, C_F are pressure and friction coefficients around the cylinder; here, the $f_{,x}$ notation is used for partial derivative $\partial f/\partial x$. Equation (7.1) is not sufficient to give the characteristics of the rivulet that will form, in terms of height and width, because only linear terms are considered. Yet, it is expected that the rivulet will saturate at the location of its initial growth. The evolution of the film thickness in equation (7.1) is seen to be controlled by two terms. The first one, which will dominate at low wind velocities ($\text{Fr} \ll 1$), represents the effect of gravity. It is maximum at $\theta = 0$, and leads to the formation of a single rivulet in the lower part of the cable section. This is the classical dripping rivulet, Figure 7.3(a) left, referred to hereafter as the gravitational rivulet. Alternatively, in the limit of large Froude numbers, $\text{Fr} \gg 1$, the second term will dominate and rivulets will grow at positions defined by the derivatives of the pressure and friction coefficients with respect to the angle θ. Two maxima are found, corresponding to two rivulets, Figure 7.3(a) right, referred to as the wind-induced rivulets. A transition between these two regimes exists at an intermediate Froude number which can be computed easily from the parameters: Figure 7.4(a) shows a comparison between measured positions of rivulets and those predicted using equation (7.1), as obtained by Lemaitre *et al.* (2006). The gravitational rivulet, at $\theta = 0$, is seen to split into two wind-induced rivulets near $\text{Fr}\sin^2\alpha/\cos\alpha = 0.1$ for supercritical values of the pressure and friction coefficients. The two rivulets are then located near the separation points.

This approach can easily be extended to the general case of arbitrary angles to derive the lower limit for the existence of rivulets, in terms of the Froude number, and the position where the wind-induced rivulets will grow (Lemaitre *et al.* 2010).

7.2.2 Tearing of rivulets

As emphasised by many authors, there is an upper limit in terms of wind velocity for the existence of RWIV. One of the issues is whether this limit is related to the destruction of rivulets or to another mechanism. At high wind velocities one may observe that rivulets are ejected by the wind forces as soon as they start to materialize. This can be modelled in the simple cross-sectional framework (Lemaitre *et al.* 2006, 2010). It is assumed that a rivulet on the cylinder surface is limited by triple points, so that the cylinder is dry outside the rivulet. The evolution with time between the initial uniform film and this limit state is certainly complex, but only the equilibrium condition is analysed now. The rivulet will exist as long as capillary forces can resist shear forces. The former are proportional to the surface tension γ, whereas the later are proportional to $\rho U^2\ell$, where ℓ is the width of the rivulet. The ratio of the two scales is $\rho U^2\ell/\gamma = \text{We}\lambda$, where $\lambda = \ell/R$. At large Weber numbers it is expected that

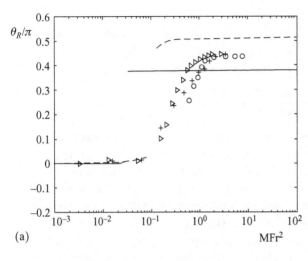

(a)

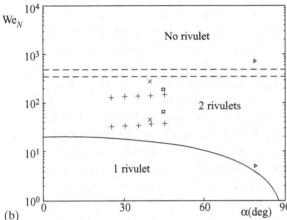

(b)

Figure 7.4. (a) Position of the rivulets on a cylinder inclined in the leeward direction, as a function of the Froude number, from Lemaitre *et al.* (2006). Data points are experimental; the lines correspond to the model of equation (7.1). (b) Domains of existence of rivulets in the space of cable inclination angle and Weber number, equation (7.3) (Lemaitre *et al.* 2010).

the rivulet will be torn off by the wind. A simple force balance yields

$$\tfrac{1}{2}\rho U^2 \ell \overline{C}_F \sin^2 \alpha = \gamma(\cos\phi_R - \cos\phi_A), \tag{7.2}$$

where $\overline{C}_F$ is an average friction coefficient over the rivulet, and ϕ_R and ϕ_A are the receding and advancing angles of the water/air/cable capillary interaction. These angles correspond to the limit shape of the rivulet before it starts moving in the windward direction. This limit may be expressed in terms of the Weber number based on the normal cross-section velocity $U_N = U\sin\alpha$, $\mathrm{We}_N = \rho U_N^2 R/\gamma$, so that rivulets are expected to exist only when

$$\mathrm{We}_N < 2(\cos\phi_R - \cos\phi_A)\frac{R}{\ell C_F}. \tag{7.3}$$

Expressing the criteria for rivulet splitting and the criteria for rivulet tearing in terms of this Weber number, one may define in the Weber-α plane three regions, as in Figure 7.4(b) (Lemaitre *et al.* 2010). For low Weber numbers and thereby low wind velocities, and low inclination angles, only one gravitational rivulet exists. For high Weber numbers, rivulets do not exist. In between, two wind-induced rivulets exist.

This is the region where RWIV is expected. Existing experiments on wind-induced rivulets are shown on the same graph: they all fall in the domain predicted by the model. Although these simple models are based on strong assumptions on the two-dimensional aspect of rivulet formation and flow dynamics, as well as on neglecting the effect of axial flow of the rivulet, they give a first answer to the questions of how rivulets appear, locate themselves and possibly disappear. This does not give any interpretation on the mechanism of RWIV, but, because these vibrations require the existence of a wind-induced rivulet, this gives some hints on the effect of geometrical and physical parameters on the possibility of RWIV.

7.3 VIV, Galloping and Drag Crisis

Several interpretations of RWIV, and in some cases quantitative models, are based on known mechanisms of cross-flow-induced instabilities. The main issue in these interpretations is, of course, why a rainwater rivulet plays a role in the mechanisms.

Matsumoto *et al.* (1995, 2003b) proposed that RWIV are actually a special case of vortex-induced vibrations, VIV. By experiments on yawed cylinders with or without artificial rivulets they showed that (a) the unsteady lift force had a low-frequency component, corresponding to a Strouhal number based on the cable diameter as low as 0.02, (b) this low-frequency component was enhanced by the presence of a rivulet near the separation point, and (c) with a significant level of upstream turbulence, this low-frequency force disappeared, except when a rivulet existed. These features also appeared on the velocity fluctuations behind the cylinder, as illustrated in Figure 7.5(b). The existence of a low-frequency component is attributed to axial vortices that are generated by the component of the flow along the cylinder axis, as shown in Figure 7.5(a). These vortices interact with Kármán vortices and amplify them periodically, leading to a subharmonic vortex-induced lift on the cylinder. Refer also to Honda *et al.* (1995) for related experiments. The corresponding Strouhal number would therefore be of the form $St = St_0/N$, N being an integer as high as 10, and St_0 the Strouhal number relevant to Kármán wake shedding. This leads to possible VIV at reduced velocities equal to $U_R = N/St_0$. Values of $N = 10$ and $St_0 = 0.2$ lead to $U_R = 50$, which is consistent with practical observations of RWIV. In those approaches, the effect of the orientation angle β is then directly related to the existence of the axial flow. Yet, the reason why a rivulet would enhance the vortex-induced vibration phenomenon remains unclear.

To answer this, a first model of the interaction between a rivulet and the oscillating wake exists in the case of classical Kármán shedding: tests on a rigid cylinder with a true rivulet (Alam & Zhou 2007) showed a significant effect of the presence of the rivulet on the characteristics of lift. The unsteady lift coefficient increased in a limited range of velocities, and the frequency of vortex shedding then deviated from the Strouhal law, as seen in Figure 7.6. It was further observed in a similar experiment that, in the range of increased lift, the rivulets did in fact move at a frequency close to that of vortex shedding, when that frequency was close to that of the capillary mode of the rivulet (Lemaitre 2006). This suggests a mechanism very similar to lock-in in VIV, as described in Chapter 3, but here between the wake and the rivulet. Preliminary models showed the possibility of significantly increased fluctuating lift, even for a small rivulet (Lemaitre 2006). This strong effect is related

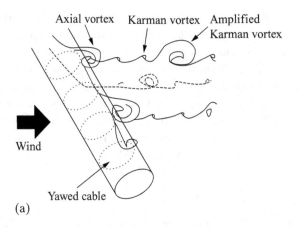

(a)

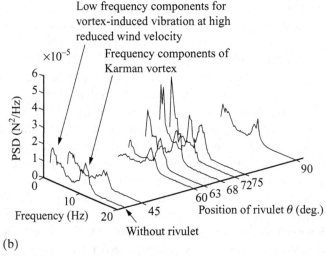

(b)

Figure 7.5. Mechanism of RWIV as described by Matsumoto *et al.* (2003b). (a) In vortex shedding from a yawed cylinder, axial flow results in amplified Kármán vortices, occurring at a low frequency. (b) Measurements of the fluctuating lift force show that the presence of a rivulet near the separation point increases this low-frequency shedding.

to the location of the rivulet at the separation point, which is the most sensitive point of the wake to external forcing. Such a coupling would generate forces in a range of frequencies corresponding to Kármán vortex shedding, and is therefore probably not the cause of what is described as RWIV. Yet, this might be an interpretation of how low-frequency shedding in relation to axial-flow effects, as described by Matsumoto *et al.* (2003b), may be enhanced by the presence of a rivulet.

A quite different interpretation is proposed by Macdonald & Larose (2008); see also some related work in Cosentino *et al.* (2003b) and Seidel & Dinkier (2006). Here again, the flow specificity of a yawed cylinder is the main cause of its motion, and the presence of rainwater rivulets is just supposed to enhance it. In the general case of inclination of the cylinder with respect to the wind (arbitrary β angle), the cross-section of the cable in the wind plane is elliptical, and its axis is not along the wind direction. A systematic experimental and analytical exploration shows that galloping of a dry yawed cylinder is in fact possible (Macdonald & Larose

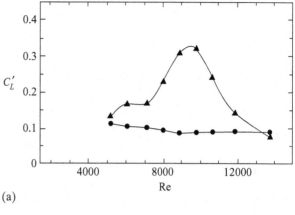

Figure 7.6. Effect of a rivulet on (a) the fluctuating lift coefficient and (b) the Strouhal number on a cylinder (Alam & Zhou 2007). Circles correspond to a cylinder without rivulets, and triangles with rivulets

(a)

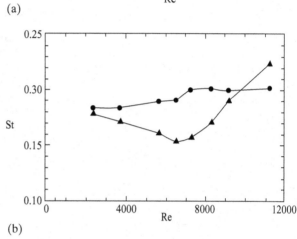

(b)

2006). This is consistent with the concepts presented in Chapter 2 of this book, but in a rather elaborate form in terms of geometry (see Figure 7.2). Moreover, in the range of critical Reynolds number, typically $Re = 2 \times 10^5$, sudden jumps in force coefficients enhance the phenomenon, consistently with the drag-crisis-induced instability presented in Section 2.6.1, but again in a more elaborate form. This defines a scenario for galloping of dry inclined cables (Macdonald & Larose 2006). There is a possible relation to RWIV, as suggested by Macdonald & Larose (2008), as follows. In the presence of water rivulets the cylinder roughness is actually modified, and consequently the critical Reynolds number, so that the possibility of the instability being enhanced by drag crisis may occur at velocities as low as 10 m/s on a real cable. This is consistent with field observations. Moreover, Flamand *et al.* (2001) did find a transitional flow behaviour caused by a rivulet, and the rivulet motion may trigger the transition. Note that if the drag crisis effect is a key feature of RWIV, then the reduced wind velocity is not a relevant parameter; Hikami & Shiraishi (1988) did in fact notice that the range of wind velocity for occurrence of RWIV did not depend much on the cable frequency. There is some evidence that part of the motion observed in RWIV is linked to galloping and Reynolds number transition, as in dry cables, but enhanced by the presence of rivulets (Macdonald & Larose 2008).

These two approaches, relating RWIV to VIV and to galloping with drag crisis, respectively, show that some of the excitation causing RWIV may not be so specific, the rain only enhancing classical phenomena. Yet, they do not allow the emergence

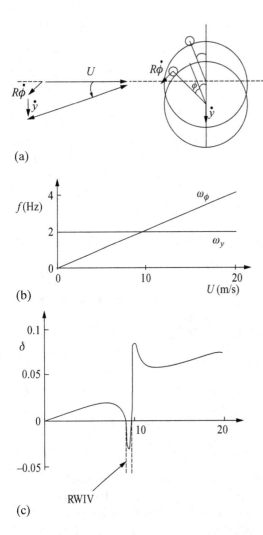

(a)

(b)

(c)

RWIV

Figure 7.7. Mechanism of RWIV as described by Yamaguchi (1990). (a) The rivulet is modelled as a moving small cylinder on the oscillating cable. (b) As the flow velocity is increased, the frequency of the rivulet mode approaches that of the cable, and coupled-mode flutter arises, with (c) a negatively damped mode.

of a satisfactory picture of the phenomenon, as the role of rainwater in all field- or wind-tunnel experiments was found to be more crucial in the existence of the instability than these two approaches suggest.

7.4 Yamaguchi's Model: A Cylinder-Rivulet-Coupled Instability

The essential, simplest and most widely used model of RWIV, at present, is that of Yamaguchi (1990). A large number of models, see for instance Van der Burgh (2004) and Cao *et al.* (2003), are based on similar basic assumptions with many refinements. The main originality of this approach is that the rivulet motion is now considered as a new degree of freedom in the system. In the simple two-dimensional case of vertical motion of a cross section of the cable, the position of the rivulet is defined by an angle $\phi(t)$, as in Figure 7.7(a). Using the quasi-steady assumption, all the forces acting on the cylinder and on the rivulet are derived from the lift and moment coefficients of a cylindrical section together with the small added cylinder. The derivation of the equations follows the standard pattern used for obtaining plunge and torsion galloping equations given in Chapter 2, but some specificities must be taken into account. First, because the rivulet is free to glide on the cylinder surface, it is assumed to have a zero stiffness. This is questionable in the case of a rivulet on a dry cylinder, where capillary forces add a stiffness, as discussed in the foregoing. Second, it is assumed that the flow-induced moment is only caused by forces acting on the rivulet. This is certainly the case for a protuberant rivulet, but less so for a flat rivulet as observed by several authors. Under these assumptions the coupled linear system reads

$$\ddot{y} + \omega_y^2 y = aU\dot{y} + bU\dot{\phi} + cU^2\phi, \tag{7.4}$$

$$\ddot{\phi} = dU\dot{\phi} + eU^2\phi + fU\dot{y}, \tag{7.5}$$

where ω_y is the natural frequency of the cable, and the dimensional coefficients a, b, c, d, e, f depend on the drag, lift and moment coefficients, and on the geometry and masses of the system (Yamaguchi 1990).

From these equations it appears that the frequency of displacement in the y direction is independent of the flow velocity (no fluid stiffness in the first equation), but the frequency of the rivulet motion is entirely flow-dependent: it is even proportional to the flow velocity. Figure 7.7(b) illustrates the resulting evolution of these two frequencies with the flow velocity. With typical values of the parameters for a cable and a rivulet (Yamaguchi 1990), these frequencies become comparable near $U = 10$ m/s. A straightforward modal analysis of these two equations shows that, at this coincidence of frequencies, the aerodynamic damping of the coupled mode becomes negative, corresponding to an instability, as shown in Figure 7.7(c). According to this model, RWIV is therefore a classical plunge-torsion-coupled instability, although the torsion is not that of the cable but the motion of the rivulet around the cable. Hence, although the model is simple in its formulation and relies on classical aeroelasticity theory, it is able to capture two important features of RWIV: the essential role of the rivulet and the limited range of instability. Yet, there are some strong limitations. First, the model is two-dimensional, ignoring all effects of the angle β, and of three-dimensional flow effects discussed in the previous subsection. Then, representing the rivulet by a small cylinder gliding on a larger one is a crude approximation in terms of geometry. Moreover, the interaction of the rivulet with the cylinder is affected by capillary forces and by friction. This model can therefore be considered as a first step to understanding RWIV, and many improvements have been introduced over of the years. For instance, Cosentino *et al.* (2003b) improved the model by taking into account the effect of the angle β on the geometry, by using a finer approximation of the load on the rivulet, based on experiments, and by adding some randomness in the load, to take into account the possibility of switching between laminar and turbulent states. The model allowed the derivation of the effect of β on the range of RWIV.

7.5 Concluding Remarks

Although the phenomenon of RWIV is now identified and experimentally documented, the question of the underlying mechanism is not fully clarified. All the models presented in this chapter reproduce or clarifiy some aspects of the phenomenon, and they probably contribute in some combination to the true mechanism. It should be emphasised that understanding RWIV poses a real challenge, because of the presence of a flowing liquid on the bluff body, and because of the three-dimensional geometry. Experiments and models are recent. Most of them have been published since the late 1990s and the field remains quite active. The dynamics of rivulet formation is now clearer, and one can expect similar progress on modelling the instability in the forthcoming years.

Epilogue

Here are some closing remarks on the topics covered, some on what is not covered and some on what remains to be done to reach "perfect understanding", fully realizing, of course, that this is an unattainable goal.*

Transverse galloping may be said to be well understood and that it can be well predicted. The same cannot be said of torsional galloping or of mixed transverse-torsional galloping/flutter, e.g. of bridge decks, or in situations involving substantial intertwining with vortex shedding. In this regard, the advent of models based on measured force and moment coefficients as in aeroelasticity has resulted in adequate predictive ability. However, this has had a deterious effect on the desire for enhanced physical understanding and on the funding opportunities towards achieving it.

Vortex-induced vibrations under lock-in conditions have received a great deal of attention over the years. Although understanding and modelling such vibrations has come a long way over the years, including the past decade, we still have a lot of ground to cover. Indeed, predictive tools are largely semi-empirical or they depend on CFD (which bypasses the need to understand physically). Thus, a great deal remains to be done.

Motivated in the 1970s and 1980s by wind-induced vibration problems on over-head transmission lines, and more recently by current-induced instabilities on off-shore structures, wake-induced flutter of small groups of cylinders has received considerable attention. For wind-induced instabilities, which typically occur at relatively high nondimensional velocities, quasi-steady theory is sufficient to give good predictions for the onset of instability. However, for cases where unsteady effects are more dominant, such as in waterflows, the aerodynamic modelling of the wake-induced forces is still an open question requiring a better understanding of the wake flow.

Fluidelastic instability of cylinder arrays in cross-flow received a lot of attention from the 1970s to the 1990s, culminating in fairly reliable design guidelines, at least for systems not radically different from those already studied. However, the basic physics requires further work, in particular, the study of near-wake dynamics and how this is related to the phase lag between cylinder motion and the forces generated thereby. Also needed is a deeper physical understanding as to fluidelastic instabilities in two-phase flows. Research in this topic has flagged over the past decade, mainly

---

* Refer, for instance, to the first footnote in Section 2.6.

because funding has become scarcer, and this because power plants of unusual design, requiring deeper understanding of possible dynamical behaviour and providing an impetus for further research, have simply not been commissioned.

The state of knowledge on ovalling instabilities of shells in cross-flow has reached a reasonable level of maturity, with a remarkable saga of the work leading to the discovery of their cause and of the battle for convincing others of its correctness. Still, some questions remain open, but because designers now know how to prevent ovalling instabilities, this area of research has been dormant over the past decade.

The opposite is true for rain-and-wind-induced vibrations (RWIV), the newest of all the phenomena covered, with serious study beginning a mere 20 years ago; ever since, this has been a hotbed of research activity, with several competing models and explanations, all with valuable facets, yet none totally satisfactory. More work remains to be done, indeed it *is* being done, to reach the level of maturity and understanding already achieved in the other topics covered in this book.

As mentioned in the Preface, several topics, which could have been included in this book, in fact have not been. An obvious example is wing aeroelasticity. Another relates to "lumpy" three-dimensional structures for which "cross-flow" is an inappropriate designation because there is no obvious long axis. Moreover, for the type of structures considered, only cross-flow-induced instabilities are treated, whereas other phenomena, such as turbulence-induced vibrations and vortex-induced vibrations with no lock-in, are left to more general textbooks on flow-induced vibrations.

It has been both interesting and rewarding to write this book. A particularly interesting aspect has been tracking the historical development of work on each topic and sorting out contributions and precedence of important ideas. Examples are contributions made by Roberts and Connors on the fluidelastic instability of cylinder arrays (Chapter 5), and those by Glauert and Den Hartog on galloping (Chapter 2).* Also important was deciphering what was really new, if anything, in some theories and what was fundamentally similar, even though appearing different.

---

* It is interesting that, conventionally, we talk about Connors' equation and the Den Hartog criterion, although a fairer notation would be the Roberts-Connors equation and the Glauert-Den Hartog criterion. But then we say America, and not Columbia-America! History and conventional usage do not always agree.

The Multiple Scales Method

This is not meant to be a formal treatment of the method, but rather an illustration of its use on a simple problem, that of the Van der Pol oscillator, the equation of motion of which may be written as

$$\frac{d^2x}{dt^2} + x = \varepsilon(1 - x^2)\frac{dx}{dt}, \tag{A.1}$$

where $\varepsilon \ll 1$; refer, e.g., to Nayfeh (1981).

Defining two timescales, that is an ordinary (fast) time and a slow time,

$$\xi = t \quad \text{and} \quad \eta = \varepsilon t, \tag{A.2}$$

and expressing $x = x(\xi, \eta)$, we can write

$$\frac{dx}{dt} = \frac{\partial x}{\partial \xi}\frac{d\xi}{dt} + \frac{\partial x}{\partial \eta}\frac{d\eta}{dt} = \frac{\partial x}{\partial \xi} + \varepsilon\frac{\partial x}{\partial \eta},$$

$$\frac{d^2x}{dt^2} = \frac{\partial^2 x}{\partial \xi^2} + 2\varepsilon\frac{\partial^2 x}{\partial \xi \partial \eta} + \varepsilon^2\frac{\partial^2 x}{\partial \eta^2}. \tag{A.3}$$

Expanding x in a Lindstedt-type perturbation series,

$$x = x_0 + \varepsilon x_1 + \varepsilon^2 x_2 + \cdots \tag{A.4}$$

and substituting into equation (A.1), we obtain

$$\frac{\partial^2 x_0}{\partial \xi^2} + 2\varepsilon\frac{\partial^2 x_0}{\partial \xi \partial \eta} + \varepsilon^2\frac{\partial^2 x_0}{\partial \eta^2} + \varepsilon\frac{\partial^2 x_1}{\partial \xi^2} + 2\varepsilon^2\frac{\partial^2 x_1}{\partial \xi \partial \eta} + \varepsilon^3\frac{\partial^2 x_1}{\partial \eta^3}$$

$$+ \cdots + x_0 + \varepsilon x_1 + \varepsilon^2 x_2 + \cdots$$

$$= \varepsilon\left(1 - x_0^2 - 2x_0 x_1\varepsilon - x_1^2\varepsilon^2 + \cdots\right)\left[\frac{\partial x_0}{\partial \xi} + \varepsilon\frac{\partial x_0}{\partial \eta} + \cdots \varepsilon\frac{\partial x_1}{\partial \xi} + \varepsilon^2\frac{\partial^2 x_1}{\partial \xi \partial \eta} + \cdots\right].$$

Collecting terms in ε^0 and ε^1, we obtain

$$\frac{\partial^2 x_0}{\partial \xi^2} + x_0 = 0, \tag{A.5}$$

$$\frac{\partial^2 x_1}{\partial \xi^2} + x_1 = -2 \frac{\partial^2 x_0}{\partial \xi \partial \eta} + \left(1 - x_0^2\right) \frac{\partial^2 x_0}{\partial \xi^2}, \tag{A.6}$$

and similar equations for ε^2, involving only x_2 on the left-hand side, *et seq.*

Equation (A.5) yields the generating solution

$$x_0 = A(\eta) \cos \xi + B(\eta) \sin \xi. \tag{A.7}$$

This solution embodies the fundamental character of the motion: it is not simple harmonic in time but one whose amplitude changes slowly with time.

Substituting into (A.6), we obtain, after expanding powers of trigonometric functions into simpler forms (e.g. $\cos^3 \xi = \frac{1}{4}(\cos 3\xi + 3 \cos \xi)$), we obtain

$$\frac{\partial^2 x_1}{\partial \xi^2} + x_1 = \left[2\dot{A} - A + \tfrac{1}{4}A(A^2 + B^2)\right] \sin \xi$$

$$+ \left[-2\dot{B} + B - \tfrac{1}{4}B(A^2 + B^2)\right] \cos \xi$$

$$+ \tfrac{1}{4}A(A^2 - 3B^2) \sin 3\xi + \tfrac{1}{4}B(B^2 - 3A^2) \cos 3\xi. \tag{A.8}$$

Eliminating the secular terms yields

$$\dot{A} = \tfrac{1}{2}A - \tfrac{1}{8}A(A^2 + B^2), \quad \dot{B} = \tfrac{1}{2}B - \tfrac{1}{8}B(A^2 + B^2). \tag{A.9}$$

Introducing polar coordinates,

$$A = \mathcal{R} \cos \theta \quad \text{and} \quad B = \mathcal{R} \sin \theta, \tag{A.10}$$

both of equations (A.9) become

$$\dot{\mathcal{R}} = \tfrac{1}{2}\mathcal{R} - \tfrac{1}{8}\mathcal{R}^3, \quad \dot{\theta} = 0. \tag{A.11}$$

To obtain the amplitude of a putative limit cycle we set $\dot{\mathcal{R}} = 0$, and we obtain

$$\mathcal{R} = \pm 2. \tag{A.12}$$

To study the stability of the limit cycle, we examine equation (A.11) and put $\mathcal{R} = 2 + r$. Linearizing about the limit cycle, we have $\dot{r} = [\mathcal{R} - \tfrac{3}{8}\mathcal{R}^2 r]_{r=2} = -\tfrac{1}{2}r$, indicating a diminishing perturbation. Hence, the limit cycle is stable.

In the foregoing the analysis was carried out to only first order; the procedure is similar if one pursues it to $\mathcal{O}(\varepsilon^2)$ and higher, but it would yield more precise estimates of the limit-cycle amplitude, as well as corrections to the frequency.

Measurement of Modal Damping for the Shells Used in Ovalling Experiments

These measurements were made *in situ*, with the shell mounted in the wind tunnel as in Figure 6.11(b), but with the wind turned off. A small shaker excited the shell, as close to one of its supports as possible, i.e. at a high-impedance point, so as to minimize coupling between it and the shell. Measurement of the input force was made via a force transducer attached to the shaker driving rod, and the shell vibration was picked up by the internally mounted fibre-optic Fotonic sensor. The resonant frequencies and modal logarithmic decrements were determined by obtaining the complex impedance of the system at different frequencies and plotting the results in the form of Nyquist plots (Ewins 1975; Ray *et al.* 1969; Ewins 1984). Sample results are shown in Figure B.1. It should be noted that this method is considered to be precise, so long as the data points fall sensibly on a circle, or a fairly large arc thereof, which may be seen to be the case in the results of Figure B.1. Once the resonant frequency, $f_{n,m}^*$, has been determined (which lies at the absolute maximum of the circle, parallel to the imaginary axis), then, given f_1 and f_2 of any of two other data points lying on the circle and the corresponding angles ϕ_1 and ϕ_2, the modal logarithmic decrement is found to be (Ewins 1975)

$$\delta_{n,m} = 2\pi(f_2 - f_1)/\left[f_{n,m}^*(\tan\phi_1 + \tan\phi_2)\right] .$$

To interpret these measurements fully, one must also determine the values of n and m associated with each $f_{n,m}$. For cantilevered shells, this could be done visually by observation from above, through the transparent ceiling of the tunnel. For clamped-clamped shells, however, this was not a trivial task. The value of n was determined by mounting the Fotonic sensor on a rotating platform and taking readings at different azimuthal locations at the same cross-sectional plane. For the dominant ovalling modes, these measurements were taken with the shell excited by the wind; for secondary modes, however, the shell was excited by the shaker at the appropriate frequencies. Typical results for the fourth and fifth mode of one shell are shown in Figure B.2. The Fotonic sensor system was not recalibrated at each azimuthal location, a very time-consuming task; as a result – as the centres of the rotating platform and the shell were probably not perfectly coincident, nor the reflectivity of the shell surface perfectly uniform – these measurements should be considered as *identifying* the mode concerned, rather than giving the precise modal shape.

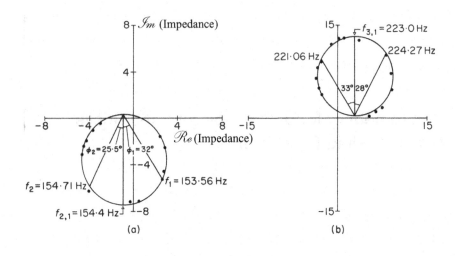

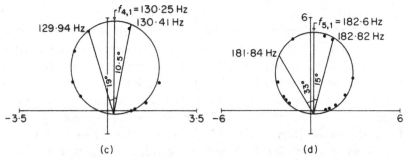

Figure B.1. Typical Nyquist plots obtained for the measurement of the natural frequencies and modal damping of clamped-clamped shells. (a) For shell B, (2,1)-mode: $f_{2,1} =$ 154.4 Hz, $\delta_{2,1} = 0.042$; (b) for shell O (with damping material applied), (3,1)-mode: $f_{3,1} =$ 223.0 Hz, $\delta_{3,1} = 0.071$; (c) for shell G, (4,1)-mode: $f_{4,1} = 130.2$ Hz, $\delta_{4,1} = 0.043$; (d) for shell G, (5,1)-mode: $f_{5,1} = 182.6$ Hz, $\delta_{5,1} = 0.037$.

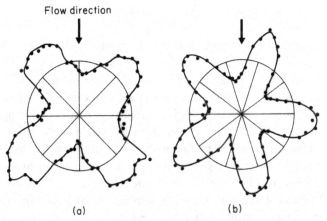

Figure B.2. Measurements of the circumferential mode shape corresponding to different resonant frequencies, obtained by rotating the Fotonic sensor inside the shell ovalling in the wind; (a) (4,1)-mode and (b) (5,1)-mode of a 147.3 mm diameter shell (shell G).

The axial modal shape was determined by traversing a Fotonic sensor along a vertical line on the surface of the shell. In some cases two Fotonic sensors were used, one trained on the upper part and the other on the lower part of the shell, so that the phase difference in cross-spectral-density analysis of their signals revealed whether the mode in question corresponded, say, to $m = 1$ or $m = 2$.

References*†

ABD-RABBO, A. & WEAVER, D. S. 1986 A flow visualization study of flow development in a staggered tube array. *Journal of Sound and Vibration* **106**, 241–256.

AFANASYEV, Y. D. & KORABEL, V. N. 2006 Wakes and vortex streets generated by translating force and force doublet: Laboratory experiments. *Journal of Fluid Mechanics* **553**, 119–141.

AGUIRRE, J. E. 1977 Flow-induced in-line vibrations of a circular cylinder. Ph.D. Dissertation, Imperial College, London, U.K.

ALAM, M. & ZHOU, Y. 2007 Turbulent wake of an inclined cylinder with water running. *Journal of Fluid Mechanics* **589**, 261–303.

ALEXANDER, C. M. 1981 The complex vibrations and implied drag of a long oceanographic wire in cross-flow. *Ocean Engineering* **8**, 379–406.

ALLNUTT, J. G., PRICE, S. J. & TUNSTALL, M. J. 1980 The control of sub-span oscillation of multi-conductor bundled transmission lines. Paper 22-01, CIGRÉ International Conference on Large High Voltage Electrical Systems, 25th Session, Vol. 1, Paris, France.

ALONSO, G., MESSEGUER, J. & PÉREZ-GRANDE, I. 2007 Galloping stability of triangular cross-sectional bodies: a systematic approach. *Journal of Wind Engineering and Industrial Aerodynamics* **95**, 928–940.

AL JAMAL, H. & DALTON, C. 2005 The contrast in phase angles between forced and self-excited oscillations of a circular cylinder. *Journal of Fluids and Structures* **20**, 467–482.

AMABILI, M. 2008 *Nonlinear Vibrations and Stability of Shells and Plates*. Cambridge: Cambridge University Press.

ANAGNOSTOPOULOS, P. 2002 *Flow-Induced Vibrations in Engineering Practice*. Boston: WIT Press.

ANDERSON, E. A. & SZEWCZYK, A. A. 1996 A look at a universal parameter for 2-D and 3-D bluff body flows. *Journal of Fluids and Structures* **10**, 543–553.

ANDJELIĆ, A. & POPP, K. 1989 Stability effects in a normal triangular cylinder array. *Journal of Fluids and Structures* **3**, 165–186.

ANDJELIĆ, A., AUSTERMANN, R. & POPP, K. 1990 Multiple stability boundaries of tubes in a normal triangular cylinder array. *ASME Journal of Pressure Vessel Technology* **114**, 336–343.

---

* The references are generally arranged alphabetically. However, for single-, double- and multiple-author papers *with the same first author*, they are listed as follows:

  (i) the single-author papers first: e.g., SMITH, A. 1990 before SMITH, A. 1991;

  (ii) the double-author papers next, according to the second author's name: e.g., SMITH, A. & BROWN, G. 1991 before SMITH, A. & GREEN, S. 1979;

  (iii) multiple-author papers, which will be cited in the text as, e.g., Smith *et al.* (1979), are listed last, *strictly chronologically*.

† All MERL Reports, e.g. Païdoussis and Denise (1970) are available on-line via www.library.mcgill.ca/psel/subguide/mecheng.htm

ANG, S.-Y. 1981 Ovalling oscillations of a clamped-clamped shell in cross-flow. Undergraduate B.Eng. (Honours) Thesis, McGill University, Montreal, Canada.

ANG, S.-Y. 1983 Ovalling oscillations of shells in cross flow: An analytical and experimental investigation. M. Eng. Thesis, McGill University, Montreal, Canada.

ANONYMOUS 1961 Wind forces on structures. Task Committee on Wind Forces of ASCE. See *Transactions ASCE* **126**, 1174.

ANTUNES, J., AXISA, F. & VENTO, M. A. 1990 Experiments on vibro-impact dynamics under fluidelastic instability. In *Flow-Induced Vibration 1990* (eds. S. S. Chen, F. Fujita & M. K. Au-Yang), PVP-Vol. 189, pp. 127–138. New York: ASME.

ANTUNES, J., AXISA, F. & VENTO, M. A. 1992a Experiments on tube/support interaction with feedback-controlled instability. *ASME Journal of Pressure Vessel Technology* **114**, 23–32.

ANTUNES, J., DE LANGRE, E., VENTO, M. A. & AXISA, F. 1992b A theoretical model for the vibro-impact motions of tube bundles under fluidelastic instability. In Proceedings International Symposium on Flow-Induced Vibration and Noise, Vol. 2: Cross-Flow Induced Vibration of Cylinder Arrays (eds. M. P. Païdoussis, S. S. Chen & D. A. Steininger), pp. 135–150. New York: ASME.

APELT, C. J. & WEST, G. S. 1975 The effects of wake splitter plates on the flow past a circular cylinder in the range $10^4 < R < 5 \times 10^4$. Part 2. *Journal of Fluid Mechanics* **71**, 145–160.

APELT, C. J., WEST, G. S. & SZEWCZYK, A. A. 1973 The effects of wake splitter plates on the flow past a circular cylinder in the range $10^4 < R < 5 \times 10^4$. *Journal of Fluid Mechanics* **61**, 187–198.

ARCHARD, J. F. & HIRST, M. E. 1956 The wear of metals under unlubricated conditions. *Proceedings of the Royal Society of London, Series A* **236**, 397–413.

ASCE 1955 1952 report of the Advisory Board on the investigation of suspension bridges. *Transactions of ASCE (American Society of Civil Engineers)* **120**, 721–781.

ASCE 1961 Wind forces on structures, Final report. *Transactions of ASCE* **126**, 1124–1198.

AUSTERMANN, R. & POPP, K. 1995 Stability behaviour of a single flexible cylinder in rigid tube arrays of different geometry subjected to cross-flow. *Journal of Fluids and Structures* **9**, 303–322.

AUSTERMANN, R., POPP, K. & KURNIK, W. 1992 Fluidelastic instabilities and bifurcations in a rotated triangular tube array subjected to cross flow. Part 2: Modelling and analysis. Proceedings International Symposium on Flow-Induced Vibration and Noise, Vol. 2: Cross-Flow Induced Vibration of Cylinder Arrays (eds. M. P. Païdoussis, S. S. Chen & D. A. Steininger), pp. 43–56. New York: ASME.

AU-YANG, M. K. 1984 Flow-induced vibration: Guidelines for design, diagnosis, and troubleshooting of common power plant components. Proceedings ASME Symposium on Flow-Induced Vibrations, Vol. 3: Vibration in Heat Exchangers (eds. M. P. Païdoussis, J. M. Chenoweth & M. D. Bernstein), pp. 119–137. New York: ASME.

AU-YANG, M. K. 2001 *Flow-Induced Vibration of Power and Process Plant Components*. New York: ASME Press.

AU-YANG, M. K., BLEVINS, R. D. & MULCAHY, T. M. 1991 Flow-induced vibration analysis of tube bundles: A proposed section III appendix N non-mandatory code. *ASME Journal of Pressure Vessel Technology* **113**, 257–267.

AXISA, F. 1993 Flow-induced vibration of nuclear system components. In *Technology for the '90s* (ed. M. K. Au-Yang et al.), pp. 899–956. New York: ASME.

AXISA, F. 2001 *Modélisation des systèmes mécaniques. Tome 4: Vibrations sous écoulements*. Paris: Hermès Science Publications.

AXISA, F., VILLARD, B., GIBERT, R. J., HETSRONI, G. & SUNDHEIMER, P. 1984 Vibration of tube bundles subjected to air-water and steam-water cross flow: Preliminary results on fluidelastic instability. Proceedings ASME Symposium on Flow-Induced Vibrations, Vol. 2: Vibration of Arrays of Cylinders in Cross Flow (eds. M. P. Païdoussis, M. K. Au-Yang & S. S. Chen), pp. 269–284. New York: ASME.

AXISA, F. ANTUNES, J. & VILLARD, B. 1988 Overview of numerical methods for predicting flow-induced vibration. *ASME Journal of Pressure Vessel Technology* **110**, 6–14.

BAJ, F. & DE LANGRE, E. 2003 Scaling of damping induced by bubbly flow across tubes. *Journal of Fluids and Structures* **17**, 351–364.

BALASUBRAMANIAN, S. & SKOP, R. A. 1996 A nonlinear oscillator model for vortex shedding from cylinders and cones in uniform and shear flows. *Journal of Fluids and Structures* **10**, 197–214.

BALSA, T. F. 1977 Potential flow interactions in an array of cylinders in cross-flow. *Journal of Sound and Vibration* **50**, 285–303.

BARRERO-GIL, A., SANZ-ANDRÉS, A. & ALONSO, G. 2009a Hysteresis in transverse galloping: The role of the inflection points. *Journal of Fluids and Structures* **25**, 1007–1020.

BARRERO-GIL, A., SANZ-ANDRÉS, A. & ROURA, M. 2009b Transverse galloping at low Reynolds numbers. *Journal of Fluids and Structures* **25**, 1236–1242.

BARTOLI, G. & MANNINI, C. 2008 A simplified approach to bridge deck flutter. *Journal of Wind Engineering and Industrial Aerodynamics* **96**, 229–256.

BATHE, K.-J., WILSON, E. L. & PETERSON, F. E. 1973 SAP IV. A structural analysis program for static and dynamic response of linear systems. Earthquake Engineering Research Center, Report EERC73-11, University of California, Berkeley, California.

BEARMAN, P. W. 1965 Investigation of the flow behind a two-dimensional model with a blunt trailing edge and fitted with splitter plates. *Journal of Fluid Mechanics* **21**, 241–255.

BEARMAN, P. W. 1969 On vortex shedding from a circular cylinder in the critical Reynolds number regime. *Journal of Fluid Mechanics* **37**, 577–585.

BEARMAN, P. W. 1972 Some recent measurements of the flow around bluff bodies in smooth and turbulent streams. *Proceedings Symposium on External Flows*, University of Bristol, UK, pp. B1–B15.

BEARMAN, P. W. & LUO, S. C. 1988 Investigation of the aerodynamic instability of a square-section cylinder by forced oscillation. *Journal of Fluids and Structures* **2**, 161–176.

BEARMAN, P. W. & MOREL, T. 1983 Effect of free stream turbulence on the flow around bluff bodies. *Progress in Aeronautical Science* **20**, 97–123.

BEARMAN, P. W. & TRUEMAN, D. M. 1971 An investigation of the flow around rectangular cylinders. *Imperial College Aero. Report* 71–15.

BEARMAN, P. W. & TRUEMAN, D. M. 1972 An investigation of the flow around rectangular cylinders. *The Aeronautical Quarterly* **23**, 229–237.

BEARMAN, P. W., GARTSHORE, I. S., MAULL, D. J. & PARKINSON, G. V. 1987 Experiments on flow-induced vibration of a square-section cylinder. *Journal of Fluids and Structures* **1**, 19–34.

BEARMAN, P. W., et al. 2006 In memory of Geoffrey Vernon Parkinson 13.12.1924–22.09.2005. *Journal of Fluids and Structures* **22**, 989–995.

BECKETT, D. 1969 *Great Buildings of the World – Bridges.* London: Paul Hamlyn.

BÉLANGER, F., PAÏDOUSSIS, M. P. & DE LANGRE, E. 1995 Time-marching analysis of fluid-coupled systems with large added mass. *AIAA Paper 1994–2267, AIAA Journal* **33**, 752–757.

BÉNARD, H. 1908 Formation des centres de giration à l'arrière d'un obstacle en mouvement. *Comptes Rendus hebdomadaires des séances de l'Académie des Sciences* (Paris) **147**, 839–842.

BENJAMIN, T. B. 1963 The threefold classification of disturbances in flexible surfaces bounding inviscid flows. *Journal of Fluid Mechanics* **16**, 436–450.

BILLAH, K. Y. & SCANLAN, R. H. 1991 Resonance, Tacoma Narrows Bridge failure and under-graduate physics textbooks. *American Journal of Physics* **59**, 118–124.

BISHOP, R. E. D. & HASSAN, A. Y. 1964 The lift and drag forces on a circular cylinder oscillating in a flowing fluid. *Proceedings of the Royal Society of London. Series A, Mathematical and Physical Sciences* **277**, 51–75.

BISHOP, R. E. D. & JOHNSON, D. C. 1960 *The Mechanics of Vibration.* Cambridge: Cambridge University Press.

BISPLINGHOFF, R. L., ASHLEY, H. & HALFMAN, R. L. 1955 *Aeroelasticity.* Cambridge, MA: Addison-Wesley (Reprinted by Dover Publications, 1996).

BLACKBURN, H. M. & HENDERSON, R. D. 1999 A study of two-dimensional flow past an oscillating cylinder. *Journal of Fluid Mechanics* **385**, 255–286.

BLEVINS, R. D. 1974 Fluidelastic whirling of a tube row. *ASME Journal of Pressure Vessel Technology* **96**, 263–267.

BLEVINS, R. D. 1977a Fluid elastic whirling of tube rows and tube arrays. *ASME Journal of Fluids Engineering* **99**, 457–460.

BLEVINS, R. D. 1977b *Flow-Induced Vibration*. New York: Van Nostrand Reinhold.

BLEVINS, R. D. 1979 Fluid damping and the whirling instability of tube arrays. In *Flow-Induced Vibrations* (eds. S. S. Chen & M. D. Bernstein), pp. 35–39, New York: ASME.

BLEVINS, R. D. 1984 A rational algorithm for predicting vibration-induced damage to tube and shell heat exchangers. In Proceedings ASME Symposium on Flow-Induced Vibrations, Vol. 3: Vibration in Heat Exchangers (eds. M. P. Païdoussis, J. M. Chenoweth & M. D. Bernstein), pp. 87–101. New York: ASME.

BLEVINS, R. D. 1990 *Flow-Induced Vibration*, 2nd edition. New York: Van Nostrand Reinhold.

BLEVINS, R. D. 2005. Forces on and stability of a cylinder in a wake. *ASME Journal of Offshore Mechanics and Arctic Engineering* **127**, 39–45.

BLEVINS, R. D. & IWAN, W. D. 1974 The galloping response of a two-degree-of-freedom system. *Journal of Applied Mechanics* **41**, 1113–1118.

BLEVINS, R. D., GIBERT, R. J. & VILLARD, B. 1981 Experiments on vibration of heat exchanger tube arrays in cross flow. 6th Conference on Structural Mechanics in Reactor Technology, Paper B6/9.

BOKAIAN, A. 1989 Galloping of a circular cylinder in the wake of another. *Journal of Sound and Vibration* **128**, 71–85.

BOKAIAN, A. R. & GEOOLA, F. 1982 Hydrodynamic galloping of rectangular cylinders. *Proceedings International Conference on Flow Induced Vibrations in Fluid Engineering*, Reading, England, pp. 105–129.

BOKAIAN, A. & GEOOLA, F. 1984a Wake-induced galloping of two interfering cylinders. *Journal of Fluid Mechanics* **146**, 383–415.

BOKAIAN, A. & GEOOLA, F. 1984b Proximity-induced galloping of two interfering circular cylinders. *Journal of Fluid Mechanics* **146**, 417–449.

BOKAIAN, A. R. & GEOOLA, F. 1984c Hydroelastic instabilities of square cylinders. *Journal of Sound and Vibration* **92**, 117–141.

BOKAIAN, A. & GEOOLA, F. 1987 Flow-induced vibrations of marine risers. *ASCE Journal of Waterway, Port, Coastal and Ocean Engineering* **113**, 22–38.

BOKAIAN, A. & GEOOLA, F. 1989 Some observations on the cross flow oscillations of a pair of interfering circular cylinders. *Aeronautical Journal* **93**, 66–72.

BOSDOGIANNI, A. & OLIVARI, D. 1996 Wind-and rain-induced oscillations of cables of stayed bridges. *Journal of Wind Engineering and Industrial Aerodynamics* **64**, 171–185.

BOURQUE, C. & NEWMAN, B. G. 1960 Reattachment of a two-dimensional incompressible jet to an adjacent flat plate. *Aeronautical Quarterly* **11**, 201–232.

BRADSHAW, P. 1965 The effect of wind tunnel screens on nominally two dimensional boundary layers. *Journal of Fluid Mechanics* **22**, 679–687.

BRIKA, D. & LANEVILLE, A. 1993 Vortex-induced vibrations of a long flexible circular cylinder. *Journal of Fluid Mechanics* **250**, 481–508.

BRIKA, D. & LANEVILLE, A. 1999 The flow interaction between a stationary cylinder and a downstream flexible cylinder. *Journal of Fluids and Structures* **13**, 579–606.

BROADBENT, E. G. & WILLIAMS, M. 1956 Effect of structural damping on binary flutter. Aeronautical Research Council, R. & M. 3169.

BROOKS, N. P. H. 1960 Experimental investigation of the aeroelastic instability of bluff cylinders. M.A.Sc. thesis, University of British Columbia, Vancouver, Canada.

BROWNJOHN, J. M. W. & BOGUNOVIC JAKOBSEN, J. 2001 Strategies for aeroelastic parameter identification from bridge deck free vibration data. *Journal of Wind Engineering and Industrial Aerodynamics* **89**, 1113–1136.

BRZOZOWSKI, V. J. & HAWKS, R. J. 1976 Wake-induced full-span instability of bundle conductor transmission lines. *AIAA Journal* **14**, 179–184.

CAI, Y. & CHEN, S. S. 1993a Chaotic vibrations of nonlinearly supported tubes in crossflow. *ASME Journal of Pressure Vessel Technology* **115**, 128–134.

CAI, Y. & CHEN, S. S. 1993b Non-linear dynamics of loosely supported tubes in crossflow. *Journal of Sound and Vibration* **168**, 449–468.

CAI, Y., CHEN, S. S. & CHANDRA, S. 1992 A theory for fluidelastic instability of tube-support-plate-inactive modes. *ASME Journal of Pressure Vessel Technology* **114**, 139–148.

CAO, D., TUCKER, R. & WANG, C. 2003 A stochastic approach to cable dynamics with moving rivulets. *Journal of Sound and Vibration* **268**, 291–304.

CARBERRY, J., SHERIDAN, J. & ROCKWELL, D. 2005 Controlled oscillations of a cylinder: Forces and wake modes. *Journal of Fluid Mechanics* **538**, 31–69.

CEGB 1966 Report of the Committee of Inquiry into the Collapse of Cooling Towers at Ferrybridge on Monday, 1 November 1965. Central Electricity Generating Board. London: H. M. Stationery Office.

CHAPLIN, J. R., BEARMAN, P. W., CHENG, Y., FONTAINE, E., GRAHAM, J. M. R., HERFJORD, K., HUARTE, F. J. H., ISHERWOOD, M., LAMBRAKOS, K., LARSEN, C. M., MENEGHINI, J. R., MOE, G., PATTENDEN, R., TRIANTAFYLLOU, M. S. & WILLDEN, R. H. J. 2005 Blind predictions of laboratory measurements of vortex-induced vibrations of a tension riser. *Journal of Fluids and Structures* **21**, 25–40.

CHEN, A., HE, X. & XIANG, H. 2002 Identification of 18 flutter derivatives of bridge decks. *Journal of Wind Engineering and Industrial Aerodynamics* **90**, 2007–2022.

CHEN, S. H. & CHEN, S. S. 1993 Chaotic vibration in fluidelastic instability of a tube row in crossflow. In *Flow-Induced Vibration and Fluid-Structure Interaction* (eds. M. K. Au-Yang & D. C. Ma), pp. 11–20. New York: ASME.

CHEN, S. S. 1975 Vibration of nuclear fuel bundles. *Nuclear Engineering and Design* **35**, 399–422.

CHEN, S. S. 1978 Crossflow-induced vibrations of heat exchangers tube banks. *Nuclear Engineering and Design* **47**, 67–86.

CHEN, S. S. 1983a Instability mechanisms and stability criteria of a group of circular cylinders subjected to cross-flow. Part I: Theory. *ASME Journal of Vibration, Acoustics, Stress and Reliability in Design* **105**, 51–58.

CHEN, S. S. 1983b Instability mechanisms and stability criteria of a group of circular cylinders subject to cross-flow. Part II: Numerical results and discussion. *ASME Journal of Vibration, Acoustics, Stress and Reliability in Design* **105**, 253–260.

CHEN, S. S. 1984 Guidelines for the instability flow velocity of tube arrays in cross-flow. *Journal of Sound and Vibration* **93**, 439–455.

CHEN, S. S. 1987 *Flow-Induced Vibration of Circular Cylindrical Structures*. Washington, DC: Hemisphere Publishing.

CHEN, S. S. 1991a A review of dynamic tube-support interaction in heat exchangers. In *Flow Induced Vibrations*, pp. 111–120. London: I.Mech.E.

CHEN, S. S. 1991b Flow-induced vibrations in two-phase flow. *ASME Journal of Pressure Vessel Technology* **113**, 234–241.

CHEN, S. S. & CHANDRA, S. 1991. Fluidelastic instabilities in tube bundles exposed to nonuniform cross-flow. *Journal of Fluids and Structures* **5**, 299–322.

CHEN, S. S. & JENDRZEJCZYK, J. A. 1981. Flow-velocity-dependence of damping in tube arrays subjected to liquid cross flow. *ASME Journal of Pressure Vessel Technology* **103**, 130–135.

CHEN, S. S. & JENDRZEJCZYK, J. A. 1983. Stability of tube arrays in crossflow. *Nuclear Engineering and Design* **75**, 351–374.

CHEN, S. S. & SRIKANTIAH, G. S. 2001. Motion-dependent fluid force coefficients for tube arrays in crossflow. *ASME Journal of Pressure Vessel Technology* **123**, 429–436.

CHEN, S. S., JENDRZEJCZYK, J. A. & WAMBSGANSS, M. W. 1984. Dynamics of tubes in fluid with tube-baffle interaction. In ASME Symposium on Flow-Induced Vibrations, Vol. 2: Vibration of Arrays of Cylinders in Cross Flow (eds. M. P. Païdoussis, M. K. Au-Yang & S. S. Chen), pp. 285–304. New York: ASME.

CHEN, S. S., ZHU, S. & CAI, Y. 1995a An unsteady-flow theory for vortex-induced vibration. *Journal of Sound and Vibration* **184**, 73–92.

CHEN, S. S., ZHU, S. & CAI, Y. 1995b Experiment of chaotic vibration of loosely supported tube rows in cross-flow. *ASME Journal of Pressure Vessel Technology* **117**, 204–212.

CHEN, S. S., CAI, Y. & SRIKANTIAH, G. S. 1998 Fluid damping controlled instability of tubes in crossflow. *Journal of Sound and Vibration* **217**, 883–907.

CHEN, X., KAREEM, A. & MATSUMOTO, M. 2001 Multimode coupled flutter and buffeting analysis of long span bridges. *Journal of Wind Engineering and Industrial Aerodynamics* **89**, 649–664.

CHEN, Y. N. 1972 Fluctuating lift forces of the Karman vortex streets on single circular cylinders and in tube bundles, Part 1: The vortex street geometry of the single circular cylinder. *ASME Journal of Engineering for Industry* **94**, 603–612.

CHEN, Y. N. 1980 Jet swing as governing factor for fluid-elastic instability of tube bundles. In Century 2 Pressure Vessel and Piping Conference, San Francisco, Paper 80-C2/PVP-103. New York: ASME.

CHENG, M. & MORETTI, P. M. 1989 Flow instabilities in tube bundles. In *Flow-Induced Vibration-1989* (eds. M. K. Au-Yang, S. S. Chen, S. Kaneko & R. Chilukuri), pp. 11–16. New York: ASME.

CHOMAZ, J.-M. 2005 Global instabilities in spatially developing flows: Non-normality and nonlinearity. *Annual Review of Fluid Mechanics* **37**, 357–392.

CHOMAZ, J.-M., HUERRE, P. & REDEKOPP, L. G. 1988 Bifurcations to local and global modes in spatially developing flows. *Physical Review Letters* **60**, 25–28.

CLAREN, R., DIANA, G., GIORDANA, F. & MASSA, E. 1971 The vibrations of transmission line conductor bundles. Transactions of the Institute of Electronic and Electrical Engineers, Paper 71 TP 158 PWR. 1796–1810, PAS-90.

CLAREN, R., DIANA, G. & NICOLINE, P. 1974 Vibrations in multiple conductor bundles. Paper 22-08, CIGRÉ International Conference on Large High Voltage Electrical Systems, 25th Session, Volume 1, Paris, France.

CONNORS, H. J., JR. 1970 Fluidelastic vibration of tube arrays excited by cross-flow. In *Flow-Induced Vibration in Heat Exchangers* (ed. D. D. Reiff), pp. 42–56. New York: ASME.

CONNORS, H. J. 1978 Fluidelastic vibration of heat exchanger tube arrays. *ASME Journal of Mechanical Design* **100**, 347–353.

CONNORS, H. J. 1980 Fluidelastic vibration of tube arrays excited by nonuniform cross flow. In *Flow-Induced Vibration in Power Plant Components* (ed. M. K. Au-Yang), PVP-Vol. 41, pp. 93–107. New York: ASME.

CONNORS, H. J. & KRAMER, F. A. 1991 U-bend shaker test investigation of tube/AVB wear potential. In *Flow Induced Vibrations*, pp. 57–67. London: I.Mech.E.

COOK, G. R. & SIMIU, E. 1990 Chaotic motions of forced and coupled galloping oscillators. *Journal of Wind Engineering and Industrial Aerodynamics* **36**, 1083–1094.

COOPER, K. R. 1974 Wind tunnel measurements of the steady aerodynamic forces on a smooth circular cylinder immersed in the wake of an identical cylinder. National Aeronautical Establishment Laboratory Technical Report LRT-LA-119, National Research Council, Ottawa, Canada.

CORLESS, R. M. & PARKINSON, G. V. 1988 A model of the combined effects of vortex-induced oscillation and galloping. *Journal of Fluids and Structures* **3**, 203–220.

CORLESS, R. M. & PARKINSON, G. V. 1992 Mathematical modelling of the combined effects of vortex-induced vibration and galloping. Part II. Proceedings International Symposium Flow-Induced Vibration and Noise, Vol. 6: Bluff-Body/Fluid and Hydraulic Machine Interactions (eds. M. P. Païdoussis, C. Dalton & D. S. Weaver), pp. 39–63. New York: ASME.

CORLESS, R. M. & PARKINSON, G. V. 1993 Mathematical modelling of the combined effects of vortex-induced vibration and galloping. Part II. *Journal of Fluids and Structures* **7**, 825–848.

COSENTINO, N., FLAMAND, O. & CECCOLI, C. 2003a Rain-wind induced vibration of inclined stay cables. Part I: Experimental investigation and physical explanation. *Wind and Structures* **6**, 471–484.

COSENTINO, N., FLAMAND, O. & CECCOLI, C. 2003b Rain-wind induced vibration of inclined stay cables. Part II: Mechanical modeling and parameter characterisation. *Wind and Structures* **6**, 485–498.

COSSU, C. & MORINO, L. 2000 On the instability of a spring-mounted circular cylinder in a viscous flow at low Reynolds numbers. *Journal of Fluids and Structures* **14**, 183–196.

COUNIHAN, J. 1963 Lift and drag measurements on stranded cables. Imperial College of Science and Technology, Report 117, London, U.K.

COURCHESNE, J. & LANEVILLE, A. 1982 An experimental evaluation of drag coefficient for rectangular cylinders exposed to grid-turbulence. *ASME Journal of Fluids Engineering* **104**, 523–528.

DAVIS, D. A., RICHARDS, D. J. W. & SCRIVEN, R. A. 1963 Investigation of conductor oscillation on the 275 kV crossing over the rivers Severn and Wye. *Proceedings Institution of Electrical Engineers* **110**, 205–219.

DAVISON, A. E. 1932 Dancing conductors. *AIEE Transactions* **49**, 1444–1449.

DE LANGRE, E. 2002 *Fluides et Solides*. Paris: Éditions de l'École polytechnique.

DE LANGRE, E. 2006 Frequency lock-in is caused by coupled-mode flutter. *Journal of Fluids and Structures,* **22**, 783–791.

DE LANGRE, E. 2008 Effects of wind on plants. *Annual Review of Fluid Mechanics* **40**, 141–168.

DE LANGRE, E., DOVEIL, F., PORCHER, G. & AXISA, F. 1990 Chaotic and periodic motion of a non-linear oscillator in relation with flow-induced vibrations of loosely supported tubes. In *Flow-Induced Vibration-1990* (eds. S. S. Chen, F. Fujita & M. K. Au-Yang) PVP-Vol. 189, pp. 119–125. New York: ASME.

DE LANGRE, E., HADJ-SADOK, C. & BEAUFILS, B. 1992 Non-linear vibrations induced by fluidelastic forces in tube bundles. In Proceedings International Symposium on Flow-Induced Vibration of Cylinder Arrays (eds. M. P. Païdoussis, S. S. Chen & D. A. Steininger), pp. 107–134. New York: ASME.

DELAIGUE, D. & PLANCHARD, J. 1986. Homogenization of potential flow models for the dynamics of cylinder arrays in transient cross-flow. In Proceedings ASME Symposium on Flow-Induced Vibrations (eds. S. S. Chen, J. C. Simonis & Y. S. Shin), PVP-Vol. 104, pp. 139–145, New York: ASME.

DELENNE, B., GAY, N., CAMPISTRON, R. & BANNER, D. 1997 Experimental determination of motion-dependent fluid forces in two-phase water-freon cross flow. In Proceedings 4th International Symposium on Fluid-Structure Interactions, Aeroelasticity, Flow-Induced Vibration and Noise, Vol. II (ed. M.P. Païdoussis), pp. 349–356. New York: ASME.

DEN HARTOG, J. P. 1932 Transmission line vibration due to sleet. *Transactions of the American Institute of Electrical Engineers* **51**, 1074–1076.

DEN HARTOG, J. P. 1956 *Mechanical Vibrations*. New York: McGraw-Hill.

DENIZ, S. & STAUBLI, T. 1997 Oscillating rectangular and octagonal profiles: Interaction of leading-and trailing-edge vortex formation. *Journal of Fluids and Structures* **11**, 3–31.

DICKEY, W. L. & WOODRUFF, G. B. 1956 The vibration of steel stacks. *Transactions ASCE* **121**, 1054–1070.

DOCKSTADER, E. A., SWIGER, W. F. & IRELAND, E. 1956 Resonant vibration of steel stacks. *Transactions ASCE* **121**, 1088–1112.

DONE, G. T. S. 1963 The effect of linear damping on flutter speed. Aeronautical Research Council, Reports & Memoranda 3396.

DONE, G. T. S. & SIMPSON, A. 1977 Dynamic stability of certain conservative and non-conservative systems. *I.Mech.E. Journal of Mechanical Engineering Science* **19**, 251–263.

DONNELL, L. H. 1933 Stability of thin-walled tubes under torsion. NACA Report 479.

DOWELL, E. H. 1995 *A Modern Course In Aeroelasticity*. Dordrecht: Kluwer Academic Publisher.

EDWARDS, A. T. & MADEYSKI, A. 1956 Progress report on the investigation of galloping of transmission line conductors. *AIEE Transactions* **75**, 666–686.

EISINGER, F. L. & RAO, M. S. M. 1998 Effect of tube-to-tube ties on fluidelastic instability of tube arrays exposed to crossflow. *ASME Journal of Pressure Vessel Technology* **120**, 179–185.

EISINGER, F. L., RAO, M. S. M. & STEININGER, D. A. 1989 Numerical simulation of fluidelastic instability of multispan tubes partially exposed to cross flow. In Proceedings 10th International Conference on Structural Mechanics in Reactor Technology (SMiRT), Vol. T: Flow Induced Dynamics, pp. 45–56.

EISINGER, F. L., RAO, M. S. M., STEININGER, D. A. & HASLINGER, K. H. 1991. Numerical calculation of fluidelastic vibration and comparison with experimental results. *Flow-Induced Vibration and Wear 1991* (eds. M. K. Au-Yang & F. Hara), PVP-Vol. 206, pp. 101–111. New York: ASME.

ERICSSON, L. E. & REDING, J. P. 1988 Fluid mechanics of dynamic stall. Part I: Unsteady flow concepts. *Journal of Fluids and Structures* **2**, 1–33.

EWINS, D. J. 1975 Measurement and application of mechanical impedance data. *Journal of Society of Environmental Engineers* **14**, 3–12.

EWINS, D. J. 1984 *Modal Testing: Theory and Practice*. Baldock, U.K.: Research Studies Press.

FACCHINETTI, M. L., DE LANGRE, E. & BIOLLEY, F. 2002 Vortex shedding modeling using diffusive Van der Pol oscillators. *Comptes Rendus Mécanique* **330**, 451–456.

FACCHINETTI, M. L., DE LANGRE, E. & BIOLLEY, F. 2004a Coupling of structure and wake oscillators in vortex-induced vibrations. *Journal of Fluids and Structures* **19**, 123–140.

FACCHINETTI, M. L., DE LANGRE, E. & BIOLLEY, F. 2004b Vortex-induced travelling waves along a cable. *European Journal of Mechanics B: Fluids* **23**, 199–208.

FARQUHARSON, F. B. 1949–1954 Aerodynamics stability of suspension bridges with special reference to the Tacoma Narrows Bridge. Parts I, III and IV. Bulletin No. 116, University of Washington Engineering Experiment Station, Seattle, WA, USA.

FEENSTRA, P., WEAVER, D. S. & JUDD, R. L. 2000 Modelling two-phase flow-excited fluid-elastic instability in tube arrays. In *Flow-Induced Vibration* (eds. S. Ziada & T. Staubli), pp. 545–554. Rotterdam: Balkema.

FENG, C. C. 1968 The measurement of vortex-induced effects in flow past stationary and oscillating circular and D-section cylinders. MASc Thesis, University of British Columbia, Vancouver, Canada.

FLAMAND, O. 1995 Rain-wind induced vibration of cables. *Journal of Wind Engineering & Industrial Aerodynamics* **57**, 353–362.

FLAMAND, O., PEUBE, J. & PAPANIKOLAS, P. 2001 An explanation of the rain-wind induced vibration of inclined stays. In *Fourth International Symposium of Cable Dynamics, Montréal, Canada*, AIM, pp. 69–76.

FLÜGGE, W. 1957 *Statik und Dynamik der Schalen*. 2nd edition, Berlin: Springer-Verlag.

FRANCE, E. R., CONNORS, H. J. 1991 Simulation of flow induced vibration characteristics of a steam generator U-tube. In *Flow Induced Vibrations*, pp. 33–43. London: I.Mech.E.

FRANDSEN, J. B. 2004 Numerical bridge deck studies using finite elements. Part I: flutter. *Journal of Fluids and Structures* **19**, 171–191.

FRANKLIN, R. E. & SOPER, B. M. H. 1977 An investigation of fluidelastic instabilities in tube banks subjected to fluid cross-flow. In Proceedings 4th International Conference on Structural Mechanics in Reactor Technology (SMiRT), Paper No. F6/7.

FRICK, T. M., SOBEK, T. E. & REAVIS, R. J. 1984 Overview on the development and implementation of methodologies to compute vibration and wear of steam generator tubes. In *ASME Symposium on Flow-Induced Vibrations, Vol. 3: Vibration in Heat Exchangers* (eds. M. P. Païdoussis, J. M. Chenoweth & M. D. Bernstein), pp. 149–161. New York: ASME.

FRICKER, A. J. 1991 Vibro-impacting of fluid-elastically unstable heat exchanger tubes with support clearances. In *Flow Induced Vibrations*, pp. 129–137. I.Mech.E: London.

FRICKER, A. J. 1992 Numerical analysis of the fluidelastic vibration of a steam generator tube with loose supports. *Journal of Fluids and Structures* **6**, 85–108.

FUJINO, Y. 2002 Vibration, control and monitoring of long-span bridges-recent research developments and practice in Japan. *Journal of Constructional Steel Research* **58**, 71–97.

FULLANA, J. M. & BEAUD, F. 1999 Identification of a quasi-unsteady model for cross-flow-induced vibration of tube arrays, *Flow-Induced Vibration*, PVP-Vol. 389 (ed. M. J. Pettigrew), pp. 65–71. New York: ASME.

FUNG, Y. C. 1955 *An Introduction to the Theory of Aeroelasticity*. New York: Wiley (reprinted by Dover Publications, 1993).

FURNES, G. K. 2000 On marine riser responses in-and depth-dependent flows. *Journal of Fluids and Structures* **14**, 257–273.

GABBAI, R. D. & BENAROYA, H. 2005 An overview of modeling and experiments of vortex-induced vibration of circular cylinders. *Journal of Sound and Vibration* **282**, 575–616.

GAWRONSKI, K. E. & HAWKS, R. J. 1977 Effect of conductor geometry on bundle conductor galloping. *Electrical Power Systems Research* **1**, 181–188.

GAY, N., DECEMBRE, P. & LAUNAY, J. 1988 Comparison of air-water to water-freon two phase cross flow effects on vibratory behavior of a tube bundle. In International Symposium on Flow-Induced Vibration and Noise, Vol. 3: Flow-Induced Vibration and Noise in Cylinder Arrays (eds. M. P. Païdoussis, S. S. Chen & M. D. Bernstein), pp. 139–158. New York: ASME.

GE, Y. J. & TANAKA, H. 2000 Aerodynamic flutter analysis of cable-supported bridges by multi-mode and full-mode approaches. *Journal of Wind Engineering and Industrial Aerodynamics* **86**, 123–153.

GIBERT, R. J. 1988 *Vibrations des structures*. Paris: Eyrolles.

GIBERT, R. J., CHABRERIE, J. & SAGNER, M. 1978 Tube bundle vibrations in transversal flow. In Proceedings BNES Conference on Vibration in Nuclear Plant, Keswick, UK, Paper 6.2.

GIBERT, R. J., SAGNER, M. & DOYEN, R. 1981 Vibration of tube arrays in transversal flow. In Transactions of 6th International Conference on Structural Mechanics in Reactor Technology (SMiRT), Paper B6/7.

GIBSON, R. D., TONKS, N. & WILKINSON, T. S. 1979 A note on the phase relationships involved in the whirling instability in tube arrays. *ASME Journal of Fluids Engineering* **101**, 530.

GIBSON, R. D., TONKS, N. & WILKINSON, T. S. 1980 A note on the phase relationships involved in the whirling instability in tube arrays. *ASME Journal of Fluids Engineering* **102**, 254–255.

GILLEN, S. & MESKELL, C. 2008 Variation of fluidelastic critical velocity with mass ratio and Reynolds number in a densely packed normal triangular tube array. In Proceedings of the 9th International Conference on Flow-Induced Vibration – FIV 2008 (eds. I. Zolotarev & J. Horaček), pp.193–198. Prague, Czech Republic.

GLAUERT, H. 1919 The rotation of an aerofoil about a fixed axis. (British) Advisory Committee for Aeronautics (ARC) R & M No. 595. In Technical Report of ARC for 1918–1919, pp. 443–447. London: H.M.S.O.

GOPALKRISHNAN, R. 1993 Vortex-induced forces on oscillating bluff cylinders. Ph.D. thesis, Massachusetts Institute of Technology, Cambridge, MA.

GORMAN, D. J. 1978 Experimental development of design criteria to limit liquid cross-flow-induced vibration in nuclear reactor heat exchange equipment. *Nuclear Science and Engineering* **99**, 324–336.

GOVARDHAN, R. & WILLIAMSON, C. H. K. 2002 Resonance forever: Existence of a critical mass and an infinite regime of resonance in vortex-induced vibration. *Journal of Fluid Mechanics* **473**, 147–166.

GOVARDHAN, R. N. & WILLIAMSON, C. H. K. 2005 Vortex-induced vibrations of a sphere. *Journal of Fluid Mechanics* **531**, 11–47.

GOVARDHAN, R. N. & WILLIAMSON, C. H. K. 2006 Defining the modified Griffin plot in vortex-induced vibration: revealing the effect of Reynolds number using controlled damping. *Journal of Fluid Mechanics* **561**, 147–180.

GOYDER, H. D. G. 1990 Fluidelastic instability in multispan tube bundles. In ASME Symposium on Flow-Induced Vibrations (eds. S. S. Chen, F. Fujita & M. K. Au-Yang), PVP-Vol. 189, pp. 51–63. New York: ASME.

GOYDER, H. D. G. & TEH, C. E. 1984 Measurement of the destabilizing forces on a vibrating tube in a fluid cross flow. In Proceedings ASME Symposium on Flow-Induced Vibrations, Vol. 2: Vibration of Arrays of Cylinders in Cross Flow (eds. M. P. Païdoussis, M. K. Au-Yang & S.-S. Chen), pp. 151–163. New York: ASME.

GOYDER, H. G. D. & TEH, C. E. 1989 A study of the impact dynamics of loosely supported heat exchanger tubes. *ASME Journal of Pressure Vessel Technology* **111**, 394–401.

GRANGER, S. 1988 Time domain method for estimation of damping in flow-Induced vibration problems. Proceedings of Pressure Vessels and Piping Conference (eds. F. Hara, P. Y. Chen, S. S. Chen & A. G. Ware), PVP-Vol. 133, pp. 27–35. New York: ASME.

GRANGER, S. 1990 A new signal processing method for investigating fluidelastic phenomena. *Journal of Fluids and Structures* **4**, 73–97.

GRANGER, S. 1991 A global model for flow-induced vibration of tube bundles in cross-flow. *ASME Journal of Pressure Vessel Technology* **113**, 446–458.

GRANGER, S. & CAMPISTRON, R. 1988 A new signal processing method to investigate fluidelastic phenomena. In Proceedings International Symposium on Flow-Induced Vibration and Noise, Vol. 3: Flow-Induced Vibration and Noise in Cylinder Arrays (eds. M. P. Païdoussis, S. S. Chen & M. D. Bernstein), pp. 255–268. New York: ASME.

GRANGER, S. & PAÏDOUSSIS, M. P. 1996 An improvement to the quasi-steady model with application to cross-flow-induced vibration of tube arrays. *Journal of Fluid Mechanics* **320**, 163–184.

GRANGER, S., CAMPISTRON, R. & LEBRET, J. 1991 An experimental analysis of the mechanisms underlying fluid-elastic coupling in a square-in-line tube bundle subject to water cross-flow. In *Flow-Induced Vibrations*, pp. 171–184. London: I.Mech.E.

GRANGER, S., CAMPISTRON, R. & LEBRET, J. 1993 Motion-dependent excitation mechanisms in a square in-line tube bundle subject to water cross-flow: An experimental modal analysis. *Journal of Fluids and Structures* **7**, 521–550.

GRIFFIN, O. M. 1985 Vortex shedding from bluff bodies in a shear flow – A review. *ASME Journal of Fluids Engineering* **107**, 298–306.

GRIFFIN, O. M. & RAMBERG, S. E. 1976 Vortex shedding from a cylinder vibrating in line with an incident flow. *Journal of Fluid Mechanics* **75**, 257–271.

GROSS, H. G. 1975 Untersuchung aerodynamischer Schwingungs-mechanismen und deren Berücksichtigung bei der Auslegung von Rohrbundelwarmetauschern. Ph.D. thesis, Technical University of Hannover, Germany.

GU, M. & DU, X. 2005 Experimental investigation of rain–wind-induced vibration of cables in cable-stayed bridges and its mitigation. *Journal of Wind Engineering & Industrial Aerodynamics* **93**, 79–95.

GUILMINEAU, E. & QUEUTEY, P. 2002 A numerical simulation of vortex shedding from an oscillating circular cylinder. *Journal of Fluids and Structures* **16**, 773–794.

HADJ-SADOK, C., PAYEN, T. & DE LANGRE, E. 1997 Non-linear vibrations of loosely supported tubes excited by fluidelastic and turbulence forces. *Proceedings of ASME 1997 International Mechanical Engineering Congress and Exposition, Dallas, TX*, pp. 193–199.

HALLAM, M. G., HEAF, N. J. & WOOTTON, L. R. 1997 Dynamics of marine structures. Ciria Underwater Engineering Group, Report UR8.

HALLE, H. & LAWRENCE, W. P. 1977 Crossflow-induced vibration of a row of circular cylinders in water. ASME Paper 77-JPGC-NE-4.

HALLE, H., CHENOWETH, J. M. & WAMBSGANSS, M. W. 1988 Shellside waterflow-induced tube vibration in heat exchanger configurations with tube pitch-to-diameter ratio of 1.42. In Proceedings International Symposium on Flow-Induced Vibration and Noise, Vol. 5; Flow-Induced Vibration in Heat-Transfer Equipment (eds. M. P. Païdoussis, J. M. Chenoweth, S. S. Chen, J. R. Stenner & W. J. Bryan), pp. 1–16. New York: ASME.

HARDY, C. & BOURDON, P. 1979 The influence of spacer dynamic properties in the control of bundle conductor motion. In Proceedings of IEEE-PAS Summer Conference, PAS 99, Paper F79 733-7, Vancouver, Canada, pp. 790–796.

HARDY, C. & VAN DYKE, P. 1995 Field observations on wind-induced conductor motions. *Journal of Fluids and Structures* **9**, 43–60.

HARTLEN, R. T. 1974 Wind tunnel determination of fluidelastic vibration thresholds for typical heat exchanger tube patterns. Ontario Hydro Report 74-309-K.

HARTLEN, R. T. & CURRIE, I. G. 1970 Lift-oscillator model of vortex-induced vibration. *ASCE Journal of the Engineering Mechanics Division* **96**, 577–591.

HEARNSHAW, D. 1974 Spacer damper performance–a function of in-span positioning. Institute of Electrical and Electronics Engineers Power Engineering Society Winter Meeting, New York, paper No. T 74 061-8. PAS 93(5), pp. 1298–1309.

HEILKER, W. K. & VINCENT, R. Q. 1981 Vibration in nuclear heat exchangers due to liquid and two-phase flow. *ASME Journal of Engineering for Power* **103**, 358–366.

HEKI, K. & HAWARA, T. 1965 Structural dynamical consideration of the ovalling deformation of steel stacks under wind. *Transactions of Architectural Institute of Japan* **100**, 22–28.

HÉMON, P. 1999a Approche du phénomène de gallop par un modèle d'effort retardé et validation expérimentale. *Comptes Rendus de l'Académie des Sciences*, Serie IIb, **327**, 679–684.

HÉMON, P. 1999b An improvement of the time delayed quasi-steady model for the oscillations of circular cylinders in cross-flow. *Journal of Fluids and Structures* **13**, 291–307.

HÉMON, P., SANTI, F., SCHNOERRINGER, B. & WOJCIECHOWSKI, J. 2001 Influence of free-stream turbulence on the movement-induced vibrations of an elongated rectangular cylinder in cross-flow. *Journal of Wind Engineering and Industrial Aerodynamics* **89**, 1383–1395.

HIKAMI, Y. & SHIRAISHI, N. 1988 Rain-wind induced vibrations of cables in cable stayed bridges. *Journal of Wind Engineering and Industrial Aerodynamics* **29**, 409–418.

HIROTA, K., NAKAMURA, T., KASAHARA, J., MUREITHI, N. W., KUSAKABE, T. & TAKAMATSU, H. 2002 Dynamics of an inline tube array subjected to steam-water cross-flow. Part III: Fluidelastic instability tests and comparison with theory. *Journal of Fluids and Structures* **16**, 153–173.

HO, C. M. & HUERRE, P. 1984 Perturbed free shear layers. *Annual Review of Fluid Mechanics* **16**, 365–424.

HODGES, D. H. & PIERCE, G. A. 2002 *Introduction to Structural Dynamics and Aeroelasticity*. New York: Cambridge University Press.

HONDA, A., YAMANAKA, T., FUJIWARA, T. & SAITO, T. 1995 Wind tunnel tests on rain-induced vibration on the stay-cable. In *Proceedings International Symposium on Cable Dynamics, Liège, Belgium*, pp. 255–262.

HOOVER, A. N. & HAWKS, R. J. 1977 Role of turbulence in wake-induced galloping of transmission lines. *AIAA Journal* **15**, 66–70.

HORI, K., LARSEN, C. M. & HALSE, K. H. 1997 Comparison of models for vortex induced vibrations of slender marine structures. *Marine Structures* **10**, 413–441.

HOVER, F. S., TECHET, A. H. & TRIANTAFYLLOU, M. S. 1998 Forces on oscillating uniform and tapered cylinders in cross flow. *Journal of Fluid Mechanics* **363**, 97–114.

HRUDEY, T. M., COWPER G. R. & LINDBERG, G. M. 1973 Finite element vibration analysis of multi-conductor electrical transmission lines. National Aeronautical Establishment Laboratory Technical Report LRT-ST-656, National Research Council, Ottawa, Canada.

HUERA HUARTE, F. J., BEARMAN, P. W. & CHAPLIN, J. R. 2006 On the force distribution along the axis of a flexible circular cylinder undergoing multi-mode vortex-induced vibrations. *Journal of Fluids and Structures* **22**, 897–903.

HUERRE, P. & MONKEWITZ, P. A. 1990 Local and global instabilities in spatially developing flows. *Annual Review of Fluid Mechanics* **22**, 473–537.

HUSE, E. 1996 Experimental investigation of deep sea interaction. In Proceedings 28th Offshore Technology Conference, Houston, Volume 2, pp. 367–372.

HUVELIN, F., LONGATTE, E., BAJ, F. & SOULI, M. 2008 Numerical prediction of fluidelastic instability threshold in tube bundles. In Proceedings of the 9th International Conference on Flow-Induced Vibration – FIV 2008 (eds. I. Zolotarev & J. Horaček), pp. 205–210. Prague, Czech Republic.

ICHIOKA, T., KAWATA, Y., IZUMI, H., NAKAMURA, T. & FUJITA, K. 1994 Two-dimensional flow analysis of fluid structure interaction around a cylinder and a row of cylinders. In *Flow-Induced Vibrations-1994* (eds. M. K. Au-Yang & K. Fujita), PVP-Vol. 273, pp. 33–41. New York: ASME.

INADA, F. KAWAMURA, K. & YASUO, A. 1996. Fluid-elastic force measurements acting on a tube bundle in two-phase cross flow. In Proceedings Flow-Induced Vibrations, PVP-Vol. 328 (ed. M. J. Pettigrew), pp. 81–87. New York: ASME.

INADA, F. KAWAMURA, K. & YASUO, A. 1997. Fluid-elastic force measurements acting on a tube bundle in two-phase cross flow. In Proceedings, 4th International Symposium on Fluid-Structure Interactions, Aeroelasticity, Flow-Induced Vibration and Noise, Vol. II, (ed. M. P. Païdoussis), pp. 357–364. New York: ASME.

INADA, F. KAWAMURA, K., YASUO, A. & YONEDA, K. 2002. An experimental study on the fluidelastic forces acting on a square tube bundle in two-phase cross-flow. *Journal of Fluids and Structures* **16**, 891–907.

IRWIN, H. P. A. H., COOPER, K. R. & WARDLAW, R. L. 1976 Application of vibration absorbers to control wind-induced vibration of I-beam truss members of the Commodore Barry bridge. National Research Council of Canada Report NAE-LA-194.

ISHIGAI, S. & NISHIKAWA, E. 1975 Experimental study on structure of gas flow in tube banks with tube axes normal to flow. Part II: On the structure of gas flow in single-column, single-row, and double-row tube banks. *Bulletin of the Japan Society of Mechanical Engineers* **18**, 528–535.

IUNGO, G. V. & BURESTI, G. 2009 Experimental investigation on the aerodynamic loads and wake flow features of low aspect-ratio triangular prisms at different wind directions. *Journal of Fluids and Structures* **25**, 1119–1135.

JAUVTIS, N. & WILLIAMSON, C. H. K. 2003 Vortex-induced vibration of a cylinder with two degrees of freedom. *Journal of Fluids and Structures* **17**, 1035–1042.

JOHNS, D. J. 1979 Discussion of the paper by Païdoussis & Helleur (1979) and authors' reply. *Journal of Sound and Vibration* **67**, 432–435.

JOHNS, D. J. 1983 Wind-induced static instability of cylindrical shells. *Journal of Wind Engineering and Industrial Aerodynamics* **13**, 261–270.

JOHNS, D. J. & ALLWOOD, R. J. 1968 Wind induced ovalling oscillations of circular cylindrical shell structures such as chimneys. *Proceedings of Symposium on Wind Effects on Buildings and Structures*, Loughborough (eds. D. J. Johns, C. Scruton, A. M. Ballantyne), Paper 28.

JOHNS, D. J. & SHARMA, C. B. 1974 On the mechanism of wind-excited ovalling vibrations on thin circular cylindrical shells. In *Flow-Induced Structural Vibrations* (ed. E. Naudascher), pp. 650–662. Berlin: Springer-Verlag.

JOHNSON, D. K. & SCHNEIDER, W. G. 1984 Flow-induced vibration of a tube array with an open lane. In Proceedings ASME Symposium on Flow-Induced Vibrations, Vol. 3: Vibration in Heat Exchangers (eds. M. P. Païdoussis, J. M. Chenoweth & M. D. Bernstein), pp. 63–72. New York: ASME.

KANEKO, S., NAKAMURA, T., INADA, F. & KATO, M. (eds) 2008 *Flow-Induced Vibrations: Classifications and Lessons from Practical Experiences*. Amsterdam: Elsevier.

KASSERA, V. & STROHMEIER, K. 1997 Simulation of tube bundle vibrations induced by cross-flow. *Journal of Fluids and Structures* **11**, 909–928.

KATSURA, S. 1985 Wind-excited ovalling vibration of a thin circular cylindrical shell. *Journal of Sound and Vibration* **100**, 527–550.

KAWAI, H. 1990 A discrete vortex analysis of flow around a vibrating cylinder with a splitter plate. *Journal of Wind Engineering and Industrial Aerodynamics* **35**, 237–273.

KHALAK, A. & WILLIAMSON, C. H. K. 1996 Dynamics of a hydroelastic cylinder with very low mass and damping. *Journal of Fluids and Structures* **10**, 455–472.

KHALAK, A. & WILLIAMSON, C. H. K. 1997 Investigation of relative effects of mass and damping in vortex-induced vibration of a circular cylinder. *Journal of Wind Engineering and Industrial Aerodynamics* **71**, 341–350.

KIM, W. & PERKINS, N. C. 2002 Two-dimensional vortex-induced vibration of cable suspensions. *Journal of Fluids and Structures* **16**, 229–245.

KING, R. & JOHNS, D. J. 1976 Wake interaction experiments with two flexible circular cylinders in flowing water. *Journal of Sound and Vibration* **45**, 259–283.

KING, R., PROSSER, M. J. & JOHNS, D. J. 1973 On vortex excitation of model piles in water. *Journal of Sound and Vibration* **29**, 169–188.

KLAMO, J. T., LEONARD, A. & ROSHKO, A. 2005 On the maximum amplitude for a freely vibrating cylinder in cross-flow. *Journal of Fluids and Structures* **21**, 429–434.

KO, R. G. 1973 A wind tunnel investigation into the aerodynamic stability of bundled conductors for Hydro-Quebec. Part III: Static Force Measurements in the 6 ft. by 9ft. wind tunnel. National Aeronautical Establishment Laboratory Technical Report LRT-LA-110, National Research Council, Ottawa, Canada.

KOBAYASHI, H. & NAGAOKA, H. 1992 Active control of a suspension bridge. *Journal of Wind Engineering and Industrial Aerodynamics* **41–44**, 143–151.

KOLKMAN, P. A. 1980 Development of vibration-free gate design. Learning from experiment and theory. In *Practical Experiences with Flow-Induced Vibrations* (eds. E. Naudascher & D. Rockwell), pp. 351–385. Berlin: Springer.

KOMATSU, S. & KOBAYASHI, H. 1980 Vortex-induced oscillation of bluff cylinders. *Journal of Wind Engineering and Industrial Aerodynamics* **6**, 335–362.

KOSKO, E. 1968 The frequency spectrum of a structural member in coupled flexural-torsional vibration. *Journal of Sound and Vibration* **7**, 143–155.

KUBO, Y., HIRATA, K. & MIKAWA, K. 1992 Mechanism of aerodynamic vibrations of shallow bridge girder sections. *Journal of Wind Engineering and Industrial Aerodynamics* **41–44**, 1297–1308.

KUMARASENA, T., SCANLAN, R. H. & EHSAN, F. 1992 Recent observations in bridge deck aeroelasticity. *Journal of Wind Engineering and Industrial Aerodynamics* **40**, 225–247.

KUNDURPI, P. S., SAMAVEDAM, G. & JOHNS, D. J. 1975 Stability of cantilever shells under wind loads. *ASCE Journal of Engineering Mechanics Division* **101**, 517–530.

KWOK, K. C. S. & MELBOURNE, W. H. 1980 Free stream turbulence effects on galloping. *ASCE Journal of Engineering Mechanics Division* **106**, 273–288.

LAMB, H. 1957 *Hydrodynamics*, 6th edition, p. 19, §20. Cambridge: Cambridge University Press.

LANCHESTER, F. W. 1907 *Aerodynamics*, pp. 43–45. London: Constable & Co.

LANEVILLE, A. 1973 Effects of turbulence on wind-induced vibrations of bluff cylinders. Ph.D. thesis, University of British Columbia, Vancouver, Canada.

LANEVILLE, A. & MAZOUZI, A. 1996 Wind-induced ovalling oscillations of cylindrical shells: critical onset velocity and mode prediction. *Journal of Fluids and Structures* **10**, 691–704.

LANEVILLE, A. & PARKINSON, G. V. 1971 Effects of turbulence on galloping of bluff cylinders. *Proceedings 3rd Int'l Conference on Wind Effects on Buildings and Structures*, Tokyo, Japan, pp. 787–797.

LANEVILLE, A., GARTSHORE, I. S. & PARKINSON, G. V. 1975 An explanation of some effects of turbulence on bluff bodies. *Proceedings 4th Int'l Conference on Wind Effects on Buildings and Structures*, London, UK, pp. 333–341.

LANGFORD, P. J. & CONNORS, H. J. 1991 Calculation of tube/AVB wear from U-tube shaker test data. In *Flow Induced Vibrations*, pp. 45–55. London: I.Mech.E.

LANGHAAR, H. L. & MILLER, R. E. 1967 Buckling of an elastic isotropic cylindrical shell subjected to wind pressure. *Proceedings of Symposium on the Theory of Shells*, University of Houston, Texas, pp. 403–429.

LARSEN, A. 1998 Advances in aeroelastic analyses of suspension and cable-stayed bridges. *Journal of Wind Engineering and Industrial Aerodynamics* **74–76**, 73–90.

LARSEN, A. 2000a Aerodynamics of the Tacoma Narrows Bridge – 60 years later. *Structural Engineering International* **10**, 243–248.

LARSEN, A. 2000b Aerodynamics of the Tacoma Narrows Bridge – Lessons learned. *Bridge Design & Engineering* **20**, 44–46.

LEISSA, A. 1993 *Vibration of Shells*. The Acoustical Society of America.

LEMAITRE, C. 2006 Dynamique d'un film d'eau de pluie sur un hauban de pont soumis au vent. Ph.D. thesis, Ecole Polytechnique, Palaiseau, France.

LEMAITRE, C., HÉMON, P. & DE LANGRE, E. 2007 Thin water film around a cable subject to wind. *Journal of Wind Engineering & Industrial Aerodynamics* **95**, 1259–1271.

LEMAITRE, C., HÉMON, P. & DE LANGRE, E. 2010 Rainwater rivulets running on a stay-cable subject to wind. *European Journal of Mechanics B/Fluids* **29**, 251–258.

LEMAITRE, C., MAHMUD ALAM, M., HÉMON, P., DE LANGRE, E. & ZHOU, Y. 2006 Rainwater rivulets on a cable subject to wind. *Comptes rendus Mécanique* **334**, 158–163.

LEONARD, A. & ROSHKO, A. 2001 Aspects of flow-induced vibration. *Journal of Fluids and Structures* **15**, 415–425.

LEVER, J. H. & RZENTKOWSKI, G. 1989 An investigation into the post-stable behaviour of a tube array in cross-flow. *ASME Journal of Pressure Vessel Technology* **111**, 457–466.

LEVER, J. H. & WEAVER, D. S. 1982 A theoretical model for the fluid-elastic instability in heat exchanger tube bundles. *ASME Journal of Pressure Vessel Technology* **104**, 147–158.

LEVER, J. H. & WEAVER, D. S. 1986a On the stability behaviour of heat exchanger tube bundles. Part 1. Modified theoretical model. *Journal of Sound and Vibration* **107**, 375–410.

LEVER, J. H. & WEAVER, D. S. 1986b On the stability behaviour of heat exchanger tube bundles. Part 2. Numerical results and comparison with experiments. *Journal of Sound and Vibration* **107**, 393–410.

LE CUNFF, C., BIOLLEY, F., FONTAINE, E., ETIENNE, S. & FACCHINETTI, M. L. 2002 Vortex-induced vibrations of risers: Theoretical, numerical and experimental investigation. *Oil and Gas Science and Technology* **57**, 59–69.

LI, L., SHERWIN, S. J. & BEARMAN, P. W. 2002 A moving frame of reference algorithm for fluid/structure interaction of rotating and translating bodies. *International Journal of Numerical Methods in Fluids* **38**, 207–223.

LI, M. & WEAVER, D. S. 1997 A fluidelastic instability model with an extension to full flexible multi-span tube arrays. In Proceedings of 4th International Symposium on Fluid-Structure Interactions, Aeroelasticity, Flow-Induced Vibration and Noise, Vol. II (ed. M. P. Païdoussis), pp. 163–172. New York: ASME.

LIANG, C. & PAPADAKIS, G. 2007 Large eddy simulation of cross-flow through a staggered tube bundle at subcritical Reynolds number. *Journal of Fluids and Structures* **23**, 1215–1230.

LIBERMAN, A. J. 1974 Oscillations and conductor galloping on HV overhead lines. Paper 22-09, CIGRÉ International Conference on Large High Voltage Electrical Systems, 25th Session, Volume 1, Paris, France.

LIDNER, H. 1992 Simulation of the turbulence influence on galloping vibrations. *Journal of Wind Engineering and Industrial Aerodynamics* **41–44**, 2023–2034.

LIGHTHILL, J. 1986 *An Informal Introduction to Theoretical Fluid Mechanics*. Oxford: Oxford University Press.

LILIEN, J.-L. & SNEGOVSKI, D. 2004 Wake-induced vibration in power transmission line: Parametric study. In Proceedings of the 8th International Conference on Flow-Induced Vibration FIV2004 (eds. E. de Langre & F. Axisa), Paris, France, Volume II, pp. 421–426.

LONGATTE, E., BENDJEDDOU, Z. & SOULI, M. 2003 Methods for numerical study of tube bundle vibrations in cross-flows. *Journal of Fluids and Structures* **18**, 513–528.

LUBIN, B. T., LETENDRE, R. P., QUINN, J. W. & KENNY, R. A. 1986 Comparison of the response of a scale model and prototype design tube bank structure to cross-flow induced fluid excitation. In Flow-Induced Vibrations-1986 (eds. S. S. Chen, J. C. Simonis & Y. S. Shin), PVP-Vol. 104, pp. 171–177. New York: ASME.

LUCOR, D., FOO, J. & KARNIADAKIS, G. E. 2005 Vortex mode selection of a rigid cylinder subject to VIV at low mass-damping. *Journal of Fluids and Structures* **20**, 483–503.

LUCOR, D., MUKUNDAN, H. & TRIANTAFYLLOU, M. S. 2006 Riser modal identification in CFD and full-scale experiments. *Journal of Fluids and Structures* **22**, 905–917.

LUGT, H. J. 1978 Autorotation of plates. David W. Taylor Naval Ship R & D Center Report 78/058, Bethesda, MD, USA.

LUGT, H. J. 1983 *Vortex Flow in Nature and Technology*. New York: Wiley.

LUO, S. C. 1985 A forced vibration study of the aerodynamic instability of a square section cylinder. Ph.D. Thesis, Imperial College, London, U.K.

LUO, S. C. & BEARMAN, P. W. 1990 Predictions of fluctuating lift on a transversely oscillating square-section cylinder. *Journal of Fluids and Structures* **4**, 219–228.

LUO, S. C., YAZDANI, M. G., LEE, T. S. & CHEW, Y. T. 1993 Aerodynamic stability of square, trapezoidal and triangular cylinders. *Proceedings 3rd International Offshore and Polar Engineering Conference*, Singapore, pp. 709–714.

LUO, S. C., YAZDANI, M. G., CHEW, Y. T. & LEE, T. S. 1994 Effects of incidence and afterbody shape on flow past bluff bodies. *Journal of Wind Engineering and Industrial Aerodynamics* **53**, 375–399.

LUO, S. C., CHEW, Y. T., LEE, T. S. & YAZDANI, M. G. 1998 Stability to translational galloping vibration of cylinders at different mean angles of attack. *Journal of Sound and Vibration* **215**, 1183–1194.

LUO, S. C., CHEW, Y. T. & NG, Y. T. 2003 Hysteresis phenomenon in the galloping oscillation of a square cylinder. *Journal of Fluids and Structures* **18**, 103–118.

MACDONALD, J. 2002 Separation of the contributions of aerodynamic and structural damping in vibrations of inclined cables. *Journal of Wind Engineering & Industrial Aerodynamics* **90**, 19–39.

MACDONALD, J. & LAROSE, G. 2006 A unified approach to aerodynamic damping and drag/lift instabilities, and its application to dry inclined cable galloping. *Journal of Fluids and Structures* **22**, 229–252.

MACDONALD, J. H. G. & LAROSE, G. L. 2008a Two-degree-of-freedom galloping. Part 1: General formulation and solution for perfectly tuned system. *Journal of Wind Engineering and Industrial Aerodynamics* **96**, 291–307.

MACDONALD, J. H. G. & LAROSE, G. L. 2008b Two-degree-of-freedom galloping. Part 2: Analysis and prevention for arbitrary frequency ratio. *Journal of Wind Engineering and Industrial Aerodynamics* **96**, 308–326.

MAEKAWA, T. 1964. Study on wind pressure against ACSR double conductor. *Electrical Engineering in Japan* **84**, 21–28.

MAGNUS, K. 1965 *Vibrations*. London: Blackie & Son.

MAIN, J. & JONES, N. 1999 Full-scale measurements of stay cable vibration. In *Wind Engineering into the 21st Century*, pp. 963–970. Rotterdam: Balkema.

MAIR, W. A. & MAUL, D. J. 1971 Aerodynamic behaviour of bodies in the wakes of other bodies. *Philosophical Transactions of the Royal Society London, A* **269**, 425–439.

MARN, J. & CATTON, I. 1991 Flow induced vibrations in cylindrical bundles: Two dimensional analysis into normal modes. In *Numerical Modelling of Basic Heat Transfer Phenomena in Nuclear Systems* (eds. F. B. Cheung & L. E. Hochreiter), HTD-Vol. 165, pp. 9–14. New York: ASME.

MARN, J. & CATTON, I. 1992 On stability analysis of a flexible cylinder in an array of rigid cylinders. *ASME Journal of Fluids Engineering* **114**, 12–19.

MARN, J. & CATTON, I. 1993 Analysis of flow induced vibration using the vorticity transport equation. *ASME Journal of Fluids Engineering* **115**, 485–492.

MARN, J. & CATTON, I. 1996 On the analysis fluidelastic instability in air-water mixture. *ASME Journal of Pressure Vessel Technology* **118**, 188–193.

MARTIN, W. W., CURRIE, I. G. & NAUDASCHER, E. 1981 Streamwise oscillations of cylinders. *ASCE Journal of Engineering Mechanics* **107**, 589–607.

MASRI, S. F. & CAUGHEY, T. K. 1979 A non-parametric identification technique for nonlinear dynamic problems. *Journal of Applied Mechanics* **54**, 918–929.

MATHELIN, L. & DE LANGRE, E. 2005 Vortex-induced vibrations and waves under shear flow with a wake oscillator model. *European Journal of Mechanics B-Fluids* **24**, 478–490.

MATSUMOTO, M. SHIRAISHI, N., KITAZAWA, M., KNISELY, C., SHIRATO, H., KIM, Y. & TSUJII, M. 1980 Aerodynamic behaviour of inclined circular cylinders – cable aerodynamics. *Journal of Wind Engineering and Industrial Aerodynamics* **33**, 63–72.

MATSUMOTO, M., YOKOYAMA, K., MIYATA, T., FUJINO, Y. & YAMAGUCHI, H. 1989 Wind-induced cable vibration of cable-stayed bridges in Japan. In *Proceedings of Japan-Canada Joint Workshop on Bridge Aerodynamics, Ottawa, Canada*, pp. 101–110.

MATSUMOTO, M., SAITOH, T., KITAZAWA, M., SHIRATO, H. & NISHIZAKI, T. 1995 Response characteristics of rain-wind induced vibration of stay-cables of cable-stayed bridges. *Journal of Wind Engineering & Industrial Aerodynamics* **57**, 323–333.

MATSUMOTO, M., SHIRATO, H., YAGI, T., SHIJO, R., EGUCHI, A. & TAMAKI, H. 2003a Effects of aerodynamic interferences between heaving and torsional vibration of bridge decks: the case of Tacoma Narrows Bridge. *Journal of Wind Engineering and Industrial Aerodynamics* **91**, 1547–1557.

MATSUMOTO, M., YAGI, T., GOTO, M. & SAKAI, S. 2003b Rain–wind-induced vibration of inclined cables at limited high reduced wind velocity region. *Journal of Wind Engineering and Industrial Aerodynamics* **91**, 1–12.

MATSUMOTO, M., MIZUNO, K., OKUBO, K. & ITO, Y. 2005 Torsional flutter and branch characteristics for 2-D rectangular cylinders. *Journal of Fluids and Structures* **21**, 597–608.

MATSUMOTO, M., SHIRATO, H., MIZUNO, K., SHIJO, R. & HIKIDA, T. 2008a Flutter characteristics of H-shaped cylinders with various side-ratios and comparisons with characteristics of rectangular cylinders. *Journal of Wind Engineering and Industrial Aerodynamics* **96**, 963–970.

MATSUMOTO, M., YAGI, T., TAMAKI, H. & TSUBOTA, T. 2008b Vortex-induced vibration and its effect on torsional flutter instability in the case of $B/D = 4$ rectangular cylinder. *Journal of Wind Engineering and Industrial Aerodynamics* **96**, 971–983.

MAVRIPLIS, D. 1982 An investigation of the limitations of potential flow in cross-flow induced vibrations of cylinder arrays. M. Eng. Thesis, Department of Mechanical Engineering, McGill University, Montreal, QC, Canada.

MAZOUZI, A. 1989 Étude sur le flottement transversal des coques cylindriques circulaires dans un écoulement uniforme. Thèse Ph.D., Département de génie mécanique, Université de Sherbrooke, Sherbrooke (Québec), Canada.

MAZOUZI, A., LANEVILLE, A. & VITTECOQ, P. 1991 An analytical model of the ovalling oscillations of clamped-free and clamped-clamped cylidrical shells in cross-flow. *Journal of Fluids and Structures* **5**, 605–626.

MESKELL, C. 2005 On the underlying fluid mechanics responsible for damping controlled fluidelastic instability in tube arrays. In Proceedings of PVP2005 2005 ASME Pressure Vessels and Piping Division Conference, Denver, CO, USA.

MESKELL, C. & FITZPATRICK, J. A. 2003 Investigation of the nonlinear behaviour of damping controlled fluidelastic instability in a normal triangular tube array. *Journal of Fluids and Structures* **18**, 573–593.

MESKELL, C., FITZPATRICK, J. A. & RICE, H. J. 2001 Application of force-state mapping to a nonlinear fluid-elastic system. *Mechanical Systems and Signal Processing* **15**, 75–85.

MEWES, D. & STOCKMEIER, D. 1991 Fluid viscosity effects on flow induced vibrations of tube bundles in cross-flow. In *Flow Induced Vibrations*, pp. 231–242. London: I.Mech.E.

MINAKAMI, K. & OHTOMI, K. 1987 Flow direction and fluid density effects on the fluidelastic vibrations of a triangular array of tubes. In *Flow-Induced Vibrations* (ed. R. King), pp. 65–75. Cranfield: BHRA.

MINORSKY, N. 1962 *Nonlinear Oscillations*. Princeton: D. Van Nostrand.

MITTAL, S. & SINGH, S. 2005 Vortex-induced vibrations at subcritical Re. *Journal of Fluid Mechanics* **534**, 185–194.

MITTAL, S. & KUMAR, V. 2001 Flow-induced oscillations of two cylinders in tandem and staggered arrangements. *Journal of Fluids and Structures* **15**, 717–736.

MIYATA, T. 2003 Historical view of long-span bridge aerodynamics. *Journal of Wind Engineering and Industrial Aerodynamics* **91**, 1393–1410.

MIZOTA, T., ZDRAVKOVICH, M. M., GRAW, K.-U. & LEDER, A. 2000 St Christopher and the vortex. A Kármán vortex in the wake of St Christopher's heels. *Nature* **404**, 226.

MODI, V. J. & SLATER, J. E. 1974 Quasi-steady analysis of torsional aeroelastic oscillators. In *Flow-Induced Structural Vibrations* (ed. E. Naudascher), pp. 355–372. Berlin: Springer-Verlag.

MODI, V. J. & SLATER, J. E. 1983 Unsteady aerodynamics and vortex induced aeroelastic instability of a structural angle section. *Journal of Wind Engineering and Industrial Aerodynamics* **11**, 321–334.

MODI, V. J., WELT, F. & SETO, M. L. 1995 Control of wind-induced instabilities through application of nutation dampers: a brief overview. *Engineering Structures* **17**, 626–638.

MOE, G. & OVERVIK, T. 1982 Current-induced motions of multiple risers. In Proceedings of 3rd International Conference on Behaviour of Offshore Structures (BOSS-82), Massachusetts Institute of Technology, Cambridge, MA, pp. 618–639.

MOLIN, B. 2002 *Hydrodynamique des Structures Offshore*. Paris: Technip.

MUREITHI, N. W., PAÏDOUSSIS, M. P. & PRICE, S. J. 1994a The post-Hopf-bifurcation of a loosely-supported cylinder in an array subjected to cross-flow. Part I: Experimental Results. *Journal of Fluids and Structures* **8**, 833–852.

MUREITHI, N. W., PAÏDOUSSIS, M. P. & PRICE, S. J. 1994b The post-Hopf-bifurcation of a loosely-supported cylinder in an array subjected to cross-flow. Part 11: Theoretical model and comparison with experiments. *Journal of Fluids and Structures* **8**, 853–876.

MUREITHI, N. W., PAÏDOUSSIS, M. P. & PRICE, S. J. 1995 Intermittency transition to chaos in the response of a loosely supported cylinder in an array in cross-flow. *Chaos, Solitons & Fractals* **5**, 847–867.

MUREITHI, N. W., NAKAMURA, T., HIROTA, K., MURATA, M., UTSUMI, S., KUSAKABE, T. & TAKAMATSU, H. 2002 Dynamics of an inline tube array subjected to steam-water cross-flow. Part II: Unsteady fluid forces. *Journal of Fluids and Structures* **16**, 137–152.

MUREITHI, N. W., SHAHRIARY, S. & PETTIGREW, M. J. 2008 Flow-induced vibration model for steam generator tubes in two-phase flow. In Proceedings of the 14th International Conference on Nuclear Engineering, ICONE16, Orlando, FL, USA.

NAGAO, F., UTSONOMIYA, H., ORYU, T. & MANABE, S. 1993 Aerodynamic efficiency of triangular faring on box girder bridge. *Journal of Wind Engineering and Industrial Aerodynamics* **49**, 565–574.

NAGASHIMA, T. & HIROSE, T. 1982 Potential flow around two dimensional isosceles triangular cylinder subjected to uniform flow from base surface. *Scientific and Engineering Reports of the National Defense Academy* **20**, 435–442 (in Japanese).

NAKAGUCHI, H., HASHIMOTO, K. & MUTO, S. 1968 An experimental study on aerodynamic drag of rectangular cylinders. *Journal of the Japan Society of Aeronautics and Space Science* **16**, 1–5 (in Japanese).

NAKAMURA, T., FUJITA, K. & HARA, F. 1992a A study on an approximate theory of fluidelastic vibration of tube array. Part 1: Basic concept of fluidelastic force and instability of tube rows. Proceedings International Symposium on Flow-Induced Vibration and Noise, Vol. 2: Cross-Flow Induced Vibration of Cylinder Arrays (eds. M. P. Païdoussis, S. S. Chen & D. A. Steininger), pp. 1–16. New York: ASME.

NAKAMURA, T., FUJITA, K., KAWANISHI, K., YAMAGUCHI, N. & TSUGE, A. 1992b Study on the vibrational characteristics of a tube array caused by two-phase flow. Part 2: Fluidelastic instability. *ASME Journal of Pressure Vessel Technology* **114**, 479–485.

NAKAMURA, T., FUJITA, K., KAWANISHI, K., YAMAGUCHI, N. & TSUGE, A. 1995 Study on the vibrational characteristics of a tube array caused by two-phase flow. Part 2: Fluidelastic instability. *Journal of Fluids and Structures* **9**, 539–562.

NAKAMURA, T., CHEN, S. S., MUREITHI, N. W. & KUSAKABE, T. 1997 Improved estimation of fluidelastic instability threshold for tube arrays in two-phase flow: A model incorporating temporal fluid force correlation along the tube axis. In Proceedings 4th International Symposium on Fluid-Structure Interactions, Aeroelasticity, Flow-Induced Vibration and Noise, Vol. II (ed. M. P. Païdoussis), pp. 365–372. New York: ASME.

NAKAMURA, Y. 1969 Vortex excitation of a circular cylinder treated as a binary flutter. *Kyushu University, Research Institute for Applied Mechanics, Report* **17**, 217–234.

NAKAMURA, Y. 1978 An analysis of binary flutter of bridge deck sections. *Journal of Sound and Vibration* **57**, 471–482.

NAKAMURA, Y. 1979 On the aerodynamic mechanism of torsional flutter of bluff structures. *Journal of Sound and Vibration* **67**, 163–177.

NAKAMURA, Y. 1980 Galloping of bundled power line conductors. *Journal of Sound and Vibration* **73**, 363–377.

NAKAMURA, Y. 1990 Recent research into bluff-body flutter. *Journal of Wind Engineering and Industrial Aerodynamics*, **33**, 1–10.

NAKAMURA, Y. 1996 Vortex shedding from bluff bodies with splitter plates. *Journal of Fluids and Structures* **10**, 147–158.

NAKAMURA, Y. & HIRATA, K. 1989 Critical geometry of oscillating bluff bodies. *Journal of Fluid Mechanics* **208**, 375–393.

NAKAMURA, Y. & HIRATA, K. 1991 Pressure fluctuations on oscillating rectangular cylinders with the long side normal to the flow. *Journal of Fluids and Structures* **5**, 165–183.

NAKAMURA, Y. & HIRATA, K. 1994 The aerodynamic mechanism of galloping. *Transactions of the Japan Society for Aeronautical and Space Sciences* **36**, 257–269.

NAKAMURA, Y. & MATSUKAWA, T. 1987 Vortex excitation of rectangular cylinders with a long side normal to the flow. *Journal of Fluid Mechanics* **180**, 171–191.

NAKAMURA, Y. & MIZOTA, T. 1975a Unsteady lifts and wakes of oscillating rectangular prisms. *ASCE Journal of the Engineering Mechanics Division* **101**, 855–871.

NAKAMURA, Y. & MIZOTA, T. 1975b Torsional flutter of rectangular prisms. *ASCE Journal of the Engineering Mechanics Division* **101**, 125–142.

NAKAMURA, Y. & NAKASHIMA, M. 1986 Vortex excitation of prisms and elongated rectangular, H and ⊢ cross-sections. *Journal of Fluid Mechanics* **163**, 149–169.

NAKAMURA, Y. & OHYA, Y. 1984 The effects of turbulence on the mean flow past two-dimensional rectangular cylinders. *Journal of Fluid Mechanics* **149**, 255–273.

NAKAMURA, Y. & TOMONARI, Y. 1977 Galloping of rectangular prisms in a smooth and in a turbulent flow. *Journal of Sound and Vibration* **52**, 233–241.

NAKAMURA, Y. & TOMONARI, Y. 1981 The aerodynamic characteristics of D-section prisms in a smooth and in a turbulent flow. *Aeronautical Quarterly* **32**, 153–168.

NAKAMURA, Y. & YOSHIMURA, T. 1982 Flutter and vortex excitation of rectangular prisms in pure torsion in smooth and turbulent flows. *Journal of Sound and Vibration* **84**, 305–317.

NAKAMURA, Y., HIRATA, K. & URABE, T. 1991 Galloping of rectangular cylinders in the presence of a splitter plate. *Journal of Fluids and Structures* **5**, 521–549.

NAKAMURA, Y., HIRATA, K. & KASHIMA, K. 1994 Galloping of a circular cylinder in the presence of a splitter plate. *Journal of Fluids and Structures* **8**, 355–365.

NATARAJA, R. & JOHNS, D. J. 1977 Random response of shell structures due to wind. *Journal of Industrial Aerodynamics* **2**, 79–97.

NAUDASCHER, E. (ed.) 1974 *Flow-Induced Structural Vibrations.* Proceedings of the 1972 IUTAM-IAHR Symposium in Karlsruhe. Berlin: Springer-Verlag.

NAUDASCHER, E. 1987 Flow-induced streamwise vibrations of structures. *Journal of Fluids and Structures* **1**, 265–298.

NAUDASCHER, E. & ROCKWELL, D. 1980 Oscillator-model approach to the identification and assessment of flow-induced vibrations in a system. *Journal of Hydraulic Research* **18**, 59–82.

NAUDASCHER, E. & ROCKWELL, D. 1994 *Flow-Induced Vibrations: An Engineering Guide.* Rotterdam: Balkema.

NAUDASCHER, E. & ROCKWELL, D. 2005 *Flow-Induced Vibrations: An Engineering Guide.* Corrected edition. Mineola: Dover Publications.

NAUDASCHER, E. & WANG, Y. 1993 Flow-induced vibrations of prismatic bodies and grids of prisms. *Journal of Fluids and Structures* **7**, 341–373.

NAUDASCHER, E., WESKE, J. R. & FEY, B. 1981 Exploratory study on damping galloping vibrations. *Journal of Wind Engineering and Industrial Aerodynamics* **8**, 211–222.

NAYFEH, A. H. 1981 *Introduction to Perturbation Techniques.* New York: John Wiley.

NAYFEH, A. H. & MOOK, D. T. 1979 *Nonlinear Oscillations.* New York: Wiley-Interscience.

NEWMAN, D. J. & KARNIADAKIS, G. E. 1997 A direct numerical simulation study of flow past a freely vibrating cable. *Journal of Fluid Mechanics* **344**, 95–136.

NI, Y., WANG, X., CHEN, Z. & KO, J. 2007 Field observations of rain-wind-induced cable vibration in cable-stayed Dongting Lake Bridge. *Journal of Wind Engineering and Industrial Aerodynamics* **95**, 303–328.

NOACK, B. R., OHLE, F. & ECKELMANN, H. 1991 On cell formation in vortex streets. *Journal of Fluid Mechanics* **227**, 293–308.

NORBERG, C. 2003 Fluctuating lift on a circular cylinder: Review and new measurements. *Journal of Fluids and Structures* **17**, 57–96.

NOVAK, M. 1969 Aeroelastic galloping of prismatic bodies. *ASCE Journal of the Engineering Mechanics Division* **96**, 115–142.

NOVAK, M. 1971 Galloping and vortex induced oscillations of structures. *Proceedings 3rd International Conference on Wind Effects on Buildings and Structures,* pp. 799–809.

NOVAK, M. 1972 Galloping oscillations of prismatic structures. *ASCE Journal of the Engineering Mechanics Division* **98**, 27–46.

NOVAK, M. 1974 Galloping oscillations of prisms in smooth and turbulent flows. In *Flow-Induced Structural Vibrations* (ed. E. Naudascher), pp. 769–774. Berlin: Springer-Verlag.

NOVAK, M. & DAVENPORT, A. G. 1970 Aeroelastic instability of prisms in turbulent flow. *ASCE Journal of the Engineering Mechanics Division* **97**, 17–39.

NOVAK, M. & TANAKA, H. 1974 Effect of turbulence on galloping instability. *ASCE Journal of the Engineering Mechanics Division* **100**, 27–47.

OHTA, K., KAGAWA, K., TANAKA, H. & TAKAHARA, S. 1982 Study on the fluidelastic vibration of tube arrays using modal analysis technique. In *Flow-Induced Vibration of Circular Cylindrical Structures* (eds. S. S. Chen, M. P. Païdoussis & M. K. Au-Yang) PVP-Vol. 63, pp. 57–69. New York: ASME.

OHYA, Y. 1997 *Yasuharu Nakamura, a Compendium of His Published Papers*. Research Institute for Applied Mechanics, Kyushu University, Japan

OLIVEIRA, A. R. E. & MANSOUR, W. M. 1983 Non-linear analysis and simulations of auto-oscillations of twin bundle conductors of transmission lines. In Proceedings 10th International Association for Mathematics and Computers in Simulation World Congress on Systems Simulation and Scientific Computation, Volume 3: Modeling and Simulation in Engineering (eds. W. F. Ames & R. Vichnevetsky), Montreal, Canada, pp. 187–190.

OTSUKI, Y., WASHIZU, K., TOMIZAWA, H. & OHYA, A. 1974 A note on the aeroelastic instability of a prismatic bar with square section. *Journal of Sound and Vibration* **34**, 233–248.

OVERVIK, T., MOE, G. & HJORT-HANSEN, E. 1983 Flow-induced motions of multiple risers. ASME Journal of Energy Resources Technology **105**, 83–89.

PAÏDOUSSIS, M. P. 1980 Flow-induced vibrations in nuclear reactors and heat exchangers: Practical experiences and state of the knowledge. In *Practical Experiences with Flow-Induced Vibrations*, (eds. E. Naudascher & D. Rockwell), pp. 1–81. Berlin: Springer-Verlag.

PAÏDOUSSIS, M. P. 1981 Fluidelastic vibration of cylinder arrays in axial and cross flow: State of the art. *Journal of Sound and Vibration* **76**, 329–360.

PAÏDOUSSIS, M. P. 1983 A review of flow-induced vibrations in reactors and reactor components. *Nuclear Engineering and Design* **74**, 31–60.

PAÏDOUSSIS, M. P. 1987 Flow-induced instabilities of cylindrical structures. *Applied Mechanics Reviews* **40**, 163–175.

PAÏDOUSSIS, M. P. 1993 *Calvin Rice Lecture*: Some curiosity-driven research in fluid–structure interactions and its current applications. *ASME Journal of Pressure Vessel Technology* **115**, 2–14.

PAÏDOUSSIS, M. P. 1998 *Fluid-Structure Interactions, Slender Structures and Axial Flows, Volume 1*. London: Academic Press.

PAÏDOUSSIS, M. P. 2004 *Fluid–Structure Interactions: Slender Structures and Axial Flow, Volume 2*. London: Elsevier Academic Press.

PAÏDOUSSIS, M. P. 2006 Real-life experiences with flow-induced vibration. *Journal of Fluids and Structures* **22**, 741–755.

PAÏDOUSSIS, M. P. & HELLEUR, C. 1979 On ovalling oscillation of cylindrical shells in cross-flow. *Journal of Sound and Vibration* **63**, 527–542.

PAÏDOUSSIS, M. P. & LI, G. X. 1992 Cross-flow-induced chaotic motions of heat-exchanger tubes impacting on loose supports. *Journal of Sound and Vibration* **152**, 305–326.

PAÏDOUSSIS, M. P. & PRICE, S. J. 1988 The mechanisms underlying flow-induced instabilities of cylinder arrays in crossflow. *Journal of Fluid Mechanics* **187**, 45–59.

PAÏDOUSSIS, M. P. & WONG, D. T.-M. 1982 Flutter of thin cylindrical shells in cross flow. *Journal of Fluids Mechanics* **115**, 411–426.

PAÏDOUSSIS, M. P., SUSS, S. & PUSTETOVSKY, M. 1977 Free vibration of cylinders in liquid-filled channels. *Journal of Sound and Vibration* **55**, 443–459.

PAÏDOUSSIS, M. P., PRICE, S. J. & SUEN, H.-C. 1982a Ovalling oscillations of cantilevered and clamped-clamped cylindrical shells in cross flow: an experimental study. *Journal of Sound and Vibration* **83**, 533–553.

PAÏDOUSSIS, M. P., PRICE, S. J. & SUEN, H.-C. 1982b An analytical model for ovalling oscillation of clamped-clamped cylindrical shells in cross flow. *Journal of Sound and Vibration* **83**, 555–572.

PAÏDOUSSIS, M. P., PRICE, S. J., FEKETE, G. I. & NEWMAN, B. G. 1983 Ovalling of chimneys: induced by vortex shedding or self-excited? *Journal of Wind Engineering and Industrial Aerodynamics* **14**, 119–128.

PAÏDOUSSIS, M. P., MAVRIPLIS, D. & PRICE, S. J. 1984 A potential flow theory for the dynamics of cylinder arrays in cross-flow. *Journal of Fluid Mechanics* **146**, 227–252.

PAÏDOUSSIS, M. P., PRICE, S. J. & MAVRIPLIS, D. 1985 A semi-potential flow theory for the dynamics of cylinder arrays in cross-flow. *ASME Journal of Fluids Engineering* **107**, 500–506.

PAÏDOUSSIS, M. P., PRICE, S. J. & ANG, S.-Y. 1988a Ovalling oscillations of cylindrical shells in cross-flow: a review and some new results. *Journal of Fluids and Structures* **2**, 95–112.

PAÏDOUSSIS, M. P., PRICE, S. J. & MARK, B. 1988b Current-induced oscillations and instabilities of a multi-tube flexible riser. *Journal of Fluids and Structures* **2**, 503–513.

PAÏDOUSSIS, M. P., PRICE, S. J., NAKAMURA, T., MARK, B. & NJUKI MUREITHI, W. 1989 Flow-induced vibrations and instabilities in a rotated-square cylinder array in cross-flow. *Journal of Fluids and Structures* **3**, 229–254.

PAÏDOUSSIS, M. P., PRICE, S. J. & MUREITHI, W. N. 1991a Non-linear dynamics of a single flexible cylinder within a rotated triangle array. In *Flow Induced Vibrations*, pp. 121–128. London: I.Mech.E.

PAÏDOUSSIS, M. P., PRICE, S. J. & ANG, S.-Y. 1991b An improved theory for flutter of cylindrical shells in cross-flow. *Journal of Sound and Vibration* **149**, 197–218

PAÏDOUSSIS, M. P., PRICE, S. J. & MUREITHI, W. N. 1992 The post-fluidelastic instability response of a loosely supported tube in an array subjected to water cross-flow. In Proceedings International Symposium on Flow-Induced Vibration and Noise, Vol. 2: Cross-Flow Induced Vibration of Cylinder Arrays (eds. M. P. Païdoussis, S. S. Chen & D. A. Steininger), pp. 243–264. New York: ASME.

PAÏDOUSSIS, M. P., PRICE, S. J. & MUREITHI, W. N. 1993 Non-linear and chaotic dynamics of a two-degree-of-freedom analytical model for a rotated triangle array in cross-flow. *Journal of Fluids and Structures* **7**, 497–520.

PAÏDOUSSIS, M. P., PRICE, S. J. & MUREITHI, W. N. 1996 On the virtual nonexistence of multiple instability regions for some heat-exchanger arrays in crossflow. *ASME Journal of Fluids Engineering* **118**, 103–109.

PANESAR, A. & JOHNS, D. J. 1985 Ovalling oscillations of thin circular cylindrical shells in cross flow – an experimental study. *Journal of Sound and Vibration* **103**, 201–209.

PANTAZOPOULOS, M. S. 1994 Vortex-induced vibration parameters: critical review. *Proceedings of the International Conference on Offshore Mechanics and Arctic Engineering, Houston, TX.* **1**, 199–255.

PAPANGELOU, A. 1992 Vortex shedding from slender cones at low Reynolds numbers. *Journal of Fluid Mechanics* **242**, 299–321.

PARKINSON, G. V. 1971 Wind-induced instability of structures. *Philosophical Transactions of the Royal Society* (London) **269**, 395–409.

PARKINSON, G. V. 1974 Mathematical models of flow-induced vibrations of bluff bodies. In *Flow-Induced Structural Vibrations* (E. Naudascher, ed.), pp. 81–127. Berlin: Springer-Verlag.

PARKINSON, G. V. 1989 Phenomena and modelling of flow-induced vibrations of bluff bodies. *Progress in Aerospace Sciences* **26**, 169–224.

PARKINSON, G. V. & BROOKS, N. P. H. 1961 On the aeroelastic instability of bluff cylinders. *Journal of Applied Mechanics* **28**, 252–258.

PARKINSON, G. V. & JANDALI, T. 1969 A wake source model for bluff body potential flow. *Journal of Fluid Mechanics* **40**, 577–596.

PARKINSON, G. V. & SMITH, J. D. 1964 The square prism as an aeroelastic non-linear oscillator. *Quarterly Journal of Mechanics and Applied Mathematics* **17**, 225–239.

PARKINSON, G. V. & SULLIVAN, P. P. 1979 Galloping response of towers. *Journal of Industrial Aerodynamics* **4**, 253–260.

PARKINSON, G. V. & WAWZONEK, M. A. 1981 Some considerations of the combined effects of galloping and vortex resonance. *Journal of Wind Engineering and Industrial Aerodynamics* **8**, 135–143.

PARRONDO, J. L., EGUSQUIZA, E. & SANTOLARIA, C. 1993 Extension of the Lever and Weaver's unsteady analytical model to the fluidelastic instability of arrays of flexible cylinders. *Journal of Wind Engineering and Industrial Aerodynamics* **49**, 177–186.

PEIL, U., NAHRATH, N. & DREYER, O. 2003 Rain-wind induced vibrations of cables–Theoretical models. In *Proceedings of the 5th International Symposium on Cable Dynamics, Santa Margherita Ligure*, pp. 345–352.

PETTIGREW, M. J. 1977 Flow-induced vibration of nuclear power station components. Atomic Energy of Canada Report AECL-5852.

PETTIGREW, M. J. & TAYLOR, C. E. 1991 Fluid-elastic instability of heat exchanger tube bundles: Review and design recommendations. *ASME Journal of Pressure Vessel Technology* **113**, 242–256.

PETTIGREW, M. J. & TAYLOR, C. E. 1994 Two-phase flow-induced vibration: An overview. *ASME Journal of Pressure Vessel Technology* **116**, 233–253.

PETTIGREW, M. J. & TAYLOR, C. E. 2003 Vibration analysis of shell-and-tube heat exchangers: an overview. Part 1: Flow, damping, fluidelastic instability. *Journal of Fluids and Structures* **18**, 469–483.

PETTIGREW, M. J. & TAYLOR, C. E. 2004 Damping of heat exchanger tubes in two-phase flow: Review and design guidelines. *ASME Journal of Pressure Vessel Technology* **126**, 523–533.

PETTIGREW, M. J., SYLVESTRE, Y. & CAMPAGNA, A. O. 1978 Vibration analysis of heat exchanger and steam generator designs. *Nuclear Engineering and Design* **48**, 97–115.

PETTIGREW, M. J., TROMP, J. H. & MASTORAKOS, J. 1985. Vibration of tube bundles subjected to two-phase cross flow. *ASME Journal of Pressure Vessel Technology* **107**, 335–343.

PETTIGREW, M. J., ROGERS, R. J. & AXISA, F. 1986 Damping of multispan heat exchanger tubes-Part 2: In liquids. In *Flow-Induced Vibrations* (eds. S. S. Chen, J. C. Simonis & Y. S. Shin), PVP-Vol. 104, pp. 89–98. New York: ASME.

PETTIGREW, M. J., TAYLOR, C. E., TROMP, J. H. & KIM, B. S. 1989 Vibration of tube bundles in two-phase cross-flow. Part 2: Fluidelastic instability. *ASME Journal of Pressure Vessel Technology* **111**, 478–487.

PICCIRILLO, P. S. & VAN ATTA, C. W. 1993 An experimental study of vortex shedding behind linearly tapered cylinders at low Reynolds number. *Journal of Fluid Mechanics* **246**, 163–195.

PIER, B. & HUERRE, P. 2001 Nonlinear self-sustained structures and fronts in spatially developing wake flows. *Journal of Fluid Mechanics* **435**, 145–174.

PIERSON, O. L. 1937 Experimental investigation of the influence of tube arrangement on convective heat transfer and flow resistance in cross flow of gases over tube banks. *Transactions of the American Society of Mechanical Engineers* **59**, 563–572.

PRICE, S. J. 1975a Wake induced flutter of power transmission conductors. *Journal of Sound and Vibration* **38**, 125–147.

PRICE, S. J. 1975b Wake induced flutter of power transmission conductors. Ph.D. thesis, University of Bristol, U.K.

PRICE, S. J. 1976 The origin and nature of the lift force on the leeward of two bluff bodies. *Aeronautical Quarterly* **27**, 154–168.

PRICE, S. J. 1982 Measurement of the mechanical characteristics of a full scale transmission conductor bundle. Proceedings 1st International Modal Analysis Conference, Orlando, Florida, USA, pp. 396–402.

PRICE, S. J. 1995 A review of theoretical models for fluidelastic instability in cross-flow. *Journal of Fluids and Structures* **9**, 463–518.

PRICE, S. J. 2001 An investigation on the use of Connor's equation to predict fluidelastic instability in cylinder arrays. *ASME Journal of Pressure Vessel Technology* **123**, 448–453.

PRICE, S. J. & ABDALLAH, R. 1990 On the effect of mechanical damping and frequency detuning in alleviating wake-induced flutter of overhead power conductors. *Journal of Fluids and Structures* **4**, 1–34.

PRICE, S. J. & KURAN, S. 1991 Fluidelastic stability of a rotated square array with multiple flexible cylinders subject to cross-flow. *Journal of Fluids and Structures* **5**, 551–572.

PRICE, S. J. & MACIEL, Y. 1990 Solution of the nonlinear equations for wake-induced flutter via the Krylov and Bogoliubov method of averaging. *Journal of Fluids and Structures* **4**, 519–540.

PRICE, S. J. & PAÏDOUSSIS, M. P. 1982 A theoretical investigation of the parameters affecting the fluidelastic instability of a double row of cylinders subject to cross-flow. In Proceedings 3rd International Conference on Vibrations in Nuclear Plant, pp. 107–119, Keswick, U.K.

PRICE, S. J. & PAÏDOUSSIS, M. P. 1983 Fluidelastic instability of a double row of circular cylinders subject to cross-flow. *ASME Journal of Vibration, Acoustics, Stress and Reliability in Design* **105**, 59–66.

PRICE, S. J. & PAÏDOUSSIS, M. P. 1984a An improved mathematical model for the stability of cylinder rows subject to cross-flow. *Journal of Sound and Vibration* **97**, 615–640.

PRICE, S. J. & PAÏDOUSSIS, M. P. 1984b The aerodynamic forces acting on groups of two and three circular cylinders when subject to a cross-flow. *Journal of Wind Engineering and Industrial Aerodynamics* **17**, 329–347.

PRICE, S. J. & PAÏDOUSSIS, M. P. 1985 Fluidelastic instability of a full array of flexible cylinders subject to cross-flow. In *Fluid-Structure Interaction and Aerodynamic Damping* (eds. E. H. Dowell & M. K. Au-Yang), pp. 171–192. New York: ASME.

PRICE, S. J. & PAÏDOUSSIS, M. P. 1986a A constrained-mode analysis of the fluid-elastic instability in a double row of flexible circular cylinders subjected to cross-flow: a theoretical investigation of system parameters. *Journal of Sound and Vibration* **105**, 121–142.

PRICE, S. J. & PAÏDOUSSIS, M. P. 1986b A single-flexible-cylinder analysis for the fluidelastic instability of an array of flexible cylinders in cross-flow. *ASME Journal of Fluids Engineering* **108**, 193–199.

PRICE, S. J. & PAÏDOUSSIS, M. P. 1989 The flow-induced response of a single flexible cylinder in an in-line array of rigid cylinders. *Journal of Fluids and Structures* **3**, 61–82.

PRICE, S. J. & PIPERNI, P. 1988 An investigation of the effect of mechanical damping to alleviate wake-induced flutter of overhead power conductors. *Journal of Fluids and Structures* **2**, 53–71.

PRICE, S. J. & SERDULA, C. D. 1995 Flow visualization of vortex shedding around multi-tube marine risers in a steady current. In Proceedings of the 6th International Conference on Flow-Induced Vibration (ed. P. W. Bearman), London, U.K., pp. 483–493.

PRICE, S. J. & VALERIO, N. R. 1990 A non-linear investigation of single-degree-of-freedom instability in cylinder arrays subject to cross-flow. *Journal of Sound and Vibration* **137**, 419–432.

PRICE, S. J. & ZAHN, M. L. 1991 Fluidelastic behaviour of a normal triangular array subject to cross-flow. *Journal of Fluids and Structures* **5**, 259–278.

PRICE, S. J., ALLNUTT, J. G. & TUNSTALL, M. J. 1979 Sub-span oscillation of bundled conductors. In Proceedings of IEE Conference on Progress in Cables and Overhead lines for 200 kV and above, London, U.K.

PRICE, S. J., PAÏDOUSSIS, M. P., MACDONALD, R. & MARK, B. 1987 The flow-induced vibration of a single flexible cylinder in a rotated square array of rigid cylinders with pitch-to-diameter ratio of 2.12. *Journal of Fluids and Structures* **1**, 359–378.

PRICE, S. J., PAÏDOUSSIS, M. P. & SYCHTERZ, P. W. 1988. An experimental investigation of the quasi-steady assumption for bluff bodies in cross-flow. In *Proceedings 1988 International Symposium on Flow-Induced Vibrations and Noise*, Vol. 1 (eds. M. P. Païdoussis, O. M. Griffin & C. Dalton), pp. 91–111. New York: ASME.

PRICE, S. J., PAÏDOUSSIS, M. P., MARK, B. & MUREITHI, W. N. 1989. Current-induced oscillations and instabilities of a multi-tube flexible riser: water tunnel experiments. Proceedings of 8th International Conference on Offshore Mechanics and Arctic Engineering, The Hague, Volume 1., pp. 447–454.

PRICE, S. J., PAÏDOUSSIS, M. P. & GIANNIAS, N. 1990 A generalised constrained-mode analysis for cylinder arrays in cross-flow. *Journal of Fluids and Structures* **4**, 171–202.

PRICE, S. J., PAÏDOUSSIS, M. P., MUREITHI, W. N. & MARK, B. 1991 Flow visualization in a 1.5 pitch-to-diameter rotated square array of cylinders subject to cross-flow. In *Flow Induced Vibrations*, pp. 243–252. London: I.Mech.E.

PRICE, S. J., PAÏDOUSSIS, M. P. & CHENG, B. 1992a A constrained-mode analysis with a one-cylinder kernel for a fully flexible in-line array in cross-flow. In Proceedings International Symposium on Flow-Induced Vibration and Noise, Vol. 2: Cross-Flow Induced Vibration of Cylinder Arrays (eds. M. P. Païdoussis, S. S. Chen & D. A. Steininger), pp. 17–30. New York: ASME.

PRICE, S. J., PAÏDOUSSIS, M. P. & MARK, B. 1992b Flow-visualization in a 1.375 pitch-to-diameter parallel triangular array subject to cross-flow. In Proceedings International Symposium on Flow-Induced Vibration and Noise, Vol. 1: FSI/JIV in Cylinder Arrays in Cross-Flow (eds. M. P. Païdoussis, W. J. Bryan, J. R. Stenner & D. A. Steininger), pp. 29–38. New York: ASME.

PRICE, S. J., PAÏDOUSSIS, M. P. & AL-JABIR, A. M. 1993. Current-induced fluidelastic instabilities of a multi-tube flexible riser: Theoretical results and comparison with experiments. *ASME Journal of Offshore Mechanics and Arctic Engineering* **115**, 206–212.

PRICE, S. J., PAÏDOUSSIS, M. P. & MARK, B. 1995 Flow-visualization of the interstitial cross-flow through parallel triangular and rotated square arrays of cylinders. *Journal of Sound and Vibration* **181**, 85–98.

PROVANSAL, M., SCHOUVEILER, L. & LEWEKE, T. 2004 From the double vortex street behind a cylinder to the wake of a sphere. *European Journal of Mechanics B: Fluids* **23**, 65–80.

RAWLINS, C. B. 1974a Discussion of Simpson and Price (1974).

RAWLINS, C. B. 1974b Effect of wind turbulence in wake-induced oscillations of bundled conductors. Institute of Electrical and Electronics Engineers Power Engineering Society Paper C 74 444-6.

RAWLINS, C. B. 1976 Fundamental concepts in the analysis of wake-induced oscillation of bundled conductors. Institute of Electrical and Electronics Engineers Transactions on Power Apparatus and Systems, PAS–95, Winter Meeting, New York. Paper F76 079-4, pp. 1377–1393.

RAWLINS, C. B. 1977 Extended analysis of wake-induced oscillation of bundled conductors. Institute of Electrical and Electronics Engineers Power Engineering Society Winter Meeting, New York, Paper F77 217-3.

RAWLINS, C. B. 1979 Galloping conductors. In *Transmission Line Reference Book*. Wind-Induced Conductor Motion, Chapter 4. Palo Alto, CA: Electric Power Research Institute.

RAY, J. D., BERT, C. W. & EGLE, D. M. 1969 The application of the Kennedy-Pancu method to experimental vibration studies of complex shell structures. *Shock and Vibration Bulletin* **39**, 107–115.

REISFELD, B. & BANKOFF, S. 1992 Non-isothermal flow of a liquid film on a horizontal cylinder. *Journal of Fluid Mechanics* **236**, 167–196.

RIABOUCHINSKY, D. P. 1935 Thirty years of theoretical and experimental research in fluid mechanics. *Journal of the Royal Aeronautical Society* **39**, 282–348.

RICHARDSON, A. S., Jr., MARTUCELLI, J. R. & PRICE, W. S. 1965 Research study on galloping of electric power transmission lines. Part I. In *Proceedings 1st Int'l Conference on Wind Effects on Buildings and Structures*, Teddington, UK, pp. 612–686.

RICHTER, A. & NAUDASCHER, E. 1976 Fluctuating forces on a rigid circular cylinder in confined flow. *Journal of Fluid Mechanics* **78**, 561–576.

RISSONE, R. F., ALDAM-HUGHES, R. R., WILTON, G. S. & ALLAN, R. A. 1968 An experimental study of bundle conductor oscillation. Paper 23-02, GIGRÉ International Conference on Large High Voltage Electrical Systems, 22nd Session, Volume 1, Paris, France.

ROBERTS, B. W. 1962 Low frequency, self-excited vibration in a row of circular cylinders mounted in an airstream, Ph.D. Thesis, University of Cambridge.

ROBERTS, B. W. 1966 Low frequency, aeroelastic vibrations in a cascade of circular cylinders, I.Mech.E. Mechanical Engineering Science Monograph No. 4.

ROBERTSON, I., LI, L., SHERWIN, S. J. & BEARMAN, P. W. 2003 A numerical study of rotational and transverse galloping rectangular bodies. *Journal of Fluids and Structures* **17**, 681–699.

ROBERTSON, A. C., TAYLOR, I. J., WILSON, S. K., DUFFY, B. R. & SULLIVAN, J. M. 2008 A model for the numerical simulation of rivulet evolution on a circular cylinder in an air flow. In *Proceedings of the 9th International Conference of Flow-Induced Vibrations, Prague*, p. 220.

ROBERTSON, A. C., TAYLOR, I. J., WILSON, S. K., DUFFY, B. R. & SULLIVAN, J. M. 2010 Numerical simulation of rivulet evolution on a horizontal cable subject to an external aerodynamic field. *Journal of Fluids and Structures* **26**, 50–73.

ROCKWELL, D. & NAUDASCHER, E. 1979 Self-sustained oscillations of impinging shear layers. *Annal Review of Fluid Mechanics* **11**, 67–94.

ROMBERG, O. & POPP, K. 1998a The influence of trip-wires on the fluidelastic instability of a flexible tube in a bundle. *Journal of Fluids and Structures* **12**, 17–32.

ROMBERG, O. & POPP, K. 1998b The influence of upstream turbulence on the stability boundaries of a flexible tube in a bundle. *Journal of Fluids and Structures* **12**, 153–169.

ROSHKO, A. 1954 A new hodograph for free streamline theory. NACA Technical Note, T. N. 3168.

ROSHKO, A. 1955 On the wake and drag of bluff bodies. *Journal of Aeronautical Sciences* **22**, 124–132.

ROSHKO, A. 1961 Experiments on the flow past a circular cylinder at very high Reynolds number. *Journal of Fluid Mechanics* **10**, 345–356.

ROTTMANN, M. & POPP, K. 2003. Influence of upstream turbulence on the fluidelastic instability of a parallel triangular tube bundle. *Journal of Fluids and Structures* **18**, 595–612.

ROWBOTTOM, M. D. & ALDAM-HUGHES, R. R. 1972 Subspan oscillation: A review of existing knowledge. Paper 22-02, GIGRÉ International Conference on Large High Voltage Electrical Systems, 24th Session, Volume 1, Paris, France.

RUSCHEWEYH, H. P. 1983 Aeroelastic interference effects between slender structures. *Journal of Wind Engineering and Industrial Aerodynamics* **14**, 129–114.

RUSCHEWEYH, H. & VERWIEBE, C. 1995 Rain-wind-induced vibrations of steel bars. In *Proceedings of the International Symposium on Cable Dynamics, Liège, Belgium*, pp. 469–472.

RUSCHEWEYH, H., HORTMANNS, M. & SCHNAKENBERG, C. 1996 Vortex-excited vibrations and galloping of slender elements. *Journal of Wind Engineering and Industrial Aerodynamics* **65**, 347–352.

RUSSELL, J. S. 1841 On the vibration of suspension bridges and other structures; and the means of preventing injury from this cause. *Transactions of the Royal Scottish Society of Arts* **1**, 304–314. Reproduced in Farquharson (1949–1954), Part I, Appendix II.

RZENTKOWSKI, G. & LEVER, J. H. 1992 Modelling the nonlinear fluidelastic behaviour of a tube bundle. In Proceedings International Conference on Flow-Induced Vibration and Noise, Vol. 2: Cross-Flow Induced Vibration of Cylinder Arrays (eds. M. P. Païdoussis, S. S. Chen & D. A. Steininger), pp. 89–106. New York: ASME.

RZENTKOWSKI, G. & LEVER, J. H. 1998 An effect of turbulence on fluidelastic instability in tube bundles: A nonlinear analysis. *Journal of Fluids and Structures* **12**, 561–590.

SAGATUN, S. I., HERFJORD, K. & HOLMÅS, T. 2002 Dynamic simulation of marine risers moving relative to each other due to vortex and wake effects. *Journal of Fluids and Structures* **16**, 375–390.

SANDSTROM, S. 1987 Vibration analysis of a heat exchanger tube row with ADINA. *Computers and Structures* **26**, 297–305.

SARKAR, P. P., CARACOGLIA, L., HAAN, F. L., SATO, H. & MURAKOSHI, J. 2009 Comparative and sensitivity study of flutter derivatives of selected bridge deck sections, Part 1: Analysis of inter-laboratory experimental data. *Engineering Structures* **31**, 158–169.

SARPKAYA, T. 1979 Vortex-induced oscillations – a selective review. *Journal of Applied Mechanics* **46**, 241–258.

SARPKAYA, T. 2001 On the force decompositions of Lighthill and Morison. *Journal of Fluids and Structures* **15**, 227–233.

SARPKAYA, T. 2004 A critical review of the intrinsic nature of vortex-induced vibrations. *Journal of Fluids and Structures* **19**, 389–447.

SARPKAYA, T. & ISAACSON, M. 1981 *Mechanics of Wave Forces on Offshore Structures*. New York: Van Nostrand Reinhold.

SAVKAR, S. D. 1970 Wake-induced oscillations in transmission line bundles. Institute of Electrical and Electronic Engineers Power Engineering Society Summer Meeting and Energy Resources Conference, Anaheim, California IEEE Summer Power Meeting and EHV Conference, Los Angeles, CA.

SAVKAR, S. D. 1977 A note on the phase relationships involved in the whirling instability in tube arrays. *ASME Journal of Fluids Engineering* **99**, 727–731.

SAWYER, R. A. 1960 The flow due to a two-dimensional jet issuing parallel to a flat plate. *Journal of Fluid Mechanics* **9**, 543–560.

SCANLAN, R. H. 1978 The action of flexible bridges under wind. I. Flutter theory. *Journal of Sound and Vibration* **60**, 201–211.

SCANLAN, R. H. 1980 On the state of stability considerations for suspended-span bridges under wind. In *Practical Experiences with Flow-Induced Vibrations* (eds. E. Naudascher & D. Rockwell), pp. 595–618. Berlin: Springer-Verlag.

SCANLAN, R. H. 1990 Bridge aeroelasticity: Present state and future challenges. *Journal of Wind Engineering and Industrial Aerodynamics* **36**, 63–74.

SCANLAN, R. H. & TOMKO, J. J. 1971 Airfoil and bridge deck flutter derivatives. *ASCE Journal of the Engineering Mechanics Division* **97**, 1717–1737.

SCANLAN, R. H., BELIVEAU, J. G. & BUDLONG, K. S. 1974 Indicial aerodynamic functions for bridge decks. *ASCE Journal of the Engineering Mechanics Division* **100**, 657–672.

SCHEWE, G. 1989 Nonlinear flow-induced resonances of an H-shaped section. *Journal of Fluids and Structures* **3**, 327–348.

SCHLICHTING, H. 1968 *Boundary Layer Theory*. New York: McGraw Hill.

SCHMID, P. J. & DE LANGRE, E. 2003 Transient growth before coupled-mode flutter. *Journal of Applied Mechanics* **70**, 894–901.

SCHRÖDER, K. & GELBE, H. 1999a Simulation of flow-induced vibration in tube arrays, In *Flow-Induced Vibration*, PVP-Vol. 389 (ed. M. J. Pettigrew), pp. 9–16. ASME: New York.

SCHRÖDER, K. & GELBE, H. 1999b New design recommendations for fluidelastic instability in heat exchanger tube bundles. *Journal of Fluids and Structures* **13**, 361–379.

SCOTT, R. 2001 *In the Wake of Tacoma: Suspension Bridges and the Quest for Aerodynamic Stability*. Reston, VA: ASCE Press.

SCRUTON, C. 1960 Use of wind tunnels in industrial aerodynamic research. AGARD Report 309, NPL Report 411.

SCRUTON, C. 1971 Wind effects on structures. James Clayton Lecture, *Proceedings Institution of Mechanical Engineers*(1970–1971) **185**, 301–317.

SEIDEL, C. & DINKIER, D. 2006 Rain-wind induced vibration: phenomenology, mechanical modelling and numerical analysis. *Computers & Structures* **84**, 1584–1595.

SHAHRIARY, S., MUREITHI, N. W. & PETTIGREW, M. J. 2007 Quasi-static forces and stability analysis in a triangular tube bundle subjected to two-phase cross-flow. In Proceedings of PVP2007 ASME Pressure Vessel and Piping Division Conference. San Antonio, TX., USA.

SHARMA, C. B. & JOHNS, D. J. 1970 Wind-induced oscillations of circular cylindrical shells: an experimental investigation. Loughborough University of Technology, Report TT7001.

SHIN, Y. S. & WAMBSGANSS, M. W. 1977 Flow-induced vibration in LMFBR steam generators: a state-of-the-art review. *Nuclear Engineering and Design* **40**, 235–284.

SHIRAISHI, N. & MATSUMOTO, M. 1983 On classification of vortex-induced oscillation and its application for bridge structures. *Journal of Wind Engineering and Industrial Aerodynamics* **14**, 419–430.

SHIRAISHI, N., MATSUMOTO, M., SHIRATO, H. & ISHIZAKI, H. 1988 On aerodynamic stability effects for bluff rectangular cylinders by their corner-cut. *Journal of Wind Engineering and Industrial Aerodynamics* **28**, 371–380.

SIMIU, E. & SCANLAN, R. H. 1978 *Wind effects on Structures*. New York: John Wiley.

SIMIU, E. & SCANLAN, R. H. 1996 *Wind Effects on Structures: Fundamentals and Applications to Design*. New York: Wiley-IEEE.

SIMPSON A. 1966 Determination of the inplane natural frequencies of multispan transmission lines by a transfer-matrix method. *Proceedings of the Institution of Electrical Engineers* **113**, 870–878.

SIMPSON, A. 1970 Stability of subconductors of smooth circular cross-section. *Proceedings of the Institution of Electrical Engineers* **117**, 741–750.

SIMPSON, A. 1971a On the flutter of a smooth circular cylinder in a wake. *The Aeronautical Quarterly* **22**, 25–41.

SIMPSON, A. 1971b Wake induced flutter of circular cylinders: Mechanical aspects. *The Aeronautical Quarterly* **22**, 101–118.

SIMPSON, A. 1972 Determination of the natural frequencies of multiconductor overhead transmission lines. *Journal of Sound and Vibration* **20**, 417–449.

SIMPSON, A. 1977 In-line flutter of tandem cylinders. *Journal of Sound and Vibration* **54**, 379–387.

SIMPSON, A. 1983 Wind-induced vibration of overhead transmission lines. *Science Progress* **68**, 285–308.

SIMPSON, A. & FLOWER, J. W. 1977 An improved mathematical model for the aerodynamic forces on tandem cylinders in motion with aeroelastic applications. *Journal of Sound and Vibration* **51**, 183–217.

SIMPSON, A. & PRICE, S. J. 1974 On the use of "damped" and "undamped" quasi-static aerodynamic models in the study of wake-induced flutter. Institute of Electrical and Electronic Engineers Power Engineering Society Summer Meeting and Energy Resources Conference, Anaheim, CA. Paper C 74 378–6.

SINGH, P., CAUSSIGNAC, P. H., FORTES, A., JOSEPH, D. D. & LUNDGREN, T. 1989 Stability of periodic arrays of cylinders across the stream by direct simulation. *Journal of Fluid Mechanics* **205**, 553–571.

SISTO, F. 1953 Stall-flutter in cascades. *Journal of Aeronautical Sciences* **20**, 598–604.

SKOP, R. A. & BALASUBRAMANIAN, S. 1997 A new twist on an old model for vortex-excited vibrations. *Journal of Fluids and Structures* **11**, 395–412.

SKOP, R. A. & GRIFFIN, O. M. 1973 A model for the vortex-excited resonant response of bluff cylinders. *Journal of Sound and Vibration* **27**, 225–233.

SLATER, J. E. 1969 Aeroelastic instability of a structural angle section. Ph.D. Thesis, University of British Columbia, Vancouver, B.C., Canada.

SMITH, F. C. & VINCENT, G. S. 1950 Aerodynamic stability of suspension bridges with special reference to the Tacoma Narrows Bridge. Part II. Bulletin No. 116, University of Washington Engineering Experiment Station, Seattle, WA, USA.

SMITH, J. D. 1962 An experimental study of the aeroelastic instability of rectangular cylinders. M.A.Sc. Thesis, University of British Columbia, Vancouver, Canada.

SOHANKAR, A., NORBERG, C. & DAVIDSON, L. 1998 Low-Reynolds number flow around a square cylinder at incidence: study of blockage, onset of vortex shedding and outlet boundary conditions. *International Journal for Numerical Methods in Fluids* **26**, 39–56.

SOPER, B. M. H. 1980 The effect of tube layout in the fluidelastic instability of tube bundles in cross flow. In *Flow-Induced Heat Exchanger Tube Vibration* (ed. J. M. Chenoweth), pp. 1–9. New York: ASME.

SOUTHWORTH, P. J. & ZDRAVKOVICH, M. M. 1975 Cross-flow-induced vibrations of finite tube banks in in-line arrangements. I. Mech. E. *Journal of Mechanical Engineering Science* **17**, 190–198.

SRIGRAROM, S. 2003 Self-excited oscillation of a equilateral triangular cylinders. *Proceedings IUTAM Symposium on Fluid-Structure Interactions*, New Brunswick, NJ, USA, pp. 145–158.

SRIGRAROM, S. & KOH, A. K. G. 2008 Flow field of self-excited rotationally oscillating equilateral triangular cylinder. *Journal of Fluids and Structures* **24**, 750–755.

STEINMAN, D. B. & WATSON, S. R. 1957 *Bridges and Their Builders*. New York: Dover.

STOKES, G. G. 1851 On the effect of the inertial friction of fluids on the motion of pendulums. *Transactions of the Cambridge Philosophical Society* **9**, 8–106.

STROUHAL, V. 1978 Über eine besondere Art der Tonerregung. *Wiedemann's Annalen der Physik und Chemie* (Leipzig, New Series) **5**, 216–251.

SUEN, H.-C. 1981 Ovalling vibration of cylindrical shells in cross flow. M.Eng. Thesis, McGill University, Montreal, Canada.

SUMER, B. & FREDSØE, J. 1997 *Hydrodynamics Around Cylindrical Structures*. Singapore: World Scientific Publishing Company.

SUMNER, D. 1999 Circular cylinders in cross-flow. Ph.D. Thesis, McGill University, Montreal, Canada.

TAMURA, T. & DIAS, P. P. N. L. 2003 Unstable aerodynamic phenomena around the resonant velocity of a rectangular cylinder with small side ratio. *Journal of Wind Engineering and Industrial Aerodynamics* **91**, 127–138.

TAMURA, T. & ITOH, Y. 1999 Unstable aerodynamic phenomena of a rectangular cylinder with critical section. *Journal of Wind Engineering and Industrial Aerodynamics* **83**, 121–133.

TAMURA, T. & MIYAGI, T. 1999 The effect of turbulence on aerodynamic forces on a square cylinder with various corner shapes. *Journal of Wind Engineering and Industrial Aerodynamics* **83**, 135–145.

TAMURA, T. & ONO, Y. 2003 LES analysis on aeroelastic instability of prisms in turbulent flow. *Journal of Wind Engineering and Industrial Aerodynamics* **91**, 1827–1846.

TAMURA, T., DIAS, P. P. N. L. & ONO, Y. 2004 On mechanisms of low- and high-speed gallopings below and over the critical side ratio of rectangular cylinder. In *Flow-Induced Vibration 2004* (eds. E. de Langre & F. Axisa), Ecole Polytechnique, Paris (not in the printed volume of Proceedings or in CD).

TANAKA, H. 1980 A study on fluid elastic vibration of a circular cylinder array (One-Row Cylinder Array). *Transactions of Japan Society of Mechanical Engineers, Section B* **46**, 1398–1407.

TANAKA, H. & TAKAHARA, S. 1980 Unsteady fluid dynamic force on tube bundle and its dynamic effect on vibration. In *Flow-Induced Vibration of Power Plant Components* (ed. M. K. Au-Yang), PVP-Vol. 41, pp. 77–92. New York: ASME.

TANAKA, H. & TAKAHARA, S. 1981 Fluid elastic vibration of tube array in cross flow. *Journal of Sound and Vibration* **77**, 19–37.

TANAKA, H., TAKAHARA, S. & OHTA, K. 1982 Flow-induced vibration of tube arrays with various pitch-to-diameter ratios. *ASME Journal of Pressure Vessel Technology* **104**, 168–174.

TANAKA, H., TANAKA, K., SHIMIZU, F. & TAKAHARA, S. 2002 Fluidelastic analysis of tube bundle vibration in cross-flow. *Journal of Fluids and Structures* **16**, 93–112.

TAYLOR, G. I. 1947 Motion of solids in fluids when the flow is irrotational. *Proceedings of the Royal Society London A* **47**, 99–113.

TEH, C. E. & GOYDER, H. G. D. 1988 Data for the fluidelastic instability of heat exchanger tube arrays. In Proceedings International Symposium on Flow-Induced Vibration and Noise, Vol. 3: Flow-Induced Vibration and Noise in Cylinder Arrays (eds. M. P. Païdoussis, S. S. Chen & M. D. Bernstein), pp. 77–94. New York: ASME.

THEODORSEN, T. 1935 General theory of aerodynamic flutter instability and the mechanism of flutter. NACA Report 496.

THOMPSON, M. C., LEWEKE, T. & PROVANSAL, M. 2001 Kinematics and dynamics of sphere wake transition. *Journal of Fluids and Structures* **15**, 575–585.

THOTHADRI, M. & MOON, F. C. 1998 Helical wave oscillations in row of cylinders in a cross-flow. *Journal of Fluids and Structures* **12**, 591–613.

TIFFANY, S. H. & ADAMS, W. H. 1988. Nonlinear programming extensions to rational function approximation methods for unsteady aerodynamic forces. NASA TP 2776.

TOKATY, G. A. 1971 *A History and Philosophy of Fluidmechanics*. Henley-on-Thames: G. T. Foulis & Co.

TRIM, A. D., BRAATEN, H. & LIE, H. & TOGNARELLI, M. A. 2005 Experimental investigation of vortex-induced vibration of long marine risers. *Journal of Fluids and Structures* **21**, 335–361.

TSUI, Y. T. 1977 On wake-induced flutter of a circular cylinder in the wake of another. *Journal of Applied Mechanics* **44**, 194–200.

TSUI, Y. T. 1978 Two-dimensional stability analyses of a circular conductor in the wake of another. *Electrical Power Systems Research* **1**, 87–95.

TSUI, Y. T. 1986 On wake-induced vibration of a conductor in the wake of another via a 3-D finite element method. *Journal of Sound and Vibration* **107**, 39–58.

TSUI, Y. T. & TSUI, C. C. 1980 Two dimensional stability analysis of two coupled conductors with one in the wake of the other. *Journal of Sound and Vibration* **69**, 361–394.

VAN DER BURGH, A. 2004 Rain-wind-induced vibrations of a simple oscillator. *International Journal of Non-Linear Mechanics* **39**, 93–100.

VAN DER HOOGT, P. J. M. & VAN CAMPEN, D. H. 1984 Self-induced instabilities of parallel tubes in potential cross-flow. In Proceedings ASME Symposium on Flow-Induced Vibrations, Vol. 2: Vibration of Arrays of Cylinders in Cross Flow (eds. M. P. Païdoussis, M. K. Au-Yang & S. S. Chen), pp. 53–66. New York: ASME.

VAN OUDHEUSDEN, B. W. 1995 On the quasi-steady analysis of one-degree-of-freedom galloping with combined translational and rotational effects. *Nonlinear Dynamics* **8**, 435–451.

VAN OUDHEUSDEN, B. W. 1996 Rotational one-degree-of-freedom galloping in the presence of viscous and frictional damping. *Journal of Fluids and Structures* **10**, 673–689.

VANDIVER, J. K. 1993 Dimensionless parameters important to the prediction of vortex-induced vibration of long, flexible cylinders in ocean currents. *Journal of Fluids and Structures* **7**, 423–455.

VENTO, M. A., ANTUNES, J. & AXISA, F. 1992 Tube/support interaction under simulated fluidelastic instability: Two-dimensional experiments and computations of the nonlinear responses of a straight tube. In Proceedings International Symposium on Flow-Induced Vibration and Noise, Vol. 2: Cross-Flow Induced Vibration of Cylinder Arrays (eds. M. P. Païdoussis, S. S. Chen & D. A. Steininger), pp. 151–166. New York: ASME.

VERWIEBE, C. & RUSCHEWEYH, H. 1998 Recent research results concerning the exciting mechanisms of rain-wind-induced vibrations. *Journal of Wind Engineering & Industrial Aerodynamics* **74**, 1005–1013.

VIKESTAD, K., VANDIVER, J. K. & LARSEN, C. M. 2000 Added mass and oscillation frequency for a circular cylinder subjected to vortex-induced vibrations and external disturbance. *Journal of Fluids and Structures* **14**, 1071–1088.

VIOLETTE, R. 2009 Modéle linéaire des vibrations induites par vortex de structures élancées. Ph.D. thesis, Ecole Polytechnique, Palaiseau, France.

VIOLETTE, R., PETTIGREW, M. J. & MUREITHI, N. 2006 Fluidelastic instability of an array of tubes preferentially flexible in the flow direction subjected to two-phase cross flow. *ASME Journal of Pressure Vessel Technology* **128**, 148–159.

VIOLETTE, R., DE LANGRE, E. & SZYDLOWSKI, J. 2007 Computation of vortex-induced vibrations of long structures using a wake oscillator model: Comparison with DNS and experiments. *Computers and Structures* **85**, 1134–1141.

VIOLETTE, R., DE LANGRE, E. & SZYDLOWSKI, J. 2010 A linear stability approach to vortex-induced vibrations and waves. *Journal of Fluids and Structures* **26**, 442–486.

VIO, G. A., DIMITRIADIS, G. & COOPER, J. E. 2007 Bifurcation analysis and limit cycle oscillation amplitude prediction methods applied to the aeroelastic galloping problem. *Journal of Fluids and Structures* **23**, 983–1011.

VON KÁRMÁN, T. 1911 Über den Mechanismus des Widerstandes, den ein bewegter Körper in einer Flüssigkeit erfährt. *Nachrichten der K. Gesellschaft der Wissenschaften zu Göttingen. Mathematischphysikalische Klasse, Göttingen,* pp. 324–338.

VON KÁRMÁN, T. 1912 *Über den Mechanismus des Widerstandes, den ein bewegter Körper in einem Flüssigkeit ehrfährt. Nachrichten von der Königlichen Gessellschaft der Wissenschaften zn Göttingen, Mathematisch-Physikalische Klasse* **5**, 547–556. See also 1911, pp. 509–517.

VON KÁRMÁN, T. 1954 *Aerodynamics.* New York: McGraw Hill.

VON KARMAN, Th. & DUNN, L. G. 1952 Wind-tunnel investigations carried out at the California Institute of Technology. Chapter VII in Part III of Farquharson (1949–1954), pp. 57–73.

WALLIS, R. P. 1939 Photographic study of fluid flow between banks of tubes. *Engineering* **148**, 423–426.

WAMBSGANSS, M. W., YANG, C. I. & HALLE, H. 1984 Fluidelastic instability in shell and tube heat exchangers – A framework for a prediction method. In Proceedings ASME Symposium on Flow-Induced Vibrations, Vol. 3: Vibration in Heat Exchangers (eds. M. P. Païdoussis, J. M. Chenoweth & M. D. Bernstein), pp. 103–118. New York: ASME.

WANG, Z., ZHOU, Y., HUANG, J. & XU, Y. 2005 Fluid dynamics around an inclined cylinder with running water rivulets. *Journal of Fluids and Structures* **21**, 49–64.

WARDLAW, R. L. 1967 Aerodynamically excited vibrations of a 3-inch × 3-inch aluminum angle in steady flow. *National Research Council of Canada* Aeronautical Report LR-482.

WARDLAW, R. L. 1971 Some approaches for improving the aerodynamic stability of bridge road decks. *Proceedings 3rd International Conference on Wind Effects on Buildings and Structures*, Tokyo, Japan, pp. 931–940.

WARDLAW, R. L. 1980 Approaches to the suppression of wind-induced vibrations of structures. In *Practical Experiences with Flow-Induced Vibrations* (eds. E. Naudascher & D. Rockwell), pp. 650–670. Berlin: Springer-Verlag.

WARDLAW, R. L. 1990 Wind effects on bridges. *Journal of Wind Engineering and Industrial Aerodynamics* **33**, 301–312.

WARDLAW, R. L., COOPER, K. R., KO, R. G. & WATTS, J. A. 1975 Wind tunnel and analytical investigations into aeroelastic behaviour of bundled conductors. *Institute of Electrical and Electronic Engineers Transactions on Power Apparatus and Systems* **94**, 642–651.

WARING, L. F. & WEAVER, D. S. 1988 Partial admission effects on the stability of a heat exchanger tube array. *ASME Journal of Pressure Vessel Technology* **110**, 194–198.

WASHIZU, K., OHYA, A., OTSUKI, Y. & FUJII, K. 1978 Aeroelastic instability of rectangular cylinders in a heaving mode. *Journal of Sound and Vibration* **59**, 195–210.

WASHIZU, K., OHYA, A., OTSUKI, Y. & FUJII, K. 1980 Aeroelastic instability of rectangular cylinders in a torsional mode due to a transverse wind. *Journal of Sound and Vibration* **72**, 507–521.

WATTS, J. A. & KO, R. G. 1973 A wind tunnel investigation into the aerodynamic stability of bundled conductors for Hydro-Quebec. Part II: Dynamic sectional model tests using a four-spring orthogonal suspension. National Aeronautical Establishment Laboratory Technical Report LRT-LA-103, National Research Council, Ottawa, Canada.

WAWZONEK, M. A. 1979 Aeroelastic behaviour of square section prisms in uniform flow. M.A.Sc. Thesis, University of British Columbia, Vancouver, Canada.

WEAVER, D. S. & ABD-RABBO, A. 1985 A flow visualization study of a square array of tubes in water crossflow. *ASME Journal of Fluids Engineering* **107**, 354–363.

WEAVER, D. S. & EL-KASHLAN, M. 1981 The effect of damping and mass ratio on the stability of a tube bank. *Journal of Sound and Vibration* **76**, 283–294.

WEAVER, D. S. & FITZPATRICK, J. A. 1988 A review of flow induced vibrations in heat exchangers. *Journal of Fluids and Structures* **2**, 73–93.

WEAVER, D. S. & GOYDER, H. G. D. 1990 An experimental study of fluidelastic instability in a three-span tube array. *Journal of Fluids and Structures* **4**, 429–442.

WEAVER, D. S. & GROVER, L. K. 1978 Cross-flow induced vibrations in a tube bank. *Journal of Sound and Vibration* **59**, 277–294.

WEAVER, D. S. & KOROYANNAKIS, D. 1982 The cross-flow response of a tube array in water – a comparison with the same array in air. *ASME Journal of Pressure Vessel Technology* **104**, 139–146.

WEAVER, D. S. & KOROYANNAKIS, D. 1983 Flow induced vibrations of heat exchanger U-tubes, a simulation to study the affects of asymmetric stiffness. *ASME Journal of Vibration, Acoustics, Stress and Reliability in Design* **105**, 67–75.

WEAVER, D. S. & PARRONDO, J. 1991 Fluidelastic instability in multispan exchanger tube arrays. *Journal of Fluids and Structures* **5**, 323–338.

WEAVER, D. S. & SCHNEIDER, W. 1983 The effect of flat bar supports on the cross flow induced response of heat exchanger U-tubes. *ASME Journal of Engineering for Power* **105**, 775–781.

WEAVER, D. S. & VELJKOVIC, I. 2005 Vortex shedding and galloping of open semi-circular and parabolic cylinders in cross-flow. *Journal of Fluids and Structures* **21**, 65–74.

WEAVER, D. S. & YEUNG, H. C. 1984 The effect of tube mass on the flow induced response of various tube arrays in water. *Journal of Sound and Vibration* **93**, 409–425.

WEAVER, D. S., FEENSTRA, P. & EISINGER, F. L. 2000 An experimental study on the effect of tube-to-tube ties on fluidelastic instability in an in-line array. *ASME Journal of Pressure Vessel Technology* **122**, 50–54.

WHISTON, G. S. & THOMAS, G. D. 1982 Whirling instabilities in heat exchanger tube arrays. *Journal of Sound and Vibration* **81**, 1–31.

WIANECKI, J. 1979 Cables wind excited vibrations of cable-stayed bridges. In *Proceedings of the 5th International Conference of Wind Engineering, Fort Collins, Colorado*, pp. 1381–1393. Oxford: Pergamon Press.

WILDE K. & FUJINO, Y. 1998 Aerodynamic control of bridge deck flutter by active surfaces. *ASCE Journal of Engineering Mechanics* **124**, 718–727.

WILDE K., FUJINO, Y. & KAWAKAMI, T. 1999 Analytical and experimental study on passive aerodynamic control of a bridge deck. *Journal of Wind Engineering and Industrial Aerodynamics* **80**, 105–119.

WILLDEN, R. H. J. & GRAHAM, J. M. R. 2004 Multi-modal vortex-induced vibrations of a vertical riser pipe subject to a uniform current profile. *European Journal of Mechanics B: Fluids* **23**, 209–218.

WILLIAMSON, C. H. K. 1996 Vortex dynamics in the cylinder wake. *Annual Review of Fluid Mechanics* **28**, 477–539.

WILLIAMSON, C. H. K. & GOVARDHAN, R. 2004 Vortex-induced vibrations. *Annual Review of Fluid Mechanics* **36**, 413–455.

WILLIAMSON, C. H. K. & ROSHKO, A. 1988 Vortex formation in the wake of an oscillating cylinder. *Journal of Fluids and Structures* **2**, 355–381.

WOOTTON, L. R. 1968 The oscillation of model circular stacks due to vortex shedding at Reynolds numbers from 10^5 to 3×10^6. National Physical Laboratory (NPL) Report 1267.

WOOTTON, L. R., WARNER, M. H. & COOPER, D. H. 1974 Some aspects of the oscillations of full-scale piles. In *Flow-Induced Structural Vibrations* (ed. E. Naudascher), pp. 587–601. Berlin: Springer.

WU, W., HUANG, S. & BARLTROP, N. 1999 Lift and drag forces on a cylinder in the wake of an upstream cylinder. 18th International Conference on Offshore Mechanics and Arctic Engineering, St John's, Newfoundland.

WU, W., HUANG, S. & BARLTROP, N. 2002 Current induced instability of two circular cylinders. *Applied Ocean Research* **24**, 287–297.

WU, W., HUANG, S. & BARLTROP, N. 2003 Multiple stable/unstable equilibria of a cylinder in the wake of an upstream cylinder. *ASME Journal of Offshore Mechanics and Arctic Engineering* **125**, 103–107.

YAMAGUCHI, H. 1990 Analytical study on growth mechanism of rain vibration of cables. *Journal of Wind Engineering and Industrial Aerodynamics* **33**, 73–80.

YETISIR, M. & WEAVER, D. S. 1985 The dynamics of heat exchanger U-bend tubes with flat bar supports. In *Fluid-Structure Interaction and Aerodynamic Damping* (eds. E. H. Dowell & M. K. Au-Yang), pp. 119–132. New York: ASME.

YETISIR, M. & WEAVER, D. S. 1988 On an unsteady theory for fluidelastic instability of heat exchanger tube arrays. In Proceedings International Symposium on Flow-Induced Vibration and Noise, Vol. 3: Flow-Induced Vibration and Noise in Cylinder Arrays (eds. M. P. Païdoussis, S. S. Chen & M. D. Bernstein), pp. 181–195. New York: ASME.

YETISIR, M. & WEAVER, D. S. 1992 A theoretical study of fluidelastic instability in a flexible array of tubes. In Proceedings International Symposium on Flow-Induced Vibration and Noise, Vol. 2: Cross-Flow Induced Vibration of Cylinder Arrays (eds. M. P. Païdoussis, S. S. Chen & D. A. Steininger), pp. 69–87. New York: ASME.

YETISIR, M. & WEAVER, D. S. 1993a An unsteady theory for fluidelastic instability in an array of flexible tubes in cross-flow. Part I: Theory. *Journal of Fluids and Structures* **7**, 751–766.

YETISIR, M. & WEAVER, D. S. 1993b An unsteady theory for fluidelastic instability in an array of flexible tubes in cross-flow. Part 11: Results and comparison with experiments. *Journal of Fluids and Structures* **7**, 767–782.

YEUNG, H. C. & WEAVER, D. S. 1983 The effect of approach flow direction on the flow-induced vibrations of a triangular tube array. *ASME Journal of Vibration, Acoustics, Stress and Reliability in Design* **105**, 76–82.

YU, M.-H., YU, C.-M. & CHEN, K.-H. 2004 Fluid elastic instability of a small circular cylinder in the shear layer of a two-dimensional jet. *Physics of Fluids* **16**, 2357–2370.

YU, P., DESAI, Y. M., POPPLEWELL, N. & SHAH, A. H. 1993a Three degrees-of-freedom model for galloping. Part I: formulation. *ASCE Journal of Engineering Mechanics* **119**, 2404–2425.

YU, P., DESAI, Y. M., POPPLEWELL, N. & SHAH, A. H. 1993b Three degrees-of-freedom model for galloping. Part II: solutions. *ASCE Journal of Engineering Mechanics* **119**, 2426–2448.

YU, P., POPPLEWELL, N. & SHAH, A. H. 1995a Instability trends of inertially coupled galloping. Part I: initiation. *Journal of Sound and Vibration* **183**, 663–678.

YU, P., POPPLEWELL, N. & SHAH, A. H. 1995b Instability trends of inertially coupled galloping. Part II: periodic vibrations. *Journal of Sound and Vibration* **183**, 679–691.

ZASSO, A. 1996 Flutter derivatives: Advantages of a representation convention. *Journal of Wind Engineering and Industrial Aerodynamics* **60**, 35–47.

ZDRAVKOVICH, M. M. 1977 Review of flow interference between two circular cylinders in various arrangements. *ASME Journal of Fluids Engineering* **99**, 618–631.

ZDRAVKOVICH, M. M. 1982 Scruton number; a proposal. *Journal of Wind Engineering and Industrial Aerodynamics* **10**, 263–265.

ZDRAVKOVICH, M. M. 1991 A comparative overview of marine risers and heat exchanger tube banks. *ASME Journal of Offshore Mechanics and Arctic Engineering* **113**, 30–36.

ZDRAVKOVICH, M. M. 1993 Interstitial flow fields and fluid forces, Chapter 2, Part III Fluid-Structure Interaction, Technology for the '90s. (ed. M. K. Au-Yang), pp. 593–658. New York: ASME.

ZDRAVKOVICH, M. M. 1997 *Flow Around Circular Cylinders*, Volume 1. Oxford: Oxford University Press.

ZDRAVKOVICH, M. M. 2003 *Flow Around Circular Cylinders, Applications*, Volume 2. Oxford: Oxford University Press.

ZDRAVKOVICH, M. M. & PRIDDEN, D. L. 1977 Interference between two circular cylinders; series of unexpected discontinuities. *Journal of Industrial Aerodynamics* **2**, 255–270.

ZDRAVKOVICH, M. M. & STONEBANKS, K. L. 1990 Intrinsically nonuniform and metastable flow in and behind tube arrays. *Journal of Fluids and Structures* **4**, 305–320.

ZHANG, Q., POPPLEWELL, N. & SHAH, A. H. 2000 Galloping of bundle conductor. *Journal of Sound and Vibration* **234**, 115–134.

ZHU, J., WANG, X. Q., XIE, W.-C. & SO, R. M. C. 2009 Turbulence effects on fluidelastic instability of a cylinder in a shear flow. *Journal of Sound and Vibration* **321**, 680–703.

ZIELINSKA, B. J. A. & WESFREID, J. E. 1995 On the spatial structure of global modes in wake flow. *Physics of Fluids* **7**, 1418.

ZUO, D., JONES, N. & MAIN, J. 2008 Field observation of vortex-and rain-wind-induced stay-cable vibrations in a three-dimensional environment. *Journal of Wind Engineering and Industrial Aerodynamics* **96**, 1124–1133.

ŽUKAUSKAS, A. & KATINAS, V. 1980 Flow-induced vibration in heat-exchanger tube banks. In *Practical Experiences with Flow-Induced Vibrations* (eds. E. Naudascher & D. Rockwell), pp. 188–196. Berlin: Springer-Verlag.

Index

Printed in the United States
By Bookmasters